Macmillan Encyclopedia of Chemistry

Editorial Board

MACMILLAN ENCYCLOPEDIA OF CHEMISTRY

Joseph J. Lagowski

Editor in Chief

Volume 4

MACMILLAN REFERENCE USA
Simon & Schuster Macmillan
NEW YORK

Simon & Schuster and Prentice Hall International
LONDON   MEXICO CITY   NEW DELHI   SINGAPORE   SYDNEY   TORONTO

Simon & Schuster Macmillan
1633 Broadway
New York, NY 10019

PRINTED IN THE UNITED STATES OF AMERICA

Printing Number

    4 5 6 7 8 9 10

LIBRARY OF CONGRESS CATALOGING-IN-PUBLICATION DATA
Macmillan encyclopedia of chemistry / edited by Joseph J. Lagowski.
    p. cm.
   Includes bibliographical references and index.
   ISBN 0-02-897225-2 (set : alk. paper). — ISBN 0-02-897221-X
(v. 1 : alk. paper). — ISBN 0-02-897222-8 (v. 2 : alk. paper).
— ISBN 0-02-897223-6 (v. 3 alk. paper). — ISBN 0-02-897224-4
(v. 4 : alk. paper)
   1. Chemistry—Encyclopedias. I. Lagowski, J. J.
QD4.M33 1997
540'.3—dc21   97-1824
                                                 CIP

COMMON ABBREVIATIONS AND SYMBOLS

′	foot; minute (of arc); single prime	∈	is an element of
″	inch; second (of arc); double prime	ε_0	dielectric constant; permittivity
+	plus	θ	plane angle
+	positive charge	λ	wavelength
−	minus	μ	magnetic moment; micro-
−	negative charge	μA	microampere
±	plus-or-minus	μC	microcoulomb
∓	minus-or-plus	μF	microfarad
×	multiplied by	μg	microgram
·	multiplied by	μg/ml	microgram per milliliter
/	divided by	μK	microkelvin
=	equals	μm	micrometer (also called micron)
≠	not equal to	μmol	micromole
≢	not equivalent to	μs, μsec	microsecond
∼	about, approximately	ν	frequency
≅	congruent to; approximately equal to	υ	velocity
≈	approximately equal to	π or pi	ratio of the circumference of a circle to its diameter; double as in double bond
≡	identical to; equivalent to		
<	less than	σ	single as in single bond; Stefan-Boltzmann constant
≤	less than or equal to		
>	greater than	Σ	summation
≥	greater than or equal to	ϕ	null set
⊂	included in	ψ	amplitude of a wave (as in *wave*, or *psi, function*)
⊃	excluded from		
%	percent	∠	angle
°	degree (temperature; angle of arc)	Ⓗ	angle of diffraction maxima (under Bragg's law)
:	ratio		
@	at	Ω	solid angle
—	single bond	→	reaction to right
=	double bond	←	reaction to left
: :	double bond	↔	connecting resonance forms
≡	triple bond	⇋	equilibrium reaction beginning at right
: : :	triple bond		
∞	infinity	⇌	equilibrium reaction beginning at left
∝	variation	⇁	reversible reaction beginning at left
∂	partial derivative or differential	⇀	reversible reaction beginning at right
α	proportional to	⇑	elimination
√	square root	⇓	absorption
∛	cube root	↺	repositioning
Δ	delta; increment of a variable	↻	ring opening

Common Abbreviations and Symbols

⌇	electrolysis
⌒	integral (may position horizontally as here or vertically)
a	acceleration; year
A	area
a_0	Bohr Unit
AA	atomic absorption
AAS	atomic absorption spectroscopy
ABS	alkylbenzene sulfate
Ac	actinium
ACE	angiotension-converting enzyme
ACM	asbestos-containing materials
ACP	acyl carrier protein
ACS	American Chemical Society
A.D.	anno domini
ADD	attention deficit disorder
ADH	alcohol dehydrogenase
ADP	adenosine diphosphate
AE	atomic emission; atomization energy
AEC	Atomic Energy Commission
AES	atomic emission spectroscopy
AF	atomic fluorescence
AFM	atomic force microscope; atomic force microscopy
AFS	atomic fluorescence spectroscopy
Ag	silver
AHERA	Asbestos Hazard Emergency Response Act
AIDS	acquired immunodeficiency syndrome
Al	aluminum
ALDH	aldehyde dehydronase
Am	americium
amp	ampere
AMP	adenosine monophosphate
cAMP	cyclic adenosine monophosphate
AMS	accelerator mass spectometry
amu	atomic mass unit
ANFO	ammonium nitrate and fuel oil
ANRORC	addition nucleophilic ring opening and ring closure
APC	air pollution control
Ar	argon
As	arsenic
ASHAA	Asbestos School Hazard Abatement Act
ASHARA	Asbestos School Hazards Abatement Reauthorization Act
ASTM	American Society for Testing and Materials
At	astatine
atm.	standard atmosphere (unit of pressure)
ATP	adenosine triphosphate
ATR	attenuated total reflectance
Au	gold
AZT	the anti-HIV drug also known as Zidovudine (trade name Retrovir)
B	boron
B	magnetic flux density
Ba	barium
BAL	British Anti-Lewisite
BAT	best available technology
BDE	bond dissociation energies
Be	beryllium
Bh	nielsbohrium, niels-bohrium, or bohrium (three forms of one of the names proposed for element 107)
BHA	butylated hydronyanisole
BHT	butylated hydroxytoluene
Bi	bismuth
Bk	berkelium
BOD	biochemical oxygen demand
b.p.	boiling point
Br	bromine
Btu	British thermal unit
c	centi-; speed of light
C	carbon; Celsius; centigrade; coulomb
C	heat capacity; electric capacitance
ca.	approximately or circa
Ca	calcium
CAA	Clean Air Act
CAM	Crassulacean acid metabolism
CAS	Chemical Abstracts Service
CASSI	Chemical Abstracts Service Source Index
C.B.E.	Commander of the Order of the British Empire
Cd	cadmium
CD	circular dichroism
CD-ROM	compact disc/read-only memory device
Ce	cerium
CERCLA	Comprehensive Environmental Response, Compensation, and Liability Act
cf.	compare, confer
Cf	californium
CFC	chlorofluorocarbon
CFR	Code of Federal Regulation
CFSE	crystal field stabilization energy
CFT	crystal field theory
CGPM	Conférence des Générale des Poids et Mesures (General Conference on Weights and Measures)
Ci	Curies

CI	configuration interaction
CIEEL	chemically initiated electron exchange luminescence
Cl	chlorine
cm	centimeter
Cm	curium
cm^3	cubic centimeter
CMC	critical micelle concentration
CMP	cytidine monophosphate
CMPES	capacitatively coupled microwave plasma emission spectroscopy
CMPS	capacitatively coupled plasma spectroscopy
CN	coordination number
CNDO	complete neglect of differential overlap
CNOC	Commission on Nomenclature of Organic Chemistry
CNS	central nervous system
Co	cobalt
COD	chemical oxygen demand
COSY	correlation spectroscopy
CPI	chemical processing industries
Cr	chromium
Cs	cesium
CSF	cerebospinal fluid; colony-stimulating factor
CT	computed tomography
Cu	copper
CVT	chemical vapor transport
cyt	cytochromes
d	day; deci-
D	debye; deuterium; dipole moment unit of measure
D	dipole moment
D	dextrorotatory
DA	dopamine
Db	dubnium (one of the names variously proposed for elements 104 or 105)
DC	direct current
DCPES	direct current plasma emission spectroscopy
DDD	dichlorodyphenyldichloroethane (a pesticide)
DDT	dichlorodiphenyltrichloroethane (a pesticide)
deg	degree
DES	diethylstilbestrol
DIFP	diisopropyl fluorophosphate
DMBA	dimethylbenzanthracene
DMF	dimethylformamide
DMPU	dimethylpropylene urea
DMSO	dimethyl sulfoxide (dimethylsulfoxide)
DNA	deoxyribonucleic acid
cDNA	complementary DNA
rDNA	recombinant DNA
DOT	Department of Transportation
doz.	dozen
DP	degree of polymerization
DPP	differential pulse polarography
DQMC	diffusion quantum Monte Carlo method
dr	diastereomer ratio
DS	degree of substitution
D.Sc.	doctor of science
dt	document type
Dy	dysprosium
DZ	double zeta
e	elementary charge
E	epinephrine
E	electric field strength; energy
ε	maximum electrical potential
E_a	activation energy
E_g	bandgap energy
ea.	each
EA	electron affinity
ECD	electron capture detector
ECG	explicitly correlated Gaussian wave function
ed.	edition, editor
EDTA	ethylenediaminetetraacetic acid
ee	enantiomeric excess
EFF	empirical force field
e.g.	for example (from Latin *exempli gratia*)
EGF	epidermal growth factor
EM	electromagnetic
e.m.f.	electromotive force
EMPD	*para*-ethoxy-*meta*-phenylediamine
EN	electronegativity
EOR	enhanced oil recovery
EP	electron pair
EPA	Environmental Protection Agency
EPR	electron paramagnetic resonance
er	enantiomer ratio
Er	erbium
erg. sec.	centimeter-gram-second (unit of measure of work)
ERT	estrogen replacement therapy
Es	einsteinium
ESCA	electron spectroscopy for chemical analysis
ESEM	environmental scanning electron microscope

Common Abbreviations and Symbols

ESR	electron spin resonance	GMP	guanosine monophosphate
esu	electrostatic unit	GnRH	gonadotropin-releasing hormone
et al.	and others (from Latin, *et alia*)	GSC	gas-solid chromatography
etc.	et cetera, and so on	GTP	guanosine diphosphate
Eu	europium	h	hour
eV	electron volts	h	Planck's constant
EWG	electron-withdrawing group	$\hbar$	h divided by 2π
f	femto-	H	enthalpy, hydrogen
F	Fahrenheit; Faraday's constant; fluorine	^{1}H or H^1	protium (most common hydrogen isotope)
F	force	^{2}H or H^2	deuterium (hydrogen isotope)
$\mathcal{F}$	Faraday's constant	^{3}H or H^3	tritium (hydrogen isotope)
FAAS	flame atomic absorption spectroscopy	Ha	hahnium (one of the names variously proposed for elements 105 or 108)
FAB	fast atom bombardment	Hb	hemoglobin
FAD	flavin adenine dinucleotide	HD	high density; "mustard gas"
FAES	flame atomic emission spectroscopy	HDPE	high-density polyethylene
FAFS	flame atomic fluorescence spectroscopy	He	helium
FAS	fatty acid synthetase	hex	hexatic
FCC	face-centered cubic; fluid-catalytic cracking	Hf	hafnium
		HF-SCF	Hartee-Fock self-consistent field
FDA	Food and Drug Administration	Hg	mercury
FES	flame emission spectroscopy	HIV	human immunodeficiency virus
FITC	fluorescein isothiocyanate	HMO	Hückel molecular orbital
Fm	fermium	HMPA	hexamethyl phosphoramide
FMN	flavin mononucleotide	Ho	holmium
Fr	francium	HOAc	acetic acid
FRET	fluorescence resonance energy transfer	HOMO	highest occupied molecular orbital
		HPLC	high-performance liquid chromatography
F.R.S.	Fellow of the Royal Society	HRE	hormone-responsive elements
FSH	follicle-stimulating hormone	HRT	hormone replacement therapy
FT	Fourier transform	Hs	hassium (one of the names proposed for element 108)
FTIR	Fourier transform infrared		
FVT	flash vacuum thermolysis	HTLV-1	human t-cell lymphotropic virus 1
g	gram	HVAO	hybrid-valence atomic orbital
g	g-orbital	Hz	hertz
G	gravity; gravitational constant	i	i-orbital
Ga	gallium	I	iodine; ionic strength; electric current
GABA	gamma aminobutyric acid	ibid.	in the same place (from Latin, *ibidem*)
GAC	granular activated carbon		
GB	Sarin (a nerve agent)	ICAPES	inductively coupled argon plasma emission spectroscopy
GC	gas chromatography		
GC-IR	gas chromatography–infrared spectrophotometry	ICPS	inductively coupled plasma spectroscopy
gc-mass spec	gas chromatography–mass spectrometry	ICRP	International Commission on Radiological Protection
GC-MS	gas chromatography–mass spectrometry	i.e.	that is (from Latin, *id est*)
		IE	ion exchange; ionization energy
Gd	gadolinium	IEEE	Institute of Electrical and Electronics Engineers
GD	Soman (nerve agent)		
Ge	germanium	IGR	insect growth regulator
g/l	grams per liter	Igs	immunoglobulins
GLC	gas-liquid chromatography	in.	inch

IP	ionization potential
IPK	international prototype kilogram
Ir	iridium
IR	infrared
IUC	International Union of Chemistry
IUPAC	International Union of Pure and Applied Chemistry
J	joule
J	electric current density
JH	juvenile hormone
Jl	joliotium (one of the names proposed for element 105)
k	*k*-orbital
K	degrees Kelvin; Kelvin; potassium
K_a	acidity constant for the dissociation of a weak acid (the weaker the acid, the lower the K_a value)
k_B	Boltzmann's constant
K.B.E.	Knight Commander of the British Empire
KE	kinetic energy
Kg	kilogram
kHz	kilohertz
kJ	kilojoule
kJ mol	kilojoule mole
km	kilometer
K_m	Michaelis constant
Kr	krypton
l	length
L	lambert; liter
L	length; Avogadro's constant
L	levorotatory
La	lanthanum
LAS	linear alkylsulfonates
laser	light amplification by stimulated emission of radiation
LC	liquid chromatography
LCAO	linear combination of atomic orbitals
LD	lethal dose
LDL	low-density lipoprotein
LDPE	low-density polyethylene
LEED	low-energy electron diffraction
LEIS	laser-enhanced ionization spectroscopy
LFER	linear free energy relation
LFP	laser flash photolysis
LFT	ligand field theory
LH	light-harvesting; loop of Henle; luteinizing hormone
Li	lithium
LIES	laser-induced emission spectroscopy
LIFS	laser-induced fluorescence spectroscopy

LIPES	laser-induced plasma emission spectroscopy
LIPFS	laser-induced plasma fluorescence spectroscopy
LLDPE	linear low-density polyethylene
LMFB	liquid-metal fast breeder (reactor)
LMO	localized molecular orbital
L/mole	liters per mole
ln	natural logarithm
log	logarithm
LP	liquefied petroleum
LPE	liquid phase epitaxy
LPG	liquefied petroleum gas
Lr	lawrencium
LSD	lysergic acid diethylamide
LTMA	lithium trimethoxyaluminium
Lu	lutetium
LUMO	lowest unoccupied molecular orbital
LVLD	very low-density lipoproteins
m	meter; milli-; molal (concentration)
m	mass
$\hat{m}$	equivalent weight
M	molar (concentration)
$\mathfrak{M}$	molar mass
m^2	square meter
m^3	cubic meter
m_e	electron mass
mA	milliamperes
M.A.	master of arts degree
MALDI	matrix-assisted laser desorption ionization
6-MAM	6-monoacetylmorphine
Mb	million base
MBE	molecular beam epitaxy
MBPT	many-body perturbation theory
MBS	minimum basis sets
MC	Monte Carlo (as in *Monte Carlo simulations*)
MCL	maximum contaminant level
MCLG	maximum contaminant level goal
Md	mendelevium
MD	molecular dynamics
mg	milligram
Mg	magnesium
mg/L	milligrams per liter
MHz	megahertz
min	minute
MINDO	modified intermediate neglect of differential overlap
MIPS	microwave-induced plasma spectroscopy
MIR	multiple internal reflectance
mix	mixture
ml	milliliter

Common Abbreviations and Symbols

mL	millilambert; milliliter	NSAID	nonsteroidal anti-inflammatory drug
MLv	multilamellar vesicle	nsec	nanosecond
mm	millimeter	NSOM	near-field scanning optical microscope
MM	molecular mechanics		
MMPD	*para*-methoxy-*meta*-phenylenediamine	NTD	neural tube defects
Mn	manganese	O	oxygen
Mo	molecular orbital; molybdenum	O.B.E.	Officer of the Order of the British Empire
mol	mole		
MOT	molecular orbital theory	OBP	odorant-binding protein
m.p.	melting point	O.E.D.	Oxford English Dictionary
MRI	magnetic resonance imaging	ORD	optical rotary dispersion
MRS	magnetic resonance spectroscopy	ORN	olfactory receptor neurons
MS	mass spectrometer; mass spectrometry; molar substitution	Os	osmium
		OSHA	Occupational Safety and Health Act; Occupational Safety and Health Administration
msec	millisecond		
MSG	monosodium glutamate		
MSW	municipal solid waste	OTC	over-the-counter
Mt	meitnerium (name proposed for element 109)	P	electron density; mass density; phosphorus; pressure; power; stress
MTBE	methyl *t*-butyl ether		
MVK	methyl vinyl ketone	P_e	critical pressure
MW	megawatt	Pa	pascal; protactinium
n	amount of substance in moles; molecular quantity; nano-; negative; neutral; neutron	PABA	*para*-aminobenzoic acid
		PAH	polycyclic aromatic hydrocarbon
		PAN	peroxyacetyl nitrate
n	neutron; number	Pb	lead
N	newton, nitrogen, nylon	PBB	polybrominated biphenyl
N	normal molecular structure; normal solute concentration	PBG	porphobilinogen
		PBI	polybenzimidazole
N_a	Avogadro's constant	PCB	polychlorinated biphenyl
Na	sodium	PCP	phencyclidine
NAAQS	National Ambient Air Quality Standards	PCR	polymerase chain reaction
		Pd	palladium
NAD	nicotinamide adenine dinucleotide	PDGF	platelet-derived growth factors
NADH	nicotinamide adenine dinucleotide (reduced form)	PDT	photodynamic therapy
		PEA	phenylethylamine
NADP	nicotinamide adenine dinucleotide phosphate	PEL	permissible exposure level
		PEM	protein energy malnutrition
NADPH	nicotinamide adenine dinucleotide phosphate (reduced form)	PET	polyethylene terephthalate; positron emission tomography
Nb	niobium	PETN	pentaerythritol tetranitrate
Nd	neodymium	pH	measure of acidity or alkalinity (from French, *pouvoir hydrogène*, meaning "hydrogen power")
Ne	neon		
NE	norepinephrine		
Ni	nickel	Ph.D.	doctor of philosophy degree
NIH	National Institutes of Health	PIC	product of incomplete combustion
NIMBY	"not in my backyard"	pK_a	the negative logarithm of the acidity constant for the dissociation of a weak acid (the pK_a value increases as the level of acidity decreases)
NIST	National Institute of Standards and Technology		
nm	nanometer		
NMR	nuclear magnetic resonance	PKU	phenylketonuria
No	nobelium	pm	picometer; pyridoxamine
Np	neptunium	Pm	promethium
NRC	Nuclear Regulatory Commission	PMMA	polymethyl methacrylate

PMT	photomultiplier tubes	RSP	reactivity-selectivity principle
Po	polonium	Ru	ruthenium
ppb	parts per billion	Rf	rutherfordium (one of the names variously proposed for element 104 or 106)
ppm	parts per million		
ppt	parts per trillion		
Pr	praseodymium	S	sulfur
PRDDO	partial retention of diatomic differential overlap	S	entropy
		SARA	Superfund Amendments and Reauthorization Act
psi	per square inch; English for ψ		
Pt	platinum	Sb	antimony
PTA	packed tower aeration	Sc	scandium
PTTH	prothoracicotropic hormone	SCF	self-consistent field
Pu	plutonium	Se	selenium
PUFA	polyunsaturated fatty acids	sec	second; secant
PVA	polyvinyl alcohol	SEM	scanning electron microscope
PVAc	polyvinyl acetate	SF	supercritical field
PVB	polyvinyl butyral	SFC	supercritical fluid chromatography
PVC	polyvinyl chloride	SFE	supercritical fluid extraction
q	quantity	Sg	seaborgium (one of the names proposed for element 106)
$\bar{q}$	specific quantity		
$\hat{q}$	equivalent quantity	SHE	standard hydrogen electrode
Q	electric charge	Si	silicon
QM	quantum mechanics	SI	Système Internationale (International System of Measurements)
QSAR	quantitative structure-activity relationships		
		SLV	single lamellar vesicle
QSPR	quantitative structure-property relationships	Sm	samarium
		Sn	tin
Ra	radium	SPF	skin protection factor
rad	radian; radiation-absorbed state	SPM	scanning probe microscope
Rb	rubidium	sr	steradian
RBE	related biological effectiveness	Sr	strontium
RC	reaction center	SRM	Standard Reference Materials
RCRA	Resource Conservation and Recovery Act	STM	scanning tunneling microscope; scanning tunneling microscopy
RDA	recommended daily allowance	STP	standard temperature and pressure (0°C, 1 atm)
RDF	radial distribution function		
rds	rate-determining step	Sv	sievert unit (1 Sv = 100 rem), used to measure radiation dose
Re	rhenium		
REM	radiation-dose unit of measure (acronym stands for "Roentgen Equivalent Man")	t	time
		T	moment of force; thermodynamic temperature (in degrees Kelvin); torque
Rf	rutherfordium	T_c	critical temperture
RF	radiofrequency	T_1, T_2, T_3	various thyroid hormones
RFLP	restriction fragment length polymorphism	Ta	tantalum
		Tb	terbium
RI	refractive index	Tc	technetium
RITC	rhodamine isthiocyanate	TCA	tricarboxylic acid
Rn	radon	TCDD	tetrachlorodibenzo-*para*-dioxin (a pesticide)
RNA	ribonucleic acid		
mRNA	messenger ribonucleic acid	Te	tellurium
tRNA	transfer ribonucleic acid	TEM	transmission electron microscope
RO	reverse osmosis	TFPIA	Textile Fiber Products Identification Act
ROMP	ring-opening metathesis polymerization		
		tg	teragram

Th	thorium
THC	tetrahydrocannabinol
THF	tetrahydrofuran
THM	trihalomethane
Ti	titanium
Tl	thalium
TL	triboluminescence
TLC	thin-layer chromatography
Tm	thulium
TNT	trinitrotoluene
TPA	tissue-plasminogen activator
TPN	triphosphopyridine nucleotide
TPNH	triphosphopyridine nucleotide (reduced form)
TRC	taste-receptor cell
TRXRF	total reflection X-ray fluorescence
TSCA	Toxic Substances Control Act
TSH	thyroid-stimulating hormone
TTHM	total trihalomethane
u	unified atomic mass unit
U	uranium
U	electric potential
dU	denatured uranium
UHF	ultrahigh frequency
Une	temporary symbol for element 109, unnilennium
Unh	temporary symbol for element 106, unnilhexium
Uno	temporary symbol for element 108, unniloctium
Unp	temporary symbol for element 105, unnilpentium
Unq	temporary symbol for element 104, unnilquadium

Uns	temporary symbol for element 107, unnilseptium
UPS	ultraviolet photoelectron spectroscopy
UV	ultraviolet
V	electric potential; vanadium; volume
V	volt
vap.	vaporization
VB	valence bond
vel.	velocity
VIS	visible
viz.	namely (from Latin, *videlicet*)
VOC	volatile organic compounds
vol.	volume (of a book)
vs.	versus
VSEPR	valence shell electron pair repulsion
VUV	vacuum ultraviolet
VX	ethyl diisopropylaminoethyl methylphosphonothiolate (a nerve agent)
W	tungsten; watt
W	mass-fraction
WORM	write once, read many
WWW	World Wide Web
X	halogen atom
Xe	Xenon
XPS	X-ray photoelectron spectroscopy
y	yocto-
Y	yttrium
Yb	ytterbium
z	zepto-
Z	atomic number
Zn	zinc
Zr	zirconium

JOURNAL ABBREVIATIONS

Acc. Chem. Res.
Accounts of Chemical Research
Acta Chim. Scand.
Acta Chimica Scandinavica
Adv. Anal. Toxicol.
Advances in Analytical Toxicology
Advances in Enzymol.
Advances in Enzymology
Adv. in Phys.
Advances in Physics
Adv. Inter. Med.
Advances in Internal Medicine
Adv. Organomet. Chem.
Advances in Organometallic Chemistry
Adv. Phys. Org. Chem.
Advances in Physical and Organic
Chemistry
Adv. Polym. Sci.
Advances in Polymer Science
Adv. Quantum Chem.
Advances in Quantum Chemistry
Am. Chem. Soc.
American Chemical Society Monographs
Amer. J. Sports. Med.
American Journal of Sports Medicine
Amer. Scient.
American Scientist
Am. Fam. Phy.
American Family Physician
Am. J. Phys.
American Journal of Physics
Anal. Biochem.
Analytical Biochemistry
Analyt. Chem.
Analytical Chemistry
Angew. Chem. Int. Ed. Engl.
Angewandte Chemie, International Edition in
English
Ann. Chim.
Annali di Chimica

Ann. Chim. Phys.
Annales de Chimie et de Physique
Ann. Rep. Med. Chem.
Annual Report of Medicinal Chemistry
Ann. Rev. Biochem.
Annual Review of Biochemistry
Ann. Rev. Phys. Chem.
Annual Review of Physical Chemistry
Ann. Sci.
Annals of Science
Antiviral Chem. & Chemother.
Antiviral Chemistry & Chemotherapy
Arch. Gesch. Naturwiss. Technik
Archiv der Geschichter der Naturwissenschaften
und Technik
Arch. Neerl. Sci. Exactes Nat.
Archive for the History of Exact Science
(Netherlands)
Atmos. Sci.
Atmospheric Science Paper
Biochem.
Biochemistry (American Chemical Society)
Biochem. Pharmacol.
Biochemical Pharmacology
Biochem. Z.
Biochemische Zeitschrift
Biochim. Biophysic. Acta
Biochimica et Biophysica Acta
Biog. Mem. Nat. Acad. Sci.
Biographical Memoirs of the National Academy
of Science
Biog. Memoirs. Roy. Soc.
Biographical Memoirs of the Royal Society
(United Kingdom)
Bull. Soc. Chem. Belg.
Bulletin of the Chemical Society of Belgium
CA
Chemical Abstracts
Cancer Res.
Cancer Research

Can. J. Chem.
   Canadian Journal of Chemistry
CAS
   Chemical Abstracts Service
CASSI
   Chemical Abstracts Service Source Index
Chem. & Biol.
   Chemistry and Biology
Chem. Brit.
   Chemistry in Britain
Chem. Commun.
   Chemical Communications
Chem. Educ.
   Chemical Education
Chem. Eng. News
   Chemical and Engineering News
Chem. Phys.
   Chemical Physics
Chem. Phys. Lett.
   Chemical Physics Letters
Chem. Rev.
   Chemical Reviews
Chemtech
   Chemical Technology (now known as Chemtech)
Clin. Chem.
   Clinical Chemistry
Comp. Rend. Acad. Sci.
   Académie des Sciences: Comptes Rendus, Série 2
Contemp. Physics
   Contemporary Physics
C. R. Hebd. Séance. Acad. Sci.
   Comptes Rendus Hebdomadaires des Séances de
      l'Académie des Sciences
Crit. Rev. Oral Biol. Med.
   Critical Reviews in Oral Biology and Medicine
Croatica Chem. Acta
   Croatica Chemica Acta
Drug Metab. Rev.
   Drug Metabolism Review
Ecol. Bull.
   Ecological Bulletin
Edin. N. Phil. J.
   Edinburgh Journal of Natural Philosophy
Educ. Chem.
   Education in Chemistry
Endo. Metab. Clinics N. Am.
   Endocrine and Metabolism Clinics of North
      America
Environ. Monit. Assess.
   Environmental Monitoring and Assessment
Environ. Pollut.
   Environmental Pollution
Exp. Clin. Pharm.
   Experimental and Clinical Pharmacology

Exp. Med.
   Experimental Medicine and Microbiology
FASEB Journal
   Federation of American Societies for
      Experimental Biology Journal
Food Tech.
   Food Technology
Helv. Chim. Acta
   Helvetica Chimica Acta (Switzerland)
Ind. Chem. Eng.
   Industrial Engineering Chemical Research
Inorg. Chem.
   Inorganic Chemistry
Int. J. Quantum Chem.
   International Journal of Quantum Chemistry
Int. J. Quantum Chem. Symp.
   International Journal of Quantum Chemistry
      Symposium
J. Am. Chem. Soc.
   Journal of American Chemical Society
J. Atmos. Sci.
   Journal of Atmospheric Science
J. Catal.
   Journal of Catalysis
J. Chem. Educ.
   Journal of Chemical Education
J. Chem. Phys.
   Journal of Physical Chemistry
J. Chem. Soc.
   Journal of the Chemical Society
J. Chem. Tech. and Biotech.
   Journal of Chemical Technology and
      Biotechnology
J. Electrochem. Soc.
   Journal of the Electrochemical Society
J. Inorg. Nucl. Chem.
   Journal of Inorganic and Nuclear
      Chemistry
J. Laryng. Otol.
   Journal of Laryngology and Otology
J. Math. Physics
   Journal of Mathematical Physics
J. Medicinal Chem.
   Journal of Medicinal Chemistry
J. Membrane Sci.
   Journal of Membrane Science
J. Molec. Struct.
   Journal of Molecular Structure
J. Mol. Med.
   Journal of Molecular Medicine
J. Org. Chem.
   Journal of Organic Chemistry
J. Pharm. Sci.
   Journal of Pharmaceutical Science

J. Phys. Chem.
   Journal of Physical Chemistry
J. Psycho. Drugs
   Journal of Psychoactive Drugs
J. R. Soc. Med.
   Journal of the Royal Society of Medicine
J. Struct. Chem.
   Journal of Structural Chemistry
Liebigs Ann. Chem.
   Liebigs Annalen der Chemie
Makromol. Chem. Symp.
   Makromoleculare Chemie Symposia
Medicinal Chem.
   Medicinal Chemistry
Med. Sci. Sports and Exercise
   Medicine and Science in Sports and Exercise
Med. Tox.
   Medical Toxicology and Adverse Drug
      Experience
Metabol.
   Metabolism: Clinical and Experimental
Mol. Pharmacol.
   Molecular Pharmacology
Nach. Chem. Tech.
   Nachrichten aus Chemie, Technik und
      Laboratorium
Nature Struct. Biol.
   Nature Structural Biology
N. Eng. J. Med.
   New England Journal of Medicine
New J. Chem.
   New Journal of Chemistry
New Sci.
   New Scientist
NIDA Res. Mono.
   National Institute on Drug Abuse Research
      Monographs
Nucl. Phys.
   Nuclear Physics
Pharmacol. & Toxicol.
   Pharmacology and Toxicology
Photochem. Photobiol.
   Photochemistry and Photobiology
Phys. Chem.
   Physiological Chemistry and Physics
Physik. Chem.
   Physikalisch-Chemische Trenn- und
      Messmethoden

Phys. Rev.
   Physiological Review
Phys. Rev. A
   Physical Review A
Phys. Rev. C
   Physical Review C: Nuclear Physics
Phys. Rev. Lett.
   Physical Review Letters
Phys. Today
   Physics Today
Polym. Lett. Ed.
   Polymer Letters Edition of the Journal of
      Polymer Science
Proc. Natl. Acad. Sci. USA
   Proceedings of the National Academy of Science
      of the United States of America
Proc. R. Soc. Edin.
   Proceedings of the Royal Society of Edinburgh
Proc. R. Soc. London
   Proceedings of the Royal Society of London
Progr. Polym. Sci.
   Progress in Polymer Science
Pure Appl. Chem.
   Pure and Applied Chemistry
Quart. J. Roy. Met. Soc.
   Quarterly Journal of the Royal Meteorological
      Society
Rev. Mod. Phys.
   Reviews of Modern Physics
Revs. Geophys. Space Phys.
   Reviews of Geophysics and Space Physics
Roy. Soc. Med.
   Royal Society of Medicine Journal
Sci. Amer.
   Scientific American
Sports Med.
   Sports Medicine
Tech. Rev.
   Technische Revue
Top. Stereochem.
   Topics in Stereochemistry
Vet. Human Toxicology
   Veterinary and Human Toxicology
West. J. Med.
   Western Journal of Medicine
Zeitschr. Chem.
   Zeitschrift für anorganische und allgemeine
      Chemie

s block

	I	II
	3 LITHIUM **Li** 6.941	4 BERYLLIUM **Be** 9.012
	11 SODIUM **Na** 22.990	12 MAGNESIUM **Mg** 24.305
	19 POTASSIUM **K** 39.1	20 CALCIUM **Ca** 40.08
	37 RUBIDIUM **Rb** 85.47	38 STRONTIUM **Sr** 87.62
	55 CESIUM **Cs** 132.90	56 BARIUM **Ba** 137.33
	87 FRANCIUM **Fr** (223)	88 RADIUM **Ra** 226.025

1 HYDROGEN **H** 1.008

d block

21 SCANDIUM **Sc** 44.966	22 TITANIUM **Ti** 47.88	23 VANADIUM **V** 50.942	24 CHROMIUM **Cr** 51.996	25 MANGANESE **Mn** 54.938	26 IRON **Fe** 55.847
39 YTTRIUM **Y** 88.906	40 ZIRCONIUM **Zr** 91.224	41 NIOBIUM **Nb** 92.908	42 MOLYBDENUM **Mo** 95.94	43 TECHNETIUM **Tc** (98)	44 RUTHENIUM **Ru** 101.07
71 LUTETIUM **Lu** 174.967	72 HAFNIUM **Hf** 178.49	73 TANTALUM **Ta** 180.948	74 TUNGSTEN **W** 183.85	75 RHENIUM **Re** 186.207	76 OSMIUM **Os** 190.2
103 LAWRENCIUM **Lr** (260)	104 RUTHERFORDIUM **Rf** (261)	105 DUBNIUM **Db** (262)	106 SEABORGIUM **Sg** (263)	107 BOHRIUM **Bh** (262)	108 HASSIUM **Hs** (265)

f block

57 LANTHANUM **La** 38.906	58 CERIUM **Ce** 140.115	59 PRAESEODYMIUM **Pr** 140.908	60 NEODYMIUM **Nd** 144.24	61 PROMETHIUM **Pm** (145)	62 SAMARIUM **Sm** 150.36	63 EUROPIUM **Eu** 151.965	64 GADOLINIUM **Gd** 157.25	65 TERBIUM **Tb** 158.925
89 ACTINIUM **Ac** 227.03	90 THORIUM **Th** 232.038	91 PROTACTINIUM **Pa** 231.036	92 URANIUM **U** 238.029	93 NEPTUNIUM **Np** 237.048	94 PLUTONIUM **Pu** (244)	95 AMERICIUM **Am** (243)	96 CURIUM **Cm** (247)	97 BERKELIUM **Bk** (247)

*Each element in the table is listed with (from top to bottom) its atomic number, its name, its symbol, and its atomic mass. Atomic mass numbers in parentheses are the mass numbers of the longest-lived isotope. Other atomic mass numbers are the average mass number of the naturally occurring isotopes.

†The names and labels for elements beyond number 103 are controversial. IUPAC initially ruled in favor of Latin names based on atomic number, but in 1994 a set of specific names and symbols was suggested. After considerable debate, a revised final list of names for elements 104-109 was issued on August 30, 1997. Temporary names were also assigned for elements 110 and 111. Various groups have suggested alternative names for some of these elements. Additional elements continue to be synthesized, though with increasing difficulty, with no definite upper atomic-number limit yet established.

						VIII
						2 HELIUM **He** 4.003
	III	IV	V	VI	VII	

p

III	IV	V	VI	VII	VIII
5 BORON **B** 10.811	**6** CARBON **C** 12.011	**7** NITROGEN **N** 14.007	**8** OXYGEN **O** 15.999	**9** FLUORINE **F** 18.998	**10** NEON **Ne** 20.180
13 ALUMINUM **Al** 26.982	**14** SILICON **Si** 28.086	**15** PHOSPHORUS **P** 30.974	**16** SULFUR **S** 32.066	**17** CHLORINE **Cl** 35.453	**18** ARGON **Ar** 39.948

27 COBALT **Co** 58.933	**28** NICKEL **Ni** 58.69	**29** COPPER **Cu** 63.546	**30** ZINC **Zn** 65.38	**31** GALLIUM **Ga** 69.723	**32** GERMANIUM **Ge** 72.61	**33** ARSENIC **As** 74.922	**34** SELENIUM **Se** 78.96	**35** BROMINE **Br** 79.904	**36** KRYPTON **Kr** 83.80
45 RHODIUM **Rh** 102.906	**46** PALLADIUM **Pd** 106.42	**47** SILVER **Ag** 107.868	**48** CADMIUM **Cd** 112.411	**49** INDIUM **In** 114.82	**50** TIN **Sn** 118.71	**51** ANTIMONY **Sb** 121.75	**52** TELLURIUM **Te** 127.60	**53** IODINE **I** 126.905	**54** XENON **Xe** 131.29
77 IRIDIUM **Ir** 192.22	**78** PLATINUM **Pt** 195.08	**79** GOLD **Au** 196.967	**80** MERCURY **Hg** 200.59	**81** THALLIUM **Tl** 204.383	**82** LEAD **Pb** 207.2	**83** BISMUTH **Bi** 208.980	**84** POLONIUM **Po** (209)	**85** ASTATINE **At** (210)	**86** RADON **Rn** (222)
109 MEITNERIUM **Mt** (266)	**110** UNUNILIUM **Uun** (269)	**111** UNUNUNIUM **Uuu** (272)	**112** (?)						

66 DYSPROSIUM **Dy** 162.50	**67** HOLMIUM **Ho** 164.93	**68** ERBIUM **Er** 167.26	**69** THULIUM **Tm** 168.934	**70** YTTERBIUM **Yb** 173.04
98 CALIFORNIUM **Cf** (251)	**99** EINSTEINIUM **Es** (252)	**100** FERMIUM **Fm** (257)	**101** MENDELEVIUM **Md** (258)	**102** NOBELIUM **No** (259)

Q

Qualitative Analysis

Qualitative analysis has evolved into a broad field of chemistry concerned with the detection, identification, and characterization of materials found in the universe. These materials may be elements, ionic or covalent compounds, or mixtures. The analytical sample may be from the environment—soil, water, rocks, the air around us—or from living systems—plant and animal tissues, blood, the foodstuffs humans eat. Both organic and inorganic compounds and substances may be included in the analytical sample.

To characterize or analyze the sample is to measure its physical properties and determine its chemical nature. Typical physical properties that are measured are melting point, boiling point, density, and heat and electrical conductivity. Color, texture, malleability, and ductility are properties that also help describe substances. Instruments have been developed to measure these physical properties, and many different techniques and methodologies have been developed to chemically analyze a substance. First, there are the "wet chemical" methods that depend on the solution chemistry of ionic compounds and the functional group chemistry of organic compounds. Also, a number of instrumental techniques, such as MASS SPECTROMETRY, NUCLEAR MAGNETIC RESONANCE (NMR) spectrometry, and INFRARED SPECTROMETRY, are utilized in the chemical analysis of compounds.

The traditional wet chemical methods are used to identify cations and anions in aqueous solutions. It would be ideal if a given reagent would specifically identify a single cation with a uniquely characteristic chemical reaction. Usually this is not possible, and cations of chemically similar groups are separated by precipitation reactions using specific group reagents. The requirements for a group reagent are that the metal ions of interest precipitate completely, free from contamination of other cations, and these precipitates can be easily filtered from solution. Extensive and elaborate qualitative analysis schemes are available to separate and identify the common cations. A brief example of such a scheme using group reagents is presented below to illustrate these ideas. Consider a solution containing the cations Ag^+, Cu^{2+}, Fe^{2+}, Pb^{2+}, and follow the scheme.

Similar schemes have been devised to separate and identify the common anions, such as nitrates, sulfates, carbonates, and chlorides. These qualitative analysis methods to identify cations and anions in a sample are based on solution equilibrium concepts.

Organic compounds can be characterized by taking advantage of functional group chemical reactions. An organic functional group is that part of the organic molecule that characterizes the molecule according to its unique chemical reactions with certain reagents. Some of the common functional groups are alcohols, aldehydes, ketones, and carboxylic acids. These can be distinguished one from the other using functional group reactions. For example, aldehydes

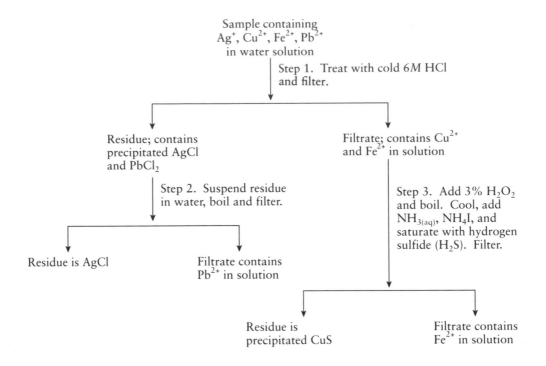

Sample containing
Ag^+, Cu^{2+}, Fe^{2+}, Pb^{2+}
in water solution

Step 1. Treat with cold 6M HCl
and filter.

Residue; contains
precipitated AgCl
and $PbCl_2$

Filtrate; contains Cu^{2+}
and Fe^{2+} in solution

Step 2. Suspend residue
in water, boil and filter.

Step 3. Add 3% H_2O_2
and boil. Cool, add
$NH_{3(aq)}$, NH_4I, and
saturate with hydrogen
sulfide (H_2S). Filter.

Residue is AgCl

Filtrate contains
Pb^{2+} in solution

Residue is
precipitated CuS

Filtrate contains
Fe^{2+} in solution

and alcohols are oxidizable, while ketones and carboxylic acids are not readily oxidizable. Also, aldehydes can be oxidized by weak oxidizing agents that will not oxidize alcohols. There are many specific chemical reactions that can be used to detect specific functional groups, and these reactions are the basis for schemes to distinguish organic compounds. It is important to be able to detect organic functional groups because they are found in the many important biological compounds in living systems.

Organic compounds have more elaborate and complex structures than do most inorganic cations and anions. Thus, elemental analysis and functional group analysis are only a part of the characterization of an organic compound. Instrumental analysis provides the information to completely define the structure of an organic molecule.

Modern instruments have been developed to do chemical analysis based on the fundamental structural properties of atoms and molecules. For example, excited atoms emit a spectrum of radiant energy unique for each metal or metalloid. Emission spectrographs are used to detect this radiant energy for qualitative elemental analysis. This technique is used to analyze the metal content of ore samples and other samples that may contain metals. Also, emission spectrographs have been used to detect elements in stars.

Before an organic or biological compound can be completely identified, it must be in a reasonably pure form. The separation of the compound from its reaction environment can be done by a combination of extraction and chromatographic techniques. Extraction techniques make use of the solubility properties of the compound in various solvents such as water, alcohols, ether, and toluene. Gas chromatography and high-performance liquid chromatography can be used to further purify the compound. Other useful separation techniques are electrophoresis, electro-chromatography, gel permeation chromatography, and supercritical fluid chromatography.

To begin to identify and characterize an organic or biological compound, elemental analysis may be done to establish the empirical formula of the compound. This analysis can be accomplished using an instrument called an automatic elemental analyzer to determine the percent of carbon, hydrogen, and nitrogen in the compound. Then the molecular weight can be determined using a mass spectrometer. Finally, the analyst can determine the functional group characteristics of the compound using instrumental techniques such as NMR, infrared, and ultraviolet spectrometries.

The mass spectrometer (MS) is an instrument that distinguishes ionized atoms and molecules of different masses. The principle of mass spectrometry was used first by J. J. THOMSON in 1913 to show that neon gas is composed of the isotopes neon 20 and neon 22. Early uses of the mass spectrometer were for the detection of elements and for the determination of atomic and molecular weights. Today, the MS is used for molecular analysis and identification of compounds by comparing the unique mass spectrum of the sample molecule with the spectra of known compounds. Another important use of the MS is for characterization of polymers. The process is also being used to determine the amino acid sequence in protein molecules and to elucidate the structure of the human genome.

Electromagnetic radiation interacts with atoms and molecules, causing them to become excited; the results of these excitations are used to elucidate the various aspects of molecular structure. NMR, infrared, and ultraviolet spectrometries are three of several techniques that make use of the interaction of radiant energy with matter to do functional group analysis.

Molecules can absorb radiant energy in the radio frequency range, and this absorption affects the spinning nuclei of the molecule. This effect is the basis of NMR spectrometry. NMR can be used to detect different functional groups and their relative positions in the molecule. Thus, NMR is a technique that can help identify and characterize the molecule. Magnetic resonance imaging (MRI) is an application of NMR used to detect cancerous cells and brain tumors in humans. MRI is a noninvasive and nondestructive analytical technique and thus is a valuable tool in medical diagnosis.

When a molecule absorbs radiant energy in the infrared (IR) region, the amplitude of the vibrational motions of the molecule increases. These molecular vibrations are characteristic of various structural features of organic functional groups. Thus, the absorption of IR energy by molecules can be used to determine which functional groups are present in the molecules. Absorption of radiant energy in the ultraviolet (UV) and visible (VIS) regions causes electronic transitions in the molecule. This is the basis of UV-VIS spectroscopy. UV spectrometry is of value in characterizing molecules with unsaturated bonds.

There exist a number of other instrumental techniques that are used to describe the properties of substances. X-ray diffraction is used to determine crystal lattice dimensions, and this capability is useful in soil, natural product, and polymer analysis. X-ray absorption locates high-atomic-weight elements in molecules and is used in bone-density studies. Optical rotary dispersion and circular dichroism are used to obtain stereochemical information that allows the determination of the absolute configuration of chiral compounds.

The application of these instrumental techniques in characterizing chemical compounds is simplified by comparing the spectrum of the unknown compound with the spectra of known compounds. The various spectra (MS, NMR, IR, etc.) of many compounds have been determined: the information is stored in computer databases and can be accessed for use in identifying compounds. In summary, qualitative analysis uses a variety of techniques to detect, identify, and describe the material substances in our world.

BIBLIOGRAPHY

McAlpine, R. K.; Soule, B. A. *Prescott and Johnson's Qualitative Chemical Analysis;* New York: D. Van Nostrand, 1933.

Moeller, T. *Qualitative Analysis;* New York: McGraw-Hill, 1958.

Robinson, J. W. *Undergraduate Instrumental Analysis,* 5th ed.: New York: Marcel Dekker, 1995.

Willard, H. H.; Merrit, L. L.; Dean, J. A.; Settle, F. A. *Instrumental Methods of Analysis,* 7th ed.; New York: Wadsworth, 1988.

JAMES E. HARDCASTLE

Quality Control

In chemistry, quality control can be considered on two levels. The first level is that of the chemist in the individual research laboratory. A chemist who synthesizes compounds must do some level of quality control to verify that the desired compounds have been made and isolated in a pure state. The second level of quality control is the more formal quality control needed by a chemical production company. Chemicals manufactured and sold by a company must meet customer specifications and, in some cases, government regulations. Similar techniques are used to establish the quality of the chemicals in individual labs and for production. The main difference is that companies need to follow carefully established protocols while individual chemists choose their methods on a case-by-case basis. In contrast to the small quantities of material prepared for use within an individual research laboratory, the large-scale production and distribution of an incorrectly identified or impure compound may have important economic, environmental, and human health consequences.

When describing a chemical reaction, a chemist may write a chemical equation such as

$$A + B \rightarrow C$$

The above equation gives the impression that all of compound A reacted with all of compound B to form only compound C. In reality, this is seldom true. Oftentimes, not all of the starting material is consumed in the reaction. Furthermore side reactions often occur, resulting in the formation of more than one product. For example, the free radical chlorination of propane results in two different chloropropanes.

$$2\,CH_3CH_2CH_3\;+\;Cl_2$$

Propane   Chlorine

$$\xrightarrow{\text{light}}\;\underset{\text{1-Chloropropane}}{\overset{\displaystyle\overset{Cl}{|}}{CH_2CH_2CH_3}}\;+\;\underset{\text{2-Chloropropane}}{\overset{\displaystyle\overset{Cl}{|}}{CH_3CHCH_3}}$$

In addition to the two products shown, there will almost certainly be some unreacted propane as well as some dichloropropane in the crude product mixture. Because chemicals in a flask seldom behave like the idealized written reaction, chemists have developed methods to verify that the compound synthesized is in fact the desired product and methods to ascertain its purity. The techniques used to follow the progress of a reaction and to identify its products constitute quality control.

Part of quality control is knowing the limitations of the assessment technique used for these evaluations. The simplest question is how does the actual weight of the product(s) formed compare to the expected or theoretical weight of the desired product? The theoretical weight is determined by calculating the maximum amount of product that would form if all of the limiting reagent were converted successfully into product.

$$\frac{\text{Actual weight of product}}{\text{Theoretical weight}} \times 100\% = \frac{\text{Percent yield}}{\text{of reaction}}$$

If the percent yield is higher than 100 percent, then clearly an impurity is present. A yield of less than 100 percent does not imply that the material is 100 percent pure, or even that the desired compound is present at all. The only information gained from comparing masses is how the mass of the crude product compares with the expected amount. It requires more time-consuming and expensive spectroscopic and chromatographic techniques to determine the purity and prove the identity of the product.

A widely used and simple test to answer the question of how many compounds are present in a flask is THIN-LAYER CHROMATOGRAPHY (TLC). A small drop of the crude reaction mixture is spotted onto an absorbent-coated glass plate. One end of the plate is immersed in solvent, and capillary action moves the solvent up the plate, carrying the spotted chemicals with it. Compounds present in the reaction mixture move to different heights on the plate depending on how strongly they bind to the absorbent and how soluble they are in the solvent. Because different compounds are carried to different heights on the plate, it is easy to determine whether more than one compound is present. A variety of techniques are used to visualize the compounds on the TLC plate. Because some compounds show up more strongly than others with the techniques used to visualize spots on the TLC plates, the relative intensities of the spots do not necessarily correspond to the relative amounts of compounds present.

After the desired compound has been separated from contaminating substances, more definitive techniques are used to prove its identity. For example, the molecular weight of a substance can be determined using MASS SPECTROMETRY. The relative amounts of different elements present in a compound can be determined using combustion analysis. The types of carbons and hydrogens in a molecule (e.g., C=O, $H_3C—CH_3$, or $H_2C=CH_2$) can be distinguished using NUCLEAR MAGNETIC RESONANCE (NMR) spectroscopy. The presence or absence of certain molecular fragments such as C=O or N—H can be ascertained using INFRARED SPECTROMETRY. Because these different techniques provide complementary information, most scientific journals require new compounds to be characterized by several different spectroscopic tests. Once known spectra are established for a compound, merely producing one or two types of spectra that match the reference spectra of that compound is considered proof that the compound in question has been successfully resynthesized.

TLC provides useful information about how many compounds are present in a flask, but, like percent yield, TLC does not provide quantitative information regarding the level of purity of a compound. Other techniques must be used to determine precisely how pure a compound is. Proton (^{1}H) NMR can detect protonated contaminants that make up at least 3–5 percent of the sample. Impurities of lower than 3 percent are not reliably detected by NMR, nor are contaminants of any percentage if that contaminant does not have any hydrogen atoms. For example, a sample of chloroform ($CHCl_3$) that is heavily contaminated by carbon tetrachloride (CCl_4) will produce the same ^{1}H NMR spectrum as pure chloroform. Since carbon tetrachloride does not have any hydrogens, it is invisible to ^{1}H NMR. Other techniques, such as specific gravity measurements, are needed to detect the presence of this contaminant.

Because of the inherent limitations of individual techniques, it is prudent to use more than one method to determine the purity of a compound.

An important aspect of quality control is ascertaining the level of quality needed. The required purity of the compound and the number of identifying techniques used for its characterization are determined in part by its intended use. Road salt will not be held to the same standards of purity as table salt, because salt used to prevent ice on roads is not designated for human consumption. While it might seem desirable to exceed the required purity, it is not always a good use of resources. It is possible to spend an hour performing a chemical reaction and then spend a week, or even months, separating the desired product from all the unreacted starting materials and side products formed in the reaction. Chemists in industry need to consider that because of their salary and benefits, their time is an important factor in the cost of an experiment.

Sometimes the purity standards are chosen by the individual chemist, while in other cases standards are externally mandated. A research chemist needs compounds pure enough for future experiments. If a chemist wants to have a research article published in a journal, the compounds that are described in that publication must meet that journal's quality control guidelines, or else that article will be rejected by the journal. A company that manufactures chemicals for sale is held to a variety of quality control standards depending on the expected use of the compound and the country in which the company is located. Some purity standards are dictated by the customer's expectations, while others are legally mandated. Companies also have fiscal incentives to maintain quality levels because mistakes in the large-scale production of chemicals are costly. To ensure that all the company products meet necessary standards, many companies have designated quality control departments that specialize in documenting that all the company's products meet all pertinent requirements.

A quality control laboratory is generally expected to be current in all regulatory requirements. For example, a pharmaceutical company must ensure that the drugs it produces meet the Food and Drug Administration's (FDA) requirements for purity. Part of this requirement is knowing and using the exact methods required by the FDA to establish the drug's purity. Pharmaceutical companies face legal action if they sell drugs that do not meet the FDA's requirements. Another important requirement is keeping appropriate records and providing requested documentation at any time. Therefore, the responsibilities of a quality control laboratory involve knowing the customer's expectations, performing the analytical chemistry, and documenting the quality control.

BIBLIOGRAPHY

BEYER, H.; WALTER, W. *Handbook of Organic Chemistry;* London: Prentice-Hall, 1996; pp. 3–9, 29–48.

KAPLAN R. "Current GMP Considerations for Biotechnology Facilities, Part 1: Applicability of GMP's, Technology Transfer, and Quality Control." *Biopharm: The Technology and Business of Biopharmaceuticals,* January–February 1995, *8,* 26–30.

LITTLE, L. E. "Keeping on Track: Current Quality Control Trends." *Biopharm: The Technology and Business of Biopharmaceuticals,* May 1996, *9,* 72–77.

SILVERSTEIN, R. M.; BASSLER, G. C.; MORRILL, T. C. *Spectrometric Identification of Organic Compounds,* 4th ed.; New York: Wiley, 1981.

WILLIG, S. H.; STOKER, J. R. *Good Manufacturing Practices for Pharmaceuticals: A Plan for Total Quality Control;* New York: Marcel Dekker, 1996.

BARBARA TUCKER

Quantum Chemistry

Early in the twentieth century it became apparent that many traditional, "classical" concepts of physical reality could no longer be regarded as fundamental. As physicists and chemists examined the behavior of matter at the atomic and subatomic level, it became apparent that there was a need to examine phenomena from a different perspective. Rather than focusing on static physical entities that are subject to linear causal transformations, there was a new emphasis placed on the energy levels and rates of change of systems that typically consisted of sets of moving particles. Matter was seen less as fixed and more as an example of a perpetual state of becoming. Energy became in a sense as integral a part of reality as matter.

The concept of the Hamiltonian, or Hamiltonian function, had been introduced by the mathematician Sir William Rowan Hamilton in 1835. The Hamiltonian of a system indicates the total energy of that system, which consists of kinetic energy (the motion exhibited by a system) plus potential energy (as determined by the positions of the particles that make up a system).

If one can write the Hamilton H, which expresses all of the interparticle interactions within a given system, SCHRÖDINGER'S WAVE EQUATION yields the energy E. Quantum chemical calculations are generally applied to a single molecule or to a small grouping of molecules. The Hamiltonian thus includes the attractions between each positively charged nucleus and all of the negatively charged electrons in the entire system. To this sum must be added the mutual repulsions between each pair of electrons. (There are also present nuclear-nuclear repulsions that must be counted. However, this procedure is greatly simplified by the Born-Oppenheimer approximation, which states that the much more massive nuclei can be treated as very nearly fixed in space in comparison to the lighter electrons.)

The difficulty in solving the Schrödinger equation for a molecule arises from the form of the wave function Ψ (see PSI FUNCTION), which depends upon all electronic positions simultaneously. Taking r_1 as a designation of the position of electron 1, r_2 for electron 2, and so forth, if Ψ $(r_1, r_2, r_3, \ldots)$ could be factored into a product of simpler one-electron wave functions $\phi_1(r_1)\phi_2(r_2)\phi_3(r_3) \ldots$, solution of the problem would be straightforward. The hurdle blocking this factorization is the presence in the Hamiltonian of interelectronic repulsions, which depend upon the position of not one but two electrons.

In an effort to circumvent this difficulty, D. R. Hartree developed in the 1920s an approximation whereby the positions of all electrons in the system, save one, are "time-averaged" in a sort of time-lapse photograph. What the first electron sees then is not a series of snapshots of instantaneous positions of the other electrons, but rather, it moves in the time-average field produced by their long-term motions. This approach reduces the problem to dimensions of one particle, which may be solved by an iterative approach, that is, the average field produced by the $n - 1$ electrons is first guessed at. Solution of the one-particle Schrödinger equation leads to a better guess for the average field. This process is repeated until the field suffers only very minimal changes from one cycle to the next and one has achieved the self-consistent-field (or SCF) solution to the problem. Because of the contributions of Z. Fock, who introduced electron antisymmetry into the original Hartree formulation, the aforementioned iterative procedure is frequently referred to as the Hartree-Fock solution of the Schrödinger equation.

A problem that arises in solving the Hartree-Fock equations is associated with the form that the one-electron wave function $\phi_1(r_1)$ should take. One answer to this question was provided by C. C. J. Roothaan, who suggested in 1951 that $\phi_1(r_1)$ might be constructed as a linear combination of atomic orbitals, the so-called LCAO treatment. This approach assumes that the orbitals of atoms are only changed to a limited extent when atoms are combined to form molecules. In other words

$$\phi_1(r_1) = c_a\chi_a + c_b\chi_b + c_c\chi_c + \cdots \quad (1)$$

where each χ_i represents a hydrogen-like atomic orbital—e.g., $1s$, $2p_y$, and so forth, centered on one of the nuclei of the molecule. The set $\{\chi_i\}$ is referred to as the basis set for the problem. Solving for self-consistency then reduces to a determination of the set of coefficients c_i.

The variation theorem of quantum mechanics provides a convenient mechanism for evaluating these coefficients. This theorem states that the energy obtained from plugging any approximate "trial" wave function into the Schrödinger equation can never be lower than the exact energy that is obtained in principle from the fully correct wave function. Knowing that there exists this "floor," below which the energy cannot sink, one takes as the best approximate wave function that which produces the lowest or minimum energy. Incorporating the Roothaan form of the wave function (equation 1) into the variation theorem yields a matric form of the Hartree-Fock equations, which may be solved by numerical determinant techniques.

At the conclusion of the aforementioned procedure, one obtains a number of molecular orbitals (MOs) of the type described by equation 1. Each MO typically contains contributions from all of the atoms present in the molecule and extends from one end of the molecule to the other. Each MO reflects the overall symmetry of the molecule by corresponding to a particular irreducible representation of the molecular point group. For a standard closed-shell molecule, the first $N/2$ MOs are occupied by 2 electrons each (where N = total number of electrons). In addition, there are a number of unoccupied, or "virtual," MOs that also conform to the symmetry of the molecule. These vacant MOs have little physical significance but play a major role in procedures for evaluating corrections to the Hartree-Fock energy (see below). The aforementioned "canonical" MOs may be transformed to a set of "localized" orbitals by various procedures, the most popular of which was developed by C. Edmiston and K. Ruedenberg in 1963. These localized MOs are concentrated in smaller regions of space and conform more closely to the chemist's traditional notions of bonding and lone electron pairs (see LONE PAIR).

An early step in the SCF computations consists of evaluating all the integrals that will be needed later. There are first a number of one-electron integrals, which are related to the kinetic energy (see KINETIC MOLECULAR THEORY) of an electron in a given atomic orbital, and potential energy of attraction to the core of nuclei. Much more numerous, however, are the two-electron integrals, which correspond to interelectronic repulsion. Not only are each of the latter more time-consuming in their evaluation, but there are a great number of them that must be evaluated and stored for later use. More specifically, the number of these integrals climbs as the fourth power of the number of atomic orbitals. It is hence the two-electron integral evaluation that is the bottleneck in most Hartree-Fock computations.

One of the key issues in quantum chemical calculations is the quality of the basis set used to represent each atom. Minimal basis sets contain one function for each orbital that would normally be occupied by one or two electrons in the ground state of a given atom. For example, the minimal basis set of hydrogen would contain a single $1s$ function while $1s, 2s, 2p_x$, $2p_y, 2p_z$ would be needed for a first-row atom like oxygen. Double-zeta basis sets represent each of these atomic orbitals by a pair of functions. Polarization functions refer to orbitals with higher quantum number l than is normally occupied—e.g., p for H, or d for O. There is consensus that the larger and more flexible the basis set used—i.e., the more functions are used to represent each atom—the more reliable will be the final result (although there are cases where smaller basis sets may better reproduce a given quantity through cancellation of error).

An important question in calculations of these types is the form of the function used to represent each atomic orbital. It is generally considered a good idea to use orbitals that resemble the functions resulting from the exact solution of the one-electron hydrogen atom Schrödinger equation. Each of these functions contains a polynomial of the distance from the nucleus r, along with a function $\exp(-\zeta r)$ that dies off exponentially with distance. One of the earliest sets of functions used to approximate these hydrogen-like orbitals in a computationally tractable manner were Slater orbitals, which replace the polynomial by a single power of r and retain the exponential. The ζ factor is typically chosen by minimizing the Hartree-Fock energy of the atom or by taking into account nuclear shielding of one set of electrons by those closer to the nucleus. The year 1960 marks the appearance of the first rigorous calculation of a polyatomic molecule using these Slater functions by A. D. McLean.

Despite the removal of the polynomial function, integral evaluation via Slater orbitals is not very efficient. One response to this problem was suggested in 1950 by S. F. Boys as the approximation of each Slater orbital in turn by a linear combination of Gaussian functions $\exp(-\alpha r^2)$, which square the radial variable in the exponential. A major advantage of Gaussians is that the product of two of them on different centers is another single Gaussian function, albeit with a different factor α and centered somewhere on a line connecting the two original centers. Although the replacement of a single Slater by a number of Gaussians increases the number of integrals that must be evaluated prior to beginning the self-consistency procedure, the much simpler form of the integrals more than offsets their larger quantity. For this reason, Gaussians are usually the functions of choice in modern quantum chemistry.

The term *closed shell* refers to a molecule wherein all electrons occur in pairs of opposite spin, and all orbitals are doubly occupied (or completely empty). Such systems can be treated by restricted Hartree-Fock (RHF) methods wherein the MOs of the electrons of one spin (α) are assumed to be identical to those occupied by β electrons of opposite spin. If, however, there are electrons that do not have a mate of opposite spin in the same orbital, the system is denoted an open shell and is typically somewhat more difficult to deal with. By relaxing the restriction that α and β MOs be identical, and optimizing each set separately, one can adopt the unrestricted Hartree-Fock (UHF) approach, as first developed by J. A. Pople and R. K. Nesbet in 1954. While computationally tractable, the drawback of this method is that the single configuration is generally not a pure spin state—doublet, triplet, etc. An alternative restricted open-shell Hartree-Fock (ROHF) approach retains the symmetry between α and β MOs but enforces strict maintenance of the spin of the state.

Even with a very large basis set, the Hartree-Fock wave function and energy can be expected to be significantly in error. The difficulty originates in the nature of the Hartree-Fock approximation itself, which time-averages the motion of the electrons. In fact, the interelectronic repulsion depends instantaneously on the distance between any pair of electrons, which in reality fluctuates very rapidly. By correlating their motions so as to maximize their distance from one another, electrons can stabilize the system via what is known as instantaneous electron correlation. The difference in energy between a Hartree-Fock calculation and another one that explicitly includes this phenomenon is termed the correlation energy.

Most quantum chemical calculations carried out prior to around 1980 were restricted to the Hartree-Fock level. While it was understood by practitioners that the total energy calculated in this manner for a given molecule was not very good, it was hoped that in many cases the errors would cancel from one calculation to the next. Taking the simple unimolecular rearrangement reaction A $\rightarrow$ B as an example, one would ordinarily not compute the reaction energy directly but rather as the difference in total energy between A and B. Both E(A) and E(B) will be in error without accounting for electron correlation. However, in many cases, the correlation energy of A will be quite similar in magnitude to that of B. Hence, the difference may be of quite high quality thanks to a fortuitous cancellation of errors.

The earliest means of computing the correlation energy was introduced by E. A. Hylleraas in 1928. The procedure was later termed configuration interaction (CI) and consists of building a super wave function as a superposition of simpler ones. The base, or reference, configuration is the Hartree-Fock wave function. Excited configurations are generated by removing one or more electrons from an occupied molecular orbital and placing them in vacant MOs. The relative contribution of each excited configuration is assessed via the variation theorem, which yields also the correlation energy. If one were to include all possible excitations and perform what is termed a complete CI, one would obtain the exact energy for a given molecule (subject, of course, to the inherent limitations of the particular basis set employed). Unfortunately, the number of configurations rises dramatically as the system studied is enlarged, and complete CI calculations are possible only for very small molecules, even with modern computers. Therefore, one is forced to limit the number of configurations in some way.

A common limitation scheme first categorizes the configurations in terms of the reference configuration. Taking one electron from an occupied to a vacant molecular orbital is termed a single excitation, a double refers to excitation of two electrons, and so on. The list of configurations to be included in a given CI calculation can be truncated at any level. One of the more typical along these lines is CISD, which refers to CI calculations including all single and double excitations from the reference.

An alternative approach to CI, and one that is non-variational, is now called "coupled cluster" and has its roots in work by J. Cizek in the mid 1960s. The basis of this approach is treatment of the correlation through an exponential operator.

A different approach to estimating the correlation energy is via perturbation theory. The latter is a general mathematical formalism for approximating the solution to a problem that is nominally insoluble if one can first obtain an exact solution to a similar problem. The difference between the real and the solvable situation is defined as a perturbation to the system, and the solution to the real situation is constructed as an expansion series of terms, which

hopefully get progressively smaller. Termination of the series at a given order can therefore provide a fairly accurate approximation.

In this case, the solvable problem is the Hartree-Fock situation. The difference between this and the exact Hamiltonian is then added as the perturbation of the system, as suggested by C. Møller and M. S. Plesset. When taken to first order, the perturbation series reproduces the Hartree-Fock solution. It is at higher orders that the electron correlation begins to appear in the wave function and energy. This means of dealing with correlation has been termed either many-body perturbation theory (MBPT) or Møller-Plesset (MP) perturbation theory. Perturbation theory is in principle a suitable treatment when the perturbation is small, i.e., when the solvable and unsolvable problems do not differ from each other by very much. Since interelectronic repulsion is not a minor effect, the perturbation series might be expected to converge slowly. However, experience has shown that in many cases second-order Møller-Plesset perturbation theory (MP2)—i.e., the first term in the expansion of the correlation energy—can furnish a very useful result, including many of the important effects of correlation.

During the early years of quantum chemistry, when computers were slower than at present and had considerably less efficient storage and retrieval capabilities, calculations on even small molecules could be prohibitively expensive. In order to provide a cheaper route to large systems, methods were developed that bypass some of the bottlenecks. These semiempirical quantum chemical methods differ from the *ab initio* approach in that the former avoid direct computation of many of the integrals that are needed in the latter (see also AB INITIO CALCULATIONS and SEMI-EMPIRICAL QUANTUM MECHANICAL CALCULATIONS). Some of the integrals are simply ignored since they are assumed to be small. Others are approximated by schemes of varying complexity. Some of the assumptions make use of experimental information—e.g., ELECTRON AFFINITIES or ionization potentials (see IONIZATION ENERGY) of the atoms concerned. Although these semiempirical calculations are formally carried out at the Hartree-Fock level, the incorporation of experimental data folds into them in some complex manner certain effects that are normally associated with correlation phenomena. Direct comparison between semiempirical and *ab initio* results can therefore resemble the apples-and-oranges paradigm. Each semiempirical method is typically identified by an acronym that helps describe some of its underlying assumptions. Some of the more popular examples include

- CNDO: Complete Neglect of Differential Overlap
- MINDO: Modified Intermediate Neglect of Differential Overlap
- PRDDO: Partial Retention of Diatomic Differential Overlap

Although each advance in computer technology enlarges the scope of systems that may be treated by *ab initio* methods, semiempirical procedures are always capable of considering still larger systems. Their biggest drawback relates to the approximations that go into their formulation, which limit their reliability. It is possible in the case of *ab initio* methods to treat a given system with a series of levels of theory that become progressively more accurate (e.g., larger basis sets or higher levels of correlation) and thereby approach the real situation in a controlled, asymptotic fashion. The same is not true for semiempirical methods, where each method is different in its basic formulation. A given calculation will yield only one answer, the accuracy of which can be judged only by the past performance of that particular method on similar systems.

Once a given *ab initio* method is chosen—e.g. Hartree-Fock or MP2—it is necessary to select a particular basis set to represent the atomic orbitals of the atoms concerned. Before the calculation can begin, one must specify the positions (coordinates) of all nuclei in the molecule. Actual designation of coordinates is usually no longer necessary but has been replaced in most quantum chemical programs by a listing instead of relative atomic positions via bond lengths and angles. Some modern graphical interfaces permit the user to build a three-dimensional model of the molecule on the display screen.

The user is faced with the question of how to choose the bond lengths and angles for a given molecule. One approach is to take these properties from experimental information. However, detailed geometries are not typically available for any molecule one wishes to study. When available, the information may not be of high enough accuracy, specifying bond lengths to perhaps only the nearest 0.01 Å. The environment in which the determination of geometry was made may not correspond to the situation of interest. For example, X-ray diffraction geometries generally correspond to a crystalline environment, whereas one may be interested in the gas phase. A common procedure to bypass these problems consists of locating that geometry which is most stable for the molecule. This geometry optimization will locate the global minimum on the potential energy surface of the molecule. It must be understood, however, that the geometry which is most stable for one basis set or level of theory will not be identical to that from another. Nonetheless, the optimized geometry is considered a valuable starting point.

Early quantum chemical calculations searched for the minimum by beginning with one parameter, e.g., a bond length, and finding the value that gave the lowest energy of the molecule. A second parameter was then minimized in the same way, followed sequentially by others. However, since the change in one parameter typically affects the minimum in another, several very tedious iterations were necessary prior to completion of the minimization. This problem has been largely overcome by modern programs that search for the minimum via a gradient algorithm that considers all parameters simultaneously.

At the minimum of the surface, it is possible to evaluate the normal modes of vibration, yielding the frequencies and intensities of the vibrational spectrum. Although not in principle necessary, these calculations usually assume a completely harmonic vibrational force field for computational efficiency. Experience has shown that Hartree-Fock–level calculations commonly overestimate bond-stretching frequencies by some 10 percent. This error can be compensated either by scaling factors or by computing the force field at a correlated level. Vibrational intensities have more stringent criteria for accurate evaluation, frequently requiring correlation for even quantative estimates.

Quantum calculations can make contributions to other areas of spectroscopy as well. Rotational (microwave) spectra are determined largely by the equilibrium geometry of the molecule, which leads to the principal moments of inertia. Electronic spectra deal with energy differences between one electronic state and another, a quantity that these calculations can address.

Quantum chemistry offers other types of information not generally accessible by experimental measurements. The molecular orbitals obtained from a calculation provide details of the distribution of electron density throughout the molecule. One can thus identify regions around the molecule that would tend to attract electropositive or negative reactants. The electronic distribution can also be partitioned into moments of varying degree—e.g., DIPOLE, quadrupole. These methods provide a convenient means to investigate extensive regions of the potential energy surface not amenable to experimental inquiry. Very short-lived species like reaction TRANSITION STATEs can be identified and characterized.

In short, quantum chemical calculations are ideal for studying the geometric and electronic structure of given species, even those that are unstable to the point of decomposition. Reactions may be followed and characterized. Potential energy surfaces may be generated that can be used by dynamic methods to extract reaction kinetic data.

BIBLIOGRAPHY

HEHRE, W. J.; RADOM, L.; SCHLEYER, P. v. R.; POPLE, J. A. *Ab Initio Molecular Oribital Theory;* New York: Wiley, 1986.

POPLE, J. A.; BEVERIDGE, D. L. *Approximate Molecular Orbital Theory;* New York: McGraw-Hill, 1970.

SADLEJ, J. *Semi-Empirical Methods of Quantum Chemistry;* New York: Wiley, 1985.

SCHAEFER, H. F., III. *Quantum Chemistry;* Oxford: Clarendon, 1984.

SZABO, A.; OSTLUND, N. S. *Modern Quantum Chemistry;* New York: Macmillan, 1982.

STEVEN SCHEINER

Quantum Mechanical Computations

It was very soon apparent following Ernest SCHRÖDINGER's presentation of quantum mechanics in 1926 that chemistry would be a principal beneficiary of this new theoretical approach to science. Perhaps the most visionary statement in this regard was that of Nobel physics laureate Paul DIRAC (1929): "The underlying physical laws necessary for the mathematical theory of a large part of physics and the whole of the chemistry are thus completely known, and the difficulty is only that the exact application of these laws leads to equations much too complicated to be soluble."

Unfortunately many years went by before quantitative quantum mechanics began to bear significant fruit of a chemical nature. In fact, it may be stated that prior to 1970, except for some important pioneering work on diatomic molecules, all of the contributions of quantum mechanics to chemistry were qualitative in nature, as typified perhaps most brilliantly by the insights of Robert Mulliken, A. D. Walsh, and Roald Hoffmann.

It began to be clear in about the year 1950 that the future of quantum mechanics in chemistry lay with automatic computing machines. Thus the heroic work of the 1950s in THEORETICAL CHEMISTRY began to focus on computational methods as well as formal advances. In particular one must note the pioneering contributions of Frank Boys (Gaussian functions and the configuration interaction method), Klaus Ruedenberg (evaluation of electron repulsion integrals), Clemens Roothaan (the matrix Hartree-Fock equations), Robert Parr (general methodology), John Pople (the treatment of unpaired spins), Robert Nesbet (configuration interaction), Albert Matsen (early work on diatomic molecules), and Per-Olov Löwdin and Harrison Shull (natural orbitals).

The decade of the 1960s again yielded many theoretical advances and the beginnings of reliable applications to atoms and diatomic molecules. But the year 1970 has frequently been identified as the origin of the "coming of age" of computational quantum chemistry. In light of the low regard in which *ab initio* quantum chemistry was held at the time (1970), it was perhaps inevitable that this fledgling field would first make its mark via some spectacular clash with experiment (see AB INITIO CALCULATIONS for a discussion of the field). That clash occurred with the unequivocal and challenging prediction that the methylene molecule was bent (bond angle 135°) rather than linear, as concluded by the brilliant spectroscopist Gerhard HERZBERG from experiments described in his Nobel Prize citation.

No discussion of the theoretical contributions to the methylene problem is complete without reference to the pioneering work of Frank Boys. Although, for reasons that will become apparent, Boys's work did not contribute to the paradigm in view here, it is a monument to the efforts of early workers in the field of *ab initio* molecular quantum mechanics. In fact, their study of CH_2 was one of the earliest *ab initio* investigations of any nonlinear polyatomic molecule. The study of Boys and his colleagues was purely predictive in nature, since there were no published spectroscopic data for CH_2 when their paper was submitted in 1959. Thirteen different geometries were investigated, resulting in a triplet ground state equilibrium geometry of 129°. Boys did not challenge experimental evidence for the simple reason that there were no experiments to challenge. Understandably, at this early stage in the development of *ab initio* theory, the authors made no particular claims concerning the quantitative reliability of their predictions.

The presentation of Herzberg's 1961 Bakerian lecture boded ill for the theoretical efforts of Boys. On the basis of his analysis of the electronic spectrum of CH_2, Herzberg concluded that the molecule in its triplet ground state is linear. Herzberg noted the work of Boys, stating, "Our experimental values are distinctly different from these predictions but not excessively so when the approximate nature of the calculations is considered." The completion of the discrediting of Boys's paper was carried out by Christopher Longuet-Higgins in 1962. At the time Longuet-Higgins was professor and head of theoretical chemistry at Cambridge, while Boys occupied a lesser post in the same department. Furthermore, Longuet-Higgins was widely known for his negative attitude toward theorists such as Boys who made serious use of computers. Longuet-Higgins prefaced his 1962 paper with the words, "It may be that future theoretical progress will require elaborate variational calculations such as those of Boys on CH_2, but until the results of such machine experiments can be interpreted physically, there would seem to be a place for more empirical theories such as that which we now describe." Longuet-Higgins then proceeded to show from his nonrigorous (i.e., choosing parameters to reproduce an experiment) scheme that the triplet ground state of CH_2 is linear. His result not only questioned Boys's 1960 paper on CH_2, but along with Herzberg's remarks quoted above, it also cast grave doubt on the overall trustworthiness of any *ab initio* procedure. Certainly if such methods were unable to predict the structure of a molecule as simple as CH_2, they would appear to be of little value.

The 1970 theoretical treatment by Charles Bender and Henry Schaefer brought to bear on the methylene problem theoretical methods that had previously been applied only to atoms and diatomic molecules. The contracted Gaussian double-zeta basis set developed by Thom Dunning was used in conjunction with self-consistent field (SCF) and configuration interaction (CI) methods. The CI treatment included the SCF wave function plus all configurations of $^3B_1$ symmetry arising from orbital occupancies differing by one or two orbitals from the SCF function, with the restriction that the $1a_1$-orbital be always doubly occupied. This yielded a total of 408 $^3B_1$ configurations, which was a rather large variational treatment for the year 1970. An iterative natural orbital procedure was then used in several subsequent CI calculations to obtain the most rapidly convergent expansion of the above type.

The CI potential energy surface for triplet methylene is displayed in Figure 1 (see page 1280) and shows that the molecule is bent. The precise bond angle predicted by Bender and Schaefer was 135.1°. This bent structure was found to lie 7 kcal per mole below the optimized linear geometry, and it was concluded that no higher level of theory could remove such a barrier. Thus methylene became the first polyatomic molecule for which it could be claimed that theory was more reliable than the conclusions of a distinguished experimentalist. And thus began methylene's role as a paradigm for quantitative quantum chemistry. An interesting aside is that Herzberg's original analysis allowed for the possibility that CH_2 was bent, but he eventually chose the linear structure because it seemed more reasonable.

Essentially concurrently with the paper of Bender and Schaefer appeared the experimental electron-spin resonance (ESR) study of Robert Bernheim and colleagues. They state, "Our measurements, which were made on CH_2 trapped in xenon at 4.2K, verify that the ground state is indeed triplet and show that triplet CH_2 is slightly bent under these conditions." These investigators did not speculate on whether the observed slight bending was due to their xenon matrix as opposed to the gas-phase conditions of Herzberg's experiments. However, there was no indication of a perceived serious disagreement with Herzberg, nor was a specific value of the CH_2 bond angle put forth.

Three months later (November 15, 1970), Edel Wasserman reported a second ESR study of triplet methylene. He approached the structural dilemma more aggressively, concluding that the CH_2 triplet bond angle lies between 128° and 143°. The most probable value was 136°, in very close agreement with Bender and Schaefer's prediction of 135.1°.

The work of Wasserman was followed by a second paper from Bernheim's group, who determined that the bond angle is 137.7° from additional observations on CH_2. At this point, the laboratory findings of the two ESR groups led Herzberg to reexamine his deductions from electronic spectroscopy. He graciously concluded, "The considerations given here point strongly toward the bent structure of the triplet ground state ($^3B_1$) of CH_2, as first suggested by the electron-spin resonance work and the *ab initio* calculations."

The experimental papers by Bernheim and Phillip Skell, by Wasserman, and by Herzberg forced many to reconsider the role of *ab initio* theory in chemistry. Instead of having been an embarrassing early failure, the work of Boys was now shown to be qualitatively correct. Equally important, state-of-the-art theory had challenged the most distinguished spectroscopic group in the world and been vindicated. A new role for theory as partner with experiment had been charted.

Of course it is now possible to study in a reliable manner much larger molecules than CH_2 using rigorous quantum mechanical methods. John Pople has argued that since the year 1960, comparable advances in quantum chemistry have resulted from (a) improvements in mathematical/theoretical/computational methods and (b) steady increases in computing power per dollar. Since the two components have a multiplicative effect on the amount of chemistry that can be accomplished by computer, the future is extremely promising.

The structure of the remainder of this contribution will be determined by the most commonly used theoretical method in quantum chemistry, namely the Hartree-Fock (HF), or self-consistent field (SCF), theory. Only *ab initio* quantum mechanical methods are discussed here. It should perhaps be noted that we use the term *ab initio* to imply that no approximations have been made in the evaluation of the one- and two-electron integrals:

$$I(i|j) = \int \phi_i^*(1) \left\{ \frac{-\nabla_1^2}{2} - \sum_A \frac{Z_A}{r_{1A}} \right\} \phi_j(1) \, dv(1) \quad (1)$$

$$(ij|kl) = j\phi_i^*(1)\phi_k^*(2) \frac{1}{r_{12}} \phi_i(1) \, \phi_1(2) \, dv(1) \, dv(2) \quad (2)$$

arising from the ordinary nonrelativistic Hamiltonian

$$H = \sum_i \left\{ \frac{-\nabla_1^2}{2} - \sum_A \frac{Z_A}{r_{1A}} \right\} + \sum_i \sum_{j>i} \frac{1}{r_{ij}} \quad (3)$$

In contrast, semiempirical methods resort to various approximate schemes, especially in evaluating the electron-repulsion integrals $(ij|kl)$.

For closed-shell molecules the Hartree-Fock wave function is of the general form

$$\Psi_{HF} = A(n) \, \phi_1(1) \, \phi_2(2) \, \ldots \, \phi_n(n) \quad (4)$$

in which $A(n)$, the antisymmetrizer for n electrons, has the effect of making a Slater determinant out of the orbital product on which it operates. The ϕ's are spin orbitals, products of a spatial orbital χ and a one-electron spin function α or β. For any given molecule system, there are an infinite number of wave functions of form (4), but the Hartree-Fock wave function is the one for which the orbitals ϕ have been varied to yield the lowest possible total energy:

$$E = \int \Psi_{HF}^* H \Psi_{HF} d\tau \quad (5)$$

The resulting Hartree-Fock equations are relatively tractable due to the simple form of the energy expression for single-determinant wave functions:

$$E_{HF} = \sum_i I(i|i) + \sum_i \sum_{j>1} [(ij|ij) - (ij|ji)] \quad (6)$$

To make this discussion more concrete we note that for singlet methylene ($^1A_1$ CH_2) the Hartree-Fock wave function is of the general form

$$\Psi_{HF} = A(8) \, 1a_1\alpha(1) \, 1a_1\beta(2) \, 2a_1\alpha(3) \, 2a_1\beta(4)$$
$$1b_2\alpha(5) \, 1b_2\beta(6) \, 3a_1\alpha(7) \, 3a_1\beta \quad (7)$$

The same energy expression (6) is also applicable to any open-shell system for which the open-shell electrons all have parallel spins. This follows from the fact that such Hartree-Fock wave functions can always be expressed as a single Slater determinant. A simple example is the case of triplet methylene ($^3B_1$) discussed earlier:

$$\Psi_{HF} = A(8) \, 1a_1\alpha(1) \, 1a_1\beta(2) \, 2a_1\alpha(3) \, 2a_1\beta(4)$$
$$1b_2\alpha(5) \, 1b_2\beta(6) \, 3a_1\alpha(7) \, 1b_1\alpha \quad (8)$$

for which the outer two $3a_1$- and $1b_1$-orbitals have parallel spins. For clarity it is often helpful to abbreviate (8):

$$\Psi_{HF} = 1a_1^2 \, 2a_1^2 \, 1b_2^2 \, 3a_1\alpha \, 1b_1\alpha \quad (9)$$

Although solution of the Hartree-Fock equations for an open-shell system such as triplet methylene is more difficult than for the analogous closed-shell system (7), the procedures are well established.

In fact, methods are available for the solution of the Hartree-Fock equations for *any* system for which the energy expression involves only coulomb and exchange integrals:

$$J_{ij} = (ij|ij) \quad (10)$$

$$K_{ij} = (ij|ji) \quad (11)$$

Open-shell singlets are a class of systems that can be treated in this way, and one such example is the first excited singlet state (of $^1B_1$ symmetry) of methylene:

$$\Psi_{HF} = \frac{1}{\sqrt{2}} 1a_1^2 \, 2a_1^2 \, 1b_2^2 \, 3a_1\alpha \, 1b_1\beta$$

$$- \frac{1}{\sqrt{2}} 1a_1^2 \, 2a_1^2 \, 1b_2^2 \, 3a_1\beta \, 1b_1\alpha \quad (12)$$

In addition, these same generalized Hartree-Fock procedures can be used for certain classes of multi-configuration Hartree-Fock wave functions.

To solve the Hartree-Fock equations exactly one must either expand the orbitals ϕ in a complete set of analytic basis functions or obtain strictly numerical (i.e., tabulated) orbitals. The former approach is impossible from a practical point of view for systems with more than two electrons, and the latter has been accomplished only for atoms and, more recently, for a few diatomic molecules. Therefore the *exact* solution of the Hartree-Fock equations is abandoned for polyatomic molecules. Instead one adopts an incomplete (but reasonable) set of analytic basis functions and solves for the best variational (i.e., lowest energy [5]) wave function of form (4). Following the University of Chicago group (Clemens Roothaan), we refer to such a wave function as being of self-consistent-field (SCF) quality. At the limit of very large basis sets, then, it is reasonable to refer to the resulting SCF wave function as near Hartree-Fock.

For very large chemical systems with no elements of symmetry, at present only minimum basis sets (MBS) can be used in all-electron *ab initio* theoretical studies. A minimum basis set includes one set of functions for each orbital occupied in the ground state of each atom included in the molecule. For the first-row atoms, B, C, N, O and F, this means that a minimum basis includes 1s-, 2s-, $2p_x$-, $2p_y$-, and $2p_z$-functions. By the term *large*, we include molecular systems with 500 or more electrons. A large system (possessing no symmetry) first treated in 1988 by the MBS SCF method is the cyclic undecapeptide (closely related to cyclosporin A) $C_{63}H_{113}N_{11}O_{12}$, with 664 electrons and 543 basis functions. Even more impressive for 1988, Robert Harrison and Martyn Guest were able to obtain an SCF wave function for the same 664 e^- system with an even larger basis set of 1,000 contracted functions. An even larger molecule conquered in 1991 with the *ab initio* MBS SCF method is the six-base-pair fragment of DNA studied by Michael Colvin. The formula for this C_1 symmetry macromolecule is $C_{116}H_{138}N_{46}O_{68}P_{10}$, and the STO-3G computation employed 1,378 contracted Gaussian basis functions.

When elements of point group symmetry are present, much larger systems may be considered with full geometry optimization. For example, Gustavo Scuseria in 1993 studied the "giant fullerene" C_{240} molecule (1,440 electrons, 1,200 basis functions) at the MBS SCF level of theory. Complete geometrical determinations were achieved for structures of I_h and D_{5h} symmetries. Moreover, for the I_h-structure, a single *ab initio* SCF wave function was determined using a double-zeta (see below) basis set of 2,400 contracted Gaussian functions.

Traditionally, minimum basis sets have been composed of Slater functions, such as those seen in Table 1 for the carbon atom. However, experience has shown that the evaluation of the molecular integrals (2) arising when Slater functions are employed is extremely time-consuming. Therefore each Slater function in an MBS is typically replaced by a linear combination of three or four Gaussian functions. The resulting chemical predictions obtained with such STO-3G or STO-4G basis sets are usually indistinguishable from the corresponding Slater function results.

As we shall see shortly, minimum basis sets are inadequate for certain types of chemical predictions. Therefore, a basis twice as large, and appropriately designated double zeta (DZ), is often used in theoreti-

cal studies. Here, however, it is not as fruitful to expand each Slater function as a linear combination of Gaussians. Instead, one uses Gaussian functions $x^p y^q z^r e^{-\alpha r^2}$ directly in atomic SCF calculations and then contracts them according to the atomic results. Perhaps the most widely used contracted Gaussian DZ basis sets are those of Thom Dunning, and his carbon atom basis is seen in Table 2. This basis has 9s

Table 1. Minimum basis set of Slater functions for the carbon atom. The orbital exponents zeta are optimum for the ^3P ground state of the carbon atom.

Label	Analytic Form	Exponent ζ
1s	$(\zeta^3/\pi)^{1/2} \exp(-\zeta r)$	5.673
2s	$(\zeta^5/\pi)^{1/2} r \exp(-\zeta r)$	1.608
$2p_x$	$(\zeta^5/\pi)^{1/2} x \exp(-\zeta r)$	1.568
$2p_y$	$(\zeta^5/\pi)^{1/2} y \exp(-\zeta r)$	1.568
$2p_z$	$(\zeta^5/\pi)^{1/2} z \exp(-\zeta r)$	1.568

Table 2. Double-zeta contracted Gaussian basis set (9s5p/4s2p) for the carbon atom ^3P ground state. Primitive Gaussian s functions are of the form $(2\alpha/\pi)^{3/4} \exp(-\alpha r^2)$.

Function Type	Exponent α	Contraction Coefficient
s	4232.61	0.002029
s	634.882	0.015535
s	146.097	0.075411
s	42.4974	0.257121
s	14.1892	0.596555
s	1.966	0.242517
s	5.1477	1.000000
s	0.4962	1.000000
s	0.1533	1.000000
p	18.1557	0.018534
p	3.9864	0.115442
p	1.1429	0.386206
p	0.3594	0.640089
p	0.1146	1.000000

and 5p original (or primitive) Gaussian functions and is contracted to 4s and 2p. Thus, the contracted basis may be designed C(9s5p/4s2p).

Just as the double-zeta basis logically follows the minimum set, the logical extension of the DZ set involves the addition of polarization functions. Polarization functions are of higher orbital angular momentum than the functions occupied in the atomic SCF wave function. That is, for carbon, d-, f-, g-, ... functions will be polarization functions. Fortunately, d-functions are far more important than f-, f-functions are far more important than g-, and so on. For most chemical applications a double-zeta plus polarization (DZP) basis including a single set of five d-functions ($d_{x^2-y^2}$, d_{z^2}, d_{xy}, d_{xz}, d_{yz}) will be adequate for first-row atoms.

Ab initio theoretical methods initially had the greatest impact on chemistry in the area of structural predictions. A good illustration of this is the methylene radical discussed earlier. For many chemists the structure of triplet methylene was the first genuine example of the usefulness of *ab initio* theoretical chemistry.

Turning from the specific to the more general, the most encouraging aspect of *ab initio* geometry predictions is their perhaps surprising reliability. Most molecular structures appear to be reliably predicted at the Hartree-Fock level of theory. Even more encouraging, many structures are accurately reproduced using only MBS SCF methods. As shown first by John Pople and coworkers, this is especially true for hydrocarbons. A fairly typical example is methylenecyclopropane, seen in Figure 1, and its MBS SCF,

Table 3. Minimum basis set (MBS) self-consistent-field (SCF) geometry prediction compared with the experiment for methylenecyclopropane (see Figure 2).

Parameter	STO-3G	Experiment
r $C_{(1)} = C_{(2)}$ (Å)	1.298	1.332
r $C_{(2)} - C_{(3)}$ (Å)	1.474	1.457
r $C_{(3)} - C_{(4)}$ (Å)	1.522	1.542
r $C_{(1)} - H_{(1)}$ (Å)	1.083	1.088
r $C_{(3)} - H_{(3)}$ (Å)	1.083	1.09
$\theta H_{(1)}C_{(1)}H_{(2)}$ (deg.)	116.0	114.3
$\theta H_{(3)}C_{(3)}H_{(4)}$ (deg.)	113.6	113.5
$\theta H_{(3)}C_{(3)}C_{(4)}$ (deg.)	149.4	150.8

structure is compared with the experiment in Table 3. Carbon-carbon bond distances differ typically by 0.02Å from experiment, and angles are rarely in error by more than a few degrees. In the same pioneering 1975 paper, Warren Hehre and Pople reported full MBS SCF geometry optimizations for thirteen other hydrocarbons containing four carbon atoms, and the results are equally impressive. Even for severely strained molecules such as bicyclo[1,1,0] butane, very reasonable agreement with the experimental structure was obtained. It is noteworthy that for only half of the C_4 hydrocarbons studied were experimental geometries available. Thus, for many purposes theory may be considered complementary or even superior to experiment in the area of structure prediction.

For molecules including atoms in addition to C and H, MBS SCF results are sometimes less reliable. This first became apparent from the work of Pople and coworkers in 1970. For example, the F_2N_2 molecule has MBS SCF bond instances $r(N—F) = 1.384$Å and $r(N=N) = 1.214$Å, which are respectively 0.107Å longer and 0.169Å shorter than the experimental values. A careful treatment of the electron correlation problem, using high-level coupled cluster methods, is needed for the quantitative prediction of difficult molecular structures such as F_2N_2.

In general, double-zeta (DZ) SCF structure predictions are considerably more reliable than those based on minimum basis sets. A noteworthy exception is

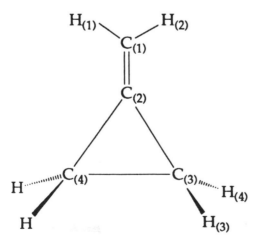

Fig. 1. The methylenecycloprane molecule, for which a complete *ab initio* structure determination was reported in 1975 and is seen in Table 3.

the water molecule, for which MBS SCF yields a bond angle of 100.0° and DZ SCF predicts 112.6°, compared to the well-known experimental value of 104.5°. The addition of oxygen-atom d-functions brings the water structure into reasonable agreement with the experiment. More typical are the HF and F_2 molecules, for which the MBS, DZ, and experimental bond distances are 0.956Å, 0.922Å, and 0.917Å (HF) and 1.315Å, 1.400Å, and 1.417Å (F_2). In fact, it can be argued that if one is not going beyond the Hartree-Fock (single configuration) approximation, DZ SCF is often a reasonable stopping point.

In recent years, methods of much higher theoretical sophistication than Hartree-Fock theory have been developed and extensively tested for small molecules, e.g., up to the size of benzene. Truly quantitative chemical predictions become possible with these new methods. Most notable in this regard are the configuration-interaction and coupled-cluster methods. As these higher-level theoretical methods are further developed and as computing speeds continue to accelerate, such theoretical techniques will become applicable to ever larger molecular systems.

BIBLIOGRAPHY

SCHAEFER, H. F. *The Electronic Structure of Atoms and Molecules: A Survey of Rigorous Quantum Mechanical Results;* Reading, MA: Addison-Wesley, 1972.

———. "Methylene: A Paradigm for Computational Quantum Chemistry." *Science* 1986, *231*, 1100–1107.

———. *Quantum Chemistry: The Development of* Ab Initio *Methods in Molecular Electronic Structure Theory;* Oxford: Clarendon, 1984.

———. "The Third Age of Quantum Chemistry." *Chimia* 1989, *43*, 1–3.

YAMAGUCHI, Y.; OSAMURA, Y.; GODDARD, J. D.; SCHAEFER, H. F. *A New Dimension to Quantum Chemistry: Analytic Derivative Methods in* Ab Initio *Molecular Electronic Structure Theory;* Oxford: Clarendon, 1994.

HENRY F. SCHAEFER

Quantum Numbers

Modern quantum mechanics characterizes the microscopic world with models that assume the existence of discrete states for a system of interest (atom, molecule, ion, or subatomic particle). Each state has specific values for the energy, ANGULAR MOMENTUM, and other quantities that describe the system. Since these values are fixed when the system is in a particular state, they are quantized. At any given time the system is in only one of the states available to it. To specify the state of a system, one can give the actual values for a sufficient number of quantities such as the energy, angular momentum, or the probability distribution for particle coordinates. A more convenient way is to use one or more numbers (integers or half integers) called quantum numbers that convey this information in a succinct form. The most common quantum mechanical method used to model atoms employs four quantum numbers to specify the state of each electron in an atom. By knowing the meaning of each quantum number, one can form a mental image of an atom that is useful in understanding its position in the periodic table, its physical properties, and its chemistry.

Niels BOHR proposed the first quantized model of any atom in 1913. The BOHR MODEL OF THE HYDROGEN ATOM drew heavily on classical mechanics and postulated fixed circular orbits for the electron. He introduced quantization in an ad hoc fashion by restricting the angular momentum of the electron to values of $n(h/2\pi)$, where h is Planck's constant and n is a quantum number with values of $n = 1, 2, 3, \ldots, \infty$. This quantum number specifies the state of the system and is a parameter in the equations for the energy of the state and the radius of the orbit. Bohr's model was useful but limited, and in 1926 a model proposed by Erwin SCHRÖDINGER replaced it. Schrödinger's model for hydrogen centers on a differential equation with an infinite number of solutions, each of which is characterized by three quantum numbers that arise during the solution (see also SCHRÖDINGER'S WAVE EQUATION). The quantum numbers indirectly specify the energy, the angular momentum, and the magnetic properties of the electron in the atom. A fourth quantum number describing the spin of the electron must still be introduced in an ad hoc fashion, but it does arise naturally in a more abstract relativistic treatment. The quantum numbers also help one to form visual images of the regions of space (orbitals) in which the electron is most likely to be found for a given state. By appropriate approximations and adjustments these solutions for hydrogen adequately describe each of the electrons in a polyelectronic atom as well and are used to rationalize periodic trends. From a chemist's perspective, one of the most powerful applications of this model is its usefulness in understanding CHEMICAL BONDING.

The first of the four quantum numbers is the principal quantum number. Its symbol is the letter n, and it has a range of $n = 1, 2, 3, \ldots, \infty$. Occasionally it is called the primary quantum number. For the hydrogen atom it alone determines the energy of the state, but for other atoms both the n and l (see below) quantum numbers determine the energy of a given electron. Since in general a number of states (orbitals) have the same value of n (or n and l for polyelectronic atoms), there are a number of orbitals with the same energy. These states are said to be degenerate. For polyelectronic atoms the energy of a state increases as $n + l$ increases, and within a given value of $n + l$, energy increases as n increases.

The second quantum number has at least four names in common use: angular momentum, azimuthal, secondary, and subshell. The preferred term is angular momentum. Its symbol is l, and its range is $0, 1, \ldots, n - 1$. Hence for a given value of n, the l quantum number can assume n different values. For convenience the symbols $s, p, d, f, g, h \ldots$ are used in place of $l = 0, 1, 2, 3, 4, 5 \ldots$ The l quantum number determines the angular momentum of an electron. More importantly, from a chemist's perspective it determines the general shape of an orbital. Orbitals with $l = 0$ (s) are spherical, those with $l = 1$ (p) are dumbbell shaped, and those with $l = 2$ (d) are mostly double dumbbell shaped. Knowing these shapes is very useful when considering a directional view of chemical bonding.

The third quantum number is the magnetic quantum number. Its symbol is m or m_l, and it has a range of $-l, -l, +1, \ldots 0 \ldots l - 1, l$. For a given value of l, m can have $2l + 1$ values. For example, when $l = 2$, m can be $-2, -1, 0, 1, 2$. This quantum number determines the magnetic properties of the electron arising from electron motion about the nucleus. It also determines the orientation of an orbital. For example, the three orbitals with $l = 1$ (p) and hence $m = -1, 0, 1$ are all dumbbell shaped but lie along three mutually perpendicular axes. This information about an atom can often be used to rationalize the shape of a molecule containing that atom.

The fourth quantum number is the electron-spin quantum number. Its symbol is m_s, and it can have the values $+\frac{1}{2}$ or $-\frac{1}{2}$. This quantum number arises because the electron behaves as if it is spinning, and consequently the spinning, charged electron creates a magnetic dipole (effectively a bar magnet), which has two quantized energy states in the presence of an external magnetic field. In the absence of an external field, the two states are degenerate. Although this quantum number does not arise naturally from the solution of the Schrödinger equation for hydrogen, there is ample experimental evidence for it.

The Schrödinger equation for polyelectronic atoms can be written but cannot be solved analytically. Fortunately the solutions for hydrogen can be used to model polyelectronic atoms in an approximate but very useful way. The same four quantum numbers apply, and the electronic structure of the atom is modeled by adding electrons to orbitals of increasing energy (increasing $n + l$; increasing n within a given value of $n + l$). The restrictions on the values of l for a given value of n and on the value of m for a given value of l still apply. However, a fundamental principle of quantum mechanics (PAULI'S EXCLUSION PRINCIPLE) states that no two electrons can have the same four quantum numbers, so two electrons with the same n, l, and m must have different m_s. Effectively, then, no more than two electrons can be in the same orbital, and if so, they must have different spins (i.e., one $m_s = +\frac{1}{2}$, one $m_s = -\frac{1}{2}$).

BIBLIOGRAPHY

ATKINS, P. W. *Physical Chemistry*, 5th ed.; New York: Freeman, 1993; Chapter 13.

BRADY, J. E.; HOLUM, J. R. *Chemistry: The Study of Matter and Its Change;* New York: John Wiley, 1993; Chapter 5.

HANNA, M. W. *Quantum Mechanics in Chemistry;* Menlo Park, CA: Benjamin/Cummings, 1981; Chapter 6.

PERRINO, C. T.; PETERSON, D. L. "Another Quantum Number." *J. Chem. Educ.* 1989, 66, 623.

J. ROBERT PIPAL

Quantum Theory

Quantum theory refers to the notion that energy and other properties of microscopic particles are "quantized"—i.e., occur in small, discrete packages. For example, the electron in a hydrogen atom has available only a number of energy levels that it can occupy, with no intervening energies allowed. This behavior differs from "classical" particles, which are permitted a continuum of allowed levels. Quantum theory can be applied to electromagnetic radiation as well due to its particle nature. The quantized units of electromagnetic energy are referred to as PHOTONS, with energy proportional to the frequency of the radiation. Application of the tenets of quantum theory to atoms and molecules leads to a subset commonly referred to as QUANTUM CHEMISTRY.

STEVEN SCHEINER

Quinone Functional Group

The quinone functional group is the conjugated diketone oxidation product of arenes. Benzene, the simplest arene, has two quinone constitutional iso-

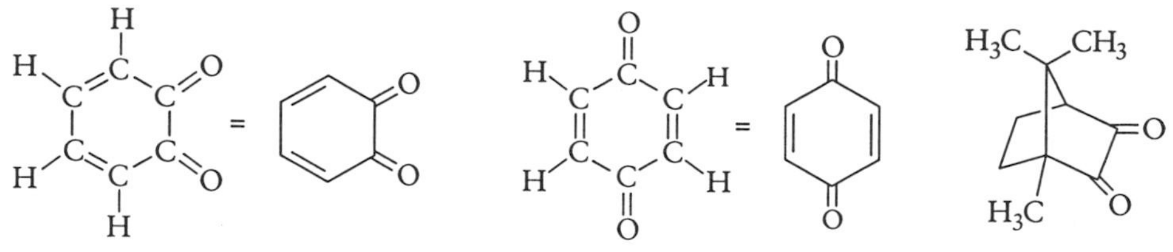

Fig. 1. (1) 1,2-Benzoquinone; (2) 1,4-benzoquinone; (3) camphor quinone.

mers: the 1,2-benzoquinone (an *ortho*-quinone) 1 and the 1,4-benzoquinone (a *para*-quinone) 2. The *ortho*-quinones, due to their adjacent electron-withdrawing ketones, are the more reactive. In the old chemical literature the term *quinone* also described derivatives of natural products having nonenolizable 1,2-diketones, such as the product from SeO$_2$ oxidation of camphor, camphor quinone 3. This usage is

archaic, yet it persists. Quinones are important molecules in organic chemical synthesis, in industry, and in nature. Their unique properties include the reactivity of their electron-deficient alkene (important in synthetic reactions and in nature); their color, which is variable depending on the quinone substituents (important in industry); and their ability to engage in reversible electron transfer (as electron carriers in nature).

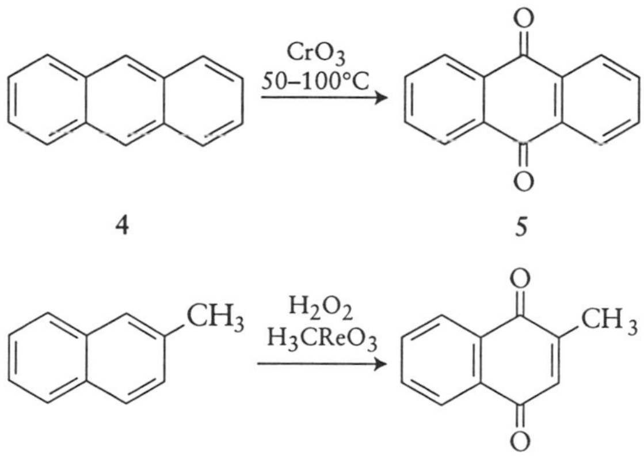

Synthesis of Quinones

The industrial synthesis of quinones, the 6e$^-$ oxidation product (corresponding to two oxygenations and a dehydrogenation) of arenes, is done by vigorous arene oxidation. Anthracene 4—an easily oxidized arene—is converted efficiently to the anthraquinone 5 (an important intermediate in dye synthesis) by reaction with CrO$_3$, a strong oxidant, at 50°–100°C. This oxidation produces, however, vast quantities of reduced chromium that are difficult to recycle. New quinone syntheses reflect improved hydrocarbon oxidation methods. An example (that formerly used Cr(VI) oxidation) is the conversion of 2-methylnaphthalene 6 to menadione (vitamin K$_3$) 7, an important nutrient of animal feed. This reaction uses the environmentally acceptable H$_2$O$_2$ (hydrogen peroxide) as the stoichiometric oxidant, and the

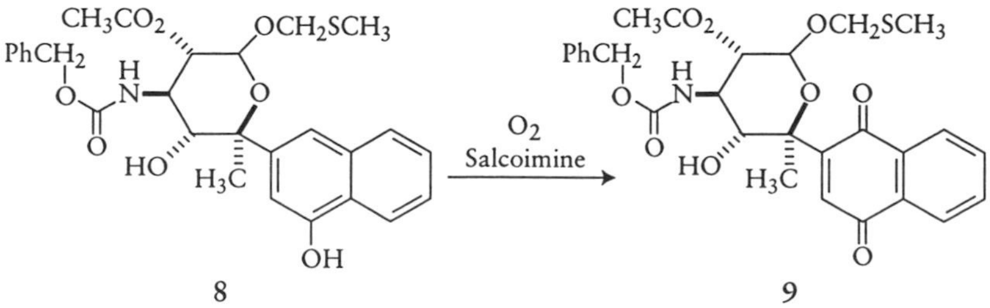

Fig. 2. (4) Anthraceine; (5) anthraquinone (5,10-anthracenedione); (6) 2-methylnaphthalene; (7) menadione (vitamin K$_3$).

Fig. 3. (8) Naphthoquinone; (9) the result of salcomine oxidation of naphthoquinone.

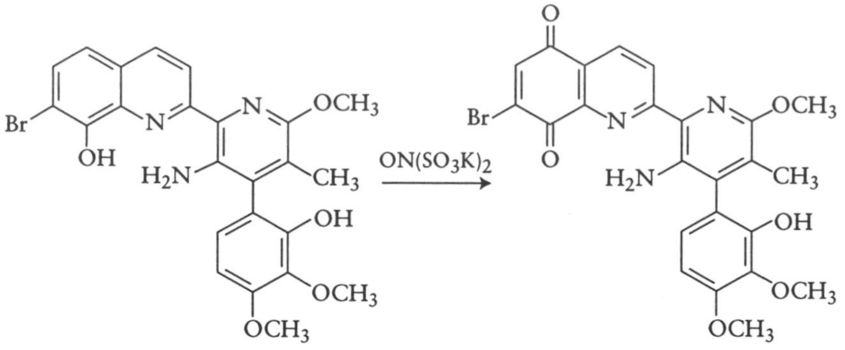

Fig. 4. (10) Hydroxyquinoline; (11) the products of the oxidation of hydroxyquinoline.

methyltrioxorhenium complex (H_3CReO_3) as the reaction catalyst.

The important methods for quinone synthesis in the laboratory are oxygenation of phenols and dehydrogenation of dihydroxyarenes (hydroquinones). Two reagents used commonly for phenol oxidation are $(KSO_3)_2NO$ (potassium nitrosodisulfonate) and N,N'-*bis*(salicylidene)ethylene-diiminocobalt(II) (salcomine) with O_2. Both oxidants generate, by $1e^-$ oxidation of the phenolate anion, the resonantly delocalized phenoxy radical. This radical then undergoes carbon atom oxygenation, to yield a hydroquinone that is oxidized easily to the quinone. These oxygenations can be remarkably selective, especially given the reactivity of these oxidants. For example, the naphthoquinone 8 undergoes salcomine-O_2 oxygenation to 9 without competitive oxygenation of the sulfur atom of the thioether. Likewise, the hydroxyquinoline 10 is oxidized to 11 without competitive amine oxygenation. In nature, quinones are biosynthesized typically by enzyme-catalyzed monooxygenation of a phenol to give the hydroquinone, which is oxidized by dehydrogenation in a second enzymatic oxidation.

Reactivity of Quinones

Two pathways dominate quinone reactivity. The quinone is one of the few functional groups capable of reversible electron transfer, a property of enormous importance in nature (as discussed below). One-electron ($1e^-$) reduction of the quinone, exemplified with 1,4-benzoquinone 2, yields the semiquinone radical intermediate 12. A second one-electron reduction provides the hydroquinone 13. The redox equilibrium for quinone electron transfer is influenced profoundly by the quinone-substituents.

Electron-withdrawing substituents (such as halides, carbonyls, and nitriles) increase the reactivity of the quinone and stabilize the hydroquinone. An example is the substituted 1,4-benzoquinone 14, known as DDQ, which is a strong oxidant useful for dehydrogenation. Electron-donating substituents (such as amines and hydroxyls) decrease the reactivity of the quinone and destabilize the hydroquinone. Reduction of the quinone 15, a degradation product of the quinone antitumor agent daunomycin, provides the hydroquinone 16. This hydroquinone, a strong reducing agent, reacts instantaneously with molecular oxygen (reducing the O_2 to H_2O_2) to regenerate the quinone. Under rigorously anaerobic conditions, however, it tautomerizes to the more stable—and oxygen unreactive—hydroquinone 17.

The other pathway for quinone reactivity is reaction of the electron-deficient alkene. Two characteristic reactions of such alkenes are as dienophiles in the

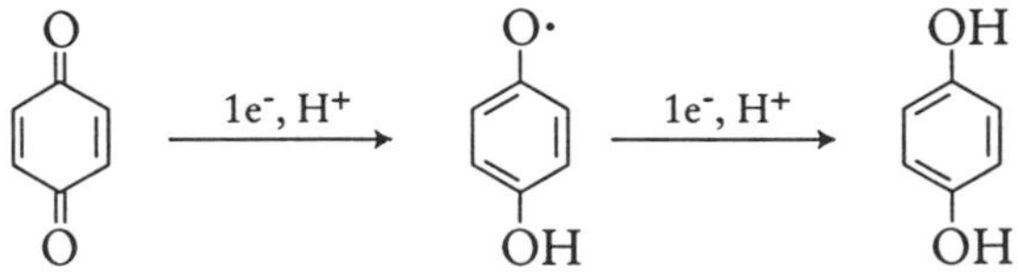

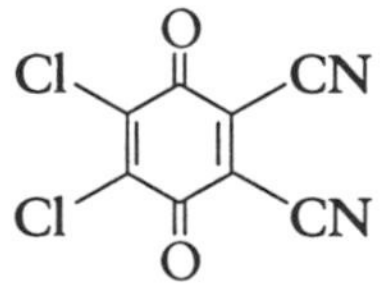

Fig. 5. (2) Quinone; (12) semiquinone; (13) hydroquinone; (14) DDQ.

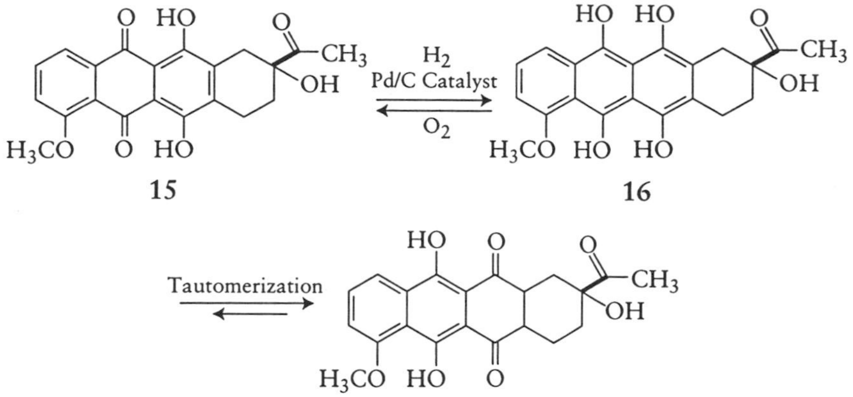

Fig. 6. (15) Quinone; (16) less stable hydroquinone; (17) more stable hydroquinone.

Diels-Alder reaction, and as electrophiles in the conjugate addition of nucleophiles. 1,4-Benzoquinone, an exemplary dienophile, undergoes rapid cycloaddition with cyclopentadiene 18 even at 0°C to give 19. At warmer temperatures significant amounts of the dimer 20 are formed. An example of conjugate addition reactivity is bromoquinone 21. Its reaction with the methoxide anion nucleophile in the presence of Ti[OCH(CH₃)₂]₄ gives (following neutralization) only the bromomethoxyquinone 22, the produce of kinetic reaction control. Methoxide addition to the less-substituted carbon of the quinone is faster (kinetic control) than addition to the other and is made irreversible by the "trapping" of the phenoxide by the titanium reagent. In the absence of this reagent, methoxide addition occurs reversibly to eventually displace the bromide by an addition/elimination sequence giving 23 (thermodynamic reaction control). Organometallic reagents are often not useful as nucleophiles toward quinones, as they may act preferentially as reducing agents.

Quinone Dyes and Pigments

The quinones occupy a pivotal position in the historical development of chemistry due to the societal and economic importance—from antiquity through the nineteenth century—given to dyes and pigments. Two important examples are the yellow dye quinazirin 24, obtained from the madder plant (which was cultivated for this purpose), and the red dye carminic acid 25, obtained from insects. In the presence of base, quinazirin deprotonates to form a red phenoxide anion, the color obtained in quinazirin dyeing of textiles. A method for the oxidation of anthracene to quinazarin was discovered independently—and virtually simultaneously—by William Perkin (in England) and Heinrich Caro (in Germany) in 1869. The economic value of the madder plant vanished shortly thereafter with dire consequences to French agriculture. The progressive color change upon aging of red wine, from ruby to brick red, is a result of oxidative polymerization of the red anthocyanin wine pig-

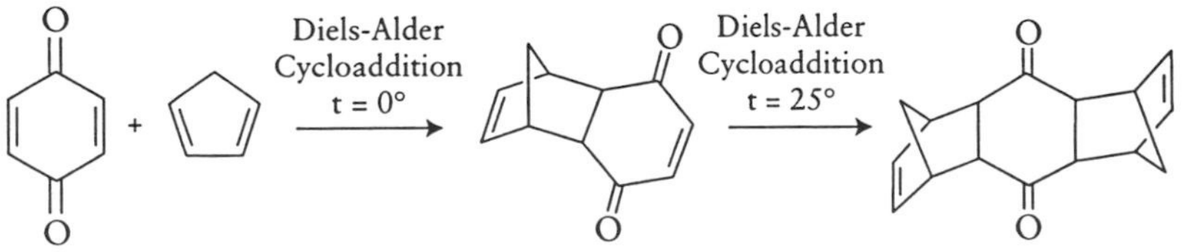

Fig. 7. (18) 1,4-Benzoquinone and cyclopentadiene; (19) results of first Diels-Alder cycloaddition to 18; (20) dimer resulting from second Diels-Alder cycloaddition.

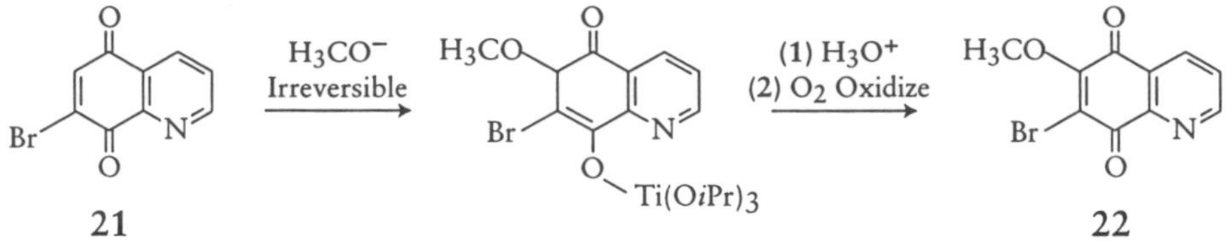

Fig. 8. (21) Bromoquinone; (22) bromomethoxyquinone.

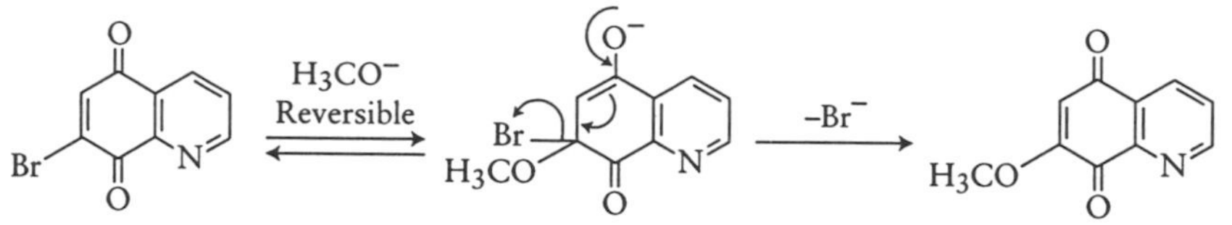

Fig. 9. (21) Bromoquinone; (23) results from a successive methoxide-addition/bromide-elimination sequence.

ments via reactive *ortho*-quinone intermediates. Similar oxidative processes, involving plant lignans, account for the yellowing of paper. A mechanism, illustrating *ortho*-quinone reactivity, has been proposed for the biosynthesis of calphostin 26. Calphostin inhibits an enzyme, protein kinase-C, important to the control of cell growth. Modern synthetic quinone dyes are used in color photography and (with additional functional groups to improve color fastness) in textiles.

The Quinone in Nature

Quinone primary metabolites are used for electron transfer in plants (during photosynthesis) and animals (during respiration). Quinones are innately cytotoxic. Consequently, quinones are expressed by plants, bacteria, and insects as defensive secretions, some of which have value as human therapeutic agents. The γ-carboxylation of the amino acid aspartate, a step required in the formation of a regulatory Ca^{+2} binding site essential to blood coagulation, requires metabolism of vitamin K.

Quinones in Biological Electron Transfer

The quinone's importance as an electron-transfer reagent derives from three characteristics: the electron transfer is reversible (thus the quinone can function as an electron carrier), the quinone can accept either one or two electrons, and lastly, the redox potential of the quinone can be adjusted (over the virtual entirety of the redox range, from strongly reducing to strongly oxidizing) by appropriate choice of substituents. It is therefore possible to create exactly the correct quinone necessary to a vital physiological electron transfer. In photosynthesis the photoexcited chlorophyll initiates an electron transfer that reduces ultimately the electron carrier ubiquinone 27. In animals, ubiquinone is biosynthesized in the mitochondria where it functions as the electron carrier to the penultimate enzyme of respiration, cytochrome *c* oxidoreductase. In its hydroquinone state (termed ubiquinol) it is found in membranes, where it terminates lipid peroxidation of the membrane. A third example of quinone-dependent electron transfer in nature is the cofactor of the enzyme lysyl oxidase. This enzyme accomplishes the oxidative deamination of the terminal amine of the amino acid lysine by the two-step process of amine dehydrogenation to an imine followed by imine hydrolysis to the aldehyde and ammonia. This reaction is essential to the formation of the elastin and collagen connective tissue. The oxidant for the dehydrogenation is an *ortho*-quinone coenzyme 28, formed within the enzyme-active site from the amino acids tyrosine and lysine.

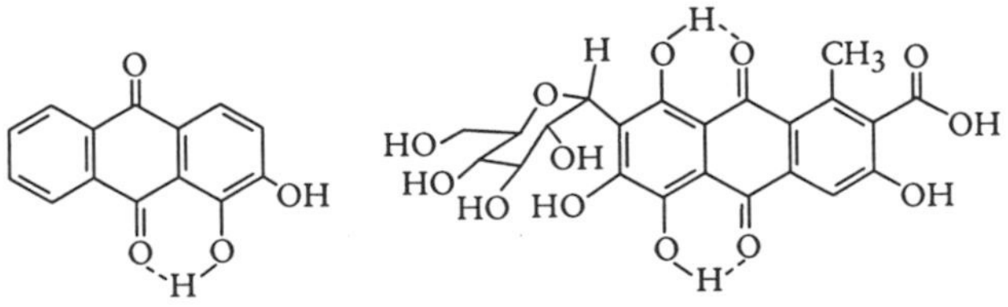

Fig. 10. (24) Alizarin; (25) carminic acid.

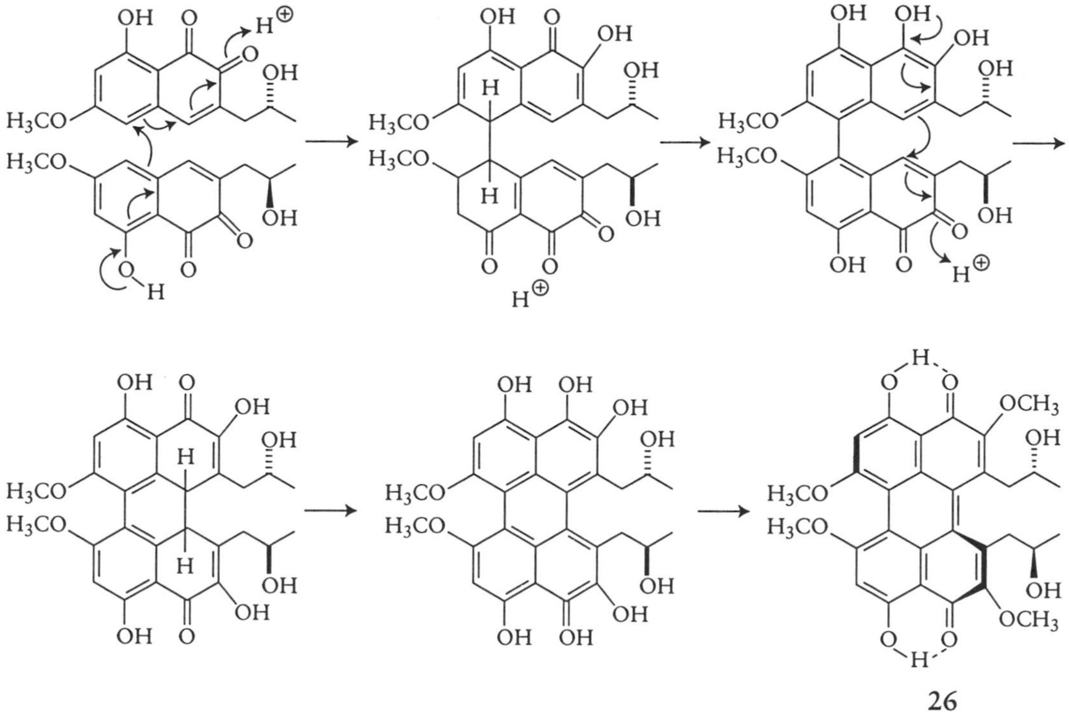

Fig. 11. Mechanism for the biosynthesis of (26) calphostin.

Quinones as Cytotoxins

Many examples of quinone defensive agents are known. The yellow naphthoquinone juglone 29 is a component of walnut wood and husks. Because juglone is toxic to many plants, it benefits walnut seed germination. It can be acutely toxic to some animals, notably horses exposed to walnut wood shavings. Curiosity as to why a certain brown alga of the Caribbean was not eaten by herbivorous fish led to the discovery, in the rust-colored ichthyotoxic secretion of the alga, of the red *ortho*-quinone stypoldione 30. This quinone is reduced by the strong reducing agent sodium dithionite ($Na_2S_2O_4$) to the colorless hydroquinone stypotriol 31, which is also present in the alga secretion. Stypotriol slowly air

oxidizes to stypolidione. Both act as narcotics at low concentrations and are toxic to fish at high concentrations. As it is highly probable that 30 and 31 interconvert in vivo, it is uncertain which one accounts for the biological activities.

Two important human anticancer agents isolated from bacteria are the red-colored adriamycin 32 and the blue-colored mitomycin 33. The electrostatic attraction of the positively charged aminosaccharide of adriamycin to the negatively charged phosphate diester of DNA enables adriamycin-DNA intercalation. Subsequent reaction with formaldehyde results in covalent bond between the saccharide amino group and the DNA base guanosine. This intercalated complex

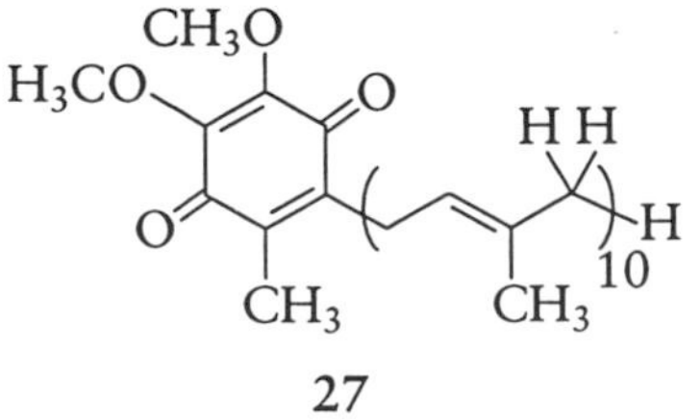

27

Fig. 12. (27) Ubiquinone (coenzyme Q_{10}).

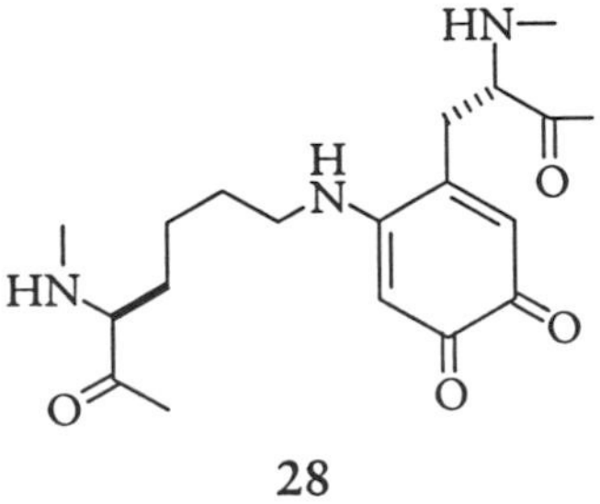

28

Fig. 13. (28) Lysine tyrosylquinone.

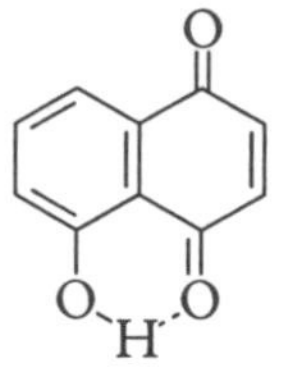

Fig. 14.　(29) Juglone.

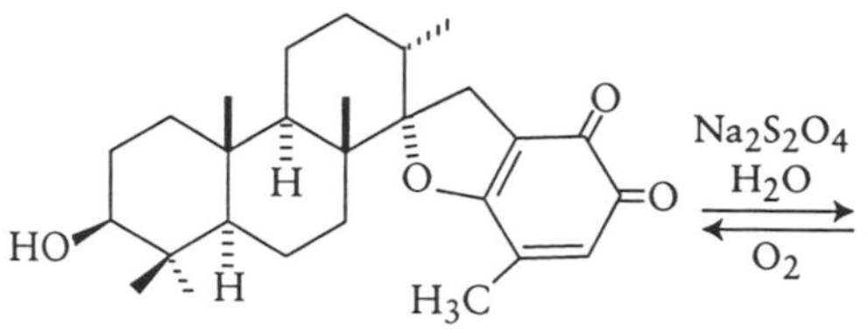

Fig. 17.　(34) Atovaquone.

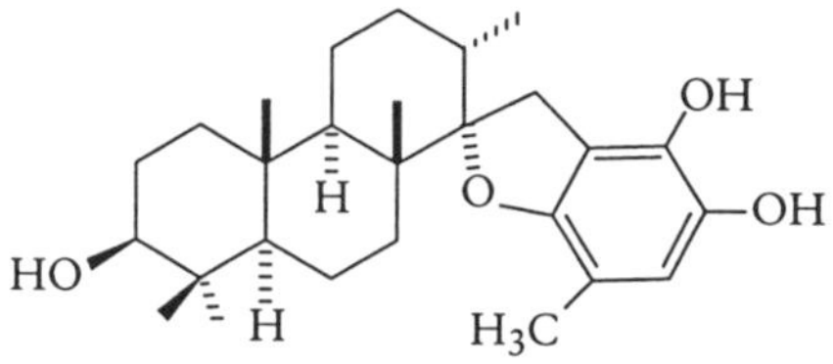

Fig. 15.　(30) Stypoldione; (31) stypoltriol.

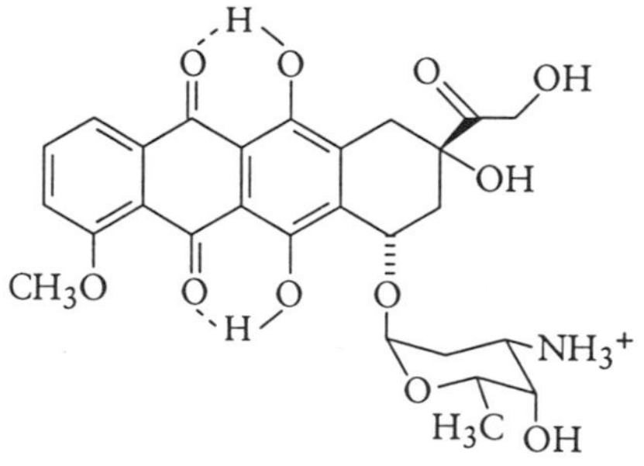

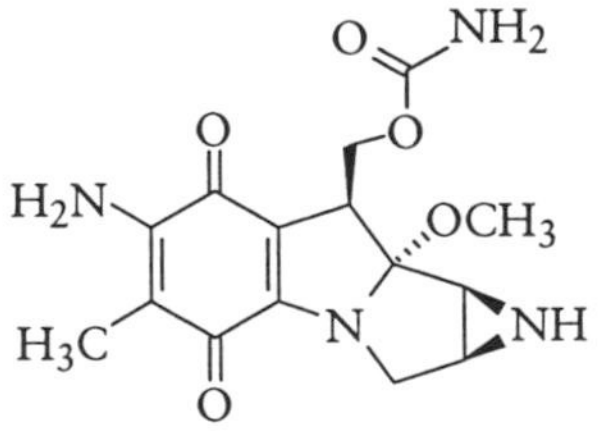

Fig. 18.　(35) Acetaminophen; (36) quinone imine.

resists DNA repair and is lethal to the cell. Mitomycin's cytotoxicity also derives from DNA damage as a consequence of reactions that ensue after quinone reduction. The naphthoquinone atovaquone **34** is used to treat human protozoal infections. The mechanism by which ethanol potentiates the toxicity of the analgesic acetaminophen **35** is induction of oxidizing enzymes. This increases ethanol metabolism, and also acetaminophen oxidation, to the highly toxic and reactive quinone imine **36**. A similar oxidation, providing a toxic quinone imine metabolite of the neurotransmitter dopamine, has been suggested as a mechanism for the progressive degeneration of the nigrostriatal dopaminergic neurons associated with Parkinson's disease.

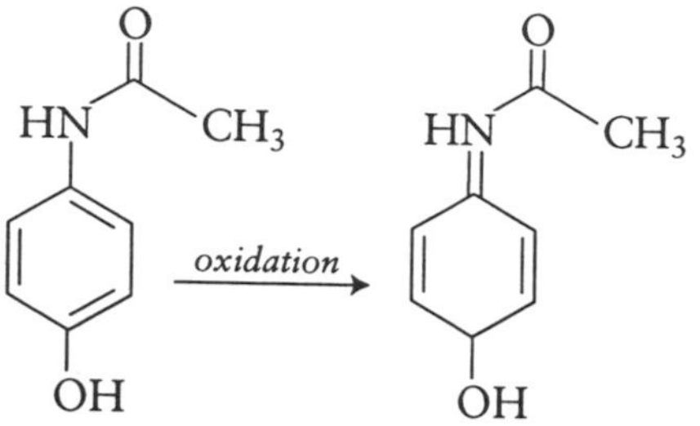

Fig. 16.　(32) Adriamycin; (33) mitomycin C.

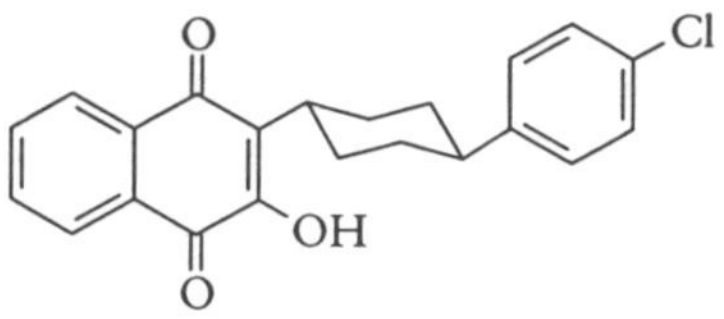

BIBLIOGRAPHY

PATAI, S.; RAPPOPORT, Z., eds. In *The Chemistry of Quinonoid Compounds;* New York: Wiley, 1988; Vols. 1 and 2.

THOMSON, R. H. In *Naturally Occurring Quinones: Recent Advances;* London: Chapman & Hall, 1987.

WANG, S. X.; MURE, M.; MEDZIHRADSZKY, K. F.; BURLINGAME, A. L.; BROWN, D. E.; DOOLEY, D. M.; SMITH, A. J.; KAGAN, H. M.; KLINMAN, J. P. "A Crosslinked Cofactor in Lysyl Oxidase." *Science* 1996, *273*, 1078–1084.

JED F. FISHER

R

Radiation, Background

Background radiation is the sum of all the RADIO-ACTIVITY in the environment, both natural and artificial.

Natural radiation in the environment can come from three general sources: (1) direct cosmic radiation, (2) primordial radionuclides, and (3) cosmogenic radionuclides.

Direct cosmic radiation has extraterrestrial origins, from both within the solar system and elsewhere in our galaxy, especially from supernova explosions. Primary cosmic rays hitting the top of Earth's atmosphere are predominantly protons with some helium nuclei and less than 1 percent electrons. These primary cosmic rays have sufficient energy to generate secondary particles that can penetrate to ground level. These secondary particles are mostly high-energy positive or negative muons and some electrons. *Muons* are subatomic particles that have a mass of about 215 times that of an electron.

Primordial radionuclides are those that were formed or present at the origin of Earth. The two categories of primordial radionuclides are series radionuclides and nonseries radionuclides. Series radionuclides consist of a long-lived parent and a decay sequence of radionuclides of wide-ranging half-lives ending with a stable isotope of lead. Nonseries radionuclides decay directly to a stable nuclide. The primordial series are named the uranium-238 (U-238) series; the uranium-235 series, often called the actinium series; and the thorium-232 (Th-232) series. The half-life of U-238 is 4.5×10^9 years, that of U-235 is 7.0×10^8 years, and that of Th-232 is 1.4×10^{10} years. The two most important nonseries radionuclides are potassium-40 and rubidium-87 (Rb-87). K-40 will decay directly to stable calcium-40 and Rb-87 will decay to stable strontium-87. The half-lives of potassium-40 (K-40) and Rb-87 are 1.27×10^9 years and 4.80×10^{10} years, respectively. These primordial radionuclides are found in Earth's crust and the underlying mantle. They are widely distributed in rocks, soil, and water, and some are found in trace amounts in foods.

Cosmogenic radionuclides are produced when cosmic rays interact with the atmospheric gases and also on the surface of Earth's crust. They are produced by nuclear spallation reactions and neutron capture reactions. A spallation reaction occurs by the high-energy bombardment of a cosmic ray with a target atom in the atmosphere, resulting in the ejection of nucleons from the target atom. Excess neutrons can be captured by a target nuclide and form a radionuclide. There are over twenty-two cosmogenic radionuclides produced by these reactions in the atmosphere. The four major widely spread radionuclides are carbon-14, hydrogen-3, sodium-22, and beryllium-7. These can contribute a measurable radiation dose to humans. Carbon-14 and hydrogen-3 contribute the highest dose to human beings because they are relatively long-lived isotopes of the elements that are major constituents of body tissue. An example of a long-lived radionuclide that is formed in Earth's crust is chlorine-36 (Cl-36). It is formed by the neutron activation of stable chlorine-35. Stray neutrons from spallation reactions will be absorbed into the nucleus of Cl-35 and become radioactive Cl-36.

The major human-made sources that contribute to background radiation include those from (1) uranium mining and processing, (2) nuclear weapons testing, and (3) releases from NUCLEAR REACTORS. Smaller amounts of background radiation can also come from medical X rays, nuclear medicines, radiopharmaceutical consumer products, and those research centers that use radioisotopes. The knowledge and measurement of background radiation are important in determining the amount of exposure to the general population. A recent study has shown that the risk of exposure to the average person in the United States from low-level sources of ionizing radiation has been determined to be 82 percent from natural sources and 18 percent from human-made sources. The single major source, among natural sources of risk to humans, is the radionuclide radon-222 gas. Radon gas is part of the uranium-238 decay series.

Background radiation measurements are also important to researchers who are doing very precise quantitative work. Analytical instrumentation must be calibrated and corrections must be made for background radiation in order to obtain precise absolute values.

BIBLIOGRAPHY

Committee on Biological Effects of Ionizing Radiation (BEIR). *Health Effects of Low Levels of Ionizing Radiation;* Washington, DC: National Academy Press, 1990.

Environmental Radiation Measurements, Report No. 50m; Bethesda, MD: National Council on Radiation Protection, 1988.

Exposure of the Population in the United States and Canada from Natural Background Radiation, Report No. 94; Bethesda, MD: National Council on Radiation Protection, 1987.

FRANK FAZIO

Radiation Chemistry

Radiation is the stream of particles or electromagnetic waves emitted by a radioactive substance as a result of nuclear decay. The actual source of the radiation is the unstable nuclei of certain isotopes. The ratio of NEUTRONs to PROTONs determines the stability of a nucleus. Because protons have a positive charge and repel each other, it is only the very strong

nuclear forces that hold neutrons and protons close together to form the nucleus. Having more neutrons increases the strong nuclear forces and helps stabilize the nucleus. However, if a nucleus contains a ratio of neutrons to protons that is too high or too low, it will emit radiation. Also, if the nucleus contains more than eighty-three protons, no number of neutrons is enough to stabilize it. Thus, any element with an atomic number greater than eighty-three is unstable and decays to form a more stable "daughter" nucleus (see DECAY, RADIOACTIVE). Often a radioactive nucleus cannot reach a stable state in a single decay process, and a decay series occurs involving several steps until a stable NUCLIDE forms.

The vast majority of nuclides found in nature are stable. The only unstable nuclides found in nature include isotopes that have a HALF-LIFE of at least 10^9 years (for example, ^{238}U), their daughter nuclides, and nuclides that are still being synthesized (for example, ^{14}C). However, bombarding stable nonradioactive nuclei with high-energy particles can transform them into other nuclei that may not be stable. Particle accelerators give the particles enough energy to overcome the repulsion and penetrate the nucleus. In this manner scientists have extended the periodic table by producing new elements with increasing numbers of protons.

The intensity of radiation that a sample emits depends on the particular isotope's activity or the number of disintegrations per second. This intensity is measured in units called curies (Ci) that are defined in terms of the number of disintegrations per second in 1 gram of radium (1 Ci = 3.7×10^{10} disintegrations/second).

There are several different types of radiation that radionuclides may emit. These include α-particles, β-particles, and γ-rays (see, respectively, ALPHA PARTICLE; BETA RADIATION; GAMMA RADIATION). An example of an alpha emitter is ^{238}U as shown in equation 1. An alpha particle consists of two protons and two neutrons and is equivalent to a helium nucleus. This type of decay is common for heavy radioisotopes. The alpha particle has relatively high mass and charge (+2) and is not very penetrating. A sheet of paper or even the dead cells on the surface of the skin are able to easily stop alpha particles. However, ingesting or inhaling an α-emitter can seriously damage cells of internal organs.

$$^{238}_{92}U \rightarrow {}^{234}_{90}Th + {}^4_2He \qquad (1)$$

An example of β-emission is the decay of ^{14}C as in equation 2. The β-particle is identical to a high-energy electron. It has a negative charge and very low mass (relative to protons or neutrons). Because of this smaller mass, the β-particle is more penetrating and can penetrate skin to a depth of approximately 4 millimeters. Prolonged exposure to the skin may cause a burn. However, a piece of aluminum foil or a thin piece of wood is able to block β-particles. Beta emission involves the conversion of an uncharged neutron to a positively charged proton plus an electron that is ejected. Thus, this type of decay occurs for nuclides whose neutron-to-proton ratios are too high to be stable.

$$^{14}_{6}C \rightarrow \, ^{14}_{7}N + \, ^{0}_{-1}e \qquad (2)$$

A new nuclide that forms in any one of these radioactive decay processes may be in an EXCITED STATE with excess energy. This nuclide may relax to its GROUND STATE by emitting γ-radiation, a form of short-wavelength (10^{-11}–10^{-14} m) electromagnetic radiation that is more energetic than X RAYS. This emission does not affect the MASS or atomic number of the nuclide. Gamma emission almost always accompanies α- and β-decay and usually occurs within 10^{-12} seconds after the α- or β-particle. However, in some cases the γ-emission is delayed slightly. An example of this is the β-decay of ^{60}Fe as in equations 3 and 4. The metastable ^{60m}Co has a half-life of 10.5 minutes. Gamma radiation is the most penetrating and damaging type of radiation. Blocking γ-rays requires thick layers of lead or concrete.

$$^{60}_{26}Fe \rightarrow \, ^{60m}_{27}Co + \, ^{0}_{-1}e \qquad (3)$$

$$^{60m}_{27}Co \rightarrow \, ^{60}_{27}Co + \, ^{0}_{0}\gamma \qquad (4)$$

Radiation damage generally occurs by knocking electrons out of molecules to form extremely reactive radicals or ions. The effects of exposure to radiation of any type can be very subtle because the energy per event is small. The effects in order of increasing doses include a decrease in the white blood cell count, nausea, and, finally, death. If an organism receives a massive dose, the effects may appear immediately. For smaller doses, damage may appear months, years, or even decades later, usually as cancer. In addition there may be genetic damage that can lead to defects in the offspring of the organism.

The effect of a dose of radiation on a particular organism depends on several factors. Higher-energy radiation can cause more damage. More penetrating radiation such as γ-radiation has the ability to cause more damage. Different types of radiation have different ionizing abilities. For example, α-particles produce a dense trail of damage, even though they do not penetrate very deeply. Basically, the more energy that radiation loses as it passes through tissue, the more damage it does to the organism. The amount of energy that the radiation deposits per kilogram of tissue is known as the radiation-absorbed dose, or rad. To take into account the effectiveness of a type of radiation on tissue, the rad is multiplied by a factor known as the relative biological effectiveness of the radiation (RBE). The result is the effective dosage in rems (see REM), as shown in equation 5. The chemical properties of the radiation source are also important because they affect the residence time of the source in the body.

$$\text{rems} = \text{rads} \times \text{RBE} \qquad (5)$$

A variety of technologies take advantage of the properties of radionuclides. Long-lived radioactive isotopes are useful in dating artifacts or rocks. For example, ^{14}C, with a half-life of 5,730 years, is used to determine the date (up to 40,000 years ago) when an organism died. This is possible because ^{14}C is constantly produced in the upper atmosphere and is incorporated in CO_2 that is taken in by plants and enters the food chain. Once an organism dies, it is no longer taking in ^{14}C, but ^{14}C continues to decay according to equation 2. Uranium-238 is much longer lived ($t_{1/2} = 4.46 \times 10^9$ years) and decays through a series of steps to ^{206}Pb. Thus, the ratio of ^{238}U to ^{206}Pb is used to determine the age of rocks. Ionizing radiation and radionuclides have important applications in the field of nuclear medicine. Radiotracers are introduced into organisms to trace pathways by monitoring radioactivity. Many different radiotracers have been employed including ^{14}C, ^{32}P, ^{131}I, ^{201}Th, and ^{99m}Tc. Therapeutic methods have been and are being developed that are very specific and cause little or no damage to healthy tissue. The use of radiolabeled compounds in chemistry has helped to elucidate mechanisms in inorganic as well as organic chemistry.

See RADIOACTIVITY, HISTORY OF for a discussion of the historical background.

BIBLIOGRAPHY

CHARLES, D. "Counting the Cost of a Lifetime's Radiation." *New Scientist* 1992, *134*, 7.

EMSLEY, J. "Superheavy Elements Could Soon Be Made." *New Scientist* 1994, *143*, 15.

EPPERLY, M. W. "Radiotherapy: The Closer the Better." *Chemtech* 1991, *21*, 744–749.

MCMURRY, J.; FAY, R. C. *Chemistry;* Englewood Cliffs, NJ: Prentice-Hall, 1995; Chapter 22.

SINGLETON, D. A.; THOMAS, A. A. "High-Precision Simultaneous Determination of Multiple Small Kinetic Isotope Effects at Natural Abundance." *J. Am. Chem. Soc.* 1995, *117*, 9357–9358.

PATRICK L. HUSTON

Radicals, Free

See FREE RADICALS.

Radioactive Tracers

See TRACERS, RADIOISOTOPE.

Radioactivity, History of

The discovery of radioactivity fundamentally changed both the world and the course of science. One measure of its importance is evidenced by the number of Nobel Prizes awarded to early investigators in the field.

The birth of nuclear science took place on January 20, 1896, when Henri BECQUEREL, a professor of physics at the Museum of Natural History in Paris, attended a lecture on the nature of X RAYS, which had been discovered by Röntgen the previous year. Becquerel was intrigued by the fact that exposure to X rays caused some chemical substances to luminesce, or glow in the dark. At the time, Becquerel was experimenting with uranium salts, which became luminescent when exposed to direct sunlight, and he wondered if the process worked both ways. In other words, if X rays cause substances to luminesce, do the same substances emit X rays while luminescing after exposure to sunlight? To test his hypothesis, Becquerel placed photographic plates, tightly wrapped in lightproof black paper, on a window ledge along with samples of uranium salts that became luminescent in sunlight. When the plates were developed, they revealed "smudges" where rays had penetrated the paper and exposed the plates. Moreover, when coins or keys, substances known to stop X rays, were placed on the plates, their silhouettes appeared just as they had in Röntgen's experiments. To Becquerel, this result was proof that luminescent substances could emit X rays when stimulated by sunlight as well as glow themselves when struck by X rays.

Then, on February 26, 1896, overcast skies forced Becquerel to postpone his experiments. After having left them for a brief period on the windowsill in the poor light, he placed the wrapped plates and uranium salts in a closed desk drawer. Two days later, curious to see if the weak sunlight had any effect on the salts, Becquerel developed the plates and was amazed to find that they were as highly exposed as if the salts had been exposed to intense sunlight. A few days later, Becquerel reported to the French Academy of Sciences that an "emanation" was coming from the salts that was not connected with their luminescent properties. Like X rays, however, the new rays could expose photographic plates and cause electroscopes to discharge. Over the next several months, Becquerel was able to show that the uranium salts continued to emit the emanations at the same intensity even after being kept in darkness for months.

Overshadowed by the worldwide interest in X rays, the new "Becquerel rays" aroused little interest except in a young Polish graduate student who was searching for a doctoral thesis topic. Marie Sklodowska CURIE and her husband Pierre CURIE read of Becquerel's work, and Marie decided to find the source of the mysterious emanations coming from uranium salts and the pitchblende ore from which they were extracted. Marie Curie was given an abandoned stockroom at the Paris School of Physics and Chemistry, where her husband, Pierre, was Director of Laboratories, in which to conduct her research. Besides chemical glassware, her main piece of research equipment was an electroscope. This is an instrument with two gold foil leaves that separate when the electroscope is electrically charged and fall back together again upon discharge. Previous work had shown that an electroscope could be discharged by either X rays or Becquerel rays. Marie Curie improved on this basic design by adding a scale so she could measure the rate of discharge.

After some initial work on uranium, Curie decided to test all elements known at the time for activity and found that thorium also gave off Becquerel rays. She coined the terms *radio elements* for minerals or elements that exhibited activity and *radioactivity* for the activity itself. The Curies also found that pitchblende ore had a higher activity, as measured by an electroscope, than either the uranium or the thorium extracted from it. In a leap of genius, Marie Curie realized what this meant: that some as yet unknown element or elements existed in pitchblende that had a higher activity than either uranium or thorium. The husband-and-wife team set out to find these unknown elements. The Curies started with about one ton of pitchblende ore and began a chemical separation process that involved repeated selective precipitations and fractional crystallizations. Marie Curie spent months dissolving the ore in hot acids, laboriously separating the salts as they crystallized out, and testing their activity with an electroscope. In July 1898, the Curies announced the existence of a new element they named polonium after her native land. Because the amount of polonium, in the form of a mixture of salts, obtained was far less than could be isolated at the time, its presence was proven by its activity, which was some four hundred times greater than that of uranium. The discovery of radium was announced four months later in December of 1898. Its activity was nearly a million times greater than that of uranium. The Curies found that the activity of polonium decreased by nearly one-half in six to nine months while radium showed no apparent loss in the same period of time.

Marie Curie obtained from Poland several tons of pitchblende ore tailings, after the uranium had been extracted, and set out to isolate polonium and radium in pure form. In 1902, the Curies were able to obtain a few milligrams of radium, which they found to be chemically similar to barium. They were never able to satisfactorily isolate polonium, which they believed to be chemically similar to bismuth. The hazards of dealing so intimately with such highly radioactive materials were, of course, unknown to the Curies. They both suffered physical burns and sores that were slow to heal from their exposure to radium. As an indication of their intense radiation exposure, Marie once commented "on the beautiful blue glow" radium gave off from its container in their darkened laboratory. Moreover, she reported that her entire laboratory was contaminated by radioactivity as it became nearly impossible for her to keep a charge on an electroscope anywhere in the room.

The doctoral thesis Marie Curie presented on the isolation and properties of polonium and radium in 1903 has come to be regarded as among the most significant in scientific history. It opened a new era in physics and chemistry that had not even been dreamed of a decade earlier. Marie and Pierre Curie shared the 1903 Nobel Prize in physics with Henri Becquerel for their studies of radioactivity. Marie Curie was awarded the 1911 Nobel Prize in chemistry for the isolation of polonium and radium. Pierre Curie was killed when he was run over by a loaded wagon in 1906, but Marie Curie spent many years in further investigations of radioactive substances. She died in 1934 of leukemia undoubtedly brought about by her long exposure to radiation.

It was first thought that Becquerel rays were weaker forms of X rays, but in 1899, utilizing magnetic and electric fields, the British physicist Ernest RUTHERFORD (1871–1937) was able to show that they contained two types of charged particles, which he named ALPHA PARTICLEs and beta particles, respectively (see also BETA RADIATION). He also found that the particles were accompanied by an extremely penetrating ray that was not affected by magnetic or electric fields. This became known as GAMMA RADIATION (Figure 1).

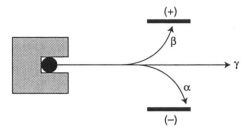

Fig. 1. The separation of radiation from a radioactive substance into alpha (α; +), beta (β; −), and gamma (γ; neutral) rays as it passes through an electric field.

Several workers were involved in determining the nature of the three types of radiation; however, most of the credit belongs to Rutherford.

The beta particles were strongly deflected toward a positive electric field, showing that they carried a negative charge. Pierre Curie found that radium took on a positive charge on emitting beta radiation, also showing the negative qualities of beta particles. Becquerel predicted them to be similar to cathode rays and, in 1903, Rutherford showed that cathode and Becquerel rays had the same charge-to-mass ratio. Beta particles were ultimately found to be high-energy electrons.

Alpha particles are deflected toward a negative field, indicating that they carry a positive charge. The suggestion was made that they were similar to the canal rays passing from anode to cathode in a helium discharge tube, and Rutherford, also in 1903, found that their charge-to-mass ratios were similar. In 1909, Rutherford isolated alpha particles and showed by spectral analysis that the gas formed from them was helium. Alpha particles turned out to be doubly positive-charged ions of helium.

Gamma radiation was unaffected by electric or magnetic fields and thus had a neutral charge. Penetrability experiments showed gamma radiation to be similar to X rays in nature but of much higher energy on the electromagnetic spectrum. Table 1 lists some of the essential features of the basic forms of radiation. The positron was found by C. D. Anderson in 1932 to be identical to the beta-particle, except that it carries a positive charge.

Ernest Rutherford was awarded the 1908 Nobel Prize in chemistry for his studies into the disintegration of radioactive elements.

Rutherford, in conjunction with McGill University chemist Fredrick SODDY, demonstrated in 1902 that a radioactive element could transform into a new element by the spontaneous emission of one or more forms of radiation, a process they called spontaneous

Table 1.

Radiation	Symbol	Mass	Charge	Penetrability
Alpha	^4_2He	4.0	+2	sheet of paper outer skin layer
Beta	$^0_{-1}e$	0.0	−1	0.25 inch of aluminum 0.40 cm of tissue
Gamma	γ	0.0	0	several inches of lead several feet of water

decay. The reaction they initially studied was the conversion of metallic radium to gaseous radon by the emission of an alpha particle (see also EMANATION).

$$^{226}_{88}Ra \Rightarrow {}^{222}_{86}Rn + {}^4_2He$$

Rutherford and Soddy found that radioactive elements decayed at a constant rate that was independent of outside factors. Rutherford coined the term HALF-LIFE as a measure of an element's activity and was instrumental in determining the exponential decay law. A contemporary wrote that "in five years, roughly from 1900 to 1905, Rutherford and Soddy set in motion a complete transformation of physics and chemistry that was breathtaking in its time."

In 1913, Soddy was able to show that an element's mass dropped by four and its atomic number by two when it emitted an alpha particle. Emission of a beta particle resulted in the same mass and an increase in the atomic number by one. Soddy studied the thorium series:

$$^{232}_{90}Th \Rightarrow {}^{228}_{88}Ra + {}^4_2He$$

$$^{228}_{88}Ra \Rightarrow {}^{228}_{89}Ac + {}^0_{-1}e$$

$$^{228}_{89}Ac \Rightarrow {}^{228}_{90}Th + {}^0_{-1}e$$

During the course of his investigations, Soddy discovered the existence of atoms with the same atomic number but different mass numbers, which were called ISOTOPES. Soddy was awarded the 1921 Nobel Prize in chemistry for his investigations into the nature and origin of isotopes.

During the course of her investigations, Marie Curie had often observed that nonradioactive substances would acquire their own activity when exposed to radium. It was left to her daughter Irène JOLIOT-CURIE, who worked in her mother's laboratory from an early age, and her husband, Frédéric Joliot-Curie (1900–1958), to determine the nature of artificial radioactivity. In 1934, they bombarded an aluminum target with alpha particles from a polonium source. For several minutes after the source was removed, the target continued to give off radiation that had been earlier identified as a positron. They proposed that aluminum was first converted to an unstable isotope of phosphorus. The ratio of protons to neutrons for stability of the nucleus was, by this time, fairly well established. The unstable phosphorus atom then ejected a positron, to become a stable isotope of silicon:

$$^{27}_{13}Al + {}^4_2He \Rightarrow {}^{30}_{15}P + {}^1_0n$$

$$^{30}_{15}P \Rightarrow {}^{30}_{14}Si + {}^0_{+1}e$$

The research of the Joliot-Curies was a modern-day culmination of the ancient alchemists' quest—that is, the search for a philosopher's stone that could change one element into another. Today many isotopes, unknown in nature but made through the process of transmutation, find use in medicine, research, and industry. The Joliot-Curies were awarded the 1935 Nobel Prize in chemistry for their synthesis of new radioactive elements.

The last missing piece in the nuclear puzzle was found in 1932 when James CHADWICK confirmed the long-suspected existence of the NEUTRON by the bombardment of beryllium with alpha particles.

$$\underset{4}{\overset{9}{Be}} + \underset{2}{\overset{4}{He}} \Rightarrow \underset{6}{\overset{12}{C}} + \underset{0}{\overset{1}{n}}$$

Chadwick received the 1935 Nobel Prize in physics for his discovery of the neutron.

The nuclear revolution in science, equal in importance to the Newtonian revolution in physics, took place in less than a single lifetime. In 1896, Becquerel first realized that something odd was going on within the atom that everyone though to be unchangeable. By 1944, Otto Hahn had received the Nobel Prize in chemistry for his discovery of fission, and Albert EINSTEIN had formulated his famous equation showing the tremendous power that could be unleashed from the atomic nucleus. There is no question that the half-century following Becquerel's discovery can be considered the most significant in the history of science.

BIBLIOGRAPHY

BADASH, L. "The Discovery of Radioactivity." *Physics Today*, February 1996, 21–26.

BORMAN, S. "Scientists Honor the Centennial of the Discovery of Radioactivity." *Chem. & Eng. News*, April 29, 1996, 55–65.

CHADWICK, J. *Radioactivity and Radioactive Substances*, 4th ed.; London: Pitman, 1953.

CURIE, E. *Madame Curie*; Garden City, NY: Doubleday, 1937.

GELLETLY, B. "Radioactivity: A Stable Foundation on Which to Build." *Physics World*, March 1996, 31–32.

RUTHERFORD, E. *Radioactive Substances and Their Radiations*; Cambridge, U.K.: Cambridge University Press, 1913.

STRUTT, R. J., *The Becquerel Rays and the Properties of Radium*; London: Edward Arnold, Ltd., 1904.

CLAIR G. WOOD

Radium

See RADIOACTIVITY, HISTORY OF; RADIATION CHEMISTRY.

Radius Ratio Rules

Radius ratio rules are derived for systems where ions (or any other spheres, even including atoms of elements) are produced around each other. The point of reference is the cation, which means the rules are developed to calculate the number of anions that can fit around, or surround, the "central" cation (see ANIONS AND CATIONS). A discussion of such rules has significance only if the cation and anion give rise to a crystalline solid having a particular structural arrangement. As noted by Linus PAULING (see Pauling 1960), there are fixed ionic radii for cations and anions in the crystalline state (see IONIC BONDING).

The existence of a crystalline ionic compound is the result of the lowest energy ascribable to the structure at hand. Two factors that lead to this lowest energy are the following:

1. The greater the number of anions in contact with a cation, the greater the degree of attractive forces between the cation and its anions. This number determines the coordination polyhedron about the cation.
2. The individual anions attached to a cation are in contact with each other tangentially in addition to being in contact with the cation. There is a repulsive force between the anions that increases as the anions are crowded together more tightly around the cation.

As a result of these two related factors, it can be said that the optimum conditions for the minimum energy state of such a structural arrangement occur when the greatest number of anions are just barely in contact with each other around an anion.

Radius ratio rules will be discussed here for some of the most commonly encountered coordination polyherons around a cation. For more exhaustive discussion see the references listed in the bibliography.

Figure 1 shows, in a general way, a cross-section illustrating the contact of a cation with anions that are themselves in contact with each other. The single small circle with the center labeled O is the cross-section of the cation, and the larger circles with centers labeled A and B are cross-sections of the identical anions. The radius of the anion is r_- and the radius of the cation is r_+. Furthermore,

$$AO = BO = r_- + r_+$$
$$AD = BD = r_-$$
$$\angle AOD = \angle BOD = \tfrac{1}{2} \angle AOB = \gamma$$
$$\angle ODA = \angle ODB = 90°$$

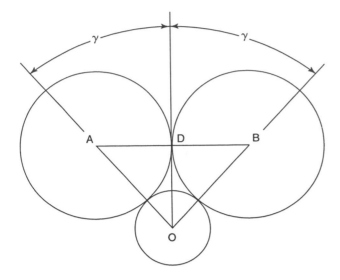

Fig. 1. Cross-section showing arrangement or cation in relation to two identical anions.

Therefore,

$$\frac{AD}{AO} = \frac{BD}{BO} = \sin \gamma = \frac{r_-}{r_- + r_+}.$$

or

$$\frac{r_- + r_+}{r_-} = \frac{1}{\sin \gamma} = 1 + \frac{r_+}{r_-}$$

Hence,

$$\frac{r_+}{r_-} = \frac{1}{\sin \gamma} - 1$$

The first case to be examined is the tetrahedral coordination of the cation as found in two representative compounds ZnS (wurtzite-zincblend) and CuCl (cuprous chloride). In these compounds the COORDINATION NUMBER about the cation is 4. The angle AOB is necessarily 109°28'. Therefore the angle is 54°44'. Hence,

$$\frac{r_+}{r_-} = \frac{1}{\sin(54°44')} - 1 = \frac{1}{0.81647} - 1 = 0.22478$$

The second case is the octahedral coordination of the cation as found in NaCl or LiF. In these compounds the coordination number is 6. The angle AOB at the cation is 90°. Therefore the angle takes a value of 45° and

$$\frac{r_+}{r_-} = \frac{1}{\sin 45°} - 1 = \frac{1}{0.70711} - 1 = 0.41421$$

The third case involves the cubic, eightfold coordination of the cation as found in CsCl and NH$_4$Cl. Here the angle at the cation, AOB, is any one of the acute angles of intersection between the body diagonals of the cube at the corners of which lie the anions (the obtuse angles of intersections between the body diagonals of the cube correspond to anions diagonally opposed in a face of the cube with no contact and have a value of 109°28'). The angle AOB at the cation then is 180° − 109°28' or 70°32'. Thus the angle γ is 35°16', which gives

$$\frac{r_+}{r_-} = \frac{1}{\sin (35°16')} - 1 = \frac{1}{0.57738} - 1 = 0.73196$$

The fourth case involves tetrahedral coordination. The coordination number is 12. (Note that even though the coordination number is 12, the coordination polyhedron is not dodecahedral. Here the angle AOB is 60° and γ is 30°.

$$\frac{r_+}{r_-} = \frac{1}{\sin 30°} - 1 = \frac{1}{0.5} - 1 = 1$$

No simple ionic compounds show this coordination. However, this is the coordination in well-known examples of cubic close packing (ccp) and hexagonal close packing (hcp) of elemental substances. Assuming equal spheres to represent elemental atoms, the ccp and hcp give rise to 12 coordination locations around each sphere, the densest packing of spheres. These packings create holes and lead to the possible use of the equation that relates the radius ratio of the cation to the anion by the sine of the angle. Here we assign r_+ as the radius of the hole in the packing of equal spheres with radius r_-. In Table 1 we give the values of the radius ratios for three types of holes.

It should be kept in mind that the rules given above are only approximate. Real cases may exhibit some variations. The respective radius ratios are limits on either side for the coordination.

Table 1. Radius of Spheres Fitting in Holes of the Closest Packing

Type of Hole	Coordination Number		r_+
Trigonal	3	60°	0.15469r_-
Tetrahedral	4	54°44'	0.22478r_-
Octahedral	6	45°	0.41421r_-

BIBLIOGRAPHY

COTTON, F. A.; WILKINS, G., *Advanced Inorganic Chemistry,* 5th ed.; New York: Wiley, 1988.

HUHEEY, J. E. *Inorganic Chemistry,* 2nd ed.; New York: Harper & Row, 1978.

JOLLY, W. L. *Modern Inorganic Chemistry,* 2nd ed.; New York: McGraw-Hill, 1991.

PAULING, L. *The Nature of the Chemical Bond,* 3rd ed.; Ithaca, NY: Cornell University Press, 1960; Chapter 13.

SHARMA, B. D., *J. Chem. Educ.* 1986, *63,* 504.

WELLS, A. F. *Structural Inorganic Chemistry,* 5th ed.; Oxford: Clarendon Press, 1989.

BRAHAMA D. SHARMA

Radon

See EMANATION.

Raman Spectrometry

Raman spectrometry, or spectroscopy, and the light-scattering phenomenon on which it is based were first demonstrated in 1928 by physicist C. V. Raman. The effect had been predicted by A. Smekal in 1923. Light scattering can be thought of as a collision between a molecule and a PHOTON. Several outcomes are possible. The most likely is called Rayleigh scattering: the photon and the molecule end up with each having the same amount of energy it had before the collision. Very often the photon changes direction. This is elastic scattering, with the particles behaving like billiard balls.

A much less likely outcome of the "collision" is a transfer of energy between the molecule and the photon so that they come away with different energies than they had before. This is an example of inelastic scattering: since the frequency of a photon is directly proportional to its energy ($E = h\nu$), it can be detected because the scattered photon has a different wavelength (and frequency) from that of the incident photon.

Raman scattering is an inelastic process, as shown in Figure 1: the scattering occurs in a very short time, due to the speed of the photon and the small size of a molecule. The molecule does not have time to absorb the photon; this distinguishes Raman scattering from fluorescence. Rather, the molecule and the photon form an "excited complex" that can have any energy, not just those few energies associated with the stable quantized states of the molecule. Such an

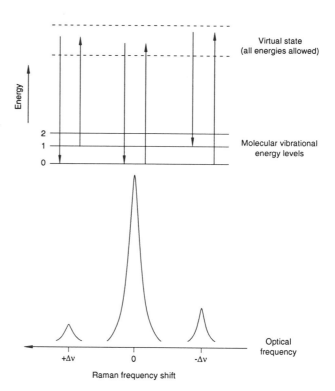

Fig. 1.   The nature of Raman scattering.

energy state is called a "virtual state," since it is not a real stationary state of the system. (See EXCITED STATE; STATIONARY STATE.) After a very short time, the molecule and photon separate and return to more stable configurations. If the molecule returns to the same energy state as prior to the encounter, it is an elastic collision and the process is Rayleigh scattering. If the molecule ends up with a different energy, then the photon must also, so that energy will be conserved: this is Raman scattering.

The molecular ENERGY LEVELS most often involved in Raman scattering are vibrational and rotational ones. While it is possible for the molecule to either gain or lose energy in the encounter, it is most likely that it will gain energy, since most molecules are in their lowest vibrational state at room temperature. We therefore normally look at that part of the spectrum in which the scattered photons have less energy than the incident ones. The energy differences between the scattered and incident photons will be exactly the energy differences between vibrational and/or rotational states of the molecule. The bottom of the diagram shows a simple Raman spectrum; since the energy (or frequency) difference between the Rayleigh line and the Raman band depends only on the molecular energies, the Raman spectrum spacing will be exactly the same, no matter what the frequency of the incident photon is. Thus the abscissa of a Raman spectrum is normally given as the frequency difference ($\Delta\nu$) between the exciting photon (the Rayleigh line) and the Raman bands. The intensities of the bands depend on the likelihood of scattering occurring; the low-frequency bands are therefore more intense. (See also VIBRATIONAL SPECTROSCOPY AND ANALYSIS.)

The spacing of the vibrational and rotational energy levels of a molecule is determined by its structure: the strengths of the bonds between the atoms, the lengths of the bonds, and the angles between bonds. Since the Raman spectrum depends on that energy-level spacing, the Raman spectrum will contain all the structural information that the infrared spectrum does. Raman spectrometry complements INFRARED SPECTROMETRY rather than duplicating it, however, because the vibrations that are active in the Raman spectrum are sometimes not the same ones as those that are infrared-active. For vibrations to be infrared-active, the molecular DIPOLE moment must change during the vibration; to be Raman-active, the polarizability of the molecule must change, and the first derivative of the polarizability must not be zero at the midpoint of the vibration. In practice this means that many vibrations that involve symmetrical motions of the molecule, and so show little or no infrared absorption, are strongly active in the Raman spectrum. One example is the vibrations of homonuclear diatomic molecules like O_2 or Cl_2; although inactive in the infrared, they are all Raman-active.

By comparison with infrared absorption, Raman spectrometry has two serious weaknesses that have prevented it from having the same wide applicability. First, Raman scattering is a very unlikely process, and the Raman spectrum is usually very weak. Under certain circumstances, its probability can be much enhanced: the most important of these is when the exciting frequency is one which the molecule can absorb. One can then observe the resonance Raman (RR) spectrum. Whereas normally the lowest concentration of scatterer that can be observed is approximately 0.01 mole/liter, a resonance spectrum can be generated by as little as 10^{-6} mole/liter or less. Where it is applicable, as in the study of many proteins, RR spectrometry has made significant contributions to our knowledge.

A second, more serious problem is caused by the fact that the fluorescence spectrum of a molecule commonly overlaps the Raman spectrum, and even weak fluorescence is likely to be stronger than Raman scattering. As a result, up to half of all materials will not show a detectable Raman spectrum.

Experimental Setup

To observe Raman spectra a bright, monochromatic light source is needed. Lasers are universally used, normally those with visible or ultraviolet output frequencies. These high frequencies take advantage of two factors: scattering intensity increases as the fourth power of the exciting frequency, and optical materials in the visible and ultraviolet are more stable and easier to use than those in other spectral regions.

The rest of the experimental setup, shown in Figure 2, resembles a fluorescence spectrometer: the scattered light is analyzed by a monochromator or interferometer with very good stray light rejection, necessitated by the relatively intense Rayleigh scattering, and the most sensitive optical detectors are used to measure the weak Raman-scattered light.

A recent important advance in Raman spectrometry is the use of near-infrared lasers to excite the Raman spectrum; because photons in this wavelength range only rarely stimulate fluorescence, many more substances yield Raman spectra under these conditions than when visible or ultraviolet exciting light is used. Because the exciting frequency is lower, the Raman intensity is also less. This loss can be largely offset by the use of an interferometer, with its greater light-gathering power, to analyze the scattered light. The instrument generates an interferogram, which is the Fourier transform (FT) of the spectrum; but the mathematical transformation required can be done rapidly by a modern microcomputer. FT-Raman has the potential to greatly expand the range of samples accessible to Raman spectrometry.

Sampling

Solids, liquids, and gases can all be sampled. Water is an excellent solvent, as it has a weak Raman spectrum and does not react with the glass and silica commonly used as sample holders. Solids can be powdered and put directly into the laser beam, or the beam can be directed onto surfaces.

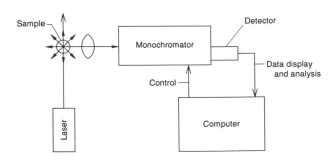

Fig. 2.   A typical Raman spectrometer.

The use of visible or near-infrared light in Raman spectrometry allows a standard optical microscope to act as a sample holder. The operator can mount the sample and choose the area to be studied, with a resolution of ca. 1 micrometer; a mirror is switched into the optical path, and the laser light is directed onto exactly the same area. The scattered light is collected by the microscope and directed into the monochromator or interferometer, as in an ordinary Raman spectrometer. The Raman microscope or microprobe, used to investigate very small samples and minute areas of larger samples, may be the single most important tool available to the Raman spectrometrist.

MOLECULAR INTERACTIONS

Many important chemical processes involve associations between molecules or between a molecule and a surface, in which no new COVALENT BONDS are formed. Such associations are often the result of hydrogen bond formation (see HYDROGEN BONDING), or dipole-dipole attractions. If the association causes changes in the structure of the molecule, it can change its vibrational spectrum. Careful interpretation of the Raman spectral changes can help determine what the structural changes are. Raman spectrometry is particularly useful for samples in aqueous solution, which are difficult to study in the infrared. Many systems of biological interest are analyzed in this way: for a current example, the formation of complexes between DNA and antitumor drugs (see ANTITUMOR AGENTS).

HOSTILE ENVIRONMENTS

The sample in Raman spectrometry can easily be located far from the rest of the instrument, because of the long distances the incident laser beam can travel without appreciable losses and the ease with which the scattered light can be collected and transmitted, again without great losses, through fiber optics. This fact makes it relatively easy to study samples that are very hot, or under high pressure, or otherwise in places where it is difficult to position a large optical instrument. Thus it is possible to study lubricants under pressure, or molten salts, which are of interest both in the industrial production of some metals and as possible high-performance battery systems (see BATTERIES).

SURFACES AND SOLIDS

As in the infrared, solids are easily sampled. Burgeoning interest in the chemical vapor deposition of thin films of diamond and other materials has led to the increased use of Raman spectrometry to determine both the physical and the chemical properties of such films and how changes in deposition conditions affect them. Catalytic systems, which also operate on the surfaces of solids, are also conveniently studied by Raman methods, as are many other surfaces and surface phenomena.

The microscope is important in geology, for example, where it has been used to classify groups of similar minerals by their structure and to study inclusions in geological materials (see GEOCHEMISTRY). It is used to determine the composition and properties of fibers and other polymers, and in the study of semiconductors, chiefly by identifying and characterizing defects and strains in the materials. It is also very useful in determining structures and stress in thin films and in many other analytical applications of Raman spectrometry.

BIBLIOGRAPHY

COLTHUP, N. B.; DALY, L. H.; WIBERLEY, S. E. *Introduction to Infrared and Raman Spectroscopy*, 3rd ed.; New York: Academic Press, 1990.

KATON, J. E.; PACEY, G. E.; O'KEEFE, J. F. "Vibrational Molecular Microspectroscopy." *Analyt. Chem.* 1986, *58*, 465A.

SKOOG, D. A.; LEARY, J. J. *Principles of Instrumental Analysis*, 4th ed.; Philadelphia: Saunders, 1992.

STROMMEN, D. P.; NAKAMOTO, K. *Laboratory Raman Spectroscopy*; New York: Wiley, 1984.

SZYMANSKI, H. A., ed. *Raman Spectroscopy: Theory and Practice*; New York: Plenum Press, 1970; Vols. 1 and 2.

TOBIN, M. C. *Laser Raman Spectroscopy*; New York: Wiley, 1971.

WILARD, H. H.; MERRITT, L. L., JR.; DEAN, J. A.; SETTLE, F. A. *Instrumental Methods of Analysis*, 7th ed.; Belmont, CA: Wadsworth, 1988.

WILLIAM E. L. GROSSMAN

Ramsay, William (1852–1916)

Sir William Ramsay was born October 2, 1852, in Glasgow, Scotland, and died July 23, 1916, at Hazlemere, England. William's grandfather, a dyer, had founded the Chemical Society of Glasgow, and the family influences were strongly oriented toward science. Ramsay entered Glasgow University in 1867 and upon graduation undertook graduate work in chemistry in Germany, first, briefly with BUNSEN in Heidleberg and, for his doctorate (1872), with Fittig at Tübingen.

At the onset of his independent research career Ramsay's interests were primarily in synthetic organic chemistry. From 1872 to 1874 he was assistant at Anderson's College, Glasgow, and in 1874 he became tutorial assistant to Professor John Fergusson at Glasgow University. In 1877 he provided the first laboratory synthesis of pyridine. His chemical and administrative abilities led to his appointment (1880) to the chair of chemistry at University College, Bristol, soon followed (1881) by his appointment as principal. By this time his research interests had broadened to included various inorganic synthetic projects and physicochemical investigations.

In 1887 he became Professor of Chemistry at University College, London, and was elected Fellow of the Royal Society in 1888. Lord RAYLEIGH, who was determining atomic weights of the gaseous elements by gas density determinations, asked Ramsay for a chemical method for the preparation of nitrogen gas, Rayleigh's earlier measurements having been made on nitrogen obtained by removing oxygen and other chemically reactive gases from air. The consequence of this was that Rayleigh found that the nitrogen from the air was denser than nitrogen prepared by oxidizing ammonia (as recommended by Ramsay). Rayleigh reported this fully at a meeting of the Royal Society on April 19, 1894. Ramsay attended this meeting and there sought Rayleigh's permission to try "to solve the mystery." Each working independently recovered an inert gaseous residue from air. In a joint paper to the Royal Society (January 18, 1895) they proved this to be a monatomic element, which they named argon. Henry Miers, Keeper of Minerals in the British Museum, who had learned of the essentials of the argon paper, wrote to Ramsay suggesting that a gas known to be liberated on heating the natural uranates might have a connection to argon. This led to Ramsay's discovery of helium (March 23, 1895). Before the end of the century Ramsay and his coworkers had discovered the remaining monatomic gases and had established these elements as a new group in the periodic table. For these achievements Ramsay was knighted in 1902 and received the Nobel Prize for chemistry in 1904.

In addition to his outstanding scientific and administrative skills, William Ramsay had an international circle of friends in science, this being enhanced by his working knowledge of several European languages, including Icelandic and Swedish. Among his friends, he was also esteemed for his bonhomie and his skill in tuneful whistling.

BIBLIOGRAPHY

TRAVERS, M. *The Life of Sir William Ramsay;* London: Edward Arnold, Ltd., 1956.

NEIL BARTLETT

Rayleigh, Lord (1842–1919)

John William Strutt, the third Baron Rayleigh, was born November 12, 1842, and died June 30, 1919. Because of delicate health as a child, he was educated in a variety of schools and privately. He entered Trinity College, Cambridge, as a fellow-commoner in 1861, where he read mathematics. He graduated (B.A., 1865) first in rank in mathematics (senior wrangler) and was a Smith prizeman. He was elected a fellow of Trinity in 1866 and turned his attention to physics. James Clerk Maxwell's papers on color vision and electromagnetism had a deep early influence, as had the work of Helmholz. The interest in color vision led to his well-known inquiry into the cause of the blue of the sky.

Rayleigh showed that the intensity of scattered, in terms of incident, light is inversely proportional to the fourth power of the wavelength (Rayleigh scattering) and thus accounts for the great preponderance of the short (blue) wavelengths of the visible spectrum in scattered light. He succeeded to the barony of Rayleigh on the death of his father in 1873 and set up a laboratory on his estate at Terling. In the 1870s he devoted his interests to a thorough theoretical treatment of the physics of sound.

On the death of Maxwell in 1879, Lord Rayleigh was invited to succeed him as the Cavendish Professor of Physics at Cambridge. This he accepted for a four-year period, during which time he focused his attention largely on the experimental tasks of standardizing the electrical units, the ohm and the ampere. Toward the end of his tenure of the Cavendish professorship, he took up the precise measurement of gas densities in an enquiry into the validity of Prout's hypothesis. This led to his best-known scientific achievement, the discovery of the element argon.

Prout had hypothesized that the atomic weights of the elements stood in simple numerical relationship to that of hydrogen. About 1882 Rayleigh set about testing the validity of this hypothesis with, as a first step, measurement of the relative densities of the elemental gases oxygen and hydrogen. He then moved to the density of nitrogen, which initially he obtained by removing oxygen, carbon dioxide, and water from air. Following a suggestion by William Ramsay, he later augmented the nitrogen production by passing air through aqueous ammonia, the entrained ammonia then being oxidized to nitrogen with hot copper oxide. Rayleigh found that the nitrogen obtained in this way was less dense than that obtained by treatment of air without added ammonia. He drew attention to this anomaly in a paper published in *Nature*, in 1892, and asked for ideas in accounting for the density discrepancy. William Ramsay undertook an investigation (using hot magnesium) to exhaustively remove nitrogen as well as oxygen and other reactive gases from air. He quickly found a residual gas differing spectroscopically from nitrogen and having a density relative to hydrogen greater than 19. This finding he communicated (August 4, 1894) to Rayleigh, who in the meantime had undertaken to repeat the 1799 experiment by Henry Cavendish. The latter had found that repeated sparking of air with oxygen over a caustic solution left an unreactive residue that was 1/120 of the volume of the original air. In August 1894 Rayleigh and Ramsay joined forces in the investigation of this gas.

From measurement of the ratio of the specific heats, Cp/Cv, for the gas, by the method of Kundt and Warburg, they showed that it was monatomic and their density measurement gave an atomic weight near 40. They reported the discovery of this inert gas, which was named argon, at a meeting of the Royal Society on January 31, 1895. For this discovery Rayleigh was awarded the Nobel Prize in physics for 1904.

Lord Rayleigh served as president of the Royal Society (London, 1905–1908) and was chancellor of Cambridge University from 1908 to 1919. Of his personal characteristics, it has been written that he looked for the good in people, had a keen sense of humor, was a good churchman, and was a hardworking public servant.

BIBLIOGRAPHY

STRUTT, R. J. *The Life of John William Strutt, Third Baron Rayleigh, O.M.,* London: Edward Arnold, 1924.

TRAVERS, M. *Sir William Ramsay;* London: Edward Arnold, Ltd., 1956.

NEIL BARTLETT

Reaction Rate

Several rates can be defined with reference to a chemical reaction. The rate of consumption of a reactant A is usually defined as the rate of decrease of its concentration:

$$v_A = -d[A]/dt$$

The rate of formation of a product Z is defined as the rate of increase of its concentration:

$$v_Z = d[Z]/dt$$

These rates are not necessarily the same; thus if a reaction is of stoichiometry

$$2A + B = Z$$

v_A is twice v_Z.

If, and only if, a reaction is of known stoichiometry, it is possible to define a rate of reaction; it is the rate of consumption of a reactant, or the rate of formation of a product, divided by the appropriate stoichiometric coefficient in the equation as written. In the above example, rate of reaction v is given by

$$v = \tfrac{1}{2}v_A = v_B = v_Z$$

It is important to note that the rate of reaction depends on the way in which the reaction is written. It is halved if the reaction is written as

$$4A + 2B = 2Z$$

See also KINETICS.

BIBLIOGRAPHY

LAIDLER, K. J. *Chemical Kinetics;* New York: Harper & Row, 1987; pp. 6–9.

KEITH J. LAIDLER

Reactivity, Theories of

Did you ever wonder why pouring water on solid calcium carbonate has no visible effect, whereas pouring glycerol on solid potassium permanganate causes, after an induction period of a couple of minutes, a violent flare-up? In the following discussion, you will see why certain chemical processes proceed spontaneously, sometimes with spectacular side effects, while others do not.

Discussions of reactivity typically consider physical and chemical processes separately. In a physical process, such as a ball rolling down a hill, the object remains unchanged at the completion of the process. In a chemical process, however, the object undergoes a chemical change and is altered in its chemical composition. An explosion is an example of a chemical process. The formulas of nitroglycerine and TNT, two common EXPLOSIVES, are on the next page.

What is the structural element common to these examples? One popular model pinpoints one or more excessively weak chemical bonds present in each. Characteristics of these weak, single bonds between N and N, N and O, or O and O are that they bridge very small atoms (note the presence of these elements

Nitroglycerine   TNT

in the extreme upper right-hand corner of the periodic table; see PERIODICITY, CHEMICAL) and include at least one LONE PAIR of electrons on each atom involved (Sanderson, 1983). Lone pairs of electrons are believed to be large, ill-focused entities, and their influence in the proper direction makes the single bonds very long and very weak by interelectronic repulsion. In the above chemical formulas, such weak bonds are labeled 1, and bonds that are equivalent to them by resonance are labeled 2. Average bond strengths are compared for these weak bonds with normal single bonds in Table 1.

Since the energies of chemical reactions are determined primarily by the balance between bond breaking (which requires energy) and bond making (which gives off energy), breaking weak bonds contributes to large energy yields. A net negative value for the energy change (more correctly, the enthalpy change) for a reaction favors spontaneity.

Actually, determining whether a reaction proceeds spontaneously is a little more complex than determining whether it gives off or takes up energy:

$$\Delta G = \Delta H - T\Delta S \qquad (1)$$

ΔG is called the free energy of the reaction, and it is the true criterion of spontaneity. The enthalpy of the reaction, ΔH, is determined, as we have seen, by the balance between bond breaking (positive) and bond making (negative). A net negative value favors spon-

Table 1. **Weak Versus Strong Bonds**

Long, Weak Bonds	Normal Bonds
N—N 163 kJ/mol	C—C 348 kJ/mol
N—O 201 kJ/mol	C—O 358 kJ/mol
O—O 146 kJ/mol	C—H 413 kJ/mol

taneity. $T\Delta S$ is the absolute temperature multiplied by the change in entropy for the reaction and is generally small, at least at low temperatures. For reactions with large negative values of the enthalpy change, the reaction is usually spontaneous. Gasforming reactions can cause ΔS to be positive and appreciable, thus contributing even further to the spontaneity:

$$2\ C_7H_3N_3O_6(s) + 7/2\ O_2(g) \rightarrow$$

TNT

$$14\ CO_2(g) + 3\ N_2(g) + 3\ H_2O(l) \qquad (2)$$

In the above equation, not only is there a net production of 13.5 moles of gas, resulting in a large, positive value for ΔS, but the process also destroys weak bonds and forms strong bonds, specifically, the O—H single bond (above), the C=O double bond (799 kJ/mol), and the N≡N triple bond (941 kJ/mol).

For example, a matter of some concern in chemical safety is the formation of peroxides from hydrocarbons upon long standing in the presence of oxygen (Committee on Hazardous Substances, 1981). Isopropyl groups are particularly susceptible.

Cumene
(isopropylbenzene)

A peroxide
of cumene

The concern arises because of the shock sensitivity and explosiveness of peroxides in pure form. Note the presence of the weak O—O bond (Sanderson, 1983) in the formula for the peroxide of cumene.

The thermodynamic criterion (ΔG, negative) is not sufficient to make the reaction proceed spontaneously. The reaction must also proceed with an appreciable rate (the kinetic criterion). In the case of explosives, the large amount of heat given off drives the temperature, and thus the rate of the reaction, up. A shock wave of product gases is formed and emanates rapidly from the center of the reaction.

Now that the basics of reactivity (thermodynamic and kinetic) have been explored with explosives and peroxides as examples, we turn to other domains of chemical reactivity. All HYDROCARBONS of molecular weight less than 100 g/mol burn spontaneously in oxygen to produce carbon dioxide and water, for example:

$$CH_4(g) + O_2(g) \rightarrow CO_2(g) + 2\,H_2O(g)$$

In this example, not only are weak bonds (O_2) replaced by strong bonds, but the entropy is favorable because of the net production of 1 mol of gas.

The reactivity of nonmetals increases as one goes toward the upper right-hand corner of the periodic chart: O_2, Cl_2, and especially F_2 are the most reactive[3].

$$CH_4(g) + 2\,F_2(g) \rightarrow CF_4(g) + 2\,H_2(g)$$

The reactivity of METALS is summed up by the activity series (Table 2).

In Table 2 reactivity increases from bottom to top. All metals above hydrogen react with acids to produce hydrogen gas and the metal ion, and although the rate is slow for the metals just above hydrogen, the reaction is dramatically rapid for elements near the top of the table. Metals near the top of the table are never found uncombined in nature; metals near the bottom frequently are.

Two other classification schemes have been found useful in describing reactivity: acid/base and reducing agent/oxidizing agent (redox) reactivity.

In Brønsted acid/base chemistry, illustrated in Table 3, acids are ranked in order of their strength from bottom to top, and their conjugate bases then vary in strength from top to bottom. An acid will react rapidly, with any base below it (see Table 3), since Brønsted ACID-BASE REACTIONS in water are kinetically fast. For example, the following reaction in aqueous solution is spontaneous:

$$HF + NH_3 \rightarrow F^- + NH_4^+ \qquad (4)$$

The activity series of metals is a portion of a more general set of redox strengths. An illustrative selection is given in Table 4.

Here the strongest oxidizing agent (F_2) is at the top left of the table, and the strongest reducing agent (Na) is at the right bottom. The weakest reducing agent (I^-) is at the top right, and the weakest oxidizing agent (Na^+) is at the bottom left. Any oxidant has the potential to react spontaneously with any reductant below it (see Table 4), although we have no information from this source regarding the rate of the reaction. For example,

$$2\,Ag^+(aq) + Fe(s) \rightarrow 2\,Ag(s) + Fe^{2+}(aq) \qquad (5)$$

does proceed, although as a heterogeneous reaction its rate is dependent on the state of subdivision (surface area) of the iron.

Table 2.   Activity Series of Metals and Hydrogen

Lithium	Nickel
Potassium	Tin
Calcium	Lead
Sodium	Hydrogen
Magnesium	Copper
Aluminum	Silver
Zinc	Mercury
Iron	Gold

Table 3.   Brønsted Acid/Base Reactivity

	Conjugate Acid		Conjugate Base
Strongest	HNO_3	NO_3^-	Negligible
Strong	H_2SO_4	HSO_4^-	Negligible
Weak	HF	F^-	Weak
Weak	NH_4^+	NH_3	Weak
Negligible	NH_3	NH_2^-	Strong
Negligible	CH_4	CH_3^-	Strongest

Table 4.   Redox Potentials

Standard Potential (V)	Reduction Half-Reaction
2.87	$F_2(g) + 2e^- \rightarrow 2\,F^-$
1.36	$Cl_2(g) + 2e^- \rightarrow 2\,Cl^-$
1.23	$O_2(g) + 4\,H^+(aq) + 4e^- \rightarrow 2\,H_2O(l)$
0.80	$Ag^+(aq) + e^- \rightarrow Ag(s)$
0.34	$Cu^{2+}(aq) + 2e^- \rightarrow Cu(s)$
0.00	$2\,H^+(aq) + 2e^- \rightarrow H_2(g)$
−0.44	$Fe^{2+}(aq) + 2e^- \rightarrow Fe(s)$
−2.71	$Na^+(aq) + e^- \rightarrow Na(s)$

BIBLIOGRAPHY

Committee on Hazardous Substances in the Laboratory; National Research Council. *Prudent Practices for Handling Hazardous Chemicals in Laboratories;* Washington, DC: National Academy Press, 1981; pp. 63ff.

COTTON, F. A.; WILKINSON, G. *Advanced Inorganic Chemistry,* 5th ed.; New York: Wiley, 1988.

GREENWOOD, N. N.; EARNSHAW, A. *Chemistry of the Elements;* Oxford: Pergamon Press, 1984.

SANDERSON, R. T. *Polar Covalence;* New York: Academic Press, 1983; pp. 20ff.

TYCODI, R. J. "Annotating Reaction Equations." *J. Chem. Educ.* 1987, 64, 243–246.

EDWARD MELLON

Reactivity-Selectivity Principle

The reactivity-selectivity principle is one of the most general postulates in chemistry. Nevertheless, it often has been formulated in a diffuse way. Perhaps the simplest formulation—and one of the most meaningful—is simply to state that the faster a reagent reacts in a chemical reaction, the less selective (in terms of product outcome) it is. Thus, the faster reagent is the one that reacts with different substrates at less differing rates. The limiting case is easily understood: the very reactive reagent reacts instantly with every molecule that it encounters, at the occasion of the first collision. Such a reagent reacts at a constant high rate and with little selectivity. The formulation of the reactivity-selectivity principle in this fashion is in agreement with the common experience that overheated reaction mixtures, and even explosive reactions, yield a mixture of different products as a result of several parallel reaction pathways. The TRANSITION STATEs for this very reactive reagent are "early"—that is to say, are reactant-like—and consequently the relative difference in the ACTIVATION ENERGY among the competing pathways is small.

A more exact definition of the reactivity-selectivity principle may be formulated with respect to a set of similar compounds (1, 2, 3, . . .), sharing a common functional group, that react with two similar reagents F (a reactive, or fast, reagent) and S (a less reactive, or slow, reagent). An example is the nucleophilic reaction of primary alkylamines (RNH_2) by an A_ND_N (S_N2) pathway, using the fast alkylating reagent iodomethane (CH_3I) and the slower alkylating reagent 1-iodopropane ($CH_3CH_2CH_2I$):

$$RNH_2 + CH_3I = RNH(CH_3) + HI$$
$$RNH_2 + CH_3CH_2CH_2I = RNH(CH_2CH_2CH_3) + HI$$

When there is sufficient similarity among the reacting compounds and the two (fast and slow) reagents, such that there is similarity in the transition state structures, then the reaction rates in the two series are approximated by the following equation (relating, for each of the series of alkylamines, the log of the rate constant for the fast reagent and the log of the rate constant for the slow reagent):

$$\log k_F = a + b \log k_S$$

This equation implies a linear dependence of the reaction rates of the two series and fulfillment of the reactivity-selectivity principle when b < 1. For the amine N-alkylation reactions described above, b is found equal to 0.63. The data for these reactions are plotted (as the open circles), as $\log k_F$ versus $\log k_S$ for a series of six alkylamines, in Figure 1.

It is evident, however, that the reactivity-selectivity principle does not have broad validity. As there are many cases where it is found, there are many where it is not. An example of its failure is the O-alkylation of substituted sodium cinnamates (also an A_ND_N mechanism) with 4-nitrophenacyl chloride (fast) and with 4-methoxyphenacyl chloride (slow); see Figure 2.

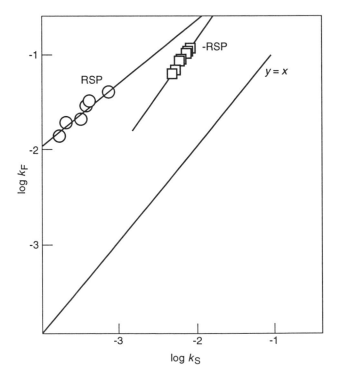

Fig. 1.   The reactivity-selectivity principle (RSP) is in evidence for the data of a series of reactions plotted with the open circles. It is not in evidence, however, for the data of another series of reactions plotted with open squares.

1304

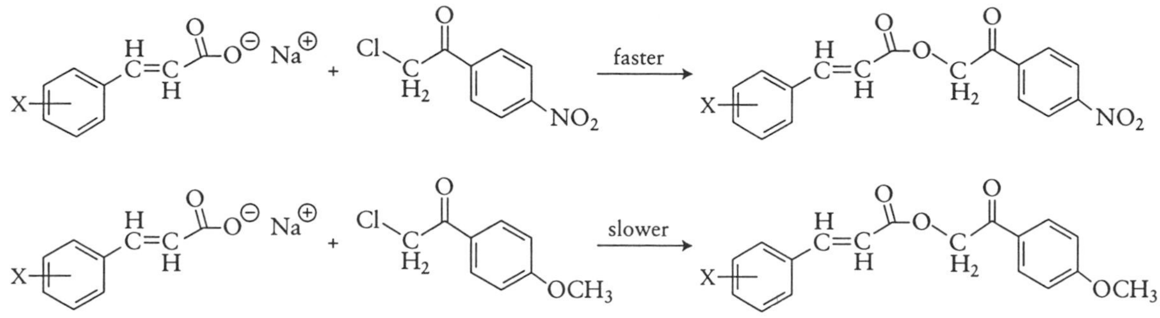

Fig. 2. Fast and slow reactions, in both of which the reactivity-selectivity principle fails to be demonstrated.

For these reactions b = 1.2 (Figure 1, open squares), implying the opposite of the reactivity-selectivity principle. There are also reagents that react with the same selectivity (b = 1). A statistical search on one hundred reactions has revealed as many failures as positive cases. The main problem is the intrinsic complexity of a reaction mechanism: for those reaction series that are sufficiently similar, proceeding by similar mechanisms, the reactivity-selectivity principle is most likely to have validity. However, it is never possible for two "similar" reactions to have the same mechanism: the identical geometry and charge distribution of transition states.

The above definition for the reactivity-selectivity principle can be extended to compare—rather than two reagents—two catalysts, two solvents, or even a long series of substrates reacting with one reagent in which the selectivity drops gradually with the increasing reaction rate. Experimental proofs are uncertain. It is proven, however, that the principle extends well to two temperatures: the reaction at the higher temperature is faster, and according to the reactivity-selectivity principle should be less selective. In other terms, the faster reaction possesses a lower activation enthalpy ($\Delta H^{\ddagger}$), and in its Arrheenius plot (log k vs. T^{-1}) the lines with the larger log k should have smaller slopes (an isokinetic relationship). This is in agreement with the common experience that at higher temperature more by-products are formed. This has been confirmed experimentally for most—but not all—reactions. An explanation for these exceptions must again be sought in the complexity of the reaction mechanism.

BIBLIOGRAPHY

EXNER, O. Correlation Analysis of Chemical Data; New York: Plenum, 1988.

———. "The Reactivity-Selectivity Principle: Redefinition and Reexamination on an Experimental Basis." V. Perkins trans. J. Chem Soc. 1993, 973–979.

JOHNSON, C. D. "The Reactivity-Selectivity Principle: Fact or Fiction." Tetrahedron 1980, 36, 3461–3480.

OTTO EXNER

Receptors

In order to maintain homeostasis and regulate metabolic activities (see METABOLISM and METABOLIC CYCLES), living cells must obtain information about their environment. Receptors are specialized PROTEINS or glycoproteins that allow cells to accomplish this. Signaling molecules that bind to receptors are called ligands. Only a small part of the receptor molecule may actually bind the LIGAND. The rest is necessary for maintaining position within the plasma membrane or binding with other regulatory proteins that are involved in signal transduction. A cell's receptors and particular metabolic machinery both determine its response to a chemical signal. Therefore, the same signal molecule may exert different effects in different cells. A cell's responsiveness to a particular signal can change with an increase or decrease in receptor number. An increase in receptors is called up regulation, and a decrease, down regulation. A signal molecule may regulate its own receptor number (homospecific regulation) or the receptors for another ligand (heterospecific regulation). Some receptors are located on the plasma membrane; others are intracellular proteins in the cytoplasm or nucleus.

Cell Surface Receptors

Cell surface receptors are protein or glycoprotein components of the plasma membrane. They are important in regulating cell processes including phagocytosis, endocytosis, antigen recognition and processing, antigen-antibody reactions, cell-to-cell communication, cell-to-matrix recognition, and responses to neurotransmitters, hormones, and other chemical messengers.

Matrix receptors bind with extracellular matrix proteins such as collagen, laminin, and fibronectin. They bind with relatively lower affinity than receptors for other signaling molecules, but they are present in higher concentration. They belong to a larger family of receptors called integrins, all of which are glycoproteins containing two distinct chains. Neural cells and nonneural cells have different integrins. The fibronectin receptor shown in Figure 1 links intracellular actin to fibronectin. It consists of alpha and beta glycoprotein chains held together by noncovalent bonds. The alpha chain has a smaller membrane-spanning portion and a larger extracellular portion held together by disulfide bonds.

Endocytosis-mediating receptors bind to certain extracellular macromolecules that accumulate in coated pits and enter the cell by endocytosis. Examples of large molecules internalized by receptor-mediated endocytosis are low density lipoprotein (LDL)-complexed cholesterol, iron-complexed transferrin, certain ANTIBODIES, and phagocytized material. The exact chemistry of the receptors and signaling mechanisms is not known, but current literature provides general information about the behavior of some of these receptors.

At least twenty-five different receptors are known to be involved in receptor-mediated endocytosis of various molecules, and all of the receptors appear to use the coated-pit pathway. For example, if a steroid-synthesizing cell needs cholesterol, the cell forms more LDL receptors and inserts them into the plasma membrane, where they associate with a coated pit. Cholesterol-containing LDL binds to the LDL receptor and is internalized. Once internalized, the receptor and LDL dissociate. Cholesterol is subsequently used, while the receptor is recycled to the plasma membrane. If there are too few LDL receptors and cholesterol uptake is deficient, cholesterol accumulates in the bloodstream and can thereby cause atherosclerotic plaques to form in blood vessels. Iron, combined in blood with the protein transferrin, is delivered to cells in a similar manner. Transport of antibodies in a mother's milk across epithelial cells of an infant's digestive tract is similar, except that in this case, the antibodies remain in transport vesicles that fuse with the abluminal epithelial cell membrane, releasing the antibodies to the extracellular fluid and bloodstream.

Phagocytosis is a special form of endocytosis. The ligand-bound receptors are linked to metabolic machinery in the cell that induces phagocytosis. Different receptors recognize different macromolecules such as complements or antibodies that coat invading microorganisms.

Receptor-mediated signal transduction involves other membrane proteins associated with the receptor. Since introduction of the receptor concept, researchers have identified cell-surface receptors for all known neurotransmitters and many hormones. Three groups or families, defined by their method of signal transduction, are ion channel-linked, G-protein-linked, and catalytic. The latter two are nonchannel-linked. Responses mediated through ion channel-linked receptors are faster than those mediated through the two nonchannel-linked, because fewer events intervene between ligand binding and the actual event that evokes the action. Receptors for neurotransmitters may be channel-linked or nonchannel-linked. Receptors for hormones are either nonchannel-linked or intracellular.

Channel-linked receptors are ion channels that open or close when bound by an appropriate ligand such as a neurotransmitter. The effect is a rapid change in the electrical properties of the responsive cell's plasma membrane. Ion channels exist for Na^+, K^+, Cl^-, Ca^{2+}, and other ions. Ions are important in maintaining the function of bioelectric membranes of nerve and muscle cells. By allowing entrance or exit of certain ions, the transmembrane potential may increase (hyperpolarize) or decrease (hypopolarize). Hyperpolarization is inhibitory, and hypopolarization, stimulatory. Acetylcholine (Ach), for example, binds to a cation channel receptor to promote Na^+ influx, subsequent hypopolarization, and stimulation of the responsive cell. Gamma amino butyric acid (GABA) and glycine, on the other hand, bind to Cl^- ion channels, allowing Cl^- influx to increase the transmembrane potential and generate an inhibitory effect. Ion channels are composed of membrane-spanning polypeptide subunits arranged around a

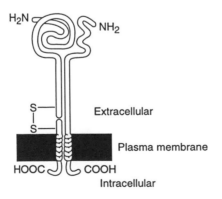

Fig. 1. Fibronectin receptor.

central aqueous pore that passes through the phospholipid bilayer of the plasma membrane. The pore is lined by polar amino acid residues in the polypeptide subunits.

Two Ach receptors are distinguished by the ability to bind nicotine or muscarine. The nicotinic receptor is the best known of all ligand-gated ion channels and has a number of alternative conformations. It was the first receptor to be purified and reconstituted into synthetic lipid bilayers. It was also the first single channel for which an electrical signal was recorded. Its gene was the first channel protein to be isolated, sequenced, and cloned. The nicotinic receptor is present on the postsynaptic membrane at vertebrate neuromuscular junctions and similar synapses in the marine ray electric organ. Binding of Ach to the receptor opens the channel to allow Na^+ influx and depolarization of the membrane.

The Ach nicotinic receptor, shown in Figure 2, has four different transmembrane polypeptide subunits designated alpha, beta, gamma, and delta. Two alpha and one each of the other three subunits are arranged around the central aqueous pore. Each subunit contains approximately 500 amino acids. All subunits have an N-terminal extracellular sequence that scientists believe crosses the plasma membrane as four alpha helices. One helix contains more polar amino acids than the others and probably forms the walls of the aqueous pore when the subunits are joined. Research indicates that the Ach-binding site is the N-terminal portion of the alpha subunits. Both alpha subunits have to be bound to Ach in order to fully open the channel for about 1 millisecond. Most of the receptor is on the extracellular surface with a smaller extension on the cytoplasmic side. The channel openings are about 25 Å in diameter, forming a 65 Å long extracellular vestibule and a 20 Å long cytoplasmic vestibule. The channel is constricted in the region of the plasma membrane.

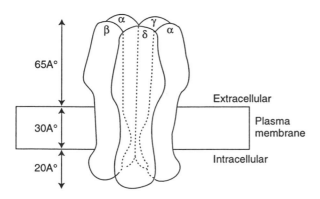

Fig. 2. Ach nicotinic receptor.

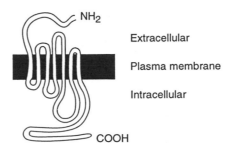

Fig. 3. A β-adrenergic receptor.

G-protein-linked receptors comprise a large family of proteins that consist of a single polypeptide chain that crosses back and forth across the plasma membrane seven times. The β-adrenergic receptor shown above in Figure 3 is an example. The N-terminus is extracellular, and the C-terminus is on the cytoplasmic side. This particular structure allows diversity in ligand-binding capacity for amines or for large and small peptides. Ligand binding to the receptor activates guanosine triphosphate-binding proteins (GTP-binding or G-proteins) that, in turn, function in signal transduction.

G-proteins comprise a large family of membrane proteins that mediate the effects of neurotransmitters, hormones, and other extracellular signals. Though G-protein-mediated responses are slower than those mediated by ion channels, they are more diverse, sensitive, flexible, and versatile. Each G-protein consists of a GTP-binding alpha subunit and beta and gamma regulatory subunits. Receptor binding activates the G-protein by adding GTP. Activated G-protein, in turn, activates specific proteins called effectors. Effector proteins are either ion channels or enzymes that catalyze reactions resulting in the formation or degradation of a second messenger. The G-protein transduction mechanism amplifies the chemical signal and directs it to the appropriate effector protein. Signals from different receptors may be integrated through G-proteins to stimulate the formation of a single second messenger, or G-proteins may balance inhibitory and stimulatory signals to provide a particular output. Information from a single receptor may be directed to more than one effector system through G-proteins.

The particular effector system that is activated categorizes the G-protein as G_s, G_i or G_p. G_s stimulates the enzyme adenylate cyclase, which in turn catalyzes formation of cyclic adenosine monophosphate (cAMP) from ATP. The cAMP regulates a cytoplasmic cAMP-dependent protein kinase enzyme that subsequently initiates a cascade of enzyme phosphorylations, ultimately producing the action attributed to the signaling molecule. Cyclic AMP can also regulate some ion channels by directly binding to them. G_i inhibits adenylate cyclase and decreases cAMP. G_p activates phosphoinositide phospholipase C. It then catalyzes release of inositol triphosphate (IP_3) and diacylglycerol (DG) from phosphoinositol diphosphate. Both IP_3 and DG are second messengers. The IP_3 pathway increases cytosol Ca^{2+} that is coupled to cell function. DG remains in the plasma membrane to activate protein kinase C, which phosphorylates proteins and activates ion channels. Obviously, once the signaling molecule binds to its particular receptor, all other events are independent of the molecule itself.

Muscurinic receptors for Ach are G-protein linked. In some of these systems, Ach binding results in the opening of K^+ channels and hyperpolarization of bioelectric membranes such as those in cardiac muscle cells. The result is inhibition. Other muscurinic receptors may close Ca^{2+} channels or activate phospholipase C.

Catalytic receptors are transmembrane alpha helix polypeptides. The most widely studied catalytic receptors in animal cells are the tyrosine-specific protein kinase enzymes. They have a ligand-binding portion located extracellularly and a cytoplasmic catalytic portion. When bound by the appropriate ligand, the enzyme autophosphorylates tyrosine residues on the catalytic portion. The exact role of tyrosine phosphorylation in signal transduction is unknown.

The insulin receptor is an example of a catalytic receptor. Though some researchers suggest an alternative form, most believe it is a glycoprotein whose suggested structure is represented in Figure 4. The receptor consists of two nonmembrane-spanning alpha chains and two membrane-spanning beta chains. The subunits are bound together by covalent disulfide bonds. Carbohydrate components include galactose, mannose, fucose, and galactosamine. Beta subunits

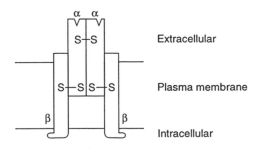

Fig. 4. The glycoprotein believed to function as an insulin receptor.

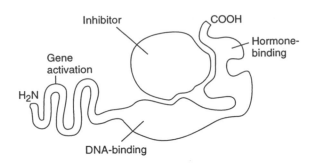

Fig. 5. Steroid hormone receptor.

have tyrosine kinase activity. When insulin binds to the extracellular portion, the beta subunits autophosphorylate tyrosine residues. This activates the kinase enzyme. Not all events in the mechanism are known, but researchers suggest that autophosphorylation is necessary for insulin's action. Following insulin binding, internalization of the receptor hormone complex occurs, suggesting a mechanism for down regulation of the insulin receptor.

Other growth factors such as epidermal growth factor (EGF), insulin-like growth factors-I (IGF-I), and platelet-derived growth factor (PDGF) are similar in possessing protein-specific tyrosine kinase activity. Amino acid sequence homologies exist in the tyrosine kinase portion of these receptors.

Intracellular Receptors

Receptors for STEROID hormones, T_3, and T_4, are intracellular. These hydrophobic hormones cross the plasma membrane of a target cell and bind with a cytoplasmic, or nuclear, receptor. Binding activates the receptor and increases its affinity to bind with nuclear genes.

A suggested model for the receptor is shown in Figure 5. It consists of a C-terminal hormone-binding domain, a DNA-binding domain, and an N-terminal domain that may be involved in gene activation. Experiments indicate that activated steroid-receptor complexes recognize certain DNA sequences and mediate their action through activation or suppression of genes. These hormone-responsive genes serve as enhancers for transcription of appropriate segments of DNA. Scientists suggest that in the inactive state the receptor is bound to an inhibitor protein that blocks the DNA-binding domain. When the hormone binds to the receptor, the inhibitor protein dissociates, allowing the receptor to interact with the hormone-responsive gene. Amino acid sequencing studies show homologous DNA-binding domains in receptors for thyroid hormones and a number of steroids.

Though interaction with surface receptors is generally thought to occur only for signaling agents other than steroids, T_3, and T_4, there is evidence for the existence of cell-surface receptors for steroids in some cases. The amphibian oocyte is an example. Progesterone, a steroid, apparently regulates meiotic maturation by interaction of cell-surface receptors coupled with adenylate cyclase inhibition.

BIBLIOGRAPHY

BLACK, J. W., et al., eds. *Perspectives on Receptor Classification;* New York: Wiley, 1987.

HADLEY, M. E. *Endocrinology,* 3rd ed.; Englewood Cliffs, NJ: Prentice Hall, 1992.

HALL, Z. W. *Molecular Biology;* Sunderland: Sinauer Associates, 1992.

HULME, E. C., ed. *Receptor-Effector Coupling: A Practical Approach;* New York: Oxford University Press, 1990.

KOHN, L. D. *Hormone Receptors;* New York: Wiley, 1982.

RUTH L. HAYS

Recommended Daily Allowance (RDA)

See DIETETIC CALORIE; NUTRITION; VITAMINS.

Recycling, Chemistry of

Recycling is the process of recovering waste material for reuse. In the United States, we generate more than 180 million tons of garbage each year. It typically contains paper (40 percent), yard waste (18 percent), metal (9 percent), plastics (8 percent), glass (7 percent), food (7 percent), and other materials (11 percent). PAPER, METALS, plastics, and glass are all recyclable; food and yard waste can be composted (converted into FERTILIZERS).

Recycling is important because it saves valuable resources from being simply thrown away. It also reduces the need for landfill space, which is becoming very scarce and expensive. Some states have passed laws mandating more recycling, so that many cities are now requiring household separation and curbside collection of newspapers, bottles, and cans. Recycling can even be a successful business operation if it includes

1. A steady source of suitable waste material
2. A reliable means for collection and separation
3. A practical technology for the recycling process
4. A dependable market for the recycled product
5. A fair margin of profit

Paper

Newspapers and corrugated cardboard are the most commonly recycled paper products. About 30 percent of all newspapers in the United States were being recycled in 1990. Standard methods for recycling waste paper usually involve (a) soaking the paper, (b) blending it into pulp, (c) adding detergent to disperse the ink, (d) screening out debris, (e) washing the pulp and draining away suspended ink, (f) bleaching the pulp, and then (g) pressure-screening the fiber slurry and sending it on to the paper machines. A newer, alternative method involves pressure-steaming the paper at about 27 atm and then flashing the soaked material back to atmospheric pressure so that the fibers are blown apart. Deinking is unnecessary in this case because the ink particles are too small to see. About one-third less energy is needed with this method.

Metals

Recycled metal scrap makes up more than 45 percent of the iron, 35 percent of the aluminum, 60 percent of the lead, 40 percent of the copper, and 20 percent of the zinc produced in the United States. Much of the recycled metal comes from automotive scrap. After the metals have been separated, they are melted in furnaces similar to those used in ordinary metal-producing plants. In fact, scrap metal is often mixed in with metal being produced from ore. Iron scrap, for example, is a common ingredient in steel furnaces, although the charge to a basic oxygen furnace cannot be more than one-third scrap iron. A ton of iron scrap can replace 1.5 tons of iron ore, with a 75 percent savings in energy and more than a 75 percent reduction in air and water pollution. Today there are many steel "minimills" that use scrap iron alone as raw material, feeding it into electric arc furnaces.

More than 60 percent of all aluminum beverage cans are being recycled. Making new aluminum cans from old ones requires only 5 percent of the energy needed to make them from aluminum ore. Old cans need only to be melted and have their impurities burned away, whereas bauxite (Al_2O_3) ore must be dissolved in cryolite (Na_3AlF_6) at a very high temperature and then subjected to electrolytic reduction—both steps are highly energy consuming. A ton of recycled aluminum also eliminates the need to mine several tons of bauxite.

Glass

Glass refuse is mostly in the form of jars and bottles, about 65 percent colorless, 20 percent amber, and 15 percent green. With beverage bottles maximum economy (about 90 percent saving) is achieved when bottles are simply collected and reused; but the bottles must be durable enough to withstand repeated use. Recycling of glass involves collection, separation by color, cleaning, and crushing into cullet. (Cullet is broken glass suitable for remelting.) Recovered cullet is a bit cheaper than glass raw materials, mainly sand (SiO_2), soda (Na_2CO_3), and limestone ($CaCO_3$); and because of its lower melting point, less energy is needed to melt it. In making a batch of glass, manufacturers typically use about 50 percent cullet, most of it from their own factory scrap. The cullet must be free of impurities, especially metals. Low-grade mixed-color cullet is sometimes used in concrete ("glascrete") or asphalt ("glasphalt").

Plastics

Recyclers usually sort plastics by type (according to stamped code numbers) and simply chop them into chips or flakes to be melted down and remolded or converted to fiber. For example, polyethylene terephthalate (PET) soda bottles are turned into carpets; high-density polyethylene (HDPE) milk jugs are remolded into detergent bottles; polystyrene containers are melted down and foamed for insulation; nylon waste is converted to tennis ball felt; and polycarbonate water bottles are made into automobile bumpers. Even mixed plastic waste can be recycled into artificial lumber or panels similar to particle board. Plastic "wood" is easy to saw; and it resists weather and insects better than real wood.

Almost any kind of plastic can be recycled. Even nonmelting thermoset plastics can be ground up for filler material. Plastics can also be degraded thermally or chemically. Some POLYMERS (such as PET) can be readily "unzipped" back to short-chain monomers from which new plastic can be made, and almost any plastics can be broken down to mixtures of HYDROCARBONS that can be used as chemical raw materials or fuels. Less than 2 percent of plastic waste was being recycled in the United States in 1990, with PET soda bottles and HDPE milk jugs accounting for most of it. But plastics recycling is a field that is only in its infancy, and it has enormous potential for growth. The fact is that landfill space is rapidly running out, and the possible uses for recycled plastics appear to be almost infinite.

BIBLIOGRAPHY

BARTON, A. F. M. *Resource Recovery and Recycling;* New York: Wiley, 1979.

BLUMBERG, L.; GOTTLIEB, R. *War on Waste;* Washington, DC: Island Press, 1989.

CONDON, J. *Recycling Paper;* New York: F. Watts, 1990.

DENISON, R. A.; RUSTON, J. *Recycling and Incineration: Evaluating the Crisis;* Washington, DC: Island Press, 1990.

Environmental Protection Agency. *The Solid Waste Dilemma: An Agenda for Action;* Washington, DC: Office of Solid Waste, 1989.

FRANKLIN, W. E. *Paper Recycling: The View to 1995;* Prairie Village, KS: Franklin Associates, 1990.

PAWLEY, M. *Building for Tomorrow: Putting Wastes to Work;* San Francisco: Sierra, Club, 1982.

DORIS KOLB

Redox Equations (Reduction Processes)

See BIOENERGETICS.

REM

The rem is a unit of radiation dose equivalent used to describe the biological effects of radiation on humans. Rem is an acronym for Roentgen Equivalent Man. The SI unit that is similar to the rem is the sievert (Sv). One sievert equals 100 rem. The rem is the unit of radiation dosage that describes not only the radiation one is exposed to but also the biological damage sustained from the exposure. This is accomplished by taking the radiation exposure in rads and multiplying it by a quality factor that depends on the type of radiation. For example, the quality factor for X and γ radiation equals 1; thus the dosage in rads equals the dosage in rems for X and γ radiation. For alpha particles the quality factor is 20. This reflects the fact that alpha particles cause twenty times more damage to human tissue than X and γ radiation, and the dose in rems for alpha particles would be twenty times greater than the dose in rads. The quality factor for beta particles is 1, for thermal neutrons 2, for fast neutrons 10, and for protons 10.

Exposures of more than 25 rem cause detectable blood changes in humans. An exposure of 500 rem is lethal to 50 percent of humans within thirty days. The average American is exposed to 0.1 to 0.2 rem per year. The largest exposure (67.6 percent of the total exposure) is from natural sources such as cosmic rays, Rn in air, transuranic elements in the soil and building materials, and internal radiation from ^{40}K in bones. The second-largest exposure (30.7 percent of the total) is from medically related procedures such as X rays, therapeutic and diagnostic procedures, and cancer treatments. For example, a typical dental X ray delivers an exposure of approximately 0.5 mrem. Therapeutic cancer treatments can produce exposures of roughly 5,000 rem or more delivered to a highly localized area, the tumor, over a period of a few months. The idea of therapeutic treatments is to kill the tumor with radiation while maintaining the patient's health.

In the general public, it is recommended by the Nuclear Regulatory Commission (NRC) and the International Commission on Radiological Protection (ICRP) that the annual exposure to an individual not exceed 0.5 rem, excluding natural radiation sources. Over an individual's lifetime, the NRC recommended exposure should average 0.17 rem per year. For workers who are occupationally exposed to radiation there is a different set of guidelines. The ICRP recommends a limit of 5 rem per year. The accumulated dose for occupational workers should not exceed $5(N - 18)$ where N is the age of the worker. Note that this formula implies that individuals below age eighteen should not be exposed to occupational radiation hazards.

BIBLIOGRAPHY

ATWOOD, C. H.; SHELINE, R. K. *J. Chem. Educ.* 1989, *66*, 389–393.

CHOPPIN, G. R.; RYDBERG, J. *Nuclear Chemistry Theory and Applications;* Elmsford, NY: Pergamon, 1980; Chapter 16.

EHMANN, W. D.; VANCE, D. E. *Radiochemistry and Nuclear Methods of Analysis;* New York: Wiley, 1991; Chapter 7.

CHARLES H. ATWOOD

Remsen, Ira (1846–1927)

Born on February 10, 1846, Ira Remsen is noted as one of the original faculty members of Johns Hopkins University in Baltimore, Maryland. He received his undergraduate degree in 1865 from the College of the City of New York. Following the wishes of his father, Remsen studied medicine at the College of Physicians and Surgeons at Columbia University and received his medical degree in 1867. During Remsen's medical studies, however, chemistry captured his curiosity,

and in the summer of 1867 he abandoned a medical career and undertook the study of chemistry.

Remsen traveled to the University of Munich, where he undertook studies with Jacob Volhard. Volhard introduced Remsen to Wohler and Fittig, whom Remsen followed to Göttingen, where he received his doctorate in 1870 for his studies of the constitution of piperic acid. Prior to returning to the United States, Remsen followed Fittig to Tübingen, where Remsen served as his laboratory assistant.

Upon returning to the United States, Remsen accepted a position at Williams College as professor of physics and chemistry from 1872 to 1876. Although he found his years at Williams College frustrating owing to his colleagues' unsympathetic attitude and the scarcity of equipment, he began writing chemistry textbooks. He translated Friedrich WÖHLER's *Organic Chemistry* into English and wrote his own *Principles of Theoretical Chemistry,* which was later translated into several languages. His textbooks and lecture style were considered marvelously clear and effective. It was fundamentally as a teacher that Remsen attained his fame.

In 1876 Remsen accepted the chairmanship of chemistry at the newly established Johns Hopkins University, where the emphasis of study was to be upon advanced graduate work. Remsen's teaching and research methods followed those he had learned while studying in Germany. Remsen's chemical research with Constantine Fahlberg at Johns Hopkins University led to the discovery of orthobenzoyl sulfimide, the sugar substitute saccharin. Remsen also began the publication of the *American Chemical Journal,* which was first issued in 1879. An immediate success, it served as the vehicle for the publication of research at Johns Hopkins University and became one of the foremost scientific publications in the world. In 1915, after the development of the American Chemical Society, the journal was merged with the *Journal of the American Chemical Society.*

Remsen also enjoyed a career as a university administrator. He served as acting president of John Hopkins University in 1889 and 1890 while the president was in Europe. Following the president's retirement in 1901, Remsen was elected president, a post he held until 1912. His tenure as president was marked by his belief that "every man does his best work when he is allowed to do it in his own way."

Remsen was honored with many awards and honorary degrees at John Hopkins University. In 1902 he became president of the American Chemical Society, and in 1903 he was elected president of the American Association for the Advancement of Science. He was elected president of the National Academy of Science in 1907 and held the post for six years. Remsen was the recipient of the Gold Medal of the Society of Chemical Industry in 1908, the Willard Gibbs Medal in 1914, and the Priestley Medal in 1923.

After resigning the presidency of Johns Hopkins University in 1913, Remsen continued his career as a consultant to the petroleum industry. He died in Carmel, California, in March 1927.

BIBLIOGRAPHY

DARROW, F. L. *The Story of Chemistry;* Indianapolis: Bobbs-Merrill, 1927.

MILES, W. D. *American Chemists and Chemical Engineers;* Washington D.C.: American Chemical Society, 1976.

PARTINGTON, J. R. *A History of Chemistry;* New York: Macmillan, 1964.

ANNA R. BERGSTROM

Research, Chemical Literature

See LITERATURE SEARCHES.

Resonance

Chemical bonding in covalent molecules and polyatomic ions is commonly represented by LEWIS STRUCTURES that locate the valence electrons of each atom in electron pairs, either in a covalent bond or a nonbonding electron pair (see CHEMICAL BONDING). Lewis structures, used with the ELECTRONEGATIVITY concept and VALENCE SHELL ELECTRON PAIR REPULSION (VSEPR) theory, are very useful in predicting chemical and physical properties such as bond polarity, molecular shape, polarity of structure and relative electron density, bond order, and BOND LENGTHS.

There are some molecules, however, whose properties are not consistent with a single Lewis structure. Representation of these molecules by a single structure will lead to inaccurate predictions. The carbonate ion is an example of this phenomenon—see Figure 1.

The structure suggests that the ion contains two carbon-oxygen single bonds and one carbon-oxygen double bond. We would expect, therefore, that one bond would be shorter than the other two, and the electron density would be higher on two of the oxy-

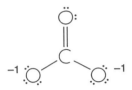

Figure 1. A proper Lewis structure of the carbonate ion.

gen atoms. We would predict that a Lewis acid, such as a proton, would preferentially react with the oxygen atoms carrying the higher electron density. In contrast, the actual carbonate ion contains three equivalent carbon-oxygen bonds of intermediate bond length, longer than a typical carbon-oxygen double bond but shorter than a single bond. The electron density is the same at each oxygen atom, lower than expected from an oxygen atom with a negative charge but higher than the neutral oxygen atom participating in the carbon-oxygen double bond. A proton would have an equal probability of reacting at any of the three oxygen atoms.

Resonance theory explains this inconsistency by proposing that not all valence electron pairs in real molecules or ions are localized in one covalent bond or on a single atom as represented in the single Lewis structure. In contrast, some are delocalized through a pi electron system, either shared by two or more atoms or participating as both bonding and nonbonding electrons. Since a single Lewis structure does not represent delocalized electrons, the actual ion is more accurately represented by a combination of several Lewis structures that, considered together, reflect its real valence-electron structure. Specifically, the carbonate ion is represented by three equivalent but different Lewis structures (Figure 2). The structures are different because each oxygen atom is a distinct atom in the ion.

Properties of the ion are consistent with an average of the three structures, as illustrated in Figure 3. The

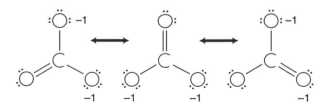

Figure 2. The carbonate ion is represented by three equivalent but different Lewis structures. (None of these resonance forms actually exist.)

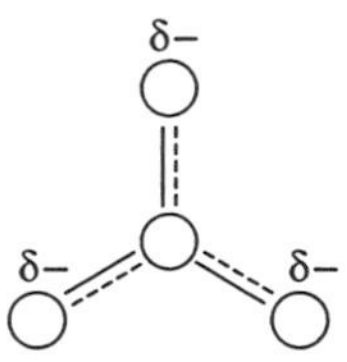

Figure 3. An average of the three structures shown in Figure 2.

actual ion is called a RESONANCE HYBRID of the three resonance structures, which are associated by the double-headed arrows in Figure 2.

The structure in Figure 3 is not a Lewis structure, since it does not illustrate nonbonded electrons, complete bonds, or full formal charges. The dashed and solid lines, however, correctly represent carbon-oxygen bonds of equal but intermediate length, and the partial negative signs illustrate the partial but equal charges. It is very important to understand that none of the resonance forms illustrated in Figure 2 actually exist. The actual ion is not an equilibrium mixture of these electronic structures but is intermediate in structure between them.

From a molecular orbital point of view, the intermediate structure represents the wave function of the actual ion. It is derived in resonance theory by adding the wave functions of each Lewis structure in Figure 2 to produce a single wave function for a molecule or ion represented in Figure 3.

BIBLIOGRAPHY

CHANG, R. *Chemistry;* New York: McGraw-Hill, 1994; pp. 353–354.

COTTON F. A.; WILKINSON, G.; GAUS, P. L. *Basic Inorganic Chemistry;* New York: John Wiley, 1987; pp. 72–74.

DEKOCK, R. L.; GRAY, H. B. *Chemical Structure and Bonding;* Mill Valley, CA: University Science Books, 1989; pp. 103–106.

FESSENDEN, R. J.; FESSENDEN, J. S. *Organic Chemistry;* Pacific Grove, CA: Brooks/Cole, 1994; pp. 72–79.

MARCH, J. *Advanced Organic Chemistry;* New York: John Wiley, 1992; pp. 26–36.

MOELLER, T. *Inorganic Chemistry: A Modern Introduction;* New York: John Wiley, 1982; pp. 198–201.

MOODY, B.; ARNOLD, E. *Comparative Inorganic Chemistry;* London: Hodder and Stoughton, 1991; pp. 81–82.

PAULING, L. *The Nature of the Chemical Bond;* Ithaca, NY: Cornell University Press, 1960; Chapters 6 and 8.

WILLIAM S. RICHARDSON III

Resonance Hybrid

A *resonance hybrid* is a single model representing a molecule for which two or more equivalent Lewis structures can be drawn. This occurs when there is a combination of single and multiple bonds between atoms in a molecule. For example, in the molecule NO_3^-, the rules for drawing Lewis structures predict three equivalent arrangements, shown in Figure 1.

None of these models, however, matches the data. Experimental measurements of the bond lengths in this molecule show that all three nitrogen-to-oxygen bonds are equal in length, shorter than a single and longer than a double bond. The bonding between nitrogen and oxygen atoms in this molecule is actually a *hybrid,* or combination, of single and double bond character, thus the word *hybrid* in the term *resonance hybrid.* The concept of resonance was introduced to account for the fact that in reality bonding electrons are not localized as shown by Lewis structures. In the case of the NO_3^- molecule, valence electrons are distributed around the central atom to form three equivalent N—O bonds. This fact is illustrated using the localized bonding model by saying that each equally possible Lewis structure (resonance structure) contributes to a resonance hybrid (combination of resonance structures). This contribution is usually represented by double-headed arrows between individual Lewis structures (Figure 2).

Both the word *resonance* and its representation with double-headed arrows very often result in the misunderstanding that the molecule actually alternates, or "resonates," between the multiple forms predicted by the rules. This leads to another misconception: the resonance hybrid is viewed as an equilibrium mixture of Lewis structures. Textbooks often show all possible Lewis structures contributing to a resonance hybrid as a way of approximating the true situation. It is important to note that in this case no single Lewis structure is correct and that the real structure is best represented by a combination of these single Lewis structures (resonance hybrid). A better resonance hybrid representation for the NO_3^- molecule would be a single structure derived from Lewis structures as shown in Figure 3.

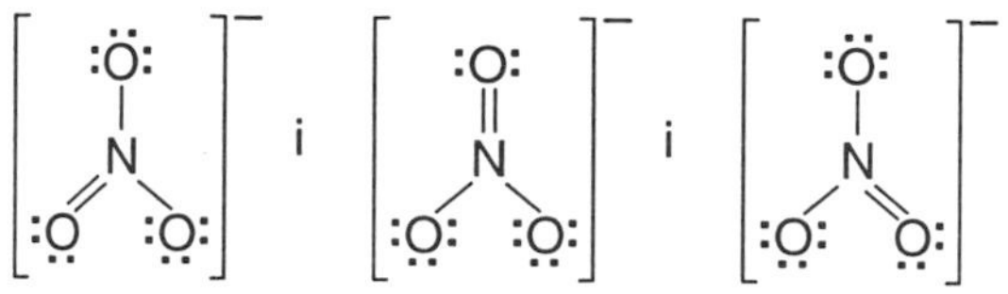

Fig. 1.

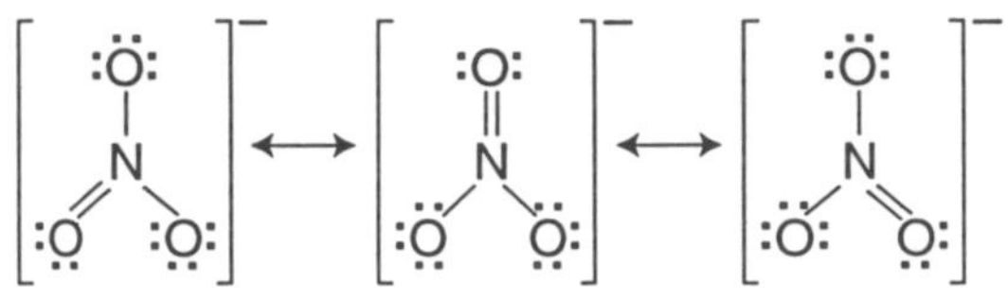

Fig. 2.

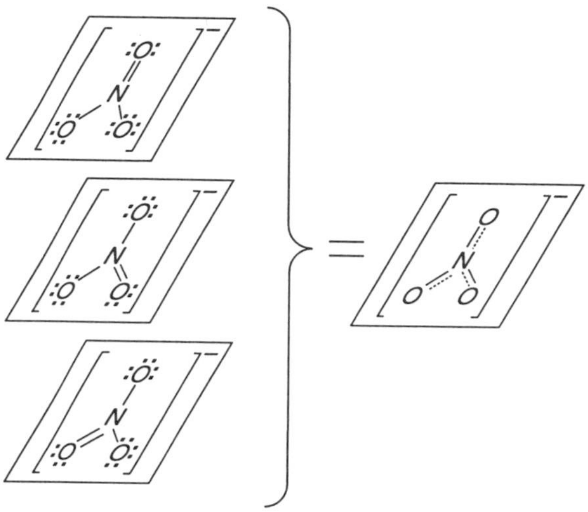

Fig. 3.

In Figure 3, all the bonds are equal in length and intermediate in character between single and double bonds, just as in the real molecule. An accurate description or model is important because the more accurately the structure of a molecule is known, the better its properties such as bond length, charge density, reactivity, and stability can be explained.

DIANE M. BUNCE
EVE LABUDA

Resonance Theory, Applications of

RESONANCE is a theoretical concept for the description of the electronic structure of molecules for which one simple Lewis structure is not appropriate (Carey and Sundberg, 1990; March, 1985; Wheland, 1955).

The LEWIS STRUCTURE is the most widely used notation for chemical compounds. It consists of a geometrical arrangement of nuclei, written as element symbols and connected by bonding electron pairs. LONE PAIRs are additional electron pairs that are localized at single atoms. For a large majority of compounds this simple model of two-center, two-electron bonds is sufficient to explain their structure and reactivity in a qualitative way. However, as will be shown below, there are certain molecules whose electronic structure cannot be reasonably described by a single Lewis formula. In order to keep the simple Lewis picture of CHEMICAL BONDING, the concept of resonance is invoked: the real molecule is described by a superposition of several resonance, or mesomeric, forms. Each of these mesomeric forms is depicted by a simple Lewis formula, and the various resonance forms are connected by double-ended arrows. This indicates that each single Lewis structure is a model that describes only part of the reality. The actual electronic structure of the molecule lies somewhere in between these limiting cases. In particular, the molecule is energetically more stable in its true state than it would be in either of the virtual resonance structures. In this way, the mesomeric forms have to be sharply distinguished from the isomeric forms of a molecule. ISOMERS are alternative geometrical structures of a given molecule; they are not model descriptions of a true state, but separate, potentially isolable compounds that are in EQUILIBRIUM with each other.

Examples of Resonance

A simple example of the necessity to invoke the concept of resonance is given by the ALLYL cation, anion, and radical (see ANIONS AND CATIONS and FREE RADICALS for an introduction to these concepts). The experiment shows that the carbon-carbon (C—C) bond lengths are equal and that the charge in the cation and anion, and the unpaired electron density in the radical, is distributed equally over the terminal carbon atoms. The π-bond is delocalized over the tree centers. It is not possible to draw a simple Lewis structure that accounts for these observations. But we can choose two model "structures" with equal weight that describe the two limiting cases of the bonding. These "structures" must not be confused with physical structures, since they are only model descriptions. By superimposing them, in the framework of the concept of resonance, a rather good description of the electronic structure of the allyl species can be achieved. A pictorial way of representing the delocalized nature of the π-bonding in the allyl system is by using dashed lines as shown in Figure 1.

Similarly, benzene is known to have six C—C bonds of equal length. Thus, its electronic structure cannot be represented by a single Lewis formula with

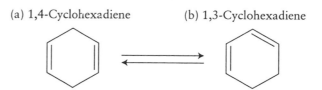

Fig. 1. Model structures approximating the nature of bonding in allyl species.

Fig. 3. Two forms of cyclohexadiene.

alternating single and double bonds. But if we use two such structures, the bonding picture should not be too far from reality (see Figure 2).

In contrast to that, the two forms of cyclohexadiene, 1,3- and 1,4-cyclohexadiene, are geometrical isomers (Figure 3). The two Lewis structures depict different molecules in equilibrium with each other, as indicated by the reaction arrows. Each molecule is well described by a single resonance form.

Of course, the concept of resonance is not restricted to organic chemistry. Two simple examples of inorganic molecules that have to be described by several Lewis structures are the nitrate ion and the phosphate ion, as shown in Figure 4.

Aromaticity as a Special Kind of Resonance

For cyclic systems, the concept of resonance has been further developed to include the concept of AROMATICITY. It is important to distinguish between the two approaches and to understand that resonance is the more general one. Cyclic structures with a system of conjugated occupied π-orbitals that show a special stabilization are called *aromatic*. Benzene is the prototype of aromatic compounds. The stabilization, as compared with one of the virtual mesomeric forms with localized single and double bonds, is attributed to the enhanced delocalization of the electrons. Accordingly, the theoretical stabilization energy, calculated as the difference between the energy of the real molecule and the theoretical energy of one of its nonexisting resonance forms (which can be estimated from quantum chemical calculations), is called delocalization energy or resonance energy. It can thus be seen that aromaticity is just another name for resonance for a certain class of molecules; no principally new effects occur. From simple MOLECULAR OR-

BITAL THEORY (MO theory), it is possible to predict the kinds of molecules for which the Lewis description of electronic structure is particularly inappropriate, leading to a specifically large resonance stabilization. The Hückel rule states that planar, monocyclic, completely conjugated hydrocarbons with $(4n + 2)$ π-electrons in the ring are such systems; they are termed *aromatic*. The concept of aromaticity shows the limit of the simple resonance model. The sole use of resonance forms cannot explain why benzene exhibits a special stabilization or why the two resonance forms should be used with equal weight to describe the bonding situation of the six-membered ring, while the two forms of cyclobutadiene shown in Figure 5 are not mesomeric forms, but rather depict two (isoenergetic) isomers. Unlike the stable aromate benzene, cyclobutadiene is a very unstable (antiaromatic) compound. Without employing the additional $(4n + 2)$ rule, which comes out of

(a) Nitrate anion

(b) Phosphate anion

Fig. 4. Representative inorganic molecules described by multiple Lewis structures.

Fig. 2. Bonding in benzene.

Fig. 5.   Two forms of cyclobutadiene.

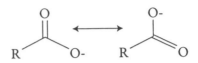

Fig. 7.   Carboxylate anion.

MO theory (see page 1319), there is no way to tell the difference between aromatic and antiaromatic compounds using the concept of resonance.

The necessity of using several resonance forms for describing the bonding situation of a molecule is not restricted to unsaturated compounds, although π-conjugated systems are the most important class of structures for which a single resonance form is often insufficient. Saturated compounds are usually well described by a single mesomeric form, but there are cases where more than one resonance form is necessary. A famous case is the norbornyl cation, which has been the focus of a controversial discussion for many years (Brown, 1977). It is well recognized now that the correct description of the bonding situation in the norbornyl cation needs two resonance forms, as shown in Figure 6. Compounds with delocalized σ-bonds such as the norbornyl cation are termed *nonclassical*.

Chemical Consequences of Resonance

The examples presented in this section will show that the notion of resonance is useful in three respects: the rationalization of structural properties, the discussion of relative stabilities, and the explanation of reaction pathways.

The two mesomeric forms of the allyl anion are mirror images of each other and must therefore contribute equally to the overall structure. The same holds for the allyl cation and the allyl radical. The carboxylate groups constitute a further example of allylic-type resonance, which is found very often in organic chemistry. Upon deprotonation of a carboxylic acid, the resulting negative charge is delocalized between the two oxygen atoms as depicted in the two resonance forms (see Figure 7). A carboxylate anion

is therefore more stable than an alcoholate ion where the negative charge must remain localized on one oxygen atom. This fact is, at least in part, responsible for the higher acidity of carboxylic acids as compared with ALCOHOLS.

The different resonance forms need not have equal weight as in the previous examples in order to be important for the description of the electronic structure of the molecule. The following examples demonstrate that resonance forms that describe different bonding situations can be helpful in explaining the structure and reactivity of molecules. The first example, Figure 8, is the AMIDEs. The amide moiety is planar and the rotational barrier of the carbon-nitrogen (C—N) bond is rather high, indicating that the second resonance form with a C—N double bond plays an important role. This effect is important in the formation of secondary and tertiary structures of proteins because it reduces their conformational mobility.

Enolate ions (see ENOLS) are an intermediate case between allyl anions and carboxylates. Here, as in the amide discussed above, the two resonance forms that are needed for their description are no longer equal in importance because they are not mirror images of each other. In a free enolate ion, one expects the left structure to be more important, as the negative charge is localized on the oxygen atom due to its higher ELECTRONEGATIVITY. The carbanionic resonance form (b) explains the nucleophilic character of the α carbon atom. One can influence the relative weight of the two resonance forms by changing the counterion of the enolate (see Figure 9).

Small, hard counterions that coordinate tightly to oxygen lead to a stabilization of the left resonance form, thereby reducing the reactivity of the enolate as a carbon NUCLEOPHILE. One can think of an enol ether as an extreme form of coordination to oxygen,

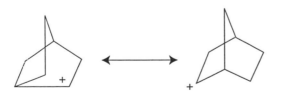

Fig. 6.   Bonding in the norbornyl cation.

Fig. 8.   Amide bonding.

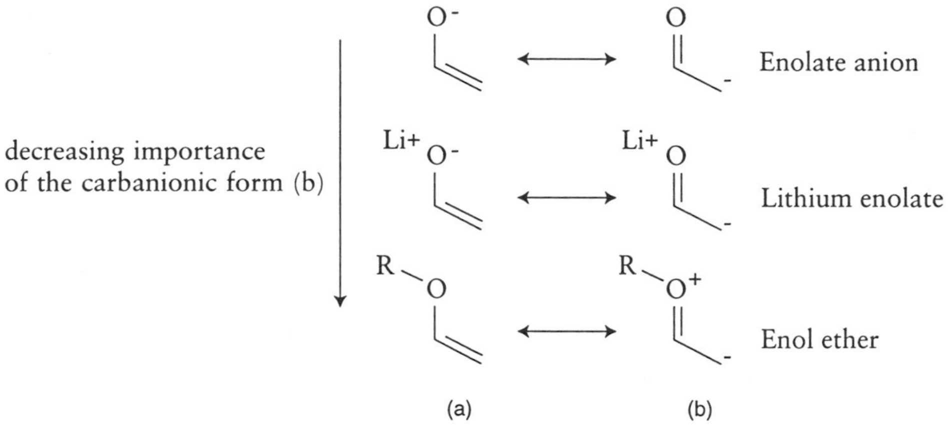

Fig. 9.   (From top to bottom) The decreasing importance of the carbanionic form (b).

where the zwitterionic form (b) is needed only to explain the reactivity of the α carbon atom. Since enamines, in which oxygen is replaced by an NR group, are more reactive than enol ethers, the dipolar resonance form must play a greater role.

When one extends the carbon chain in the allyl anion or cation by one double bond, one obtains a pentenyl ion, which can be described by three resonance forms, (a), (b), and (c). Structure (b) should contribute somewhat less than resonance forms (a) and (c) to the anion because of the greater stability of primary carbanions. The opposite is true for the pentenyl cation, the extended allyl cation. Here, resonance form (b) is more important than the other two. Both pentenyl systems, however, should have C—C bonds of approximately equal length. In general, linear conjugated π-systems with an odd number of car-

bon atoms are strongly resonance-stabilized and show nearly identical C—C bond lengths. The formal insertion of double bonds into the conjugated chain produces so-called vinylogues, which often have similar properties to their parent compounds. Many representatives of this class are used as dyes (see DYES AND PIGMENTS). Polymethine dyes such as Pinacyanol, for example, can also be looked at as vinylogues of the amidine group. Phenolphthalein, in its anionic form, contains a vinylogue of a carboxylate group. Such conjugated π-systems have two main resonance structures with the charge on either end of the conjugated chain. They are sometimes called *push-pull* systems. See Figure 10.

Butadiene is reasonably well described by only one Lewis structure with two double bonds. The two dipolar forms have been added to explain the shorten-

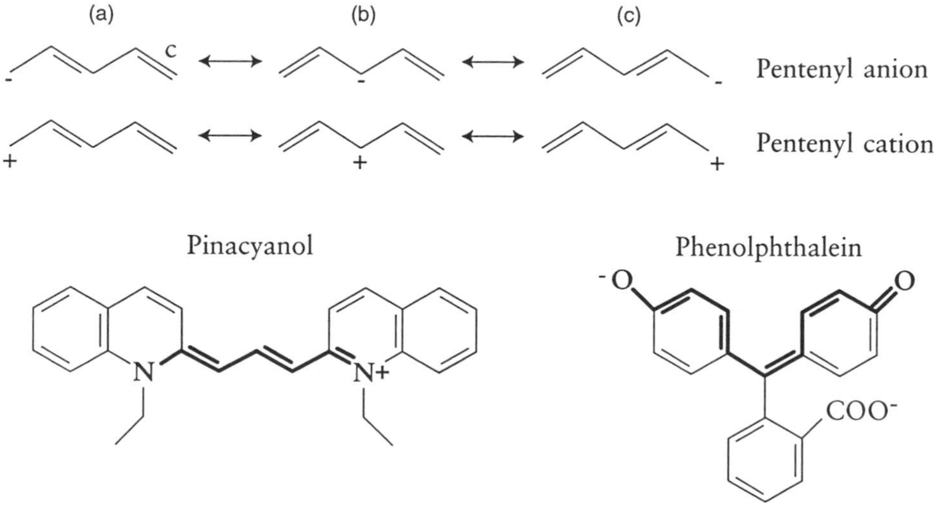

Fig. 10.   Representative push-pull systems.

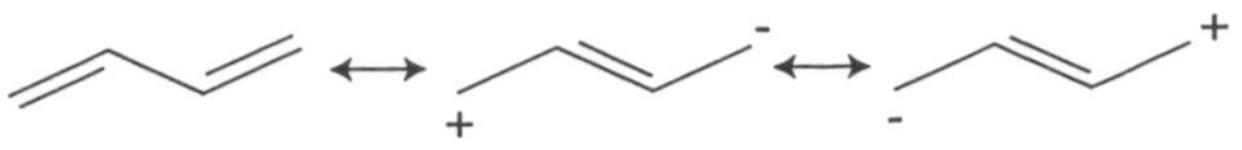

Fig. 11.   Butadiene.

ing of the central C—C bond with respect to a normal C—C single bond (see Figure 11). The effect is small, however, and might also be attributed to the fact that the central bond is a single bond between two sp^2 centers. Note that the central C—C bond is still markedly longer than the terminal C—C bonds. Generally, π-systems with an even number of conjugated double bonds have alternating bond lengths.

The description of the bonding situation using resonance forms not only is helpful in explaining the structure of a molecule but may also be used to rationalize its reactivity in a chemical reaction. The electrophilic substitution reaction of aromatic compounds may serve as representative. See Figure 12 for two examples.

In general, the rate-determining step of electrophilic aromatic substitution reactions is the attack of the electrophile on the aromatic compound, leading to a σ-complex as a high-energy intermediate. The TRANSITION STATE leading to this complex is product-like. The σ-complex can, therefore, serve as a model to understand the relative substitution rates at

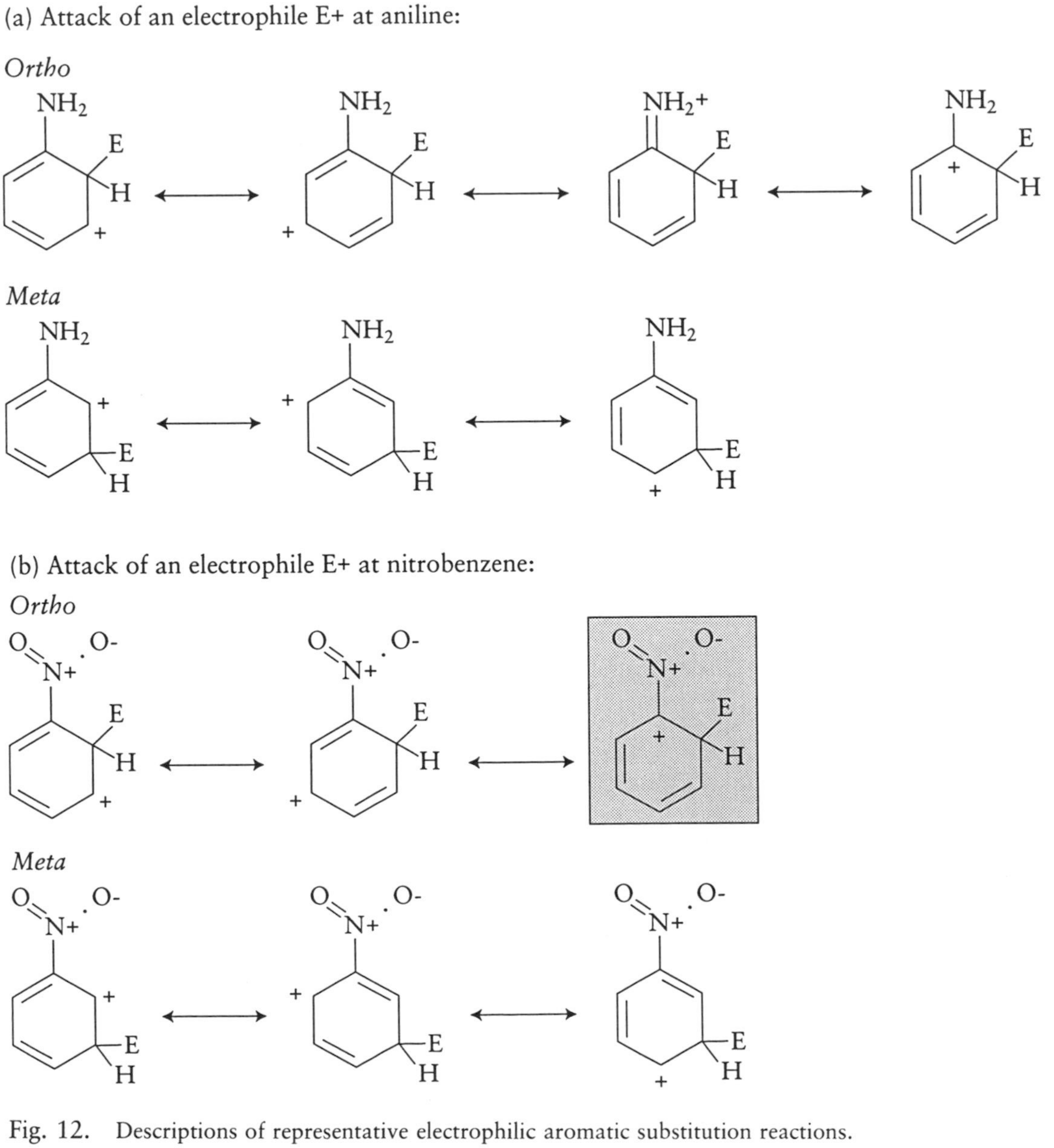

(a) Attack of an electrophile E+ at aniline:

Ortho

Meta

(b) Attack of an electrophile E+ at nitrobenzene:

Ortho

Meta

Fig. 12.   Descriptions of representative electrophilic aromatic substitution reactions.

different positions of substituted benzenes. The σ-complex is a pentadienyl cation that forms relatively easily if resonance structures exist that spread the positive charge over a large part of the molecule. Substituents with nonbonding electron pairs can stabilize positive charges resulting from attack at the *ortho* and *para* positions, reflecting the greater reactivity of these positions in electrophilic aromatic substitutions. This is illustrated by the resonance structures of the σ-complex formed upon the attack of an electrophile at an *ortho* position in aniline. Substituents of this type have been classified as (+M)-substituents because of their ability to donate electrons in mesomeric forms. Substituents that can act as electron-withdrawing groups in resonance structures are called (−M)-substituents and strongly decrease the reactivity of *ortho* and *para* positions because there exist resonance forms with two neighboring positively charged centers in the corresponding σ-complex. In molecules like nitrobenzene, the *meta* position is thus the most reactive one.

Theoretical Aspects of Resonance

The pictorial Lewis description of the electronic structure of molecules has its theoretical foundation in the VALENCE BOND THEORY (VB theory) developed by Linus PAULING et al. (Pauling, 1960). In VB theory the wave function Ψ of a molecule is constructed from products of atomic orbitals of the atoms involved. Consider the hydrogen molecule H_2 as the most simple example. It is built up by two hydrogen atoms A and B that can be described by wave functions Ψ_A and Ψ_B. As a VB wave function for the hydrogen molecule we might thus choose

$$\Psi = \Psi_A(1) \cdot \Psi_B(2)$$

where the numbers (1) and (2) designate electrons 1 and 2. In this description, electron 1 is located at atom A and electron 2 at atom B. But we know that the two electrons cannot be distinguished in the H_2 molecule. We thus have to add a second electronic configuration where the two electrons are exchanged:

$$\Psi = \Psi_A(1) \cdot \Psi_B(2) + \Psi_A(2) \cdot \Psi_B(1)$$

This wave function Ψ gives us a purely covalent description of the H_2 molecule in the framework of VB theory. If we want to further improve the description, we have to include additional resonance forms. In the case of H_2 these can only be ionic contributions:

$$H\text{—}H \leftrightarrow H^+ H^- \leftrightarrow H^- H^+$$

Covalent   Ionic

Of course, for H_2 the ionic forms will not play a significant role, but there are cases where several resonance structures of equal or nearly equal weight have to be considered. These additional electronic configurations are included in the VB wave function by simply adding them, weighted by a factor λ:

$$\begin{aligned} \Psi = \Psi_A(1) \cdot \Psi_B(2) &+ \Psi_A(2) \cdot \Psi_B(1) \\ &+ \lambda \cdot \Psi_A(1) \cdot \Psi_A(2) + \lambda \cdot \Psi_B(1) \cdot \Psi_B(2) \end{aligned}$$

In this way, the concept of resonance occurs naturally in the VB theory of chemical bonding. The real molecule is described by a wave function Ψ that includes several electronic structures as limiting cases. These resonance forms are just mathematical products of atomic orbitals—they don't have any further physical significance. From the VB wave function the relative importance of the various mesomeric forms can immediately be extracted: it is simply given by the corresponding weighting factor λ.

There is a second, competing approach for the description of chemical bonding: molecular orbital (MO) theory. This method uses molecular orbitals (MOs) instead of atomic orbitals (AOs) to construct the wave function Ψ of a molecule. One important consequence is that it is no longer necessary to invoke the somehow artificial concept of resonance for a proper description of electronic structure. This clearly indicates that the notion of resonance has no physical foundation, but is just a by-product of a special way of looking at chemical bonding. Similarly, the frequently discussed resonance stabilization energy is not a physical observable, but rather a theoretical quantity that should be treated with some caution. Molecular orbital theory does not have such a strictly defined quantity as the resonance energy. The term *conjugation* is often used in discussions of bonding situations in which the VB term *resonance* would be appropriate. It is not possible to extract a clear-cut definition for the term *conjugation* from its current use in the literature. We speak of π-conjugated compounds in molecules that contain π-systems separated by only one single bond. For example, 1,3-cyclohexadiene is a π-conjugated molecule, while 1,4-cyclohexadiene is not π-conjugated, because the double bonds are separated by two σ-bonds. The norbornyl cation is an example of σ-conjugation. If there is considerable conjugative interaction between double bonds that are separated by more than one single bond, the term *homoconjugation* is used (see CONJUGATION).

How conjugative effects are studied in the context of MO theoretical methods can be shown by taking 1,3-butadiene as an example. In the VB picture, the

(a) In the ground state conformation   (b) When rotated 90° around the central C–C bond

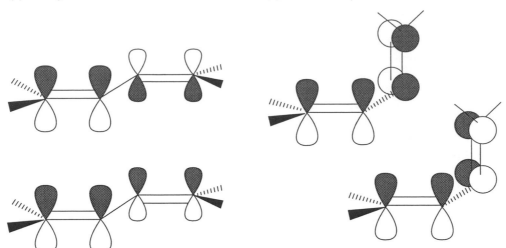

Fig. 13.   The two energetically lowest-lying occupied π-orbitals of 1,3-butadiene.

mesomeric Lewis structures that have been shown above are a shorthand notation for two π-bonds that interact by overlap of the $p(\pi)$-orbitals of C2 and C3. Molecular orbital calculations, on the other hand, yield four delocalized π molecular orbitals that are formed by a linear combination of the $p(\pi)$ atomic orbitals of the carbon atoms. The two energetically lowest orbitals, occupied with two electrons each, are shown in Figure 13. They describe the π-bonding in 1,3-butadiene. An analysis of the total wave function then yields bond orders that can be interpreted in terms of the mesomeric structures shown above. When the same calculation is carried out with a different geometry in which the C2-C3 bond is rotated by 90°, the $p(\pi)$ AOs of the C2 and C3 carbon atoms are orthogonal to each other. No linear combination of the four π MOs results in bonding between all four centers (see Figure 13). The conjugation between the two double bonds is switched off. This example shows that, although MO theory does not provide the traditional localized picture of chemical bonds, it can be used in a convenient way to study conjugative effects.

Resonance in Modern Research

The concept of resonance and the use of mesomeric forms to describe the bonding situation in molecules is still found in modern research, although it is often mixed with arguments using MO theory. Unfortunately, the terms *resonance* and *conjugation* are sometimes used in an indiscriminate way. Resonance forms are easy to write down, while the discussion of

conjugative effects in MO theory is more difficult to visualize. However, modern quantum chemical calculations are seldom based upon VB theory; most of them use MO theory as theoretical background. Nevertheless, the pictorial description of conjugative effects is then sometimes presented by using resonance forms. For this purpose, mathematical tools have to be applied to extract chemical information from the somewhat abstract wave function of the molecule. The development of such tools is one of the most prominent tasks of applied quantum chemistry.

One recent application of modern quantum chemical methods to the problem of conjugation concerns the controversy about the importance of the resonance forms of the carboxylate anion. It had long been accepted that the higher acidity of carboxylic acids compared with alcohols could be completely described by the high conjugative stabilization of the carboxylate anion. This view has been challenged by several authors who proposed that the inductive effect of the carbonyl group is more important than resonance stabilization. The latter argument was mainly based on the discussion of atomic charges that do not differ strongly for the neutral acid and its anion. Valence bond calculations on model compounds revealed, however, that inductive and resonance effects must be accounted for about equally.

BIBLIOGRAPHY

BROWN, H. C., with comments by P. V. R. SCHLEYER. *The Nonclassical Ion Problem;* New York: Plenum Press, 1977.

CAREY, F. A.; SUNDBERG, R. J. *Advanced Organic Chemistry;* New York: Plenum Press, 1990.

MARCH, J. *Advanced Organic Chemistry;* New York: John Wiley, 1985.

PAULING, L. *The Nature of the Chemical Bond;* Ithaca, NY: Cornell University Press, 1960.

WHELAND, G. W. *Resonance in Organic Chemistry;* New York: John Wiley, 1955.

GERNOT FRENKING

MARTIN STAHL

ULRICH PIDUM

Resorcinols

See PHENOL FUNCTIONAL GROUP.

Ribonucleic Acid (RNA)

See NUCLEIC ACIDS.

Richards, Theodore William (1868–1928)

Theodore William Richards was born in Germantown, Pennsylvania. His father was a successful painter, his mother a published poet. He was educated at home under his mother's direction until, at age fourteen, he entered Haverford College. Upon graduation three years later, he went on to Harvard University as a senior; he received a chemistry degree in 1886, when he graduated summa cum laude. Except for brief periods spent in Europe, he was to be part of the Harvard community for the rest of his life.

Richards's sponsor in graduate school was Josiah Parsons Cooke, Jr., whom Richards had known since 1874, and who was his reason for enrolling at Harvard. His research involved the accurate determination of the relative atomic weights of hydrogen and oxygen. The hydrogen gas was weighed in large glass containers and then reacted with hot copper oxide. From the weight of the resulting water, the ratio of the atomic weights was obtained. The value he calculated for hydrogen differs from today's accepted value by only four parts in 10,000. He was awarded a Ph.D. degree in 1888 and spent the next year abroad on a fellowship because of his outstanding dissertation. Upon his return, at age twenty-one, he joined the faculty at Harvard.

His work on oxygen and hydrogen led to a careful study of the composition of copper oxide and resulted in his correcting the accepted atomic weight of copper. Studies of the weights of numerous other metals followed; they proved that cobalt and nickel had slightly different atomic weights, with nickel the lighter of the two, and that the value for iron was much lower than had been supposed. The usual method involved the addition of silver nitrate to solutions of the metal halide, precipitating silver halide. Strontium chloride and bromide seemed to give slightly different results; careful work eventually proved that the problem lay with the accepted value for the atomic weight of chlorine. Correcting chlorine led to other changes, all determined with great care and precision. Nitrogen, for example, was checked both by synthesizing silver nitrate and by decomposing ammonium chloride. With improved accuracy, Richards and his coworkers determined atomic weights for twenty-five elements including sodium, silver, and potassium. It is this body of work for which he is best known and for which he received the 1914 Nobel Prize in chemistry.

The same factors that led to success in atomic-weight determinations—thoughtful experiment design, parallel experiments to cross-check results and eliminate sources of error, meticulous precision in measurements—also resulted in the production of accurate physical constants of other sorts. Richards's group developed an improved calorimeter and measured specific heats of acids, bases, and salts, along with heats of neutralization. Experiments with galvanic cells demonstrated the precision of Faraday's law of deposition; identical currents produced identical amounts of silver from solutions of silver nitrate in water at room temperature and solutions in molten sodium and potassium nitrates at 250°C. Studies on the compressibility of elements led to measurements of many other physical properties.

In addition to his research efforts, which produced a large number of prominent chemists in the early part of this century, Richards was known as an excellent teacher. He offered courses in quantitative analysis until Cooke's death in 1894, and in physical chemistry thereafter. He was still active in both teaching and research until just before his death on April 2, 1928.

BIBLIOGRAPHY

FORBES, G. S. "Investigations of Atomic Weights by Theodore William Richards." *J. Chem. Educ.* 1932, 9, 453–458.

KOPPERL, S. J. *The Scientific Work of Theodore William Richards;* Ph.D. dissertation, University of Wisconsin; Madison, WI, 1970.

RICHARDS, T. W. "Ideals of Chemical Investigation." *J. Chem. Educ.* 1929, 6, 2239–2245.

BENJAMIN B. CHASTAIN

Roasting

Roasting is a process in which an ore is heated to create chemical or physical changes. Usually roasting involves heating an ore containing a desired metal or metals, usually as sulfides, in excess air to a temperature that is below the melting point of the ore but at which oxidation can take place. This oxidation by reaction with oxygen in air converts the ore to an oxide or an oxy-sulfide (sulfate). The following examples of these reactions are for processing of copper ores where an initial stage can be considered to be

$$2\,CuS + O_2 = Cu_2S + SO_2$$

and subsequent stages as

$$Cu_2S + 2\,O_2 = 2\,CuO + SO_2$$

or

$$Cu_2S + 2\,O_2 = Cu_2SO_4$$

The complete oxidizing roast burns out the sulfur from sulfides, converting the desired metallic element to an oxide, which can be reduced by carbon or hydrogen or hydrocarbon gases directly to metal. A roasting operation that converts the desired metal to a sulfate provides the opportunity to leach the sulfate from the roasted ore for subsequent electrolysis to produce a pure metal. An intermediate roast will provide a charge material for conversion to a sulfide matte in a subsequent SMELTING process.

Other roasting operations are carried out by heating under oxidizing conditions to create volatile oxides, which are driven off from the roasting ore and recovered as fine, condensed solid particles or fume in the gas cleaning system. Arsenic, antimony, and zinc have volatile oxides and can be recovered from ores in this manner.

Chloridizing roast processes can be conducted under oxidizing or reducing conditions. In an early process to create a water-soluble chloride, sulfide ores were mixed with about 10 percent salt (NaCl) and roasted with air at 500–600°C. This process can be represented by the example reaction

$$CuS + 2\,NaCl + 2\,O_2 = Na_2SO_4 + CuCl_2$$

The chloridizing roast can be utilized to create volatile chlorides from the nonferrous metal oxides or sulfides. These reactions are carried out at relatively high temperatures using a chloride salt or gaseous chlorine. At about 1,200°C, reactions involving a salt are, for example,

$$ZnO + CaCl_2 = ZnCl_2 + CaO$$

$$ZnS + CaCl_2 + 3/2\,O_2 = ZnCl_2 + CaO + SO_2$$

or in the case of gaseous chlorine, at 900–1,000°C,

$$ZnO + Cl_2 = ZnCl_2 + 1/2\,O_2$$

$$ZnS + Cl_2 + O_2 = ZnCl_2 + SO_2$$

Chloridizing roast processes are also conducted under conditions where oxides are reacted with carbon or hydrogen in the presence of chlorine, often to create a volatile chloride that can be condensed and recovered as a pure compound. For example, titanium tetrachloride is produced in the following reaction at about 200°C:

$$TiO_2 + 2C + 2Cl_2 = TiCl_4 + 2CO$$

Reducing roast processes involves exposing a metal oxide to a reducing gas at high temperature to produce a metal as a material to be charged to a smelting or leaching process. In the case of iron ore processing, after crushing and grinding, exposure of an ore containing hematite, Fe_2O_3, to a moderately reducing gas, as, for example, a carbon monoxide–carbon dioxide-bearing gas, to produce magnetite (Fe_3O_4) enables magnetic separation of the iron oxide from silicate gangue material. The overall process reaction is

$$3\,Fe_2O_3 + CO = 2\,Fe_3O_4 + CO_2$$

Other roasting processes can be carried out on sintering strands or in fluidized or packed bed reactors with the purpose of changing the physical condition of the ore and agglomerating particulate materials.

In order to create a suitable chemical state as well as the desired physical condition for the material, roasting is carried out in several processes. Early primitive oxidation roasting of sulfide ore in heaps has been replaced by mechanical roasting, flash roasting, and fluidized bed roasting.

Mechanical or hearth roasting is carried out in a vertical multihearth furnace in which eight to twelve hearths are contained in a large cylindrical shell, e.g., 20 to 25 feet in diameter and 30 feet high. The ore is charged at the top and falls from hearth to hearth, driven by a rotating rabble attached to a central shaft that moves the ore across each hearth. The ore is moved to openings alternating on the inner and outer edges of each hearth, from which it falls to the hearth below. Production rates for sulfide ores have been reported to be 100 to 200 tons of material per day in this type of roaster. The thermal energy required by the system is usually supplied by combustion of the sulfide ore, i.e., the process is autogenous, but supplemental energy can be supplied by combustion of fuels.

Flash roasting was developed in a modification of the multihearth roaster. Oxidation was observed to occur on the surface of the ore exposed to roasting gases, particularly as the ore fell from one hearth to the hearth below. This process is carried out in a roasting furnace similar to the multihearth roaster, except that several intermediate hearths are absent. Upper hearths are utilized for drying and preheating and lower hearths collect the settled calcine. Control of gases in the reactor allows for use of auxiliary and for injection of SO_2 to control the atmosphere for conversion to sulfate, e.g., in the case of zinc roasting.

Fluidized bed roasting has become widely adopted. Through control of particle size and gas velocities, a bed of ore can be fluidized, providing excellent gas–solid contact, high reaction rates, and a high production throughput for moderately sized reactors. Off-gas collection and cleaning are effectively incorporated in the fluidized bed reactor system. For larger installations processing sulfide ores, the discharged SO_2 and SO_3 gases are converted to sulfuric acid, a valuable by-product. Temperatures in the fluidized bed can be controlled by proportioning the relative amounts of feed and air. The air is often enriched with pure oxygen.

Sinter-roasting offers an advantage over the previously described roasting processes in that agglomeration of the roasted material is accomplished. The blast furnace requires a suitably large particle size, and iron and lead sulfide ores are usually sinter-roasted. Roasting is carried out on a traveling grate sintering machine, and temperatures are allowed to rise to the point where incipient fusion occurs, producing a porous, cinderlike material called sinter. The sulfur in the ore acts as a fuel that can be supplemented with an added fuel such as powdered coal or coke. The high temperatures and oxidizing conditions can provide low sulfur content, which is important, particularly for roasting of pyrite (FeS_2) or pyrrhotite (FeS).

BIBLIOGRAPHY

Evans, J. W.; DeJonghe, L. C. *The Production of Inorganic Materials;* New York: Macmillan, 1991; Chapter 8.4, pp. 248–256.

Gilchrist, J. D. *Extraction Metallurgy;* London: Pergamon, 1967; pp. 204–210.

Pehlke, R. D. *Unit Processes of Extractive Metallurgy;* New York: American Elsevier, 1973; pp. 7–16.

Rosenqvist, T. *Principles of Extractive Metallurgy;* New York: McGraw-Hill, 1974; Chapter 8, pp. 245–255.

R. D. PEHLKE

Röntgen Rays

See X RAYS.

Rubber

See ELASTOMERS.

Rumford, Count (1753–1814)

Benjamin Thompson, Count Rumford, was born in Woburn, Massachusetts, on March 26, 1753. After a life of extraordinary variety, he was to die in Paris, France, on August 21, 1814. His claim to have attended Harvard College was characteristically spurious, and he seems to have been largely self-educated. He early displayed his vast ambition and gift for social climbing. He moved to Concord (previously called Rumford), New Hampshire, where he found employment as a schoolteacher. Here he married a wealthy widow, Sarah Rolfe, by whom he had a daughter, also named Sarah. Both were to be abandoned when Thompson fled to England in March, 1776. The reasons for flight were Thompson's loyalist sentiments and his open friendship with Governor Wentworth, the Royal Governor of New Hampshire. Only many years later was it revealed that Benjamin Thompson had also been a paid informer of the British.

Thompson's fortunes in England rose rapidly. Within seven years he had served as secretary of the Province of Georgia, been elected Fellow of the Royal Society, led his own King's American Dragoons in the closing campaigns of the War of Independence, and been dubbed Sir Benjamin by King George III. He also had begun the series of investigations that would lead to his reputation as one of the founders of applied science. With the signing of the Treaty of Paris in September 1783, Colonel Sir Benjamin Thompson sought his fortune as aide-de-camp to the Elector of Bavaria in Munich. The Munich period was the most varied and productive of Thompson's career. He reformed the army, banished beggary, founded reform schools and workhouses, replanned the city, and laid out the English Gardens—and in doing so he made countless enemies. He investigated thermal conductivity (for army blankets), illuminating efficiency (for workhouses), heating efficiency (for fireplaces and stoves), nourishing soups (for feeding the poor), and continuous heat generation (in the boring of army cannon). The last of these was his most important scientific discovery, and it was acknowledged by

Joule as a semiquantitative foreshadowing of "heat as a form of motion." For these and other services he was elected Count (Rumford) of the Holy Roman Empire.

Having worn out his welcome in Munich, Rumford returned to London in 1799. Here he proposed and helped found the Royal Institution of Great Britain. Two of its first professors were Thomas Young and Humphry DAVY. It flourishes to this day, still occupying the premises Rumford helped to redesign. Once again Rumford fell out with his colleagues and moved to Paris. Rumford's last years were something of an anticlimax. His marriage to the widow of Antoine LAVOISIER quickly led to acrimony and divorce. He continued his researches in applied science and published copiously. His behavior became increasingly eccentric—in winter he dressed entirely in white to protect himself from "frigorific rays"—though his final days were warmed by a visit from Sir Humphry Davy and his young traveling companion, Michael FARADAY. In his eulogy on Rumford, Baron Cuvier cogently remarked, "Nothing would have been wanting to his happiness had the amenity of his behaviour equalled his ardor for public utility."

BIBLIOGRAPHY

BROWN, S. C. Count Rumford: Physicist Extraordinary; Garden City, NY: Anchor/Doubleday Books, 1962.

DEREK DAVENPORT

Rutherford, Ernest (1871–1937)

Ernest Rutherford was born on August 30, 1871, near Nelson on New Zealand's South Island. He was the fourth of the twelve children of James and Martha Rutherford, who had migrated to New Zealand from the British Isles. He obtained a scholarship to attend Nelson College (a secondary school) and then went on to Canterbury College, Christchurch, where he received a B.A. in 1892, with "first class honours" in mathematics and physics, and an M.A. in 1893. Rutherford stayed at Canterbury College for a fifth year to do research on the detection of radio waves. He became engaged to Mary Newton, his landlady's daughter; they were married when Rutherford returned to New Zealand for the summer of 1900. Their only child, Eileen, was born in 1901.

In 1895 Rutherford went to work under J. J. Thomson at the Cavendish Laboratory of Cambridge University, supported by an 1851 Exhibition Scholarship. At first he continued his research on the magnetic detection of radio waves, but he soon switched, at Thomson's direction, to the ionization of gases by X rays. Shortly after the discovery of radioactivity, Rutherford investigated the ionization of gases by the radiation from radioactive sources. He soon discovered two types of radiation, which he named ALPHA and BETA, and eventually proved that alpha particles are doubly charged helium. Rutherford became professor of physics at McGill University in Montreal in 1898. He continued to study alpha particles, working with Frederick SODDY, a chemist. This work led Rutherford to the conclusion, in 1902, that radioactivity is the result of the disintegration or transmutation of elements. For this discovery he was awarded the 1908 Nobel Prize in chemistry.

In 1907 Rutherford moved back to England to the University of Manchester. He continued to study the alpha particle (which he considered *his* alpha particles). Based on measurements of the scattering of alpha particles by thin foils, mostly done by Hans GEIGER and Ernest Marsden under his direction, Rutherford demonstrated in 1911 that atoms consist of a tiny, very dense, positively charged nucleus surrounded by electrons. Niels BOHR worked with Rutherford for a year before returning to Copenhagen to develop his model of the electronic structure of atoms.

Rutherford was knighted in 1914. He had become world-famous as the large and boisterous but friendly and extremely effective leader of one of the most important physics research groups. During World War I he studied underwater acoustics for submarine detection. This work laid the foundation for the later development of sonar. Continued work with alpha particles led, in 1919, to the discovery of the first artificial transmutation—the bombardment of nitrogen by alpha particles resulted in oxygen and hydrogen. In 1919 he succeeded J. J. THOMSON as director of the Cavendish Laboratory. Much of his time in the 1920s and 1930s was spent in seeking government support of basic research and in being a leader of the international scientific community. Rutherford speculated on the existence of the neutron, which James CHADWICK discovered at the Cavendish Laboratory in 1932. It was under Rutherford's leadership that Cockcroff and Walton first observed a transmutation induced by artificially accelerated protons.

In 1931 Rutherford was made a baron, Lord Rutherford of Nelson. He was quite healthy most of his life but did have a small umbilical hernia. A strangulation of the hernia led to his death at Cambridge on October 19, 1937.

BIBLIOGRAPHY

CHADWICK, J., ed. *The Collected Papers of Lord Rutherford;* 3 vols.; London: Allen and Unwin, 1915.

EVE, A. S. *Rutherford;* Cambridge, U.K.: Cambridge University Press, 1939.

WILSON, D. *Rutherford, Simple Genius;* Cambridge, MA: MIT Press, 1983.

JACK C. NORMAN

Rutherford-Bohr-Sommerfeld Model of the Atom

In May 1909, two of Ernest RUTHERFORD's students, Hans W. GEIGER and Ernest Marsden, submitted a paper reporting that a small percentage of a beam of positively charged alpha (α) particles (α = helium nucleus = $_2^4$He = He^{2+}) directed at a gold foil scattered from the foil at angles greater than 90 degrees. Rutherford recognized that the prevailing model of the atom proposed by J. J. THOMSON (see ATOM, CLASSICAL)—electrons stuck like raisins into a "pudding" positive charge—was incompatible with the Geiger-Marsden experiment. To account for the ALPHA PARTICLE deflections, Rutherford proposed in 1911 that the atom is composed of a dense, compact central nucleus of positive charge containing most of the atomic mass. Around this nucleus orbits a sufficient number of very-low-mass, negatively charged electrons to balance the positive nuclear charge Z (Figure 1).

In 1911, the young Danish student Niels BOHR went to work with Rutherford in Manchester on problems involving alpha particle absorption and Rutherford's new atomic model. Despite compelling

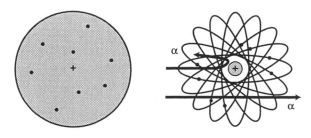

Fig. 1. (a) The Thomson "plum pudding" atom with negatively charged electrons embedded in a pudding of positive charge and (b) the Rutherford atom showing an α-particle backscattered from the dense, positively charged nucleus and another α-particle passing through the electron orbits essentially undeflected. Atoms with atomic number $Z = 8$ are shown.

evidence for the Rutherford atom derived from alpha-scattering experiments, the model appeared to suffer from a fundamental defect. Classical electrodynamics clearly predicted that the electrons orbiting the atomic nucleus must dissipate their energy as light, which would in turn cause their orbits to collapse. To resolve this difficulty, Bohr, in three papers published in 1913, took an imaginative leap and simply assumed that certain orbits are stable or stationary and therefore not subject to radiative instability. Moreover, he explained the sharp line spectra of atoms—very narrow characteristic bands of color at selected, well-defined regions of the spectrum rather than a continuous rainbow of colors—in terms of transitions from one stationary orbit to another stationary orbit. Electronic charge in the atom only accelerated in the classical sense and produced or absorbed light when the electrons jumped from one Bohr orbit to another. Each orbit was assumed to have a well-defined energy, $E = -R/n^2$, where R is a numerical constant called the RYDBERG CONSTANT and n is a QUANTUM NUMBER (the principal quantum number) that could take on *only* the positive integer values 1, 2, 3, . . . This orbit energy formula was derived by requiring the ANGULAR MOMENTUM of the electron orbiting the nucleus to be quantized. For an electron of mass m_e orbiting the nucleus at radius r and speed v, the orbital angular momentum is $m_e v r$. Bohr simply required $m_e v r$ to be equal to $n \cdot (h/2\pi)$, where n is the principal quantum number and h is PLANCK's numerical constant. The frequency v of the electromagnetic radiation (light) required to induce a transition from one orbit to the other could be found by dividing the difference in energy between the two orbits by Planck's constant, since $E_{(\text{orbit 2})} - E_{(\text{orbit 1})} = hv$ (Figure 2). Bohr's imaginative leap became the foundation for quantum mechanics (see QUANTUM CHEMISTRY; QUANTUM THEORY).

To the basic Bohr model, Arnold SOMMERFELD in 1916 added a notable refinement. In Bohr's original model only circular electron orbits were considered and only one quantum number, the principal quantum number n, was needed to describe the atom. Sommerfeld was prompted by the appearance of fine structure or splittings in atomic spectra and by the appearance of still more splittings when atoms were placed in a magnetic field, the Zeeman splittings (see ZEEMAN EFFECT), to generalize Bohr's original circular orbits to include ellipses as well (Figure 3).

From these ideas came the azimuthal number, k, which in more modern formulations becomes the orbital angular momentum quantum number, l, with a slightly different interpretation (l can take on the

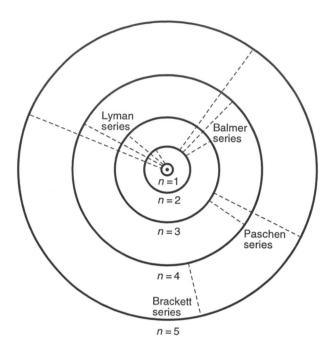

Fig. 2. The first five Bohr orbits of the hydrogen atom, drawn to scale. The radius of the $n = 1$ orbit is 52.9 pm, the Bohr radius. Transitions between orbits leading to well-known atomic hydrogen line spectra are noted.

The Rutherford-Bohr-Sommerfeld model of the atom stands as the high-water mark of what is now called the "old quantum mechanics." To other creative scientists—P. A. M. DIRAC, Werner HEISENBERG, Erwin SCHRÖDINGER, Max Born, and Wolfgang PAULI, to name but a few—would fall the task of developing the mathematical and conceptual structure of the "new quantum mechanics." With the new quantum mechanics, not only could the results of the Rutherford-Bohr-Sommerfeld atom model be obtained, but the quantum mechanical behaviors of complicated atomic, molecular, and nuclear systems also could be addressed in a highly sophisticated and systematic way.

BIBLIOGRAPHY

HAKEN, H.; WOLF, H. C. *Atomic and Quantum Physics;* Berlin: Springer-Verlag, 1984; pp. 96–98.

HERZBERG, G. *Atomic Spectra and Atomic Structure;* New York: Dover, 1944.

EISBERG, R. M. *Fundamentals of Modern Physics;* New York: Wiley, 1961; pp. 110–137.

PAIS, A. *Niels Bohr's Times, in Physics, Philosophy, and Polity;* Oxford: Clarendon, 1991.

JOHN W. KENNEY III

values 0, 1, 2, 3, . . ., $n - 1$). Sommerfeld also introduced Einstein's relativistic effects into the orbital motion of the electron, from which he predicted that the elliptical orbits would themselves slowly rotate or precess around the nucleus.

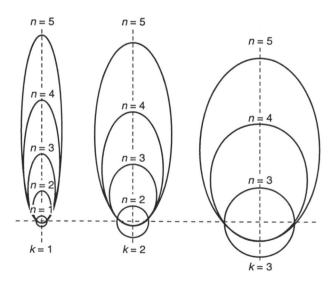

Fig. 3. The Bohr-Sommerfeld elliptical electron orbits where n is the principal quantum number and k is the azimuthal quantum number. For a given value of n, k can take on the values 1, 2, 3, . . . , n.

Ruzicka, Leopold (1887–1967)

Born in Vukovar, Croatia, in 1887, Leopold Ruzicka was originally interested in entering the priesthood; however, his academic career steered him instead in the direction of classic organic chemistry. After studying at the Technical University of Karlsruhe from 1906 to 1910 under Hermann Staudinger, Ruzicka received his doctorate following his studies of phenyl methyl ketene. When the influential Staudinger moved to Zurich to take a position at the Swiss Federal Institute of Technology, Ruzicka followed as his assistant. Although he lost the support of Staudinger when he wanted to pursue his own research interests, Ruzika obtained a lectureship at the Swiss Federal Institute of Technology in 1918, followed by a lectureship at the University of Zurich in 1920. At the same time he collaborated with a Geneva perfume company on the synthesis of the sesquiterpene perfumes farnesoll and nerolidol from animal and plant resources. He also studied the structures of several odorous ketones (see ODORS).

During the 1930s, Ruzicka moved to a position with the Naef chemical company in Geneva, followed by a chairmanship of organic chemistry at the University of Utrecht, and finally he returned to the Swiss Federal Institute of Technology as the chair of inorganic and organic chemistry. During this fruitful period, he reorganized and developed the laboratories at the Swiss Federal Institute of Technology to create a first-class research facility as well as advance his career. As a result of his research findings, Ruzicka challenged Adolph BAEYER's ring structure theory. Baeyer had predicted that large-numbered ring structures were unstable, but Ruzicka showed that his 15- and 17-carbon ring structures were stable.

Another major area of interest for Ruzicka was the synthesis of male sex HORMONES. He successfully converted cholesterol into androsterone, which is a sterodial male sex hormone. Perhaps Ruzicka's most important work was his biogenetic isoprene rule. Ruzicka's biogenetic isoprene rule states that out of all the possible structures of a particular terpene, the favored one can be derived mechanistically by the cyclization of an alphatic precursor.

For his research on polymethylenes and higher terpenes, Ruzicka was awarded the 1939 Nobel Prize, which he shared with Adolf BUTENANDT. During his career, Ruzicka published almost six hundred papers, was a member of the Swiss Chemical Society, and received many honorary degrees and awards, including the Marcel Benoist Prize. In addition, Ruzicka was active in the chemistry publishing circle. He served as an editor of *Experientia* and *Ergebnisse der Vitamin und Hormonforschung* and was a member of the editorial board of *Helvetica Chimica Acta*. In addition to his contributions to classical organic chemistry, Ruzicka will be remembered for his stand on human rights. During wartime, he supported Nazi refugees and allowed them to work in his laboratories. He also resigned from his position as a chief editor of *Ergebnisse der Vitamin und Hormonforschung* when the publisher fired two Jewish executives. Ruzicka died in 1967.

BIBLIOGRAPHY

COBB, C.; GOLDWHITE, H. *Creations of Fire: Chemistry's Lively History from Alchemy to the Atomic Age;* New York: Plenum Press, 1995.

ESCHENMOSER, A. "Leopold Ruzicka—from the Isoprene Rule to the Question of Life's Origin"; E. Loewenthal, trans.; *Chimia* 1990, 44, 1–21.

FARBER, E. *Nobel Prize Winners in Chemistry, 1901–1961;* New York: Abelard-Schuman, 1963.

JAMES, L. *Nobel Laureates in Chemistry, 1901–1992;* Washington, DC: American Chemical Society and Chemical Heritage Foundation, 1993.

ANNA R. BERGSTROM

Rydberg, Johannes Robert (1854–1919)

The Swedish physicist Johannes Robert Rydberg was born on November 8, 1854, in Halmstad. After studying at the University of Lund, Rydberg received his doctorate in mathematics in 1879. He then joined the university's faculty on attaining professorial status in 1897. Rydberg's research focused on SPECTROSCOPY, and he labored to make sense of the spectral lines produced by elements that suggested a harmonic oscillation in atoms. Unaware that Johann Balmer had published a paper on the empirical relationship between the four prominent lines of the hydrogen spectrum, Rydberg developed his own relationship. After learning of Balmer's work, Rydberg was able to demonstrate that Balmer's equation was merely a special case of a more general relationship. Rydberg's own equation was purely empirical, as he did not manage to work out why the equation existed (see RYDBERG CONSTANT). It was not until Niels BOHR's application of quantum notions of atomic structure that the why was explained. Rydberg died at the age of sixty-five on December 28, 1919, in Lund, Malmohus.

BIBLIOGRAPHY

FARBER, E. *The Evolution of Chemistry: A History of Its Ideas, Methods, and Materials;* New York: Ronald Press, 1952.

IHDE, A. J. *The Development of Modern Chemistry;* New York: Harper & Row, 1964.

ANNA R. BERGSTROM

Rydberg Constant

The Rydberg number, R_H, is the only constant in a simple equation giving all the wavelengths observed in the line spectrum of the hydrogen atom. The number and equation were discovered empirically by Balmer (1885) and RYDBERG (1890) for the H atom and derived theoretically by BOHR (1913) and SCHRÖDINGER (1926) for one-electron atoms A—i.e., H, D, He$^+$, Li^{+2} ... (see SCHRÖDINGER'S WAVE EQUATION).

Theory shows that $P_A = 2\pi^2\mu_A e^4 c^{-1} h^{-3}$, where $\mu_A^{-1} = m_e^{-1} + m_A^{-1}$ is the reduced mass of the electron and A nuclear masses. This small mass dependence of R_A led to the discovery of DEUTERIUM (heavy hydrogen) by Urey and colleagues (1931), through its line spectrum. The symbols e, c, and h

represent the electron charge, speed of light, and Planck's constant, respectively. For infinite nuclear mass, $\mu_A = m_e$ and $R_A = R_\infty = 1.0973731534 \times 10^7 \ m^{-1}$. This extremely precise universal constant (which is about 0.05 percent smaller than R_H) relates e, c, h, and m_e and can be used in work to improve the precision of their values.

Gaseous atoms in excited electronic states (e.g., H atoms produced from H_2 in an electric discharge tube) spontaneously lose energy by emitting electromagnetic radiation (light) at a set of unique wavelength values constituting their "line spectrum" (see EXCITED STATE). Balmer's formula for a series of H atom spectral lines was generalized by the Swedish physicist Johannes Rydberg to

$$\lambda_{nm}{}^{-1} = -R_H Z_H^2 (n^{-2} - m^{-2}) = E_n - E_m$$

where n and m are positive integers, $Z_H = +1$ is the nuclear charge, and λ_{nm} is the wavelength of radiation emitted by the H atom when it drops from high-energy state E_n to low-energy state E_m. This Rydberg-Ritz equation expresses the Ritz combination principle (1908) that spectral lines always occur as a difference between two energy "term" values. The line spectrum of the H atom is composed of several spectral series, e.g., the Balmer series, which connects the upper-energy states with $n = 3, 4, 5, \ldots$ to the same lower-energy state with $m = 2$. An atom can also *absorb* radiant energy with the wavelength λ_{nm}

from a light beam, and the atom is ionized if the excitation is to an upper state located above E_∞ (i.e., E_m with $m = \infty$). The normal ionization potential (i.e., excitation from the "ground" state E_1) is $E_\infty - E_1 = R_A Z_A^2$, showing that R_H is the ionization potential of the H atom.

The Rydberg-Ritz equation is specific for H-like atoms, but its form is useful for sodiumlike atoms with a single electron in the outer electron shell. Thus, Rydberg (in 1890) analyzed spectral series for neutral atoms by using $-R_H(n - \delta)^{-2}$ for the running term E_n, where an empirical constant δ corrects for energy level shifting due to the atomic core. Electronic states of any atom (or molecule) with a single electron excited to energies near the ionization limit are now called "Rydberg states" because of the approximate validity of this expression for their term values.

BIBLIOGRAPHY

BARROW, G. M. *Introduction to Molecular Spectroscopy;* New York: McGraw-Hill, 1962.

HERZBERG, G. *Atomic Spectra and Atomic Structure,* 2nd ed.; New York: Dover, 1944.

MILLS, I. M.; CVITAS, T.; HOMANN, K.; KALLAY, N.; KUCHITSU, K., eds. *Quantities, Units and Symbols in Physical Chemistry;* Oxford: Blackwell, 1988.

RICHARD REDINGTON

S

Sanger, Frederick (1918–)

Frederick Sanger, born on August 13, 1918, in Rendcombe, Gloucestershire, England, was the son of Frederick Sanger, a prosperous country doctor, and Cicely Crewson Sanger.

After receiving a B.A. degree in 1939 from Cambridge, Sanger continued at Cambridge during the years of World War II, having been exempted from military service as a conscientious objector. He received his Ph.D. in 1943, working with A. Neuberger, and later studied under Albert C. Chibnall.

Frederick Sanger joined the ranks of such illustrious scientists as Linus PAULING, Marie CURIE, and John Bardeen by winning two Nobel Prizes.

The 1958 Nobel Prize recognized Sanger's method for sequencing AMINO ACIDS in PROTEINS, especially in the protein hormone insulin. Sanger's method depended upon the reaction of fluorodinitrobenzene with free amino groups in small polypeptides. Following the oxidation of the cystine disulfide bridges of insulin, treatment with fluorodinitrobenzene and hydrolysis of all peptide bonds gave yellow dinitrophenyl (DNP) derivatives of glycine and phenylalanine and of the e-NH_2 of lysine, which were identified by chromatographic techniques. N-terminal DNP derivatives of aspartic acid and asparagine did not survive hydrolysis. Sanger reasoned that insulin contained two types of chains, one terminating in phenylalanine and one terminating in glycine.

Sanger and Hans Tuppy identified the sequence of thirty residues in the phenylalanine chain, and Sanger identified the sequence of twenty-one amino acid residues in the glycine chain.

Performic acid oxidation of the cystine disulfide (—S—S—), cross-links between the two peptide chains of insulin, gave four cysteic acid (CH_2SO_3H) residues in the glycine chain and two in the phenylalanine chain, indicating that two bridges connect the two chains and one connects different parts of the glycine chain. Sanger was fortunate that insulin could be obtained in highly purified form and that methionine and tryptophan, which react with performic acid, were absent from insulin. By 1953 Sanger had determined the complete structure for insulin.

Sanger showed that whereas the amino acid sequence in a given polypeptide chain is random—not ordered and highly organized, as previously thought—each position in that chain is always occupied by one specific amino acid residue, completely unique for a given biological species. A given protein exhibited little variation in composition among different species—not a mixture of different substances. This implied that the sequence of the amino acids governs the biological function.

Sanger's Nobel Prize in chemistry in 1980, shared with Walter Gilbert and Paul Berg, mainly recognized the impact on genetic engineering of his methods for sequencing the nucleotides in the nucleic acids. The chemical basis of Sanger's method was the addition of dideoxynuleoside triphosphates, which, because they lack a 3'-OH group for addition of the next nucleoside, terminate the growing chain. Using cloning techniques, he inserted a desired section of template DNA into a single-stranded vector and annealed an oligonucleotide primer to the single-stranded DNA. In four different reactions, the extension of the primer proceeded in a mixture of four P^{32}-labeled nucleoside triphosphates, small amounts of a single chain-terminating dideoxynucleoside triphosphate (of either adenine, guanine, cytosine, or thymine), and DNA polymerase. When RNA was being sequenced, reverse transcriptase was used to make a DNA copy of the RNA. The polymerase generates a DNA fragment complementary to the insert and releases a series of fragments ending in a specific dideoxynucleotide. Sanger was able to identify the sites of specific nucleotides and compare the complementary sites on the original DNA that had been the template of the clone. He established the sequence in the DNA of approximately two hundred nucleotides by separating the fragments on gel electrophoresis and by analyzing by autoradiography. He used computer programs to store, overlap, and arrange data for increasingly longer sequences.

Sanger devoted his career to the belief that knowledge of molecular sequences is a key to understanding biological systems. He was head of the Division of Protein Chemistry in the Medical Research Council Laboratory for Molecular Biology at Cambridge until he retired in 1983 after a forty-year tenure at Cambridge.

He credited his wife, Margaret Joan Howe, with contributing more to his work than anyone else by providing a peaceful, happy home. Students said he was unpretentious and seemed more like a laboratory caretaker than its star researcher. Colleagues praised his work as elegant and efficient because of his painstaking methods.

BIBLIOGRAPHY

Laszlo, P. "A History of Biochemistry: Molecular Correlates of Biological Concepts, Sequence Analysis." In *Comprehensive Biochemistry;* M. Florkin, E. H. Stotz, eds.; Amsterdam: Elsevier, 1986; Vol. 34A; Chapter 16, pp. 296–307.

McGill, F. N., ed. *The Nobel Prize Winners: Chemistry 1967–1989;* Pasadena, CA: Salem Press, 1990; Vol. 3, pp. 1049–1057.

Sanger, F. *Chemistry;* Amsterdam: Elsevier, 1958; pp. 541–557.

———. *Nobel Lectures, Chemistry, 1942–1962;* Amsterdam: Elsevier, 1964.

MURIEL B. BISHOP

Scanning Tunneling Microscopy

Several technologies come together to give unprecedented details of the atomic structure of surfaces. In this relatively new technique, a microscopic tip guides a scanning mechanism over the contours of a surface by using the effect of the "tunneling current."

Tunneling

Have you had the experience of trying to throw something—a ball, perhaps—over a tall fence, but not quite clearing it? It's frustrating because all you really want to accomplish is to move the ball a few feet sideways, except for that fence! If only the ball could somehow ooze through it. Strictly speaking, the laws of physics don't forbid that from happening, but the odds in favor of it are overwhelmingly small. On the other hand, if the ball were small enough—about the size of an ELECTRON—and if the fence were very thin—a few atomic diameters—it could happen. In fact, it would be pretty common. When two metals are brought to within a few atoms' width in diameter, large amounts of electrons can flow through the space separating them, in spite of the fact that they don't have enough energy to actually escape either of the metals. Like a ball oozing through a fence, the electrons are said to tunnel through the space between. The flow of electrons is called the tunneling current.

An important aspect of the tunneling current is that it is very sensitive to the distance separating the two metals; moving them apart by one atom's width in diameter can easily cause the current to fall by a factor of ten. Because the current falls by such a factor for every extra atom's width in diameter the two metals separate, the dependence of current with separation must be decaying exponentially. Because the current changes so quickly with distance, we can use it to monitor a sharp metal probe as it scans across a surface. Imagine scanning across a smooth surface and passing over a ledge to a region one sheet of atoms lower, or even passing over a pit caused by a missing atom. If you are monitoring the tunneling current, you get a big drop in current; and if the metal probe is mounted on a positioning device called a piezoelectric device, you can easily order it to compensate. Changing the voltage slightly on the piezoelectric device will bring the metal probe back to the original distance from the surface. Keeping track of this control voltage then gives us a map of the shape of the surface, since the volts per nanometer property of the piezoelectric element is known.

Some bad news comes along with the extreme sensitivity of the tunneling current with distance that we find so useful above: You get very little warning as you bring a probe to the surface before it crashes into it and destroys your sharp probe. Avoiding this occurrence requires a careful approach. Initially, the tip is advanced toward the surface by turning on a screw separating the tip and the sample until they are as close as can be managed, while the operator watches through an optical microscope. Then the tip inches forward under electronic control.

Scanning

The probe tip must be able to move toward, away from, and across a sample surface with very fine motions. Unless movement can be directed with a precision of better than 1 nanometer (about a ten-thousandth of the thickness of a human hair), either the sample surface or the probe tip may be damaged. This controlled movement is usually accomplished with piezoelectric materials. In the piezoelectric effect, certain crystalline materials have an unsymmetric placement of the positive and negative charges. When that is the case, applying an electric field to the crystal will cause the crystal to change shape slightly. A common material that will do this is quartz. A common amount of change for some materials is 2 nanometers per applied volt of electricity. Just the sort of movement that is needed! In some instruments, three bars of piezoelectric are glued at right angles to supply the three different directions of movement needed, and then a probe tip is attached at the corner.

More recent instruments manage to make the three movements—up/down and two sideways movements—by using a single hollow cylinder of piezoelectric material. The electrodes can be placed so that a voltage to extend the length of the cylinder is applied to only one side of the cylinder, and a voltage is applied to the other side of the cylinder that will make it become shorter in length. The result of this combination of voltages is that the cylinder is forced to bend to the side. In this way, sideways motion is achieved. Keep in mind that we require bends of only a few thousandths of a millimeter (micron), while the cylinder may be about 50 millimeters long, so there is no danger of breaking the cylinder.

With the proper program of voltages, it is possible to make the piezoelectric driver sweep across the sample surface line by line in a "raster pattern" similar to that used in TV screens. All the while, it is using the tunneling current to sense the height of the surface and to keep the probe just above the surface.

Scanning Probe Tips

To put the MICROSCOPY in scanning tunneling microscopy, we must approach the surface with a very sharp tip. Common sense and theoretical calculations indicate the need for a tip that is only a handful of atoms at the end. This is the part of this field where the greatest amount of uncertainty and art is found. There are two approaches: One can work to build such a tip methodically, checking the work along the way, or one can try empirically something that hasn't any apparent need to work, to see the results.

In the spirit of the first approach, it is possible to make very fine needles of tungsten and other metals by dipping wire in caustic solutions and electrochemically etching away the metal. Currents are set up in the solution that cause the etching to be fastest near the surface of the bath. The wire forms a very narrow "waist" until the lower portion of the wire drops away, leaving a sharp tip. These tips, if made properly, can be measured by electron microscopy to be as small as about 20 nanometers. However, that is not quite good enough, since a tip of 1 nanometer or smaller is wanted. Various treatments with high-voltage pulses and other imaginative treatments can yield tips with just 1 atom, and this has been checked with the specialized technique of field electron microscopy.

This last technique is one of the very few ways to directly check the quality of a needle with atomic detail, and few laboratories are set up to perform this fairly difficult procedure. So more often, empirical methods prevail. In fact, scanning tips are used that are judged good because they give the expected results on well-understood materials. Sharp enough tips are made by cutting metal wire with scissors "just so"!

A puzzle arises when these tips are examined with microscopy; they don't appear to be all that sharp. Scientists believe that extending from the end of the fairly dull tips are sharp filaments, or asperities, that are too fine to be seen by ordinary microscopy but that allow the tips to achieve atomic resolution when operating in the microscope.

Atomic pictures of surfaces are now routinely obtained around the world. They reveal complex arrangements and rearrangements of the surfaces of materials. They even reveal the resting places of gas atoms on surfaces. Many of these results could only be inferred, speculated, or argued about in the past. The first spectacular success of the method was to clearly reveal the complex way silicon atoms can rearrange on some silicon crystals to form a "7 × 7" surface. For bringing this new form of microscopy into the world, Gerd Binnig and Heinrich Rohrer became two of the three recipients of the Nobel Prize in physics in 1986.

See also ATOMIC FORCE MICROSCOPY; MICROSCOPY.

BIBLIOGRAPHY

BINNIG, G.; ROHRER, H. "The Scanning Tunneling Microscope." *Sci. Amer.*, 1985, *253*, 50.

BINNIG, G.; ROHRER, H.; GERBER, C.; WEIBEL, E. "7 × 7 Reconstruction on Si(111) Resolved in Real Space." *Phys. Rev. Lett.* 1983, *50*, 120.

DENLEY, D. R. "Practical Applications of Scanning Tunneling Microscopy." *Ultramicroscopy* 1990, *33*, 83.

GOLOVCHENKO, J. A. "The Tunneling Microscope: A New Look at the Atomic World." *Science* 1986, *232*, 48.

DAVID DENLEY

Schrödinger, Erwin (1887–1961)

The Austrian physicist Erwin Schrödinger was born in Vienna on August 12, 1887, the only child of Georgine and Rudolf Schrödinger. As was common practice for upper-middle-class children, Erwin was tutored at home as a child in preparation for the entrance examination to a gymnasium. He passed the examination easily and entered the Akademisches Gymnasium in 1898 at age eleven.

In the fall of 1906 Schrödinger entered the University of Vienna, where he quickly gained a reputation as an outstanding student and performed brilliantly in physics and mathematics. He was greatly influenced by Professor Fritz Hasenohrl's courses on theoretical physics and became focused mainly on theory while at the university. Schrödinger stated that no one except his father had a more profound influence on his life than Hasenohrl. In 1910 he received a doctor of philosophy degree, roughly equivalent to a master's degree in American universities.

After a one-year stint as a volunteer in the Austria-Hungary Army reserves, he returned to the University of Vienna, where he was appointed to an assistantship in experimental physics under Franz Exner and Fritz Kohlrausch. He missed an opportunity to take an assistantship with Fritz Hasenohrl in theoretical physics because of his military service, but he was happy to have had the practical laboratory experience, which influenced what he would eventually accept as possible in the realm of theoretical physics.

Schrödinger presented his first theoretical paper in 1912 at the Vienna Academy as part of his *habilitation*, a process similar to obtaining a Ph.D. in American universities. In 1914 he completed his *habilitation* at the University of Vienna and was paid to teach courses there. Schrödinger then spent 1914 to 1918 in military service during World War I. He was able to continue some scientific work while in uniform. After the war, he resumed his research at the Institute of Physics at the University of Vienna. He continued there until 1920, when he married Annemarie Bertel and took a position teaching physics at the University of Jena in Germany. He went from the University of Jena to other teaching and research positions at the universities of Stuttgart (Germany), Breslau (Poland), Zürich (Switzerland), and Berlin (Germany). At Berlin he was Max PLANCK's successor. Out of disgust for the Nazis, he left Berlin in 1933 and took a fellowship at Oxford University in England. He then taught at the University of Graz in Austria before becoming director of theoretical physics at the Institute for Advanced Studies in Dublin from 1940 until his retirement in 1955. He returned to the University of Vienna as a professor emeritus in 1958.

Schrödinger's important research included work on atomic spectra and statistical thermodynamics. However, he is best known for the development of wave mechanics, a mathematical description of the wave behavior of electrons as they orbit atomic nuclei. His theory of wave mechanics published in 1926 was considered the mathematical equivalent of the theory of matrix mechanics proposed by Werner HEISENBERG in 1925 (see also SCHRÖDINGER'S WAVE EQUATION). These contributions of Schrödinger and Heisenberg are considered to be the foundation of quantum mechanics (see QUANTUM THEORY). Schrödinger received a Nobel Prize in 1933 for his contribution to the development of quantum mechanics.

Schrödinger died in Vienna on January 4, 1961, and is buried in Alpbach, Tirol.

BIBLIOGRAPHY

MOORE, W. *Schrödinger: His Life and Thought*; Cambridge, U.K.: Cambridge University Press, 1967.

SCHRÖDINGER, E. *Space-Time Structure*; Cambridge, U.K.: Cambridge University Press, 1950.

———. *What Is Life? and Mind and Matter*; Cambridge, U.K.: Cambridge University Press, 1967.

MARY GEE

Schrödinger's Wave Equation

Whereas the trajectory of a classical particle can be described by Newtonian mechanics to offer a precise prediction of its state at any future time, the Heisenberg uncertainty principle precludes precision of this sort for submicroscopic particles such as electrons. In 1926 the Austrian physicist Erwin Schrödinger developed an equation that describes the position and velocity of a microscopic particle in terms of a "wave function," which recognizes the wave character of such particles. Schrödinger lived between 1887 and 1961, and his accomplishments have been commemorated by placing his likeness on one denomination of the Austrian unit of currency, the schilling.

When expressed in one dimension x of a single particle, the wave function $\Psi(x, t)$ conforms to the Schrödinger wave equation:

$$-\hbar/i \; \partial\Psi(x, t)/\partial t = -(\hbar^2/2m) \; \partial^2\Psi(x, t)/\partial x^2 + V(x, t)\Psi(x, t)$$

where $\hbar$ is Planck's constant, which is related to the intrinsic degree of uncertainty. $V(x, t)$ expresses in terms of position and time the potential energy field in which the particle of mass m moves.

This equation relates the time evolution of the wave function to its rate of change with respect to position. There are many situations where the potential energy V does not change over time. In such a case, the wave function $\Psi(x, t)$ can be written as $\Psi(x) \exp[-Et/\hbar]$, where E refers to the energy of the system. In such a case, the Schrödinger wave equation reduces to its time-independent form:

$$-(\hbar^2/2m) \; d^2\Psi(x)/dx^2 + V(x)\Psi(x) = E\Psi(x)$$

The imaginary part of the exponential time dependence of $\Psi(x, t)$ is eliminated when this wave function is multiplied by its complex conjugate, yielding the product $\Psi(x, t)\Psi^*(x, t)$, which is the probability density for the particle.

BIBLIOGRAPHY

Levine, I. N. *Quantum Chemistry,* 4th ed.; Englewood Cliffs, NJ: Prentice-Hall, 1991.

STEVEN I. SCHEINER

Seaborg, Glenn Theodore (1912–)

Glenn Theodore Seaborg was born in Ishpeming, Michigan, on April 19, 1912, to parents of Swedish descent. His parents moved their small family from the coal mining town of Ishpeming to Los Angeles in 1922 because they wanted their son to have a better education. Although Seaborg had no particular interest in science, he was already showing signs of intellectual promise. Seaborg soon fell under the influence of his gifted chemistry and physics teacher, Dwight Reid, at David Starr Jordan High School in Watts and began one of the most distinguished careers in science, education, and public service.

Seaborg earned an A.B. degree in chemistry in 1934 from UCLA and a Ph.D. in 1937 from UC Berkeley where he immediately joined the faculty, first as a research associate and later as an assistant professor. Seaborg and his colleagues had access to a small cyclotron, a machine that gets subatomic particles moving at nearly the speed of light; they believed they could use it to create new, heavier elements. Seaborg stated that they just wanted to see if they could expand the periodic table; however, that was only the beginning. The periodic table has guided many scientists; it guided Seaborg into the MAN-HATTAN PROJECT.

Seaborg was needed for war work when the weapon potential for plutonium was evident. In 1942 the thirty-year-old Seaborg went to Chicago to work on the atomic bomb project, returning to Berkeley only long enough to marry Helen Griggs. Seaborg was co-winner of the 1951 Nobel Prize in chemistry for his pioneering research in the chemistry of the TRANSURANIC ELEMENTS, ten of which he and his colleagues discovered.

Seaborg has served with distinction as the chair of the Atomic Energy Commission and as chancellor of the University of California at Berkeley. As chancellor of UC Berkeley, he presided over the years 1958 to 1961, during which a high point was bringing President John F. Kennedy to speak at Memorial Stadium. He headed the Atomic Energy Commission from 1961 to 1971; one of his proudest accomplishments was laying the groundwork for the Limited Test Ban Treaty in 1963.

Seaborg, his wife, and their six children were hikers and enjoyed the outdoors. The elder Seaborgs spent every weekend of 1980, for example, laying out the trail across California that was followed by the forty-odd members of Hike-A-Nation in their fundraising shore-to-shore walk.

Seaborg's continuing dedication to education is reflected in his work as chair of the Lawrence Hall of Science in Berkeley, his service on the National Commission on Excellence in Education, which published the 1983 report "A Nation at Risk," and his active membership on the Seaborg Center National Advisory Board. In the past decade, Seaborg's other great passion, the science education of children, has become a priority. Seaborg has always been involved with children: As a guest on the old radio show "Quiz Kids," he blurted out the discovery of einsteinum and fermium when a thirteen-year-old asked whether the periodic table was still expanding. Teachers everywhere respect Seaborg, who has made a practice of showing up at science fairs and class-rooms, including a college-freshman chemistry lecture class for Berkeley students every spring.

BIBLIOGRAPHY

SEABORG, G. T. *Science and Liberal Education in the Space Age;* Rock Island, IL: Augustana College Library, 1960.

SEABORG, G. T.; LOVELAND, W. *The Elements beyond Uranium;* New York: Wiley, 1990.

MARCIA J. TALKMITT

Selenium

34
Se
78.96

Melting Point: 490 K (gray); **Boiling Point:** 958 K (gray); **Density:** 4.81 g/cm^3 (gray); **Most Common Ion:** -2

Selenium is sixty-sixth in abundance in the Earth's crust (0.05 ppm). Of its analogues, sulfur (sixteenth), is far more abundant (340 ppm), and tellurium (seventy-second) a bit less abundant (0.001 ppm). Selenium's discovery is shared by two Swedish chemists, J. J. BERZELIUS and J. G. Gahn, who in 1817 found a residue following the oxidation of sulfur dioxide from copper pyrites. Elemental tellurium had been discovered thirty-five years earlier, and they initially believed this residue to be tellurium. Further work proved otherwise. Having found selenium associated with tellurium (named after the Latin word *tellus* for earth), Berzelius named the element selenium (after the Greek word *selene,* moon).

Selenium is sandwiched on the periodic table between sulfur and tellurium in the group and between arsenic and bromine in period four. Selenium reacts with active metals to form selenides containing the Se^{-2} ion. The selenides are chemically and physically similar to the sulfides. Selenium reacts with oxygen to form several oxides. Several COVALENT species, for example, SeO$_4$$^{-2}$, exist.

Selenium forms compounds with sixteen other elements and is a main constituent in thirty-nine mineral species, but these are widely dispersed and in low concentrations. Crooksite and clausthalite are two selenium minerals, but these are rare. Consequently, selenium is not mined alone but always as a by-product of other materials. There are several plant species that accumulate high selenium concentrations from soils and are helpful in locating regions with high selenium content. Some species have been found containing up to 1.5 weight percent selenium.

The element has been obtained from flue deposits in copper sulfide processing but is now predominantly obtained from electrolytic copper slimes. Recovery involves the formation of a water-soluble selenium specie with subsequent reduction to the element. Major producers of selenium in 1980 included Japan, Canada, the United States, the CIS (the former USSR), and Mexico. That year these sources accounted for 1,392 tons of refined selenium.

Selenium consists of several allotropes that exhibit the ring-chain behavior associated with sulfur. Amorphous selenium is obtained by the rapid cooling of liquid selenium. There are also three crystalline modifications. Red selenium exists in two monoclinic forms and β-selenium (or gray selenium) exists in a hexagonal form.

Selenium exhibits both photovoltaic and photoconductive properties. As such, selenium is used in photoelectric cells (electric eyes), rectifiers, and photographic exposure meters. Xerography is perhaps the most important application of the photoconductive property of selenium. Selenium has also been used in several metallurgical processes. Small amounts of selenium decrease the surface tension and the porosity of some steels. Selenium is also used to improve the machinability of nickel/iron and cobalt/iron alloys. Selenium has been used by the glass industry since about 1915 to decolorize glass. When mixed with other materials—cadmium sulfide, for example—selenium glasses are used to make signal glasses, the types of glass used in airport and railroad lighting. A suspension of selenium sulfide has been marketed for several years in a dandruff-control shampoo.

Selenium is a micronutrient for livestock and poultry. In selenium-deficient diets, white muscle disease develops in animals. Selenium also appears to play a role in dietary health for humans, although large doses of selenium have produced poisoning in both humans and livestock.

BIBLIOGRAPHY

ELKIN, E. M. "Selenium and Selenium Compounds." In *Encyclopedia of Chemical Technology,* 3rd ed.; R. E. Kirk, D. F. Othmer, M. Grayson, eds.; New York: Wiley-Interscience, 1981; Vol. 20, pp. 575–601.

GREENWOOD, N. N.; EARNSHAW, A. *Chemistry of the Elements;* New York: Pergamon, 1984; pp. 882–889.

LIDE, D. R., ed. *Handbook of Chemistry & Physcis,* 71st ed.; Cleveland, OH: Chemical Rubber Co., 1990–1991.

ALTON J. BANKS

Semiempirical Quantum Mechanical Calculations

The electronic structure of a molecule determines its geometry, physical properties, spectroscopic behavior, and reactivity. The electronic state function, $\Psi(r_1, r_2, r_3, \ldots)$, a function of the coordinates of the n electrons in the molecule, defines the electronic properties of the molecule. Application of the appropriate quantum mechanical operators to the molecular electronic state function generates corresponding molecular properties. It follows that determination of the electronic state function lies at the heart of modern computational chemistry (see also QUANTUM MECHANICAL COMPUTATIONS).

The function Ψ satisfies the time-independent Schrödinger equation:

$$-\frac{\hbar^2}{2m} \sum_i^n \frac{\partial^2 \Psi}{\partial r_i^2} + V(r_1, r_2, r_3, \ldots)\Psi = E\Psi$$

where V is the electronic potential energy of the molecule. The determination of Ψ thus amounts to a solution of this deceptively simple differential equation. Unfortunately, the potential energy depends on the coordinates of all the electrons, an example of the many-body problem often encountered in physics. Such problems can be solved analytically only for the simplest atomic and molecular systems; for virtually all molecules we must settle for approximate solutions to the Schrödinger equation.

While the simultaneous repulsion of all the electrons in a molecule precludes an exact solution, reducing the many-electron problem to a set of one-electron problems can produce chemically reasonable approximations to Ψ. In this approach we treat each electron in a molecule as if it moves in an average electric field due to all the other electrons. We characterize each electron by its own one-electron function, called a molecular orbital. Constructing these molecular orbitals as linear combinations of atomic orbitals (LCAO) provides an additional simplification to this problem:

$$\Phi = \sum_i c_i \phi_i$$

We conveniently represent the many-electron Schrödinger equation as $H\Psi = E\Psi$, where H is the many-electron "operator" on the function Ψ. A one-electron operator, F, treats the one-electron orbitals similarly: $F\Phi = \varepsilon\Phi$. The LCAO approximation results in n simultaneous linear equations:

$$\sum_j^n c_{ij}(F_{ij} - \varepsilon S_{ij}) = 0, i = 1 \text{ to } n$$

requiring that $\det(F_{ij} - \varepsilon S_{ij}) = 0$, where $F_{ij} = \int \Phi_i{}^* F\Phi_j dv$ and $S_{ij} = \int \Phi_i{}^* \Phi_j dv$. Evaluating S_{ij} and F_{ij} over all the atomic functions and solving the polynomial equation resulting from the expanded determinant provide one-electron energies, ε. Solution of the system of linear equations then produces the coefficients, c_{ij}, which define the molecular orbitals. Finally, combination of the molecular orbitals creates an approximate molecular electronic state function.

The field, due to all the electrons, depends on Φ for each electron, requiring an iterative approach to the problem. The process begins with an estimation of the initial average field encountered by each electron. A solution based on these fields results in molecular orbital functions and new average fields, which yield new functions, and so forth. Eventually the field calculated in one cycle will be indistinguishable from the field calculated in the previous cycle; the process has created a self-consistent field (SCF). Although the method is approximate, it contains no empirical parameters. We base the calculations solely on first principle, or *ab initio*, considerations (see also AB INITIO CALCULATIONS).

For a given molecular geometry we must evaluate the overlap integrals, S_{ij}, only once. In contrast, we must compute every F_{ij} in each cycle of the iteration, since it depends on potential energy. Each F_{ij} contains integrals for electron-electron repulsion (Coulomb integrals) and integrals to correct for the difference in interaction between electron pairs with the same and opposite spins (exchange integrals). The calculation of F_{ij} requires the evaluation of a large number of two-electron integrals of the general form for electron 1 and electron 2:

$$\int \phi_i(1)\phi_l(1)r_{12}{}^{-1}\phi_k(2)\phi_1(2)$$

The number of these integrals that must be evaluated becomes formidable as the number of atomic functions (i.e., the size of the molecule) increases. A molecular orbital calculation to determine the electronic structure of a compound containing twenty carbon atoms requires the evaluation of more than twelve million two-electron integrals for the carbon atoms alone. Thus, for molecules containing more than a few atoms, the computational effort becomes extreme and further approximations become necessary to reduce the number of two-electron integrals.

Semiempirical molecular orbital methods cover a wide array of approaches incorporating various schemes to approximate two-electron integrals.

1335

These schemes either ignore the integrals or parameterize them with experimental data (hence the term *semiempirical*). The simplest (and crudest) strategy is to avoid the SCF procedure altogether, determining each F_{ij} directly from experimentally measured ionization potentials. This method, known as the extended Hückel method, appears often in the literature, chiefly because of its ease of use and minimal computational requirements. Extended Hückel calculations involve valence electrons only. While results can differ drastically from those determined from more rigorous calculations, the method often describes useful chemical trends.

The SCF procedure can be maintained with only a small increase in effort by noting that at a given location in space the product of a function with itself is likely to be larger than the product of two different functions at the same location. On the basis of this somewhat shaky premise, we set all two-electron integrals for which $i \neq j$ or $k \neq l$ to zero and evaluate all those with $i = j$ and $k = l$. As with the extended Hückel method, we consider only valence electrons. We consider the product of two different functions, so-called functional overlap, to be negligible in such cases, rendering the integral equal to zero (neglected). The differential of an integral is the function itself, and the method described is deemed "complete neglect of differential overlap," abbreviated CNDO. The CNDO approximation also ignores the overlap integrals, S_{ij}. In this approximation, the omission of all exchange integrals gives molecular states with the same electronic configuration the same energy, regardless of differences in electron spin (e.g., singlet versus triplet states).

A significant portion of the exchange interaction can be retained by including exchange integrals in which the functions are on the same atomic centers. The inclusion of these one-center exchange integrals increases the computational effort modestly but neglects the much larger set of multicenter two-electron integrals. This modification of the CNDO method is labeled INDO (intermediate neglect of differential overlap). Modifications of INDO using experimentally parameterized integrals in place of computed ones result in schemes such as MINDO (modified intermediate neglect of differential overlap). The examples cited are representative of a number of methods that employ similar methodology, each identified by its own acronym.

A more rigorous approach results from an estimation of the complete contribution of the exchange integrals to F_{ij}. The exchange contribution can be shown to be proportional to the cube root of the electron density, $\rho^{1/3}$. We can account for the effect of *all* the exchange integrals if we know the electron density distribution. This knowledge allows us to replace the exchange integrals with terms containing $\rho^{1/3}$, scaled with a proportionality constant, α. We determine the value of α from a fit to specific atomic properties obtained from *ab initio* calculations or experiment. Since the state function that we seek depends on the electron density used to determine it, we call methods that use this approach density functional methods. We call the replacement of the exchange interaction with an alpha-scaled electron-density term the X-α approximation. In contrast to other semiempirical methods, density functional techniques usually include all electrons in the calculation. Depending on the computational approach and selection of atomic functions (called basis functions), X-α self-consistent field methods can yield electronic structures and properties that parallel those from *ab initio* calculations.

Semiempirical quantum mechanical calculations increase in computational effort as the approximations become less restrictive. Extended Hückel calculations require the evaluation of only a few integrals, but the results border on the qualitative. On the other end of the scale, density functional methods require the evaluation of a large number of integrals. Nevertheless, the number is at least an order of magnitude lower than the number required in *ab initio* methods, while properties computed from density functional calculations are often as accurate as those obtained from *ab initio* calculations.

BIBLIOGRAPHY

ATKINS, P. W. *Molecular Quantum Mechanics*; Oxford: Oxford University Press, 1983.

ATKINS, P. W. *Quanta*; Oxford: Oxford University Press, 1991; pp. 329–330.

HIRST, D. M. *A Computational Approach to Chemistry*; Oxford: Blackwell Scientific Publications, 1990; pp. 84–97.

LEVINE, I. N. *Quantum Chemistry*; Boston: Allyn and Bacon, 1974; pp. 450–455.

MURREL, J. N.; HARGET, A. J. *Semi-Empirical Self-Consistent Field Molecular Orbital Theory of Molecules*; London: Chapman and Hall, 1972.

SEGAL, G. A., ed. *Semiempirical Methods of Electronic Structure Calculations*; New York: Plenum, 1977.

SLATER, J. C. *The Calculation of Molecular Orbitals*; New York: Wiley, 1979.

SNOW, R. L.; BILLS, J. L. "A Simple Illustration of the SCF-LAO-MO Method." *J. Chem. Educ.* 1975, *8*, 506–509.

DENNIS W. BENNETT

Separation Science

Processes for separating materials and chemicals into their constituents may be classified, as shown in Table 1, according to phase states involved and basic mechanisms and driving forces used to accomplish the separation. Chemical separations are based on attaining different compositions in different phase states in contact and then physically separating the

Table 1. Table of Industrial Separations

Based on physical separations
  Fluid-fluid
    Size-based
      Filtration (demisting)
    Density-based
      Gravity settling
      Cyclone
      Centrifugation
    Surface forces
      Foam fractionation
  Fluid-solid
    Size-based
      Filtration
        Microfiltration
      Scrubbers
    Density-based
      Gravity
        Settling
        Fluidization
      Cyclone
      Centrifugation
    Magnetic field
    Electric field
      Electrostatic precipitator
    Surface forces
      Flocculation
      Flotation

Based on diffusional separations
  Equilibrium interphase
    Fluid-fluid
      Absorption/desorption/stripping
      Humidification/dehumidification
      Evaporation
      Distillation
        Azeotropic
        Extractive
        Steam
      Extraction
        Single solvent

        Double solvent
        Supercritical solvent
        Pyrometallurgical
    Fluid-solid
      Adsorption/desorption
        Leaching (extraction), washing, expression
        Supercritical extraction
        Chromatography
      Sublimation
      Drying
        Freeze drying
      Crystallization
        Zone melting
  Rate-governed
    Thermal diffusion
    Molecular distillation
    Ultracentrifuge
    Membrane
      Gaseous diffusion
      Ultrafiltration
      Reverse osmosis
      Dialysis
      Gas permeation
      Pervaporation
      Liquid membrane
      Electrodialysis
    Electrophoresis

Based on chemical reactions
  Fluid-fluid
    Gas-liquid heterogeneous reactors
      Gas scrubbers
      Chemisorption
      Oxidation/reduction
    Solvent extraction with reactive or complexing agents
    Hydrometallurgical extraction
    Pyrometallurgical extraction
  Fluid-solid
    Fluid-solid heterogeneous reactors
      Precipitation, clathration
      Leaching
      Chemisorption
      Ion exchange
    Catalytic reactors
      Catalytic converters
      Reactive distillation
      Catalytic membranes
  Biochemical oxidation/reduction
  Electrolysis
    Electrometallurgy
  Electromagnetic

---

phases to isolate products. Phase states may be any combination of gas, liquid, and solid (see PHASES AND PHASE DIAGRAMS; STATES OF MATTER). Different compositions may be attained by combinations of physical forces, molecular diffusion, and/or chemical reactions. Diffusional separations may be based on rate-governed intraphase and/or equilibrium interphase and segregation of constituents. The separating agent added to the process may be energy (as in distillation) or mass plus energy (as in extractive distillation, where a solvent is added).

Physical Separations

Physical separational processes are generally effective only for species or particles that are at least macromolecular (>0.1 μm) in size. Force fields may be inertial, gravitational, electrostatic, or magnetic. Particle (or species) size, density, and polarizability are important parameters.

Filtration processes are ubiquitous in industry for removing solid particulates from gases, solid particulates from liquids, and liquid aerosols from gas (demisting). Examples are dust from flue gas originating in a lime kiln, paint pigment from glycol grinding fluid, and machine-oil mist from air in a metalworking facility. Filtrations are predominantly size-based; microfilters are needed for particulates <2 μm in size (e.g., filters in hood vents of nuclear and biological facilities and in air-supply ducts for cleaning rooms). Separation mechanisms include simple screening (trapping at interstices), inertial impaction, diffusion, and interception followed by (electrostatic) adhesion on surfaces of fibrous mesh or membrane medium (and in previously collected beds of particulate solids). Particulates smaller than the effective trapping size for the medium pass through with the fluid; collection efficiency becomes exponentially lower as particle size decreases. In solid-liquid filtrations, some liquid inevitably remains occluded in the bed of solid particles, further limiting the efficiency of separation. Filtration is a batch process (see BATCH AND CONTINUOUS PROCESSES) that is stopped when the filter chamber becomes filled or the filter medium becomes plugged, except in demisting, where coalescence of the adsorbed liquid into separable drops enables continuous operation.

Wet scrubbers use liquid to provide renewable collection surface for removing particulates from gas streams in continuous operation. Simple spray chambers are not efficient; packed-bed and venturi contactors have high efficiency at 12 and 4 μm particle size, respectively. Scrubbers are employed for cleaning flue gas.

Electrostatic precipitators use a high-voltage electrostatic field to generate gas ions for charging aerosol particles; this causes them to migrate to a collecting surface, from which agglomerated particles are occasionally vibrated free and collected. These filters have high efficiency at 1 μm particle size. One use is removing fly ash from flue gas in coal-fired power plants.

Magnetic separators and filters use strong magnetic fields to cause ferromagnetic particles, and to a lesser extent paramagnetic particles, to migrate to a collection surface in order to separate these materials from the suspending fluid or from diamagnetic particles. An industrial example is removal of tramp iron and rust in ceramic processing equipment.

Density-based separations are used to physically separate liquid or solid aerosols from gas, solid particulates from liquids, and liquid-liquid dispersions. Inertial and gravitational force fields act on dispersed-phase particles and continuous-phase fluid according to the density of the respective phases. The greater the density difference, the greater the driving force that causes the dispersed-phase particles to migrate to a collection surface or interface. Particle size is important in setting the migration rate and in determining effects of surface forces.

Gravity settlers are efficient only for aerosols and particles >100 μm in size. Examples are primary fly ash collection from flue gas, settling of entrained liquid above aerated or boiling liquid, sedimentation in thickeners of $Ca(OH)_2$ sludge from acid mine water neutralization treatment, and settling steam-distillation condensates to decant organic and water layers. Coalescing baffles or mesh is used in liquid-liquid settlers to aid breaking emulsions of finer droplets.

In fluidization, gas or liquid flows up through a bed of particles to suspend or transport them. Particles of different sizes or densities may be segregated, so smaller or lighter particles are selectively elutriated with the exiting fluid. Examples of fluidization separations are steam fluidization to strip oil from catalyst collected from riser reactor in catalytic cracking and backflushing of dirt from sand filters with water.

In cyclones for gas-solid and liquid-solid separations, inertial forces add to gravitational forces. Single cyclones have high efficiency at 100 μm particle size, multistage cyclones at 20 μm. Some fluid remains occluded in the solid-phase product. Examples are recovery for recycling of the fluidized catalyst particles from gas at the top of petroleum catalytic crackers and separation of rock particles from coal particles with a dense-media cyclone.

In centrifuges for liquid-liquid and liquid-solid separations, inertial forces are 1,000–15,000 times gravitational acceleration; efficiencies are high at 10 μm particle size. Some liquid remains occluded in the recovered solid phase. Representative uses of continuous units are the removal of emulsified water from steam-distilled essential oils and the concentration of sludge from bio-oxidation reactors for treating wastewater from textile dying.

Surface forces may cause compositions of materials to be different at phase interfaces than in the bulk phase. In foam fractionation, gas is bubbled through liquid to cause surfactants in the liquid to concentrate at the top in a foam layer to be skimmed off. An example is removal of detergents from laundry waste. In FLOTATION, gas is bubbled through a liquid-mixed solid suspension, containing surfactant that preferentially adsorbs on one solid species, to concentrate this species at the surface. An example is the recovery of zinc sulfide from carbonate gangue. In flocculation, polyelectrolyte water-soluble polymers are added to fine suspensions to agglomerate finer particles up to perhaps 20 μm in size to make them easier to separate by physical methods. An example is adding 20 ppm polyelectrolyte to gold ore slimes to increase sedimentation rate.

Diffusional Separations

EQUILIBRIUM INTERPHASE

In the equilibrium interphase category, the diffusion of chemical species occurs across the interface between the phases and is followed by physical separation of the phases to attain separation of chemicals. Thermodynamic equilibrium is reached at the interface. Compositions in the bulk fluids are different from the equilibrium values and thus set up the concentration driving forces that cause intraphase diffusion of chemical species in each phase to feed the interphase transfer process. Phase contact mode may be stagewise (e.g., mixer-settler) or differential countercurrent (e.g., gas flow up, liquid flow down in packed tower). A cascade of stages, usually connected in countercurrent phase-flow fashion or extending the length of countercurrent contractors, is used to increase the efficiency of separations. Operating modes may be continuous or batch (again, see BATCH AND CONTINUOUS PROCESS).

In fluid-fluid diffusional separations involving non-condensable gas and liquid solvent, transfer of the solute (being separated) from the gas phase to the liquid phase is called *absorption;* the transfer of the solute from the liquid phase to the gas phase is called *desorption* or *stripping.* Important factors affecting distribution of solutes between gas and liquid phases, and thus ease of separation, are the solubilities and Henry's law coefficients for the solutes in the liquid phase. Operating temperature and pressure affect equilibrium distribution of solutes. For dilute solutions, heat effects are small and operation may be at preset temperature. Number of contact stages or column efficiency and gas-liquid flow ratio affect overall separation efficiency. Examples of separations are removal of H_2S from natural gas by absorption into ethanolamines, desorption of volatile organic contaminants (VOCs) from drinking water supplies by air stripping followed by capture on carbon, and steam stripping of dissolved and suspended essential oil from process wastewater in flavor oil production plants.

Humidification/dehumidification is a special case of desorption/absorption where the liquid phase is pure solvent, which is the transferred component. The driving force is the difference between the partial pressure of the solvent in the gas phase and the VAPOR PRESSURE of the solvent at the interface temperature; heat effects are extensive. Examples are the cooling of water by evaporation, and thus humidification of air, in water-cooling towers for power plants and the dehumidification of air in conventional air conditioners (air cooling units).

Evaporation involves the separation of solvent vapor from liquid mixtures containing dissolved non-volatile solutes. The driving force is the difference between vapor-phase pressure and partial pressure exerted by solvent in the liquid phase at the interface temperature; heat effects are large. Solvent partial pressure, like vapor pressure, is a very significant function of temperature, solvent composition, and the ionic nature of the solutes. Multistage cascades operated at progressively lower temperature, and thus progressively lower pressure, are used to increase heat recovery efficiency. An example is multi-effect evaporation of freshwater from seawater.

Distillation, a method used for separation of components of liquid mixtures, depends on the distribution of constituents between the liquid mixture and the vapor in equilibrium with the mixture. The two phases exist by formation of the vapor phase through partial evaporation of the liquid mixture. Each phase can be recovered separately with the more volatile components concentrated in the vapor, while the less volatile ones are concentrated in the liquid. Concentration offsets at the interface are based on the differences in volatilities, and thus effective vapor pressures, of the constituents. System pressure is present, and boiling temperature decreases on proceeding from the stillpot where vapor is generated to the vapor condenser at the top. The number of contact stages or column efficiency and the vapor-liquid flow ratio, and thus the reflux ratio and energy input, affect the overall separation efficiency. An example is the fractional distillation of propylene from propane and other C_3 hydrocarbons to prepare a monomer for manufacturing polypropylene.

In the distillation of highly nonideal polar-nonpolar mixtures, constant-boiling blends called azeotropes may occur that become inseparable by distillation. Azeotropes are usually lower-boiling than their pure constituents. If the constituents are sparingly soluble (e.g., benzene-water), a heterogeneous azeotrope of two liquid phases and a vapor phase can form. An azeotrope-forming entrainer, or an azeotrope-breaking solvent, may be purposely added to a distillation to facilitate separation. Subsequent separation processes are then needed to recover the added entrainer or solvent. Examples occur in the manufacturing process for polycarbonate polymer; the water-pyridine azeotrope is distilled from aqueous NaCl (salt) solution in one distillation column and is broken in the next column by adding a desiccating agent to hold back the water while pure pyridine is distilled overhead.

In extractive distillation, a high-boiling solvent is added to change the equilibrium characteristics of the constituents to facilitate their separation or to break an azeotrope. An example is the use of furfural solvent in distillation of C_4 hydrocarbons to recover 1,3-butadiene.

In steam distillation, steam is injected directly into the material in the stillpot to vaporize the liquid; vaporization occurs at a lower temperature than the normal boiling point of the constituents because the steam displaces much of the partial pressure space in the vapor phase. Temperature lowering also occurs because water forms low-boiling azeotropes with many organic compounds that have normal boiling points up to 200 K higher than that of water. Examples are steam distillation of peppermint oil from peppermint hay and steam distillation of dimethylsulfide from peppermint oil.

Liquid-liquid extraction is a process for separating components in solution by their distribution between two immiscible phases. The process is analogous to distillation except that differences in solubility rather than differences in volatility govern the equilibrium concentration offsets at the phase interface. A single solvent or two immiscible solvents may be added to the mixture being separated. Each phase can be recovered separately with the more soluble components concentrated in the extract phase, while the less soluble ones are concentrated in the raffinate phase. The solvents need to be recovered from the extract and raffinate products in downstream separational processes such as distillation. The operating temperature and nature of the solvents affect the equilibrium distribution of the solutes. The number of extraction stages or column efficiency and the liquid-liquid flow ratio (solvent-raffinate flow ratio) also affect the overall separation efficiency.

An example of single-solvent liquid extraction is the recovery of C_6-C_8 aromatics from catalytically reformed naphtha using mixed-glycol solvent. An example of double-solvent extraction is found in the ketene process for manufacturing acetic anhydride from ACETIC ACID, wherein an immiscible organic solvent is used to extract acetic acid from an aqueous acid stream. One use of a supercritical solvent is in the propane deasphalting process of petroleum lube oil; resins are selectively precipitated from oil extracts in propane above its critical temperature and pressure (see CRITICAL PHENOMENA). An example of pyrometallurgical extraction is the use of molten lead to absorb zinc from zinc-laden furnace gas, followed by cooling and the settling of a molten zinc layer from the molten lead.

In fluid-solid diffusional separations involving gas or liquid with solid adsorbent, the transfer of the solute (being separated) from the fluid phase to the solid phase is called ADSORPTION; the transfer of the solute from the solid phase to the fluid phase is called desorption. Equilibrium adsorption isotherms are used to relate the concentration of the solute in the fluid at the fluid-solid interface to the amount (bulk concentration) of solute adsorbed in the solid phase. Operating temperature and pressure affect the equilibrium distribution of solutes. Solid nature and porosity and heat effects are important parameters. The operating mode is cyclical-batch or semicontinuous. Breakthrough curves in adsorption indicate when the solid becomes saturated. Retention of occluded liquid in the solid decreases the sharpness of the separation. Regeneration follows the adsorption step, the adsorbed material being recovered by pressure or thermal swings or by using a displacement fluid (which latter requires another downstream separation process). Some separations are the separation of oxygen from nitrogen by pressure-swing adsorption on zeolites, the adsorption of chlorine from drinking water in an activated carbon filter, and the separation of p-xylene from C_8-aromatics on zeolites in a simulated moving bed.

Leaching is batch desorption where a solvent is used to extract solutes found in solid matrices. It is also called extraction, or supercritical extraction if the solvent is above its critical temperature and pressure. Washing refers to the displacement of liquid, not tightly bound, from pores in solid material. Expression refers to the use of mechanical force to push liquid from pores of a compressible solid matrix. Examples are the leaching of vanilla resin from ground vanilla beans by aqueous ethanol, the decaffeination of coffee beans with supercritical carbon dioxide, the washing of occluded aqueous ethylene glycol from paint pigment cake in a filter press using water, and the chopping and pressing of alfalfa plants to recover juice from which protein is agglomerated and dried to prepare animal feed.

In chromatography, carrier fluid (gas, liquid, or supercritical) is used to transport solute mixtures through fixed adsorbent beds in a semibatch mode. Differential adsorption/desorption causes the solutes to be eluted from the bed in bands, allowing separation of pure-solute fractions depending on their retention times. The high carrier fluid/solute flow ratio and high pressure drop through packed beds make preparative chromatography an expensive process. One application is the production of single enantiomers from racemic mixtures of chiral drugs.

Sublimation is the vaporization of a substance from the solid into the vapor state without the formation of an intermediate liquid phase. Desublimation is the direct preferential condensation of vapor to solid, perhaps with preferential participation in crystal structure. Examples are purification of sulfur by sublimation from impurities and purification of phthalic anhydride by desublimation.

The drying of solids is a batch-desorption operation wherein a solvent is removed from a solid matrix into a gas. It resembles a humidification operation when the solid surface is wet with solvent; it likewise resembles a sublimation operation when the temperature is below the freezing point of the solvent. A vacuum may be used in lieu of gas to lower the partial pressure of the solvent at the interface. The drying of gases with desiccants is an adsorption operation. Heat effects are considerable in drying. Examples are spray drying of detergent powders, freeze-drying for food dehydration and preservation, and dehydration of cracked petroleum gas with alumina.

Crystallization involves the purification of a chemical species by solidification from a liquid mixture. In solution crystallization a diluent is added to the mixture. The solution is then directly or indirectly cooled and/or a solvent is evaporated to effect crystallization. In melt crystallization no diluent is added to the mixture, and the solid phase is formed by direct or indirect cooling of the melt. Solubilities and equilibrium compositions are functions of temperature. Crystals precipitated are theoretically pure chemical species or eutectic mixtures. Occluded liquor inevitably remains with the solids during subse-

quent solid-liquid phase separation; this limits the chemical separating efficiency of the crystallization process. An example of solution crystallization is the dewaxing of lubricating oil by chilling the oil in ketone diluent to form wax crystals, which are removed by filtration. An example of melt crystallization is the separation of *para*-dichlorobenzene from its isomeric mixtures.

In zone melting, a small molten zone moves through a long-charge solid material. Multiple passes are made through the ingot. Distribution of a solute between the liquid and solid phases effects a separation greater than that attained in simple melt crystallization. Zone refining is employed in the semiconductor industry for the refining of germanium and silicon.

RATE-GOVERNED PROCESSES

In the rate-governed category, separational processes work by virtue of differences in transport rate through some medium under the impetus of an imposed force, resulting from a gradient in pressure, temperature, composition, electric field, or the like. Usually rate-governed processes give product phases that would be fully miscible if mixed with each other, whereas ordinary equilibration processes necessarily generate products that partition.

Thermal diffusion relies on the difference in migration rates of different molecules or isotopically substituted materials between regions of different temperature in the gas phase. Thermal diffusion is currently used to enrich ^{15}N and noble gases on a production scale.

Molecular distillation is performed under high-vacuum conditions to increase the mean free path (see STATISTICAL MECHANICS) of the vapor molecules. The rate of distillation is controlled by the rate of evaporation of atoms or molecules from the surface of the liquid; the evaporating species are removed from the system by immediate condensation on a nearby surface. Therefore, it is not valid to assume that an equilibrium is maintained between the concentrations in the vapor and in the liquid. Separation factors theoretically are larger by a factor of the square root of the mass ratio than in ordinary distillation under equilibrium conditions. An example is vacuum dezincing for the purification of lead.

In ultracentrifugation, molecules in a fluid are separated by the density gradient brought about by diffusion under high centrifugal force; heavier molecules migrate to the highest gravity zone. Examples are the separation of biological substances in liquids and isotope separations in a gas centrifuge. Acceleration of 2×10^5 times gravitational acceleration is used to separate $^{235}UF_6/^{238}UF_6$ in a multistage isotope enrichment plant, with a separation factor of 1.005 per stage.

In the membrane separational processes, two fluids are separated by a thin membrane barrier that prevents hydrodynamic flow, so that transport through the membrane is by sorption and diffusion. The driving force may be concentration, pressure, and/or an electric field. The membrane may be polymeric, ceramic, or metallic or an immiscible liquid. The property of the membrane describing its rate of transport is known as its permeability. A membrane is semipermeable if under identical conditions, it transports different molecular species at different rates (see also OSMOSIS).

Gaseous diffusion is based on the difference in rates of Knudsen or surface diffusion through a porous barrier. It is used to separate $^{235}UF_6$ from $^{238}UF_6$ in multistage cascades of diffusion cells. Stage-feed gas, UF_6, flows past a diffusion barrier made of porous material with holes smaller than the mean free path of UF_6 molecules. About half of the feed gas flows through the barrier to a lower-pressure region. The gas passing through the membrane is slightly richer in $^{235}UF_6$ than the gas remaining on the high-pressure side, because the mean speed of $^{235}UF_6$ molecules is 1.004 higher (square root of molecular mass ratio) than that of $^{238}UF_6$ molecules. Viscous flow, surface effects, and back-pressure diffusion reduce separation efficiency. Interstage gas streams need to be recompressed and cooled.

Ultrafiltration is a pressure-driven membrane process capable of separating solution components on the basis of molecular size and shape. It is effective for polymeric-size (<0.5 μm) molecules, colloids, or fine particulates. The purpose is usually to concentrate the high-molecular-mass material by removing some solvent, such as in concentrating proteins with cellulose acetate membranes.

Osmosis occurs when solutions of two different concentrations in the same solvent are separated from one another by a membrane. If the membrane is more permeable (because of higher combined solubility and diffusivity) to the solvent than to the solute(s), solvent flow occurs from the more dilute to the more concentrated solution. This solvent flow continues until the concentrations are equalized or the pressure on the more concentrated side of the membrane rises to the osmotic level. If a pressure in excess of the OSMOTIC PRESSURE is applied to the more concentrated solution, the solvent can be caused to flow to the more dilute solution. This reverse osmosis process can be used to separate solution flowing over the membrane at high pressure into a solvent-rich stream that flows through the membrane and a more concentrated solution rejected downstream. Membrane integrity at high pressure, plugging, ion rejection efficiencies, low solvent permeability, and inadequate concentration of rejected solution are problem areas. An application is the production of drinking water from NaCl brine using cellulose acetate membrane modules.

Dialysis uses membranes to separate small solutes (<0.002 μm) preferentially from solution by diffusion from a concentrated solution to a dilute solution. The process takes advantage of the fact that low-molecular-mass solutes have a higher diffusion coefficient in the membrane material than higher-molecular-mass solutes. Simultaneous osmosis of solvent occurs in the opposite direction. Bulk flow of solvent through the membrane may be reduced by balancing the osmotic pressure of the feed solution by using a flowing solution with the same osmotic pressure on the other side of the membrane. Examples are recovery of NaOH in rayon manufacture and artificial kidneys.

Gas permeation involves gases on the high-pressure side of a membrane permeating through the membrane to its low-pressure side. An industrial application is recovery of hydrogen from refinery gas using polysulfone hollow fibers coated with silicone rubber.

Pervaporation maintains a liquid phase on the high-pressure side of a membrane and a vapor phase on the low-pressure side (sustained with a vacuum pump). The driving force for the permeating component is the difference between its partial pressure in the liquid feed and in the vapor phase permeate. A commercial application is dehydration of azeotropic organic distillate (e.g., ethanol-water) using a polyvinyl alcohol membrane.

Liquid membrane separations transfer solutes from one liquid phase to another liquid phase through an immiscible liquid film; the separation is based on different rates of permeation through the liquid film. In a contactor that provides high interfacial area, small vector droplets emulsified in larger organic drops rise through a counterflowing continuous water phase. Solutes transfer between the counterflowing aqueous phases. The process does require stabilization (e.g., with surfactant) of the aqueous droplets within the organic drops, as well as facilities for separating both levels of liquid dispersion after the contacting. An example of the process is the extraction of phenol from wastewater into drops containing aqueous NaOH.

For electrolyte solutions, migration of charged species in an electric field constitutes an additional mechanism for mass transfer. In electrodialysis, the concentration or composition of electrolyte solutions is altered as a result of electromigration through membranes in contact with these solutions. In electrodialysis units, feed solution is pumped through compartments separated by alternating cation-exchange and anion-exchange membranes, which are selectively permeable to positive and negative ions, respectively. With electric current turned on, ion migration causes salt depletion in alternate compartments and salt enrichment in adjacent ones. Its major use is demineralization of brackish waters.

Electrophoresis is separation by differential migration of charged particles, macromolecules, or colloids in an electric field. In continuous-flow zone apparatus, the field is applied transversely to the direction of flow of a delta-shaped laminar layer of fluid. A practical example is protein separation.

Processes Based on Chemical Reactions

The capabilities of the diffusional methods described above can be further expanded by chemical derivitization, in which components to be separated are subjected to some form of chemical reaction. Either some components react and others do not, or else all components react, such that the products are more readily separable than the initial unreacted mixture. The chemical reaction energies involved are usually higher than the phase transition energies (<40 kJ/mol) of the diffusional unit operations. For reversible chemical complexing when using a mass separating agent, bond energies are likely to be in the 20–60 kJ/mol range. This is substantially less than the 200^+ kJ/mol for covalent bonds because of the need for regeneration and avoiding decomposition of the complexing agent itself.

In the fluid-fluid contacting category, gas-liquid heterogeneous reactors are used to amplify the separating power of absorption/desorption. In gas scrubbers, adding a reactive chemical to the liquid solvent provides a chemical sink for removing selected solutes transferred from the gas phase while increasing the solute-concentration-difference driving force and reducing the mass-transfer resistance in the liquid phase. An example is the irreversible absorption of chlorine from vent gas by scrubbing with aqueous sodium hydroxide to form Cl^- and ClO^- in solution.

Chemisorption refers to the use of a chemically reactive adsorbent to increase selectivity and capacity; this is usually a reversible chemical complexing agent to enable recovery of transferred solutes and reagent. Examples are absorption of CO_2 from natural gas with ethanolamines and the use of cuprous ammonium acetate for extractive distillation of butadiene and butenes.

Oxidation-reduction reactions are performed in the liquid phase by contacting with oxidizing or reducing gas, or directly in the gas phase, in order to convert chemical species to separable forms. Examples are air oxidation of ferrous iron in acid coal mine drainage to enable precipitation as ferric hydroxide when neutralized with lime; reduction of Cr^{+6} to Cr^{+3} in metal-finishing wastewaters using sulfur dioxide to enable precipitation as chromic hydroxide when neutralized with lime; and incineration of chlorinated organics in vent gas to produce CO_2, H_2O, and HCl, which is then scrubbed out with hydroxide solution.

Solvent extraction with reactive or complexing agents involves use of chemical reactions to provide selectivity and capacity for selected solutes. An example is caustic washing of petroleum to extract mercaptans.

For a metallic element to be extractable by an organic solvent immiscible with water, the element must be capable of forming an organic-soluble, electrically neutral complex compound with the solvent or with added complexing agent. The oxidation-reduction potential for the aqueous phase is adjusted to control the proportion of an element in different valence states in order to vary its distribution coefficient. An example of such hydrometallurgical extraction is the Purex process for the separation of plutonium, uranium, and fission products: the elements are partitioned between nitric acid and their complexes with tributylphosphate dissolved in dodecane using ferrous salt to control valence states.

Reactive pyrometallurgical extraction between molten phases involves changing between metallic and ionic valence states of the transferring element. An example of pyrometallurgical extraction is the distribution of uranium, plutonium, and fission products between a zinc alloy phase and a molten chloride salt phase in the nuclear breeder reactor program.

In the fluid-solid contacting category, fluid-solid heterogeneous reactors are used to selectively precipitate chemical species or amplify the separating power of adsorption/desorption. Examples of precipitation separation are desilverization of liquid lead with zinc in which a zinc-silver intermetallic compound floats into the crust layer and neutralization of acid mine water with lime to raise the pH and precipitate iron and aluminum as hydroxides.

Clathration involves precipitation of chemicals as hydrates. For example, formation of $C_3H_8 \cdot 18H_2O$

and $CO_2 \cdot 7H_2O$ is troublesome in high-pressure natural gas pipelines.

Leaching of chemically bonded species requires the use of reactive solvents. An example is dissolving copper oxide from ore with dilute sulfuric acid to form soluble copper sulfate.

Chemisorption refers to use of chemically reactive adsorbent to increase selectivity and capacity while maintaining reversibility; this reactivity increases the energy required to regenerate the adsorbent and recover adsorbate. An example is recovering 1,3-butadiene from liquefied, cracked refinery gas by reacting with copper chloride in a slurry.

Ion exchange involves passing aqueous solution through solid resin that contains bound groups carrying an ionic charge (either basic − or acidic +) accompanied by displaceable ions of opposite charge. Respectively, cations or anions from the aqueous solution replace the dissimilar ions of the same charge type initially in the resin. The process is cyclical—each resin must be regenerated by displacing the adsorbed ions with a concentrated solution of acid, base, or salt. Water can be deionized with dual ion-retardation resin in which H^+ and OH^- are the displaceable species. Regenerant and rinse waters need to be treated by other downstream separational processes. Plugging resin pores may be a problem. A major use of ion exchange is in the softening of hard water, a process in which calcium ions in the water are exchanged with sodium ions in polymeric resin containing bound sulfonic acid groups; regeneration is with saturated NaCl brine.

Catalytic reactors use a catalyst to increase the reaction rate without affecting the reaction equilibrium concentrations. Catalytic converters are reactors that convert chemical species into another chemical form that is separable or innocuous. Examples are recovery of H_2S from natural gas plant H_2S/CO_2 vent gas by partial combustion of the hydrogen sulfide to sulfur dioxide and catalytic combination of the hydrogen sulfide and sulfur dioxide to produce molten sulfur; and conversion of CO in automotive exhaust to CO_2 by oxidation with excess oxygen over platinum catalyst.

Reactive distillation separations involve simultaneous chemical reaction and multistage distillation. This combined unit operation suits those chemical reactions where equilibrium limits conversion in a fixed-bed reactor to a low-to-moderate level. By continuously separating products from reactants while the reaction is in progress, the reaction can proceed to a much higher level of conversion than would otherwise be possible. An example is fractional distillation of a C₄-methanol azeotrope from methyl-t-iso-butylether formed by the reaction of methanol with isobutene over distillation packing containing an acidic ion-exchange catalyst.

Catalytic membranes use reversible chemical reactions between the transferring solute and the membrane material to form chemical intermediate species that increase selectivity and the permeation rate. An example is the diffusion of hydrogen through a palladium membrane, facilitated by the formation of Pd_2H and diffusion of hydrogen between metal atoms by formation and breaking of the hydride bonds.

Biochemical reactors use aerobic (or anaerobic) chemically tolerant microorganisms to oxidize (or reduce) chemical species to remove them from polluted water. The amount of energy required is low because operation is at ambient temperature. An example is passive treatment of acid coal mine drainage employing air oxidation of ferrous iron in a trickling bed reactor using *Ferrobacillus* and *Thiobacillus* bacteria along with neutralization in a bed of crushed limestone.

Electrolysis separation of chemical species is based on the difference in rates of production of an element by oxidation or reduction of an ionic species or, inversely, different rates of discharge of ions at an electrode. Electric energy is used to drive the ionization reactions in an electrolytic cell; an electrolyte supports charge transfer by ionic conductance between the anode and cathode. An example is the enrichment of deuterium ($^2H_1$) over hydrogen ($^1H_1$) in the hydrogen gas produced at the cathode in a cell electrolyzing water containing deuterium (HDO), with potassium hydroxide added as an electrolyte.

In electrometallurgy, electrolysis is performed using either an aqueous or a fused-salt electrolyte. Examples are electrolytic refining of copper by dissolving it in sulfuric acid; plating purified copper metal from the aqueous copper sulfate, leaving an anode mud containing selenium and other impurities; and the recovery of aluminum by electrolysis of alumina in a bath of fused cryolite.

Electromagnetic separation is based on the same principles as mass spectroscopy, using a device consisting of an ion source, an accelerating electric field, a magnetic field, and collecting apparatus in an ultra-high-vacuum enclosure. Metals are usually introduced as chlorides. An example is uranium isotope enrichment in a calutron.

BIBLIOGRAPHY

BARTON, P. "The Acid Mine Drainage." In *Sulfur in the Environment*; J. O. Nriagu, ed.; New York: Wiley, 1978; Part 2, Chapter 8.

EGGERSTEDT, P. M.; ZIEVERS, E. F.; ZIEVERS, E. C. *Chem. Eng. Progress* 1993, *1*, 62–68.

KING, C. J. *Separation Processes,* 2nd ed.; New York: McGraw-Hill, 1980.

PEHLKE, R. D. *Unit Processes of Extractive Metallurgy;* New York: Elsevier, 1973.

PERRY, R. H.; GREEN, D. W., eds. *Perry's Chemical Engineers' Handbook,* 6th ed.; New York: McGraw-Hill, 1984.

SCHWEITZER, P. A., ed. *Handbook of Separation Techniques,* 3rd ed.; New York: McGraw-Hill, 1996.

PAUL BARTON

Sewage Treatment

See WASTE TREATMENT.

Sidgwick, Nevil Vincent (1873–1952)

Nevil Vincent Sidgwick was born in Oxford, England, May 8, 1873. Except for graduate studies in Germany under the direction of Wilhelm OSTWALD, Sidgwick was educated at Oxford and remained there as a member of its faculty for all his professional life.

G. N. LEWIS had established the symbolism for electronic bonding in compounds. Although his method worked for a large number of compounds, it had its inadequacies. In 1923 Sidgwick showed that in particular compounds the electrons of a shared pair are contributed by the same atom, a characteristic he termed *coordinate covalency.* This causes a polarity to be established. The donor atom becomes slightly positive and the acceptor atom slightly negative. This allowed Lewis's method to extend beyond organic chemistry into coordination compounds and chelation complexes and to organic molecules in which nitrogen atoms are involved.

Sidgwick's specialization in the organic chemistry of nitrogen led to a book on the subject. Considered a classic, it was first published in 1910, and in 1947 it was expanded to two volumes. Sidgwick was a fellow of the Royal Society of Chemistry in Great Britain. He died in Oxford, March 15, 1952.

BIBLIOGRAPHY

IHDE, A. J. *The Development of Modern Chemistry;* New York: Harper & Row, 1964.

SUTTON, L. E. Obituary. *Proceedings of the Chemical Society;* 1958; pp. 310–319.

ANNA R. BERGSTROM

Sight, Sense of

See VISION, CHEMISTRY OF.

Sigma Bond

See SINGLE BONDS.

Signal Transduction

See MEMBRANES, BIOLOGICAL.

Silicon

14
Si
28.0855

Melting Point: 1683 K; **Boiling Point:** 2628 K; **Density:** 2.33 g/cm^3

Silicon is the second most abundant element in the Earth's crust, following oxygen. The name is derived from the Latin word *silex,* meaning "flint." Given the use of flint (a silicon oxide) in the construction of prehistoric weapons and tools, silicon has been used for centuries. The element does not occur free, and its proclivity for reaction with oxygen retarded isolation of the element until the early 1800s. DAVY reported the substance silica in 1800, but he thought it was a compound. GAY-LUSSAC and Thenard are reported to have isolated silicon in 1811 by reducing silicon tetrafluoride with potassium. BERZELIUS is frequently credited with the first isolation of the element using a similar technique—reduction of K_2SiF_6 with potassium.

Silicon is a gray, brittle material belonging to the same group as carbon. Silicon's chemistry is that of the COVALENT BOND, with the major, naturally occurring compounds being either oxides or silicates. Notable oxides of silicon in addition to flint are sand, quartz, amethysts, agates, and opals. Silicates include clay, mica, granite, and ASBESTOS.

Silicon has become an element in great demand since the discovery of the transistor in 1947 and the integrated circuits that form the basis of modern electronics. This growth is evident in the fact that Bureau of Mines data indicate that in 1955 5–10 metric tons of semiconductor-grade silicon were consumed compared with 1980 levels of 2,800 metric tons.

Production of silicon consists of conversion of the silicon compound to an easily reduced substance. Frequently $SiCl_4$ is produced and subsequently reduced by magnesium. Alternatively, quartzite or sand is reduced with coke in an electric arc furnace.

Silicon is used as a major component of many carbon and steel alloys along with aluminum, manganese, strontium, titanium, vanadium, and zirconium. Silicon has been shown to increase tensile strength and improve corrosion resistance of steels. Silicon also has been used to reduce metal oxides in metal slags, permitting the pure metal to be recovered and alloyed in steels. Silicon carbide, also known as carborundum, is a widely used abrasive, with a hardness only slightly less than that of diamond. SiO_2 is used as high-purity quartz for oscillators and transducers. Silica gel, an amorphous form of SiO_2, is used as a desiccant and as an anticaking agents in foods. The glass industry uses large amounts of SiO_2 to make many different glasses. Silicates such as clays form many types of pottery and bricks. Asbestos has been used in many insulating applications but is currently out of favor owing to the health concerns over asbestosis.

Silicon forms many organic compounds. The first compound was tetraethyl silicon, made in 1863 by Friedel and Crafts. Silicones are organosilicon compounds and exist as oils, greases, elastomers, and resins. They are used as heat-transfer agents, lubricants, gaskets, sealants, adhesives, and insulators for electrical equipment.

BIBLIOGRAPHY

GREENWOOD, N. N.; EARNSHAW, A. *Chemistry of the Elements;* New York: Pergamon, 1984; pp. 379–426.

LIDE, D. R., ed. *Handbook of Chemistry & Physics,* 71st ed.; Cleveland: Chemical Rubber Co., 1990–1991.

RUNYAN, W. "Silicon and Alloys." In *Encyclopedia of Chemical Technology,* 3rd ed.; R. E. Kirk, D. F. Othmer, M. Grayson, eds.; New York: Wiley-Interscience, 1981, Vol. 20, pp. 826–845.

ALTON J. BANKS

Silver

47
Ag
107.8682

Melting Point: 1235 K; **Boiling Point:** 2485 K; **Density:** 10.50 g/cm³; **Most Common Ion:** +1

Silver is the sixty-fourth most abundant element in the Earth's crust, with an abundance equal to that of mercury (0.08 ppm). The elemental symbol is derived from the Latin word *argentum,* meaning "silver." The element occurs free in nature, albeit rarely, and has been used throughout human history. Silver utensils dating to 3000 B.C. prove the existence of metallurgical techniques that permitted separation of silver from other metals.

Silver occurs in ores such as argentite, Ag_2S, and horn silver, $AgCl$. There are fifty-five silver-containing minerals, most of which occur with copper, gold, lead, or zinc, and the metallurgy of silver is that of extraction from these metals. Silver is mined in open pits and, to a lesser extent, subsurface mines and shafts. Major producers of silver are Canada, Mexico, Peru, and the United States. Several processes have been used for silver recovery, the most recent of which involves conversion of finely ground silver ore into a cyano-silver complex—a process known as cyanidation.

Silver is a shiny white metal and has the greatest thermal and electrical conductivity of all the metals. While not quite as malleable as gold, silver can be worked into many forms—for example, fine wires, foil, and leaf.

The uses of silver have changed dramatically over time. Silver is one of the well-known "coinage metal" trio (with gold and copper), and in 1971 coinage uses of the metal accounted for 879 tons of silver. In 1980 that number fell to 491 tons. The ease of working with it and its luster made silver an important resource for making jewelry, silverware, and other decorative objects. Those uses are no longer the major consumer of silver. Photographic emulsions contain silver halides (bromide and chloride) and have been the largest source of silver consumption in the United States for several years. High thermal and electrical conductivity make silver an important element in electrical contacts and in paints. Silver oxide batteries are used in watches and radios and have military applications (in submarines and torpedoes). Silver-copper alloys are used in brazing and soldering. Steel bearings that are lead-coated and silver-plated have been used in aircraft engine bearings. Silver catalysts are used in reactions, for example, the production of formaldehyde. Dental fillings are alloys containing silver and tin with small amounts of copper and zinc. Silver salts, specifically AgI, have been used in seeding clouds. Some silver salts have also been used as topical antiseptics.

BIBLIOGRAPHY

GREENWOOD, N. N.; EARNSHAW, A. *Chemistry of the Elements;* New York: Pergamon, 1984; pp. 1364–1367.

LIDE, D. R., ed. *Handbook of Chemistry & Physics,* 71st ed.; Cleveland: Chemical Rubber Co., 1990–1991.

SISTARE, G. H. "Silver and Silver Alloys." In *Encyclopedia of Chemical Technology,* 3rd ed.; R. E. Kirk, D. F. Othmer, M. Grayson, eds.; New York: Wiley-Interscience, 1981; Vol. 21, pp. 1–15.

ALTON J. BANKS

Single Bonds

A single bond is a COVALENT BOND holding two atoms together by the sharing of a pair of electrons with opposite spins, in accordance with well-established laws of quantum mechanics. Remarkably, chemical bonds in many compounds had been recognized and correctly classified as single or multiple by 1868, a half-century before QUANTUM THEORY appeared, long before ELECTRONs had been discovered, and well before leading philosophers had accepted the atomic nature of matter. This creative feat had been accomplished by meticulous measurements of the relative proportions of elements in diverse compounds and by studies of volumes of the vaporized compounds. The simplest coherent interpretation was to introduce the concept of chemical bonds between atoms.

By 1939, heats of reaction had shown that a given type of single bond (e.g., C—H or Si—Cl) is associated with its own characteristic stabilization energy irrespective of the compound in which the bond is found. Likewise, experimental diffraction and spectroscopic investigations revealed that each type of bond has its own distinctive interatomic distance and frequency of vibration. These characteristic features of linkages between atoms put the concept of the single bond on a firm empirical footing one-third of a century before a fully satisfactory theoretical substantiation was feasible. Accurate quantum calculations had to wait for major advances in computer technology (see QUANTUM MECHANICAL COMPUTATIONS). Electron pairs, which are charged negatively, bind together the positively charged nuclei of atoms by electrostatic forces. What is inherently quantal about the phenomenon is the peculiarly nonclassical character of the spatial distribution of the electron pair.

BIBLIOGRAPHY

MACKLE, H. "The Evolution of Valence Theory and Bond Symbolism." *J. Chem. Educ.* 1954, *31*, 618–625.

PARTINGTON, J. R. *A History of Chemistry;* London: Macmillan, 1964; Vol. 2.

PAULING, L. *Nature of the Chemical Bond*, 3rd ed.; Ithaca, NY: Cornell University Press, 1960.

L. S. BARTELL

SI Units

See INTERNATIONAL SYSTEM OF UNITS.

Smell, Sense of

See ODORS.

Smelting

Smelting is a process involving the melting or fusing of ore to provide for separation of metal from a remaining oxide melt called a *slag*. Smelting is usually carried out under reducing conditions to produce a liquid metallic phase, or a liquid sulfide phase called a *matte*.

In copper smelting, copper in the charge in the form of oxide, sulfide, or sulfate is converted to cuprous sulfide, Cu_2S. The affinity of oxygen for iron and sulfur and of sulfur for copper results in the overall reactions

$$2\,CuO + 2\,FeS_2 = Cu_2S + 2\,FeS + SO_2$$

$$Cu_2O + FeS = Cu_2S + FeO$$

In the case of iron smelting, the overall process may be represented by the reactions

$$Fe_2O_3 + 3\,C = 2\,Fe + 3\,CO$$

and

$$2\,Fe_2O_3 + 3\,C = 4\,Fe + 3\,CO_2$$

The principal smelting processes are for the production of matte or of metal.

Matte Smelting

Matte smelting follows roasting, a process in which the sulfur content of the ore is reduced to a level that will form a high-metal-concentration matte (e.g., one that is rich in copper, copper and nickel, or lead) and a slag containing most of the gangue (minerals of no value that contaminate the ore). This slag forms from the gangue in the roasted ore (calcine) when the roaster calcine is melted with a suitable flux. The flux is usually silica, SiO_2, often with a small amount of limestone, which provides CaO to the slag on calcining and increases slag fluidity. The matte contains enough FeS to protect the more valuable metallic sulfides from oxidation, since FeS is one of the most readily oxidized common sulfides. The control of matte composition is based on the roaster calcine, and except for a limited amount of oxidation, the matte smelting process involves melting and separation of the molten slag, which floats on the liquid matte. The temperature of the process must be high enough to keep both phases molten and easily separable by flotation owing to their density differences.

Matte smelting of lump ores is carried out in a blast furnace, or in a reverberatory furnace if the concentrate consists of fine particles. The reverberatory furnace for smelting copper or copper-nickel calcines is a long (often up to 120 feet in length), narrow (30 feet wide) chamber that is 10 feet high. It is lined with magnesite or chrome-magnesite refractories. The reverberatory furnace is equipped with charging facilities, often feed hoppers set in the roof, which are fed by a fettling conveyor. The molten matte and slag phases are generated as the calcines and fluxes are exposed to long, intense flames from burners at one end of the furnace. Matte and slag are tapped almost continuously into ladles at the lower end of the hearth.

In copper smelting, copper matte content varies between 20 and 70 percent. The sulfur content of the matte is in the range of 20 to 25 percent. The remainder of the matte is composed of iron and minor constituents, often including precious metals and other metallic elements that are later recovered. The matte can be assumed to be a mixture of Cu_2S and FeS, although the sulfur content is usually slightly less than the concentration required to account for stoichiometric formation of these compounds. The slag is a complex silicate, primarily $2\,FeO \cdot SiO_2$, although substantial quantities of other oxides may be present. The slag contains some sulfur and some copper, resulting in a copper loss. The concentration of copper is usually a few tenths of a percent in the slag and may be present both as dissolved copper or copper sulfide and as entrained matte.

The flash smelter developed by Outokumpu in Finland represents an attempt to achieve a higher SO_2 level in the process off-gases by minimizing the amount of gas generated by combustion of fuel. It is possible to eliminate added fuels by combusting the fine concentrate with gaseous oxygen, which increases combustion temperatures; this also increases the SO_2 concentration in the process gases. A market for sulfuric acid is an important economic aspect of this smelting process.

Electric smelters minimize the generation of SO_2. The smelting operation is carried out in furnaces similar to submerged arc furnaces, where the carbon electrodes resistively heat the slag rather than arcing on the bath.

Matte smelting is followed by a converting process in which the liquid matte is oxidized to generate an impure metal, an oxide slag, and SO_2-bearing gas. In copper converting, the reactions are

$$FeS + 3/2\ O_2 = FeO + SO_2$$

$$Cu_2S + O_2 = 2\,Cu + SO_2$$

This further step generates additional SO_2-containing gases, which are collected and converted to sulfuric acid, or in other cases are dispersed from very tall chimneys, often creating "acid rain."

Attempts have been made to combine smelting and converting of mattes while generating SO_2-rich off-gases for economic conversion to sulfuric acid. The Noranda process is based on this concept, but early difficulties occurred because smelting and converting reactions do not occur separately in the reactor, and an exiting slag high in copper is generated. In the Mitsubishi continuous smelter there is concurrent flow of slag and matte through three furnaces for smelting, slag cleaning, and converting. This process has accomplished the simultaneous smelting-converting for copper, as has the QSL process for lead.

Metal Smelting

Tin smelting is carried out in a reverberatory furnace in which a reducing agent, such as powdered coal or coke, is charged with tin oxide concentrates to produce metallic tin.

The production of iron, and subsequent conversion to steel, from iron ores is the largest-scale smelting process. Iron smelting is carried out in the iron blast furnace, a large (over 100 feet in height by 35–40 feet in diameter) vertical-shaft furnace. Larger furnaces can produce 12,000 to 15,000 tons of iron per day. Iron ore or iron oxide pellets made from iron oxide concentrate, coke made by heating coal, and limestone as a flux to produce a fluid-refining slag are charged to the top of the furnace. Preheated air is blown through the charge materials, reducing the iron oxide in the ore to liquid metallic iron saturated with carbon (about 4.0–4.5 percent) that contains minor amounts of silicon, manganese, phosphorus, and sulfur.

Significant improvements have occurred over the past several years that have increased performance in terms of higher productivity and reduced coke rates. These advances include higher blast temperature; oxygen enrichment of the blast; fuel injection in the form of hydrocarbon gases, oil, and powdered coal; higher pressure in the furnace; and improvements in charge materials and charging practices.

Recent investigations and large-scale pilot plants are leading to direct smelting processes for iron. The charge materials for these processes are powdered iron ores, coal, and other fuels with gaseous oxygen injection. The products are liquid iron similar in composition to blast furnace iron, slag, and a carbon monoxide–rich off-gas. The great advantages of these processes are smaller equipment, which allows smaller-scale plants and a range of plant sizes, and the use of lowest-cost raw materials.

The lead blast furnace is much smaller than the iron blast furnace and is rectangular in cross section. The principles of operation are similar, but since lead oxide is more easily reduced to metal, it is not necessary to operate the furnace at as high a temperature or under reducing conditions as strong as for the iron blast furnace. Consequently, the blast is not preheated. The charge is lead oxide, which is about 10 percent coke, and a small amount of scrap iron, which reduces any lead sulfide and minimizes the lead content of the slag.

The smelting of lead-zinc ores in the blast furnace for recovery of both lead and zinc metals was considered an impossibility for many years. The imperial smelting process, which is carried out in a blast furnace resembling the lead blast furnace, reduces lead in the standard manner and volatilizes zinc into the stack gases. The zinc is subsequently condensed from these off-gases by absorption in liquid lead. This process offers a considerable advantage in processing ores containing both lead and zinc, a common source for these metals.

The submerged arc furnace is commonly used for the production of ferroalloys of silicon, manganese, or chromium. The reduction of oxides of these elements requires high temperatures and this process is ideal. Coke and wood, which provide some porosity to the bed of charge materials, are the reducing agents, and steel scrap is added to produce the desired alloy concentration in a ferroalloy, as well as lowering the melting temperature of the ferroalloy produced and hence reducing the required process temperature.

BIBLIOGRAPHY

CHOI, K. C.; KIM, H. J.; LEE, Y. H. "The New Q.S.L. Lead Smelter at Korea Zinc's Onsan Plant." In *Metallurgical Processes for the Early Twenty-First Century;* H. Y. Sohn, ed.; Warrendale, PA: Minerals, Metals & Materials Society, 1994; Vol. 2, pp. 317–331.

EVANS, J. W.; DEJONGHE, L. C. *The Production of Inorganic Materials;* New York: Macmillan, 1991; Chapter 8.5, pp. 256–268; Chapter 8.6, pp. 269–273.

GILCHRIST, J. D. *Extraction Metallurgy;* London: Pergamon, 1967; Chapter 10, pp. 210–215.

PEHLKE, R. D. *Unit Processes of Extractive Metallurgy;* New York: American Elsevier, 1973; Chapter 4, pp. 67–82.

ROSENQVIST, T. *Principles of Extractive Metallurgy;* New York: McGraw-Hill, 1974; Chapter 9, pp. 264–297; Chapter 10, pp. 299–321.

THEMELIS, N. J.; ZHAO, B. "Continuous Flow Reactors in Pyrometallurgy." In *Metallurgical Processes for the Early Twenty-First Century;* H. Y. Sohn, ed.; Warrendale, PA: Minerals, Metals & Materials Society, 1994; Vol. 1, pp. 879–896.

R. D. PEHLKE

Soaps and Detergents

The term *soap* is generally considered to refer to the sodium (or potassium) soaps of the natural fatty acids derived from triglycerides. When other cations are involved, it is usual to prefix the cation to the term, i.e., calcium soaps. The term *detergent*, on the other hand, is taken to refer to so-called synthetic detergents. Strictly speaking, detergent means a substance that deterges, i.e., washes or cleanses. Properly speaking, these substances are all called surface-active compounds or surface-active agents. This term has been shortened to *surfactant*.

In the older literature, it was not uncommon to refer to the synthetic surface-active agents as "soaps," a somewhat confusing usage.

Surface-active agents are characterized by a structure in which the molecule is more or less clearly divided into distinct moieties. One moiety is hydrophilic, or water-soluble, the other hydrophobic. In most cases this hydrophobic portion will, in fact, be lipophilic, i.e., exhibit solubility in organic liquids.

Owing to the hydrophilic-lipophilic (or polar) character of these molecules, they possess certain properties that are not characteristic of other types of molecules. In particular, they adsorb at interfaces (gas/liquid, liquid/liquid, liquid/solid, and so forth) to form oriented monolayers. In addition, in solution they may aggregate to form micelles. The usefulness of these surface-active materials stems largely from these two effects.

Classes of Surface-Active Compounds

Surface-active compounds may be divided into the following five principal classes:

1. Anionic
2. Cationic
3. Nonionic
4. Amphoteric
5. Water-insoluble

Only classes 1 through 4 are, strictly speaking, surface-active agents; the last type may include finely divided solids and vegetable gums, e.g., tragacanth and gum arabic. They find their principal application as emulsifying agents, often in conjunction with the more conventional compounds.

The simple classifications may be expanded as follows:

1. Anionic
   a. Carboxylic acids
      (1) Carboxyl joined directly to hydrophobic group
      (2) Carboxyl joined through an intermediate linkage
   b. Sulfuric esters (sulfates)
      (1) Sulfate group joined directly to hydrophobic group
      (2) Sulfate group joined through intermediate linkage
   c. Alkane sulfonic acids
      (1) Sulfonic group directly linked to hydrophobic group
      (2) Sulfonic group joined through intermediate linkage
   d. Alkyl aromatic sulfonic acids
      (1) Hydrophobic group joined directly to sulfonated aromatic nucleus
      (2) Hydrophobic group joined to sulfonated aromatic nucleus through intermediate linkage
   e. Miscellaneous anionic hydrophilic groups
      (1) Phosphates and phosphoric acids
      (2) Persulfates, thiosulfates, etc.
      (3) Sulfonamides
      (4) Sulfamic acids, etc.
2. Cationic
   a. Amine salts (primary, secondary, tertiary)
      (1) Amino group joined directly to hydrophobic group
      (2) Amino group joined through intermediate link
   b. Ammonium compounds
      (1) Nitrogen joined directly to hydrophilic group
      (2) Nitrogen joined through an intermediate group
   c. Other nitrogenous bases
      (1) Nonquaternary bases (e.g., guanidine, thiuronium salts, etc.)
      (2) Quaternary bases
   d. Nonnitrogenous bases
      (1) Phosphonium compounds
      (2) Sulfonium compounds, etc.
3. Nonionic
   a. ether linkage to solubilizing groups
   b. Ester linkage
   c. Amide linkage
   d. Miscellaneous linkages
   e. Multiple linkages
4. Amphoteric
   a. Amino and carboxy
      (1) Nonquaternary
      (2) Quaternary
   b. Amino and sulfuric ester
      (1) Nonquaternary
      (2) Quaternary
   c. Amino and alkane sulfonic acid
   d. Amino and aromatic sulfonic acid
   e. Amino and aromatic sulfonic acid
   f. Miscellaneous combinations of basic and acidic groups
5. Water-insoluble emulsifying agents
   a. Ionic hydrophilic group
   b. Nonionic hydrophilic group

Figure 1 gives examples of various types of surface-active agents; the dashed line separates the lipophilic and hydrophilic moieties of the molecule. Commercially, the most important surfactants are the anionic and nonionic agents, with the cationic agents a somewhat distant third. Amphoteric agents, although of considerable interest, have more restricted uses.

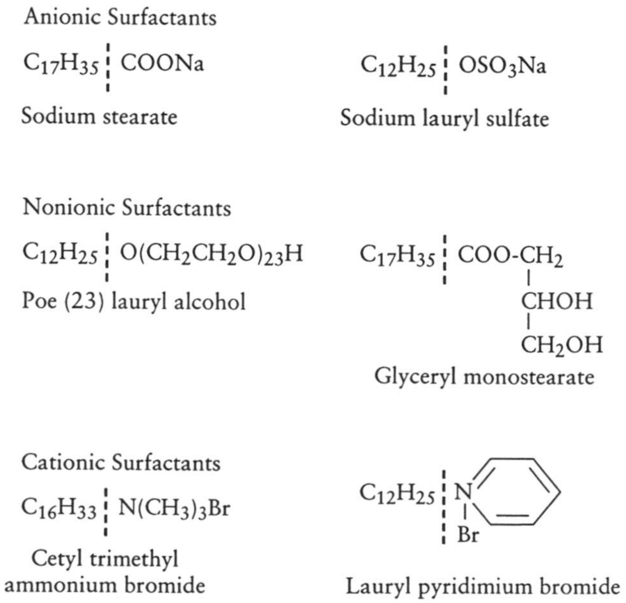

Anionic Surfactants

$C_{17}H_{35}$ ┊ COONa

Sodium stearate

$C_{12}H_{25}$ ┊ OSO_3Na

Sodium lauryl sulfate

Nonionic Surfactants

$C_{12}H_{25}$ ┊ $O(CH_2CH_2O)_{23}H$

Poe (23) lauryl alcohol

$C_{17}H_{35}$ ┊ COO-CH$_2$
                   |
                 CHOH
                   |
                 CH$_2$OH

Glyceryl monostearate

Cationic Surfactants

$C_{16}H_{33}$ ┊ $N(CH_3)_3Br$

Cetyl trimethyl ammonium bromide

$C_{12}H_{25}$ ┊ N⟨⟩
                  |
              ┊ Br

Lauryl pyridimium bromide

Fig. 1. Structural formulas of the various types of surfactant. The portion of the molecule to the left of the dashed line is the lipophilic moiety; that to the right, the hydrophilic.

Surface Properties

As is well known, the surface of a liquid or the interface between two liquids is characterized by the existence of a surface (or interfacial) tension. The surface tension may be thought of as arising from the unbalanced forces acting on the liquid molecules in the neighborhood of the surface or interface, as represented schematically in Figure 2.

The addition of a solute to a liquid may affect the surface tension in varying ways. In the first instance, there may be a small increase in the surface tension. This will be evidenced with nonpolar organic molecules (e.g., glycerine). A second effect, found usually with simple inorganic electrolytes, is a small depression of the surface tension. Finally, a dramatic effect, resulting in a substantial depression of surface or interfacial tension, is found when the surfactant compounds are the solute. Figures 3 and 4 show schematically the range of surface and interfacial tension lowering for the usual types of surface-active agents. It should be noted that the break in the curves is more pronounced with actual surfactants.

The lowering of the surface tension arises from the strong surface adsorption of the polar surface-active molecules (i.e., the concentration of surface-active molecules in the surface layer is much higher than in the bulk of the solution). This occurs because of the ability of the polar molecules to orient at the surface or interface with the lipophilic portion in the nonaqueous phase (air or oil). As a consequence of this, the original water surface—which is a surface of high energy, as evidenced by its high surface tension (72.8 mN/m)—is reduced to the normally much lower energy of the hydrocarbon (18–20 mN/m).

The extent of the adsorption of the surfactant at the interface is given by the well-known adsorption isotherm of Gibbs:

$$\Gamma = \frac{c}{RT}\frac{d\gamma}{dc}$$

where γ is the surface tension, c is the bulk concentration of surface-active agent (more precisely, the activity a), R and T have their usual meanings, and Γ is the surface concentration, in moles per square centimeter.

Thus, the more steeply the surface tension decreases with increasing concentration of surface-active agent, the greater the surface concentration. The Gibbs equation governs the portion of the surface tension curve in which the surface tension is decreasing (i.e., negative $d\gamma/dc$), as shown in the typical surface tension/concentration curve in Figure 3. It will be noted that the slope of the curve abruptly changes

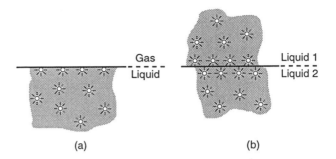

Fig. 2. Forces acting on the molecules at the interface: (a) gas-liquid; (b) liquid-liquid.

to zero (approximately) at some definite concentration characteristic of the particular surface-active agent being studied. At this point, the Gibbs equation does not apply because the nature of the interface and of the surfactant in the bulk has changed. At this concentration, the layer of adsorbed surfactant covers the surface in a monomolecular layer (or monolayer). Having calculated the value of Γ at this point

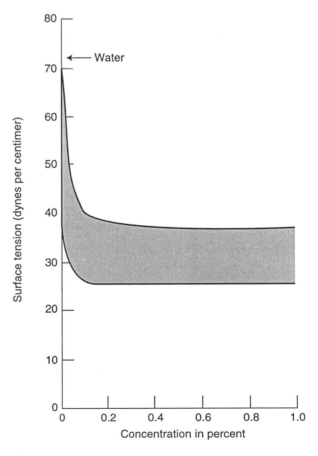

Fig. 3. Range of values of surface tension found in aqueous solutions of most surfactants. Note that in real surface tension curves the break is generally more abrupt.

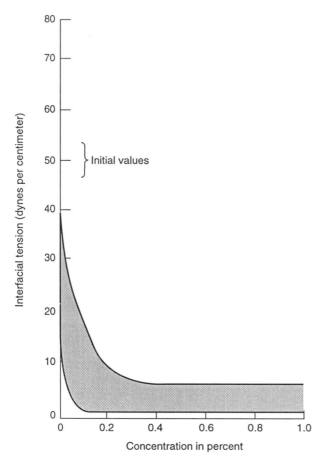

Fig. 4. Range of values of interfacial tension found in aqueous versus nonaqueous phases containing surfactant for most surfactants. Note that in real interfacial tension curves the break is generally more abrupt.

of their polar structure, surfactants possess only a limited solubility in either water or organic liquids. Thus, when their solubility limit is reached, rather than precipitating, the surfactant molecules produce oriented aggregates, called micelles, in which the portion of the molecule that is compatible with the solvent is directed outward. In recent years, this type of action has been recognized as an example of a more general phenomenon known as self-assembly.

Initially, it was assumed that micelles were essentially spherical, but it has been shown that a number of other, more asymmetric forms are possible.

As noted above, the concentration at which micelle formation begins, corresponding to the break in the surface-tension curve, is characteristic of the particular surface-active compound. This concentration is referred to as the *critical micelle concentration*. It is usually represented by cmc or CMC. The range of concentrations over which the CMC values vary is suggested by the small number of characteristic values (chosen at random) in Table 1.

Applications of Soaps and Detergents

The phenomena of adsorption and micellization are responsible for the well-known behavior of surfac-

from the Gibbs equation, it is possible to determine the cross-sectional area of the surfactant at the surface or interface.

The break concentration also corresponds to the formation of molecular aggregates by the surfactant, or micelle concentration. This topic is discussed in the following section.

In some cases the surface-tension concentration has been found to show a minimum, rather than a sharp, break. In the past, an attempt was made to explain this phenomenon as a failure of the Gibbs equation, but it is now known to be due to the presence of small amounts of impurities in the surface-active agent.

MICELLE FORMATION

Micelle formation is another example of the ability of polar surfactants to orient themselves. At very low concentrations, the surface-active agent is in true solution, like any simple compound. However, because

Table 1. Critical Micelle Concentrations of Selected Surface-Active Agents

Compound	CMC (mol/L $\times$ 10^{-3})
Anionic	
Sodium laurate	24.0
Sodium myristate	6.0
Sodium lauryl sulfate	6.5
Sodium myristyl sulfate	1.6
Cationic	
Lauryl trimethylammonium chloride	16.0
Lauryl trimethylammonium bromide	16.0
Cetyl trimethylammonium bromide	1.3
Nonionic	
Polyoxyethylene (23) lauryl alcohol	0.091
Polyoxyethylene (9.5) nonyl phenol	0.085
Polyoxyethylene (30) nonyl phenol	0.280

Table 2. Projected U.S. Soap and Detergent Demand 1994–2000

Millions of lb	1994	2000	% annual growth (1994–2000)
Anionic	4,804	5,595	2.6
Nonionic	1,890	2,240	2.9
Cationic	700	850	3.5
Amphoteric	53	75	6.0
Total	7,447	8,770	2.8

tants—e.g., detergency or cleansing, solubilization, foaming, stabilization of suspensions and emulsions, and so on. There is almost no area in industry where surface-active agents do not have an application; the widespread literature supports this.

INDUSTRIAL PRODUCTION

The worldwide production of soaps and detergents is a major industry. Table 2 indicates that between the years 1994 and 2000 the U.S. demand will increase at an annual rate of 2.8 percent. Note that, as suggested above, anionics constitute the major component, staying constant at 6.4 percent over the interval.

BIBLIOGRAPHY

KARSA, D. R., ed. *Industrial Applications of Surfactants;* London: Royal Society of Chemistry, 1987.

PORTER, M. R. *Handbook of Surfactants;* New York: Chapman & Hall, 1991.

ROSEN, M. J., ed. *Surfactants in Emerging Technologies (Surfactant Science 26);* New York: Dekker, 1987.

SAMPERSTONE, V. J. *Soaps and Surface Active Agents: Index of New Information;* Annandale, VA: ABBE Publications Association of Washington, D.C., 1992.

SPITZ, L., ed. *Soap Technology for the Nineteen Nineties;* Champaign, IL: American Oil Chemists' Society, 1990.

WASAN, D. T.; GINN, M. E.; SHAH, D. O., eds. *Surfactants in Chemical/Process Engineering (Surfactant Science 28);* New York: Dekker, 1988.

PAUL BECHER

Soddy, Frederick (1877–1956)

The originator of the concept of ISOTOPES, Frederick Soddy, was born on September 2, 1877, in Eastbourne, England. His mother died when he was two years old, and he was raised by his half-sister. He was interested in science early in life, and his science master at Eastbourne College encouraged him to study chemistry at Oxford. He studied under William RAMSAY and earned a first-class honors degree in chemistry in 1898. He spent the next two years in independent chemical research at Oxford.

In 1900, Soddy accepted a position as demonstrator at McGill University in Montreal, Canada. This brought him together with Ernest RUTHERFORD, and together they developed the disintegration theory of radioactive elements. The idea that one radioactive element could change itself into another element by emission of particles was revolutionary. It meant abandoning one of the principal assumptions of traditional chemistry: that one element could not be changed into another.

In 1903, Soddy returned to England to work with Ramsay on predictions of the disintegration theory. He showed that the disintegration of radium produced helium, which was the first time that the production of one element by another was documented. In 1904, he was appointed lecturer at the University of Glasgow, where he further developed and tested the disintegration theory. Going against traditional chemical assumptions once again, he stated in 1910 that "elements of different atomic weight may possess identical chemical properties." Further research with nonradioactive elements prompted him to propose, in 1913, the concept of isotopes: atoms of the same element that have different atomic weights.

While at Glasgow, Soddy formulated his displacement law, which used the type of emission (alpha or beta) to predict the products of a radioactive element's decay. In 1914, Soddy became professor of chemistry at the University of Aberdeen. There he demonstrated that lead actually consisted of a number of isotopes, as predicted by the displacement law.

In 1919, Soddy was appointed professor of chemistry at Oxford with the expectation that he would establish a school of radiochemistry at the university. Instead, he worked hard to improve teaching standards and modernize laboratory facilities. He received the 1921 Nobel Prize for chemistry "for his important contributions to our knowledge of the chemistry of radioactive substances and his investigations into the origins and nature of isotopes."

After winning the Nobel Prize, Soddy turned away from chemistry and toward economic, social, and political theory, after coming to believe that scientists were not considering the social impact of their research. He wrote that the blame for the downfall of civilization "must rest on scientific men, equally with others, for being incapable of accepting the responsibility for the profound social upheavals which their own work primarily has brought about in human relationships."

In 1908, Soddy married Winifred Bilby, and they shared a happy and stable relationship. Her death in 1936 prompted him to resign from Oxford, and he retired to Brighton, England. He died there on September 22, 1956.

BIBLIOGRAPHY

PANETH, F. "A Tribute to Frederick Soddy." *Nature* 1957, *180*, 1085–1087.

TRENN, T. In *Dictionary of Scientific Biography*, C. C. Gillispie, ed.; New York: Scribner, 1975; Vol. 10, pp. 504–509.

KATHLEEN HOLLEY

Sodium

11
Na
22.9898

Melting Point: 371 K; **Boiling Point:** 1156 K; **Density:** 0.971 g/cm^3; **Most Common Ion:** +1

Sodium is the seventh most abundant element in the Earth's crust. The elemental symbol derives from the Latin word *natrium*, for sodium. The element does not occur free in nature, but has been used throughout human history. The word *salary* derives from the use of salt (sodium chloride) by Romans to pay wages. The element was isolated in 1807 by Humphry DAVY shortly after he isolated potassium, using electrolysis with fused sodium hydroxide.

Sodium occurs widely in deposits as rock salt ($NaCl$), trona (Na_2CO_3), saltpeter ($NaNO_3$), mirabilite (Na_2SO_4), and borax. Additionally, sodium occurs in ocean waters in essentially unlimited amounts. Sodium can be produced through chemical reduction or electrolytic techniques. The first industrial process by Deville involved reduction of sodium carbonate by carbon at high temperatures. Since the late 1890s, sodium has been produced by electrolytic techniques. The Downs Cell, patented in 1924 by J. C. Downs, utilizes sodium chloride and also produces elemental Cl_2.

Sodium is a soft, shiny gray metal and is quite reactive. The element is used in the reduction of certain metal halides—for example, titanium or zirconium. The most important alloy of sodium is NaK, used as a heat-transfer agent in nuclear reactors (see POTASSIUM).

Most of sodium's importance arises from its compounds. Sodium carbonate is used in pulp and paper mills and in manufacturing glass. Sodium hydrogen carbonate is used in foods, fire extinguishers, and pharmaceuticals. Sodium compounds are used in the sodium vapor lamps used for street lighting. Sodium chloride has hundreds of uses including the melting of snow and ice on roads, the preservation of meats, and the delivery of minerals in animal diets. Soaps are typically sodium salts of fatty acids. Sodium nitrate is used as a fertilizer. Sodium iodide has been used to treat certain animal conditions, for example, chronic bronchitis and goiter. The sodium ion has also been shown to be essential to human life.

BIBLIOGRAPHY

GREENWOOD, N. N.; EARNSHAW, A. *Chemistry of the Elements*; New York: Pergamon, 1984; pp. 75–85.

LEMKE, C. H. "Sodium and Sodium Alloys." In *Encyclopedia of Chemical Technology*, 3rd ed.; R. E. Kirk, D. F. Othmer, M. Grayson, eds.; New York: Wiley-Interscience, 1981; Vol. 21, pp. 181–204.

LIDE, D. R., ed. *Handbook of Chemistry & Physics*, 71st ed.; Cleveland: Chemical Rubber Co., 1990–1991.

ALTON J. BANKS

Solubility

Solubility refers to the extent to which two pure components mix to form a homogeneous system (SOLUTION). The component present in lesser amount is referred to as the *solute*, while the one in greater amount is called the *solvent*. The most common solubilities of interest are those of solids, liquids, or gases in a liquid solvent. However, solutions of gases in other gases (e.g., air), gases in solids (e.g., H_2 in Pd), liquids in solids (e.g., Hg in Ag), and solids in solids (e.g., Cu in Ag) are well known. Since no irreversible chemical reactions occur, solution components can be separated by physical means such as by evaporating or distilling the more volatile component. Like other chemical and physical phenomena, solubility is controlled (at fixed T) by the free-energy changes: $\Delta G_{sol'n} = \Delta H_{sol'n} - T\Delta S_{sol'n}$. The more negative $\Delta G_{sol'n}$, the more soluble is the solute in the solvent. The enthalpy term $\Delta H_{sol'n}$ reflects the change in potential energy of interaction between particles in solution compared with those in the pure components, while $\Delta S_{sol'n}$ measures the increase in disorder of the solution as compared with the pure components. So-called ideal solutions have $\Delta H_{sol'n} = 0$ and are formed by substances whose attraction for each other is the same as attractions among the identical particles of the pure substances; such substances are generally nonpolar molecular species for which only van der Waals forces operate. An example is a solution of benzene in toluene. Liquids that form IDEAL SOLUTIONS are miscible in all proportions. For ideal solutions, solubility is driven only by the positive value of $\Delta S_{sol'n}$. It can be shown that for an ideal solution of two liquids,

$$\Delta G_{sol'n} = RT(x_1 \ln x_1 + x_2 \ln x_2) = -T\Delta S_{sol'n}$$

where x_i is the mole fraction of the ith component. $\Delta G_{sol'n}$ (at 298 K) is minimized at -1674 kJ for $x_1 = x_2 = 0.5$. For solids that form ideal solutions, solubility is independent of the identity of the solvent; its variation with temperature is given by

$$\ln x = \frac{\Delta H_{fusion}}{R} \frac{(T_{fusion} - T)}{TT_{fusion}}$$

where x is the mole fraction of solute (approximating the activity a), assuming that ΔH_{fusion} is a constant. Since the heat of fusion is always positive, solubility is predicted to increase with temperature for $T < T_{fusion}$. The ideal solubility of gases is given by an analogous equation,

$$\ln x = \frac{-\Delta H_{vap}}{R} \frac{(T_b - T)}{TT_b}$$

where T_b is the boiling temperature and ΔH_{vap} the heat of vaporization, a positive quantity. The ideal solubility of a gas is predicted to decrease for $T > T_b$. This is generally true, even for nonideal solutions, since any favorable interaction energies between dissolved gas molecules and solvent do not outweigh the entropy decrease on passing from the gaseous state to confinement in solution (see also NONIDEAL GASES).

More commonly, interactions between species in solution differ from those in each separate pure substance, resulting in nonideal solutions having $\Delta H_{sol'n} \neq 0$. Hildebrand derived an expression for $\Delta G_{sol'n}$ of a nonideal solution of two pure liquids:

$$\Delta G_{sol'n} = RT(x_1 \ln x_1 + x_2 \ln x_2) + V(\delta_1 - \delta_2)^2 \phi_1 \phi_2$$

where V is the solution volume, ϕ_i the volume fraction, and δ_i a solubility parameter of the ith component. δ is related to the cohesive energy of the liquid and is defined as

$$\delta = \left(\frac{\Delta E_{vap}}{molar\ volume}\right)^{1/2} = \left(\frac{\Delta H_{vap} - RT}{molar\ volume}\right)^{1/2}$$

δ is related to intermolecular forces in the liquid. In Hildebrand's expression $\Delta H_{sol'n} = V(\delta_1 - \delta_2)^2 \phi_1 \phi_2$,

it reflects the energy expenditure to make holes in one liquid to accommodate molecules of the other and takes no account of specific interactions between the two different types of molecules. The more different the liquids (as expressed in the different values of δ) the more positive will be $\Delta H_{sol'n}$ and the less negative $\Delta G_{sol'n}$. Typical values of δ range from 33–60 (J/mL)$^{1/2}$, so the components need not be very different in order that complete miscibility be decreased to only partial miscibility. Because T does appear in the entropy term but not in the enthalpy term, heating increases solubility. Similar considerations apply to solutions of nonpolar solids such as SnI_4 in nonpolar solvents such as CCl_4. However, no independent way of determining δ for solids is available, and the values must be derived from measured solubilities.

In most solutions encountered, additional specific interactions such as SOLVATION by polar solvents, H-bonding, and dissociation of ionic solutes are important. (Attempts have been made to modify Hildebrand's equation to take account of specific interactions in solution.) Generally, solubilities are low when the forces experienced by species in solution are very different from the forces experienced in the pure components. For example, a polar solvent dissolves little of a nonpolar solute. The energy required to disrupt the mutual attractions of polar molecules is not compensated for by their interactions with nonpolar solute molecules introduced into the matrix. Similarly, a polar solute has low solubility in a nonpolar solvent. On the other hand, if both solute and solvent are polar (or both nonpolar), the energies of interaction in solution may be quite comparable to energies required to disrupt the structures of the pure components. These considerations are the basis for the solubility rule "like dissolves like."

An especially important polar solvent is water, in which geochemical and physiological reactions occur as well as many industrial and laboratory reactions. Pure water has an ordered structure on account of H-bonds that continually migrate; the energy involved is ~25 kJ/mol of H-bonds. Solutes that disrupt H-bonding are not very water-soluble; for example, only about 0.285 g/L of I_2 dissolves in water at 20°C. Nonpolar organic compounds such as CH_4 are also not appreciably soluble in water. A rule of thumb is that one O atom (capable of forming H-bonds) for every two or three C atoms leads to water solubility. For example, dioxane ($C_4H_8O_2$) is completely miscible with water, while diethyl ether [$(C_2H_5)_2O$] is only moderately soluble (75 g/L). Because water is so polar, it readily dissolves many ionic substances. The BORN-HABER CYCLE is helpful in understanding factors contributing to $\Delta H_{sol'n}$:

$$MX(s) \xrightarrow{U_0} M^+(g) + X^-(g)$$

$$\Delta H_{sol'n} \searrow \quad \downarrow \Delta H_{M^+} \quad \downarrow \Delta H_{X^-}$$

$$M^+(aq) + X^-(aq)$$

$$\Delta H_{sol'n} = U_0 + \Delta H_{M^+} + \Delta H_{X^-}$$

where U_0 is the lattice energy of the salt; ΔH_{M^+} and ΔH_{X^-} are the heats of solvation of the cation and anion, respectively. Highly polar water molecules orient themselves around ions; the smaller and more highly charged the ion, the more highly concentrated its charge and the more negative the heat of solvation. Lattice energies follow a similar trend, being larger the smaller the interionic distance (all other things being equal). U_0 is > 0 while ΔH_{M^+} and ΔH_{X^-} are < 0; so $\Delta H_{sol'n}$ may turn out either positive or negative in a particular case. For example, ammonium salts are very water-soluble; but the solution feels cold because $\Delta H_{sol'n} > 0$. In a series of related salts such as the sodium halides, both U_0 and the solvation energies display regular trends; however, $\Delta H_{sol'n}$ involves small difference between large numbers and does not necessarily give a parallel trend. In addition to $\Delta H_{sol'n}$, solubilities depend on $\Delta S_{sol'n}$. Destruction of the CRYSTAL LATTICE always involves a large favorable entropy contribution; on the other hand, ions orient solvent molecules giving a negative contribution. The net result is generally that $\Delta S_{sol'n} > 0$ for hard-hard cation-anion pairs (except for small highly charged ions) and $\Delta S_{sol'n} < 0$ for soft-soft pairs. Thus, we can understand why NaCl is soluble, but MgO, which has the same solid structure, is insoluble. Smaller, more highly charged ions make U_0 for MgO much greater than for NaCl, making $\Delta H_{sol'n}$ more positive while they also better orient water molecules, making $\Delta S_{sol'n}$ less positive.

Finally, the nature of the solvent environment can be changed very drastically by the presence of other dissolved species. Large quantities of dissolved salts provide ions that may pair with oppositely charged ions of a newly dissolving salt, thus increasing its solubility. In fact, the picture of ions totally surrounded by water molecules is valid only for very dilute solutions of salts with low ionic charges. In other cases appreciable ion pairing exists. For example, in a 0.44M solution of $MgCl_2$, 70 percent of the cations are present as Mg^{2+} and 30 percent as $MgCl^+$.

BIBLIOGRAPHY

DASENT, W. E. *Inorganic Energetics: An Introduction*, 2nd ed.; Cambridge, U.K.: Cambridge University Press, 1982; Chapter 5.

Handbook of Chemistry and Physics; Cleveland, OH: Chemical Rubber Publishing Co., 1995, 1996; pp. 3-1ff, 4-35ff.

HAWKES, S. J. "Salts Are Mostly NOT Ionized." *J. Chem. Educ.* 1996, *73*, 421–423.

JANDER, J.; LAFRENZ, C. *Ionizing Solvents;* London: Wiley, 1970.

SHINODA, K. *Principles of Solution and Solubility;* New York: Marcel Dekker, 1978.

JOHN J. ALEXANDER

Solutions

A solution is a mixture of one or more pure substances (elements or compounds) in which the particles of those substances, the constituents, are of molecular or ionic size. In contrast to other mixtures, the constituents of a solution will not spontaneously separate. Liquid solutions may be colored but are commonly clear rather than cloudy or milky. Solutions are homogeneous: the composition and properties do not change throughout a sample. However, the composition of a solution is variable; that is, the amount of any constituent may be changed, at least within limits. When one substance, X, mixes with another substance, Y, to form a solution, X is said to dissolve in Y and to be soluble in Y. If the amount of a substance that can dissolve in another is limited, that limit is called its solubility.

Solutions may be gaseous, liquid, or solid, as may their constituents. All gases are soluble in one another in all proportions. Some classes of solutions have special names. Solutions of metallic elements in one another are called alloys. Solutions in which the major constituent is water are called aqueous. Most of the complex reactions that make life possible take place in aqueous solutions, within cells or in various body fluids.

The constituents of a solution are often described as a solvent and one or more solutes. This distinction is somewhat arbitrary. Usually the most abundant constituent is regarded as the solvent, but there are exceptions. For example, in liquid solutions where water is a major constituent it is usually considered the solvent regardless of the amount of another material, say a salt or acid, present.

In the laboratory, in industry, and in nature, chemical reactions can take place in liquid solution that do not occur between the pure substances. One reason for this is that dissolved particles, being in constant motion, can easily come into contact and can rotate so as to bring reactive parts of the molecules of different species together. When these species do react, excess energy may be released. This energy may then be taken up and carried away by surrounding molecules and is not available for the reverse reaction.

Another way the solution state makes some chemical processes feasible is by the moderation or slowing down of very rapid or violent reactions. A reaction is often (but not always) slowed if the concentration of one or more reactants is decreased. The extent to which a reaction occurs in solution and the reaction rate are both affected by a host of other conditions including temperature and precise solution composition. By careful control of concentrations and other conditions a chemist can cause one reaction to be favored over competing ones.

The solution state is also useful in chemical analysis. Reactions in solution are commonly used for separation and identification of substances. In addition, many instrumental methods—e.g., absorption and nuclear resonance spectrometry—often are best applied to dilute solutions.

H. BRADFORD THOMPSON

Solvation

When a substance called a solute is placed in a solvent, the solvent molecules interact with the solute and the solute is solvated. This generalized interaction is termed *solvation*, whether it occurs in a highly polar solvent like water or a nonpolar solvent like a hydrocarbon. When water is the solvent, the term HYDRATION is used instead of solvation. The forces involved in the interaction are a function of both the solute and the solvent and can range from very weak interactions to very strong interactions. In many systems, Lewis acid-base electron pair donation/acceptor interactions predominate. For example, a polar substance added to a highly polar solvent like water will become solvated because of the interaction between the permanent DIPOLEs of the solute and solvent. These interactions may be quite strong. When a nonpolar species dissolves in a nonpolar solvent, solvation may occur because of the weaker covalent interactions between the molecules. These two examples may be considered the limiting cases. The solvation of the solute is in competition with the interactions among solvent molecules themselves, and the solute may first have to weaken the solvent-solvent structure before solvation can occur. The observation that polar molecules tend to solvate polar species and nonpolar molecules tend to solvate non-

polar species leads to the common observation that "like dissolves like" in describing the solubilities of solutes in solvents.

Although solvation occurs in any solvent system, much more is known about solvation in water SOLUTIONS. When an ionic substance such as a salt is added to water, energy from the solvent is required to break the ionic interactions holding the salt ANIONS AND CATIONS together (the lattice energy). Because the formation of solvated ions releases energy, whether or not a salt dissolves is determined by the relative magnitudes of the quantities. For example, NaCl dissolves because the energy released by the formation of the hydrated sodium ions and the hydrated chloride ions is more than sufficient to overcome the ionic interactions in the solid. For AgCl, however, the reverse is true, and AgCl is not very soluble in water. If we compare the results in liquid AMMONIA, another solvent system, NaCl is still more soluble than AgCl, but the solubility of AgCl increases by four thousand compared with that in water, whereas the solubility of NaCl decreases by a factor of three in liquid ammonia. The enhanced solubility of AgCl in ammonia occurs because of the solvation of the silver ion by the NH_3 molecules to form stable species called *complexes*.

Dissolving a material like $Cr(NO_3)_3$ in water results in the formation of a six-coordinate octahedral complex, $Cr(H_2O)_6^{3+}$, and solvated nitrate ions. Although the subsequent reaction with another species like SCN^- is written in the form of an apparent addition reaction, $Cr^{3+} + SCN^- = CrSCN^{2+}$, this is actually incorrect. The chemical reaction that occurs is a substitution reaction between the solvated Cr^{3+} ion and the thiocyanate ion: $[Cr(H_2O)_6]^{3+} + SCN^-$ (aq) = $[Cr(H_2O)_5SCN]^{2+}$, where the symbol (aq) is used for the solvated thiocyanate ion and the square brackets denote a complex. This example is chosen because we know the structure of the solvated chromium ion to be a six-coordinate octahedral species. The water molecules bound to the Cr^{3+} are exchanged with the solvent in a process called the cation-solvent exchange reaction, $[Cr(H_2O)_6]^{3+} = [Cr(H_2O)_5]^{3+} + H_2O$, and this reaction has a HALF-LIFE of more than a day at room temperature. Other cations undergo this process more rapidly, sometimes in as short a time as 10^{-9} seconds. In the case of Cr^{3+} the slow step in the reaction with thiocyanate is the rate of solvent exchange, and hence the overall substitution reaction cannot be more rapid than the cation-solvent exchange process. When the solvent exchange rate is slow, it is sometimes possible to actually determine the structure of the solvated ions by using modern techniques like NUCLEAR MAG-

NETIC RESONANCE or neutron diffraction. The interaction between a metal ion and solvent molecules not only is important in simple systems but also plays a role of the mechanisms by which metalloenzymes operate in living organisms.

Solvation is also important when nonionic but polar molecules are added to water. For example, simple sugars like glucose are polar covalent substances containing carbon, hydrogen, and oxygen atoms. The oxygen atoms interact with the hydrogen atoms in the water molecules via hydrogen bonding, and the sugar molecules become solvated. These solvation interactions are quite strong, and sugar is very soluble in water.

Although the solute-solvent interactions are usually weaker in nonpolar solvents, solvation is still important. For example, modern gasoline is not just a liquid hydrocarbon, octane, but rather a complex mixture of solvated species held in solution by the interactions between the additives and the solvent molecules. Because the additives are put into gasoline for particular reasons, it is critical that they dissolve in the nonpolar mixture—or they would cause breakdowns in the engine during combustion. The choice of a nonaqueous solvent system in carrying out reactions is very complex, depending not only upon the solvation to selectively dissolve or precipitate out the reactants and products but also on the acidity or basicity of the solvent molecules.

The description of how soap or detergents work is a good summary of the principles involved in solvation. Much of the dirt on our clothes is nonpolar, such as grease and grime, and these materials are not very soluble in water, but they are soluble in nonpolar solvents. A soap can be the sodium salt of a large-chain organic acid, which is mostly a nonpolar organic species coupled to an ionic tail (the acid salt portion of the molecule). The long-chain hydrocarbon portion solvates the nonpolar dirt, whereas the ionic end dissolves in water, and when the water is drained out, the dirt-soap complex is removed. In the presence of hard water containing calcium or magnesium ions, the soap-dirt entity is insoluble in water and remains behind.

BIBLIOGRAPHY

HUHEEY, J. E.; KEITER, E. A.; KEITER, R. L. *Inorganic Chemistry: Principles of Structure and Reactivity,* 4th ed.; New York: HarperCollins, 1993.

ROBINSON, R. A.; STOKES, R. H. *Electrolyte Solutions,* 2nd ed.; London: Butterworths, 1965.

HERBERT B. SILBER

Solvents

By convention, the solvent is the component present in the greatest proportion in the homogeneous mixtures of pure substances called solutions. Components of such mixtures present in minor proportions are called solutes. Thus, technically, homogeneous mixtures are possible with liquids, solids, or gases dissolved in liquids, solids in solids, and gases in gases. In common practice this terminology is applied mostly to liquid mixtures for which the solvent is a liquid and the solute can be a liquid, solid, or gas.

Three broad classes of solvents are recognized— aqueous, nonaqueous, and organic. Formalistically, the nonaqueous and organic classifications are both not aqueous, but the term *organic solvents* is generally applied to a large body of carbon-based compounds that find use industrially and as media for chemical synthesis. Organic solvents are generally classified by the functional groups that are present in the molecule, for example, alcohols (isopropyl and ethyl alcohol), halogenated hydrocarbons (chloroform and carbon tetrachloride), or hydrocarbons (hexane and benzene); such groups give an indication of the types of physical or chemical interactions that can occur between solute and solvent molecules. Nonaqueous solvents are generally taken to be inorganic substances and a few of the lower-molecular-weight, carbon-containing substances such as acetic acid, methanol, and dimethylsulfoxide. Nonaqueous solvents can be solids (e.g., fused LiI), liquids (H_2SO_4), or gases (NH_3) at ambient conditions; the solvent properties of the first-named substance are manifested in the molten state, whereas the last-named substance must be liquefied to act as a solvent.

Nonaqueous Solvents

The classification, often arbitrary, of nonaqueous solvents can be made on the basis of a variety of factors. Excluding utilitarian considerations and classifications based on chemical character, that is, the presence of distinctive groups such as a carbonyl function in the molecule, useful classification schemes involve the protophilic nature of the solvent and its solvating power. Classification schemes involving nonaquous solvents inevitably involve considerations of acid-base phenomena. Of the two major theories, the BRØNSTED-LOWRY protonic concept has been the most useful because the early nonaqueous solvents of interest were invariably potential proton donors. However, the LEWIS theory of acidity has become very useful for understanding solution phenomena in aprotic systems such as SO_2.

The ability of a solvent to form a so-called onium (positively charged) species is an important factor in defining the nature of the solution phenomena it will support. Onium species can be formed by reaction with a potential proton donor or by self-ionization of the solvent. Four classes of solvents are generally recognized according to their ability to coordinate with the proton—basic, acidic, aprotic, and amphiprotic solvents.

Basic Solvents

Such solvents form the onium species most readily and are generally derivatives of ammonia (amines, hydrazines, pyridine, and so on) or water (alcohols and ethers). Hydrogen-containing species in this classification can undergo self-ionization to form the onium species, for example,

$$2\,NH_3 \rightleftarrows NH_4^+ + NH_2^-$$

but the other substances in this class require the presence of a proton donor to do so,

$$R_2O + HX \rightleftarrows R_2OH^+ + X^-$$

Acidic Solvents

These solvents exhibit a greater tendency to release protons than do basic solvents, and they form onium species with great reluctance. However, it is possible for acidic solvents to undergo self-ionization, for example,

$$2\,CH_3CO_2H \rightleftarrows CH_3CO_2H_2^+ + CH_3CO_2^-$$

to form the corresponding onium species. Even though a solvent may be classified as acidic, for example, CH_3CO_2H, it is possible to protonate such molecules with a more strictly acidic substance. Thus, the strong mineral acids such as HCl ionize in anhydrous acetic acid to form the onium species,

$$CH_3CO_2H + HCl \rightleftarrows CH_3CO_2H_2^+ + Cl^-$$

Aprotic Solvents

Solvents such as SO_2, also commonly called inert, have very little affinity for protons, and they are incapable of dissociating to give protons. Such aprotic solvents are also called indifferent, nondissociating, or nonionizing.

Amphiprotic Solvents

These solvents are capable of either adding or donating protons. Thus ammonia is amphiprotic because it can lose or accept a proton as shown in the following reactions:

$$(C_6H_5)_3C^- + NH_3 \rightleftarrows (C_6H_5)_3CH + NH_2^-$$
$$NH_3 + HX \rightleftarrows NH_4^+ + X^-$$

The classification of a substance as amphiprotic adds little to understanding solvent phenomena because this designation depends only on the relative strengths of the aids or bases involved. It has always been possible to produce a medium sufficiently acidic to protonate even acidic solvents, which, of course,

Table 1. Autoprotolysis Constants of Some Common Solvents at 25°C

Solvent	Log K_s	Dielectric constant (ε)
H_2O	14.0	78.4
H_2O_2	13	84.2
CH_3OH	16.9	32.6
H_2S	34.5 (−78°C)	8.99 (−78°C)
H_2SO_4	3.6	101
HCO_2H	6.2	58.5 (16°C)
$H_2NC_2H_4OH$	5.1 (20°C)	37.7
$(CH_3)_2SO$	32	46.6
$HCON(CH_3)_2$	>21	36.7
CH_3CN	28.5	36.0
C_2H_5OH	19.1	24.3
$H_2NC_2H_4NH_2$	15.3	14.2 (20°C)
CH_3CO_2H	14.5	6.1 (20°C)

Solvents

also would be expected to be good proton donors to bases. From this point of view, anhydrous acetic acid is considered amphiprotic. Whether a substance is an acid or base depends upon the character of the solvent in which it is dissolved. For example, urea, which is very weakly basic in water, is a weak acid in the more basic solvent ammonia and a strong base in the acidic solvent CH_3CO_2H.

Amphiprotic solvents exhibit a well-defined self-dissociation, as in the case of H_2S,

$$2\,H_2S \rightleftarrows H_3S^+ + HS^- \; (K_s = a_{H_2S^+} \cdot a_{HS^-})$$

for which an autoprotolysis constant K_S can be measured (see Table 1). It is generally difficult to measure K_S for nonaqueous solvents because apparent self-ionization can arise from the presence of traces of water or other amphiprotic contaminants.

Chemical Behavior

Solvents intervene in chemical process by producing species from solutes that are more reactive than if the solvent were not present. If the solute consists of ions (such as NaCl), the energy for the dissolution process is supplied almost entirely from the solvation of ions; this is primarily an electrostatic process. However, two processes may occur when a covalent solute is dissolved: the solvation of

$$NaCl_{(S)} + S \rightarrow Na^+(S)_x + Cl^-(S)y$$

molecules (such as CO_2) or the formation of ions. The solvation of molecular species usually involves

$$CO_{2(g)} + H_2O \rightleftarrows CO_2\,(aq)$$
$$CO_2(aq) + H_2O \rightleftarrows H_2CO_3$$
$$H_2CO_3 + H_2O \rightleftarrows HCO_3^- + H_3O^+$$
$$HCO_3^- + H_2O \rightleftarrows CO_3^{2-} + H_3O^+$$

dipolar interactions or, specific interactions such as hydrogen-bond formation, and formation of covalent bonded species via coordinate covalent-bond formation. In many instances it is possible for such intermediate species to undergo ionization. The formation of ions from covalent molecules is a measure of the ionizing power of the solvent; attempts have been made to correlate this with the dielectric constant of the solvent. Unfortunately, there are a sufficient number of contradictions to make such relationships unreliable in the light of our present knowledge.

The extent to which a substance is ionized by neutral donor (solvent) molecules should increase with

Table 2. Donicity (D_n) and Dielectric Constant (ε) of Some Common Solvents

Solvent	D_n	ε
1,2-Dichloroethane	0.0	10.1
Sulfurylchloride	0.1	10.5
Tetrachloroethylene carbonate	0.2	9.2
Thionyl chloride	0.4	9.2
Acetyl chloride	0.7	15.8
Benzoyl chloride	2.3	23.0
Nitromethane	2.7	35.9
Dichloroethylene carbonate	3.2	31.8
Nitrobenzene	4.4	34.8
Acetic anhydride	10.5	20.7
Phosphorous oxide chloride	11.7	14.0
Benzonitrile	11.9	25.2
Selenium oxide chloride	12.2	46.0
Monochloroethylene carbonate	12.7	62.0
Acetonitrile	14.1	38.0
Sulfolane	14.8	42.0
Propylene carbonate	15.1	69.0
Benzyl cyanide	15.1	18.4
Ethylene sulfite	15.3	41.0
Isobutyronitrile	15.4	20.4
Propionitrile	16.1	27.7
Ethylene carbonate	16.4	89.1
Phenylphosphorous oxide difluoride	16.4	27.0
Methyl acetate	16.5	6.7
Butyronitrile	16.6	20.3
Acetone	17.0	20.7
Ethyl acetate	17.1	6.0
Water	18.0	81.0
Phenylphosphorous oxide dichloride	18.5	26.0
Diethyl ether	19.2	4.3
Tetrahydrofuran	20.0	7.6
Diphenylphosphorous oxide chloride	22.4	—
Trimethyl phosphate	23.7	6.8
Tributyl phosphate	23.7	6.8
Dimethylformamide	26.6	36.1
Dimethylacetamide	27.8	28.9
Dimethyl sulfoxide	29.8	45.0
Diethylformamide	30.9	—
Diethylacetamide	32.2	—
Pyridine	33.1	12.3
Hexamethylphosphoramide	(38.8)	30.0

increasing stability of the cation resulting from nucleophilic attack,

$$S: +X - Y \rightarrow S - X^+ + Y^-$$

compared with that of the un-ionized solute. The strength of the coordinate covalent bond formed in the species SX$^+$ is related to the donor ability of S, to acceptor ability of X in the species XY, to steric effects, and to the magnitude of specific solvent-solute interactions, such as hydrogen bonding. The donor strengths, expressed as donicity, D_n, of solvent molecules, have been defined relative to the reference acceptor SbCl$_5$; the enthalpy of the reaction between SbCl$_5$ and a series of donors in an inert medium is taken as the measure of donicity. The donicity of a solvent has been interpreted as a measure of its donor intensity, nucleophilicity, or Lewis-base strength and can be a useful guide in assessing the ionizing power of the solvent. The donicities of some common nonaqueous solvents appear in Table 2. Inspection of the data in Table 2 indicates that there is no necessary correlation between the doncity of the solvent and its dielectric constant ε.

Supercritical Fluids

The combination of physical and chemical properties described above that determine the dissolving power of a liquid can be affected by pressure. Fluids (i.e., gases and liquids) near their critical point have dissolving power comparable to that of liquids. A substance is supercritical when its temperature (T) and pressure (P) are higher than the corresponding critical values (T$_c$, P$_c$). When the temperature and pressure of supercritical fluids (SCFs) are in the range $1 < T/T_c < 1.1$ and $1 < P/P_c < 2$, the distinction between gases and liquids disappears, and the substance exists as a single phase—the fluid phase. The SCF has many of the advantageous properties of both a liquid and a gas. Because of the high compressibility of fluids near the critical point, their density and dissolving power can be fine-tuned through small changes in pressure. Although many details concerned with kinetics and thermodynamics in SCFs need to be developed, practical applications of SCFs have appeared. For example, large-scale decaffeination of coffee with supercritical CO$_2$ is well established, and the ROSE process is used for separating the components of the heavy fraction of petroleum using supercritical pentane.

BIBLIOGRAPHY

ECKERT, C.; KNUDSON, B. L.; DEBENEDELTE, P. G. "Supercritical Fluids As Solvents for Chemical and Materials Processing." *Nature 383*, 313–317.

McHUGH, M.; KRUKONIS, V. "Supercritical Fluid Extraction." Oxford, U.K.: Butterworths-Heinemann, 1994.

MEEK, D. W. "Lewis Acid-Base Interactions in Polar Non-Aqueous Solvents." In *The Chemistry of Non-Aqueous Solvents;* J. J. Lagowski, ed.; San Diego, CA: Academic Press, 1966; Vol. 1, p. 1062.

PRICE, E. "Solvation of Electrolytes and Solution Equilibria." *The Chemistry of Non-Aqueous Solvents;* J. J. Lagowski, ed.; San Diego, CA: Academic Press, 1966; Vol. 1, pp. 67–94.

JOSEPH J. LAGOWSKI

Sommerfeld, Arnold Johannes Wilhelm (1868–1951)

Arnold Sommerfeld was born in Kaliningrad (formerly Königsberg), Russia, on December 5, 1868. His father, Franz Sommerfeld, was a physician. His mother was Cäcile Matthias Sommerfeld, whom he described as energetic and intellectually vigorous. Sommerfeld graduated from high school in 1886 and attended the University of Königsberg. Although interested in literature and history, he gravitated toward mathematics and science in his major studies.

Sommerfeld became *Privatdozent* in mathematics in March 1886 at Göttingen, where he developed a numerical method for solving integrals associated with diffraction theory. With this accomplishment, he provided a means for later scientists to solve complex integrals in physical calculations. The notable Henri Poincaré used this method in relation to his own work, which was an important precursor for the development of Albert EINSTEIN's relativity theory. After a few years at Göttingen, Sommerfeld became a full professor of mathematics at the Bergakademie at Clausthal in 1897. While there, he married Johanna Hopfner, whose father was the head of the University of Göttingen.

In 1900 Sommerfeld became full professor at the Technical Hochschule in Aachen. He investigated fluid dynamics and developed explanations for the onset of turbulent flow. This so impressed his engineering colleagues that Sommerfeld was offered a prestigious position at the Berlin mining academy (which he declined).

Sommerfeld became associated with Peter DEBYE while at Aachen, where the two of them collaborated on many areas of interest. His growing reputation as a theoretical physicist landed him the position as chair of theoretical physics at the University of Munich in 1906. The university established an institute for Sommerfeld's work, and one of his students, Walter Friedreich, discovered X-RAY DIFFRACTION by crystals while working there. Sommerfeld was also an early supporter of Einstein's theory of special relatively. He used it to extend into vector form and applied it to certain fundamental problems in physics. He became a colleague of Einstein, although he resisted Einstein's view that PHOTONs existed and were "particles of light."

He became interested in the problems of atomic structure and was well-versed on the work of Niels BOHR, who had applied Max PLANCK's quantum ideas to the hydrogen atom structure with considerable success. Since the ZEEMAN EFFECT required more than a single quantum number to explain observed emission lines, Sommerfeld tried to generalize the BOHR MODEL of the atom by introducing elliptical orbits for electrons instead of the circular ones suggested by Bohr. This extended the Bohr model but was still too limited to explain additional emissions that had been observed. This work and a book he wrote made him one of the leading theoretical physicists of the time, until the development of quantum theory in 1926. Thanks to these efforts, he turned out more doctoral degree students than any other theoretical physicist in his day. Einstein admired Sommerfeld greatly for his proficiency and success in physics.

Sommerfeld's awards, while not including a Nobel Prize (which was won by some of his students such as Hans Bethe and Peter Debye), included many honorary doctorates.

Sommerfeld did not support the political movements in Germany during the 1930s and 1940s and was targeted by the Nazis as part of the "Jewish physics" that they so despised even though he was not Jewish. He ultimately resumed the directorship of the Institute for Theoretical Physics. He was struck by an automobile in 1951 and died on March 26 of his injuries.

BIBLIOGRAPHY

FORMAN, P.; HERMANN, A. "Arnold Sommerfeld," In *Dictionary of Scientific Biography;* C. C. Gillispie, ed.; New York: Scribner, 1975; pp. 525–532.

HERMANN, A. "Sommerfeld's Role in the Development of Early Quantum Theory." In *Physics of One- and Two-Electron Atoms;* F. Bopp, H. Kleinpopper, eds.; New York and Amsterdam: Elsevier, 1969.

PAUL RIDER

Space Chemistry

The space between stars in not empty. It contains detectable quantities of gas and dust grouped into structures known as intersteller clouds with diameters ranging from 10^{13}–10^{15} km. Together these clouds contain several percent of the mass of our Milky Way Galaxy. Astronomers study these clouds by analyzing their spectra, the pattern made by the strength of the light at different wavelengths. In this way over 100 different molecules have been identified, ranging from diatomic species such as hydrochloric acid to complex molecules with nine carbons in a chain. These molecules form in space by chemical reactions. The raw materials for the molecules are the atoms of stellar origin. Two major sources of energy for chemical processing are ultraviolet (UV) photons (energies ≥ 6 eV) and cosmic ray protons with energies near 1–2×10^6 eV. Ultraviolet photons come from hot young stars and have enough energy to break a molecular bond in most small molecules. Cosmic rays are mainly high-energy protons (energies $> 10^9$ eV have been measured) formed in events such as supernovae; they can penetrate dense interstellar clouds whereas UV photons cannot.

Interstellar clouds can be classified as diffuse or dense. Diffuse clouds are thin enough that some visible light from distant stars passes through them. These thin clouds have gas densities in the range of 10–$1,000$ molecules cm^{-3}, which by Earth's standards is a vacuum (at standard temperature and pressure [STP], the density of Earth's atmosphere is $\sim 3 \times 10^{19}$ molecules cm^{-3}). Diffuse cloud temperatures range from ~ 50–100 K. Hydrogen, the dominant species, remains as atomic hydrogen because abundant energetic UV photons break H_2 bonds. Along with atomic hydrogen, some simple diatomic molecules identified in diffuse clouds include CH, CH^+, CN, H_2, CO, OH, and C_2. However, molecular abundances are low because intense UV photons destroy molecules rapidly.

Dense clouds have 10^3–10^6 molecules cm^{-3}. In dark dense clouds where there is no evidence of star formation, cloud temperatures are near 10 K; these regions are where molecules are most often detected and they are referred to as molecular clouds. Giant molecular clouds are more complex and are warmer because they have regions in which star formation occurs. The largest of these are huge and contain material equal to a million solar masses. About 1 percent of the mass of a dense cloud is due to small dust grains with diameters $< 1 \times 10^{-5}$ cm. Astronomers think these dust grains form in the extended shell of gas surrounding older stars and then move out into space by the push of the star's light. Dust acts as a shield blocking UV photons from dissociating molecules. The abundance of interstellar molecules becomes appreciable in these shielded, dusty regions, since the rate of formation is greater than the

Table 1. Identified Interstellar Molecules

Molecules with two atoms		Molecules with three atoms	
H_2	Hydrogen	C_2S	
CO	Carbon monoxide	SiC_2	
CSi	Carbon monosilicide		
CP	Carbon monophosphide	Molecules with four atoms	
CS	Carbon monosulfide	NH_3	Ammonia
NO	Nitric oxide	H_2CO	Formaldehyde
NS	Nitrogen monosulfide	H_2CS	Thioformaldehyde
SO	Sulfur monoxide	C_2H_2	Acetylene
HCl	Hydrochloric acid	HNCO	Isocyanic acid
NaCl	Sodium chloride	HNCS	Thioisocyanic acid
KCl	Potassium chloride	$HCNH^*$	
AlCl	Aluminum monochloride	H_3O^+	
AlF	Aluminum monofluoride	$HOCO^*$	
PN	Phosphorous mononitride	C_3S	
SiN	Silicon mononitride	H_2CN	
SiO	Silicon monoxide	C_3H-linear	
SiS	Silicon monosulfide	Cyclic-C_3h	
CO^+		HCCH	
SO^+		H_2CO^+	
CH		C_2CN	
CH^+		C_3O	
NH			
OH		Molecules with five atoms	
C_2		CH_4	Methane
CN		SiH_4	Silane
		CH_2NH	Methyleneimine
Molecules with three atoms		NH_2CN	Cyanamide
H_2O	Water	CH_2CO	Ketene
H_2S	Hydrogen sulfide	HCOOH	Formic acid
HCN	Hydrogen cyanide	HC_2CN	Cyanoacetylene
HNC	Hydrogen isocyanide	HCCNC	Isocyanoacetylene
CO_2	Carbon dioxide (seen	H_2C_3-linear	
	only as an ice)	cyclic-C_3H_2	
SO_2	Sulfur dioxide	CH_2CN	
OCS	Carbonyl sulfide	C_4H	
MgNC	Magnesium isocyanide	HNCCC	
NaCN	Sodium cyanide	C_4Si	
N_2O	Nitrous oxide	C_5	
$?H_2D^+$			
CH_2		Molecules with six atoms	
NH_2		CH_3OH	Methanol
C_2H		CH_3SH	Methanethiol
HCO		C_2H_4	Ethylene
HCO^+		CH_3CN	Methylcyanide
HOC^+		CH_3NC	Methylisocyanide
NH_4^+		HC_2CHO	Propynal
HNO		NH_2CHO	Formamide
HCS^+		H_2C_4-linear	Butadiyne
C_3		HC_3NH^+	
C_2O		C_3H	

Table 1. (Continued)

Molecules with seven atoms		Molecules with nine atoms	
CH_3C_2H	Methylacetylene	$(CH_3)_2O$	Dimethyl ether
CH_3CHO	Acetaldehyde	C_2H_5OH	Ethanol
CH_3NH_2	Methylamine	C_2H_5CN	Ethylcyanide
H_2C_2HCN	Acrylonitrile	CH_3C_4H	Methylbutadiyne
HC_4CN	Cyanobutadiyne	HC_6CN	Cyanohexatriyne
C_6H			
		Molecules with ten atoms	
Molecules with eight atoms		$(CH_3)_2CO?$	Acetone
CH_3COOH	Acetic acid	$CH_3C_4CN?$	Cyanomethylbutadiyn
$HCOOCH_3$	Methyl formate	$NH_2CH_2COOH?$	Aminoacetic acid
CH_3C_2CN	Cyanomethylacetylene		
		Molecule with eleven atoms	
		HC_8CN	

* Molecules detected in gas around older stars are included.
? Denotes uncertain detection.

rate of destruction. Table 1 is a list of molecules identified in molecular clouds (including molecules, ions, and radicals). The identification of most molecules has been made by matching their radio emissions with emission spectra of gases measured by scientists in the laboratory. On Earth, ions and radicals (neutral fragments with an unpaired electron) are extremely unstable, since they undergo $\sim 10^{10}$ collisions sec^{-1}. In a molecular cloud, ions and radicals appear stable because the time between collisions is many years.

Although the formation mechanisms for interstellar molecules are not well understood, two types of reactions are probably most important in interstellar clouds: gas phase reactions and reactions occurring on dust grains. Ion-neutral molecule reactions play an important role in the gas phase. The process starts with the ion H_3^+, formed by cosmic rays from H_2. It reacts in a variety of ion-molecule chemistry pathways to make molecules such as H_2O, CH_4, and NH_3. CO, the second most abundant molecule in interstellar regions (there is ~ 1 CO for every 10,000 H_2), probably forms by several processes, including ion-molecule reactions. More complex carbon molecules build up by a succession of reactions.

Grain surfaces appear to play a crucial role in the formation of H_2, the seed molecule on which much of interstellar gas phase chemistry is based. H atoms stick on cold (T < 10 K) grains for $\sim 10^5$ years. Since H atoms are light, they can move on the grain through a process called tunneling until they are close enough to another H atom to form a chemical bond,

creating H_2. This is an exothermic reaction and the energy released is absorbed by the grain. The resulting temperature rise is enough to unstick H_2, releasing it into the gas phase. The importance of surface reactions to form more complex molecules is not known. Big molecules that do form on grains probably remain stuck, since the Van der Waals binding energy between the molecule and the grain increases with the number of atoms in contact with the grain.

Complex molecules can be made from simple molecules (e.g., H_2O + CO + CH_4) condensed as icy coatings on grains. Ultraviolet photons and cosmic ray particles are both energetic enough to break molecular bonds and cause ionizations in ices. The relative importance of the UV and the cosmic rays depends on the location of the icy grain in the cloud. Laboratory experiments try to simulate the low-temperature and -pressure space environment and study chemical reactions in an ice after UV photons and particles are absorbed. Laboratory spectra show that many new molecules form when the molecular fragments recombine in the ice. Other laboratory results show that a regular molecular structure becomes more random (damaged), and lighter atoms like H are lost from the solid, resulting in a compositional change. Similar processes may occur on ice-coated interstellar grains, although on longer time scales and with smaller radiation fluxes than in the laboratory experiments.

In the neighborhood of our Sun, there is a local excess of high-energy particles from the Sun called the solar wind. The study of planetary atmospheres and of planetary surfaces is helped by an understanding of chemical reactions caused by the combined UV and solar wind environment in interplanetary space. Understanding the degradation processes of human-made materials in space is crucial to the success of manned and unmanned space missions. The effect of radiation on most metal structures over periods of tens of years is not severe. However, spacecraft electronic semiconducting materials suffer damage from ionizing particles that can cause logic changes and performance degradation. These effects are minimized by shielding electronics with other spacecraft components and by using specially processed materials. Both UV photons and particle radiation change the mechanical and physical properties of plastic polymers by breaking C—C bonds. Some plastics become soft and weak, while others become brittle and weak. The strength of many materials diminishes and paints can discolor. The study of methods to improve the performance and useful lifetime of these materials continues.

BIBLIOGRAPHY

HERBST, E. "The Chemistry of Interstellar Space." *Angew. Chem. Int. Ed. Engl.* 1990, 29, 595–608.

TAYLOR, S.; WILLIAMS, D. "Star-Studded Chemistry." *Chem. in Britain,* August 1993, 680–683.

MARLA H. MOORE

Specific Heat

The specific heat or specific heat capacity, c, of a substance is the amount of heat needed to raise the temperature of one gram of that substance by one degree Celsius (C). It is heat capacity, C, on a per gram basis. Specific heats are listed as J/C-g or as dimensionless values relative to the HEAT CAPACITY of water (4.18 J/C-g).

Different materials have different heat capacities. For example, one gram of water requires 4.18 joules of energy to raise the temperature by one degree Celsius, while one gram of aluminum requires only 0.90 J of energy to raise the temperature one degree Celsius.

Specific heat is approximately constant over the temperature range of the state of a substance. Thus the specific heat of aluminum at 25°C is about the same as the specific heat of aluminum at 400°C. When aluminum melts at 660°C, the heat capacity changes for liquid aluminum compared with solid aluminum.

Specific heat is an intensive property; heat capacity is an extensive property. For example, two blocks of solid copper that have different masses both have a

Table 1. Specific Heat for Six Common Substances

Substance	Specific Heat
Aluminum	0.22
Carbon	0.17
Ethanol	0.59
Iron	0.11
Lead	0.03
Water	0.50

specific heat of 0.38 J/g-C, but the larger block will have a larger heat capacity. The specific heat (J/C-g) multiplied by the mass (g) gives the heat capacity (J/C).

The specific heat for some representative substances is listed in Table 1.

DAVID C. FINSTER

Specific Quantity

If quantity, symbol q, is some physical quantity capable of quantitative expression associated with a chemical substance (energy, volume, absorptivity, conductivity, etc.), then the corresponding specific quantity is the value of that quantity divided by its mass. The symbol for the specific quantity is $\tilde{q}$. Thus, $\tilde{q} = q/m$, where m is the mass.

The corresponding molar quantity is $\tilde{q} = q/n$, where n is the molar amount. (See MOLE CONCEPT; MOLAR.) The corresponding equivalent quantity is $\hat{q} = q/e$, where e is the equivalent amount. (See EQUIVALENTS.)

The obsolete term *specific gravity* was formerly used for the dimensionless ratio between the density of a substance and that of water, both densities being obtained by weighing in air at the same temperature.

TRUMAN P. KOHMAN

Spectrometry

See SPECTROSCOPY.

Spectroscopy

Spectroscopy is the measurement of energy absorption of emission, which reveals separations between energy levels. The measurement produces a spectrum, which is a plot (graph) in which the x-axis typically is energy and the y-axis is intensity detected at a particular energy level. The equipment or instrument that is used to make the measurement is usually called a spectrometer (the older name was spectroscope).

Spectroscopy takes advantage of the fact that atomic and molecular energies are quantized and electromagnetic radiation is quantized (see also QUANTUM CHEMISTRY; QUANTUM THEORY). If a PHOTON with energy E is absorbed, two energy levels in the absorbing material must differ by the amount E. By monitoring the absorption or emission of energy it is possible, for example, to determine the energies of the vibrations of a molecule (infrared or Raman spectra), the separations between electronic energy levels (ultraviolet and visible spectra), or the energies of nuclear spin states in the presence of a magnetic field (see NUCLEAR MAGNETIC RESONANCE). To provide an overview of spectroscopy this article first provides a brief historical perspective, then discusses types of spectroscopy and the corresponding timescales, and concludes with comments on how chemists use spectroscopy.

Historical Perspective

Early scientists were fascinated by the behavior of visible light. In 1621, Snell discovered that the path of a beam of light was refracted (bent) upon passage through a prism. Building upon this work, Sir Isaac Newton, in 1666, passed a narrow beam of white light from the sun through a prism and displayed the result on a screen. He observed a series of colors (the colors of a rainbow: red, orange, yellow, green, blue, indigo, and violet), which he called a spectrum. The positions of different colors along the spectrum were due to the varying amounts of refraction that occurred for the different colors. Although prior workers had observed these colors, he was the first to recognize that all the colors had been present in the white light. The prism had separated the colors, not created them. On the basis of this experiment, Newton is credited as the father of spectroscopy. Newton's prism was of relatively low quality, so his spectrum appeared to be continuous. As others examined the solar spectrum at higher resolution, they found that the intensity of light with different colors was not constant—there were in fact dark lines where there was little or no light in the spectrum. Many of these lines were carefully measured by Fraunhofer, an optician in Munich, in the early 1800s and became known as FRAUNHOFER LINES.

The earliest work on spectroscopy was limited to the visible region of the spectrum—the part that could be detected with the human eye. In 1800, Sir William Herschel examined the heating power of light with different colors and found that maximum heating occurred beyond the red end of the spectrum in what is now known as the infrared region. Detection in this region was done with temperature sensors rather than with the human eye. The observation that the spectrum extended beyond the red end spurred others to examine the spectrum beyond the violet end. Experiments soon showed that the spectrum extended into what became known as the ultraviolet region. Fraunhofer lines were observed in both the infrared and the ultraviolet regions.

Early in the twentieth century quantum mechanics provided a framework to interpret the Fraunhofer lines. Colors of the spectrum correspond to electromagnetic radiation with different wavelengths and frequencies. The fundamental relationships are $E = h\nu$, where E is the energy per photon, h is Planck's constant (6.626×10^{-34} joules seconds), and ν is the frequency of the radiation. The frequency (ν) and wavelength (λ) of the radiation are related by $\lambda\nu = c$, where c is the speed of light (2.9979×10^8 meters/second). The spectrum of the radiant energy from the sun is essentially continuous. However, atoms and molecules in the atmosphere absorb energy at characteristic values. Therefore, the solar spectrum detected at Earth has dark lines (decreased intensity) at positions in the ultraviolet, visible, and infrared that correspond to absorption energies for molecules in the atmosphere.

Types of Spectroscopy

ELECTROMAGNETIC RADIATION SPECTROSCOPIES

Modern sources and detectors permit generation and measurement of electromagnetic radiation at energies orders of magnitude higher and lower than those represented by the ultraviolet, visible, and infrared ranges that were first studied. This range of energies provides the opportunity to probe a wide variety of energy spacings. Table 1 gives examples of the frequencies and wavelengths of electromagnetic radiation that provide the basis of some common spectroscopies.

RESONANT ENERGY ABSORPTION OR EMISSION. Several of the techniques listed in Table 1 measure resonant absorption of energy by the sample. A source generates radiation that is incident upon the sample, and the detector measures the amount of radiation of a particular energy that was not absorbed. Energy absorption is observed only when there is a match between the energy supplied and the spacing between two energy levels of the atom or molecule.

Table 1. Spectroscopies Based on Electromagnetic Radiation

Name of Spectroscopy*	Frequency† (sec^{-1})	Wavelength (m)	Timescale‡ (sec)
Nuclear Magnetic Resonance	3×10^8	1.0	10^{-1} to 10^{-6}
Electron Paramagnetic Resonance = Electron Spin Resonance, Microwave	9×10^9	3.3×10^{-2}	10^{-4} to 10^{-8}
Infrared, Raman	1.5×10^{14}	2.0×10^{-6}	10^{-13}
Visible, Atomic Absorption, Atomic Emission, Fluorescence	6×10^{14}	5×10^{-7}	10^{-14}
Ultraviolet	1.2×10^{15}	2.5×10^{-7}	10^{-15}
Ultraviolet Photoelectron	5×10^{15}	6.2×10^{-8}	10^{-15}
X-Ray Photoelectron = Electron Spectroscopy for Chemical Analysis	1×10^{17}	3×10^{-9}	10^{-17}
Mössbauer, gamma-ray	1×10^{19}	3×10^{-11}	10^{-7}

* Each name is typically followed by the word "spectroscopy" unless that word is embedded in the name. When two names are separated by commas they refer to two distinct types of spectroscopy. When two names are separated by an equal sign, the two names are used approximately interchangeably.

† Typical values are cited. Most spectroscopies encompass a range of frequencies. The table is arranged in order of increasing energy.

‡ Order-of-magnitude values are shown. Timescales for a particular system will vary.

These techniques include MICROWAVE SPECTROS-COPY; ULTRAVIOLET SPECTROSCOPY; and visible and infrared spectroscopy.

A few of the techniques measure emission of energy by a sample that has previously absorbed energy. These techniques include FLUORESCENCE AND PHOSPHORESCENCE and atomic emission spectroscopy.

Thermal energy at room temperature (298 K) is 2.48 kJ/mole. Since this value is expressed per mole, it must be divided by AVOGADRO'S NUMBER of particles per mole (6.02×10^{23}) to find the corresponding energy per particle, 4.12×10^{-21} J. If we use $E = h\nu$, this energy corresponds to $\nu = 6.2 \times 10^{12}$ sec^{-1}. When energy levels have spacings less than thermal energy, more than one energy level is significantly populated at room temperature at thermal equilibrium. The most common spectroscopies that observe transitions between energy levels in this category are nuclear magnetic resonance and electron spin resonance (also known as electron paramagnetic resonance). In these cases both absorption and stimulated emission occur, and the spectroscopy measures the net absorption of energy.

There are quantum mechanical restrictions on the transitions that can be observed in each type of spectroscopy. These are called selection rules. Transitions forbidden by these selection rules can be observed, although usually with lower probability and therefore lower intensity in the spectrum.

When energy is absorbed, the atom or molecule goes into an EXCITED STATE. The process by which it returns to its original state (the ground state) is called relaxation. The energy could be given up to the radiation field by stimulated emission, which has the same selection rules and the same probability as stimulated absorption, or by spontaneous emission. The rate of spontaneous emission increases rapidly as the energy of the radiation increases. Thus, it is so fast in the visible region of the spectrum that slower processes, such as fluorescence and phosphorescence, have separately been identified and form distinguishable types of spectroscopy. In the radiofrequency region, as used, for example, in nuclear magnetic resonance, spontaneous emission is so slow that it is relatively easy to saturate the energy levels—i.e., to equally populate the ground and excited states. In the visible region very high photon intensity, as with a high-power laser, is necessary to achieve saturation.

NONRESONANT ENERGY ABSORPTION. Unlike the resonant technique described above, some spectroscopies supply radiation to a sample at a single energy that is greater than the spacings between energy levels to be studied. The spectroscopy measures the energy that is "left over" after the molecule is excited. Two techniques that demonstrate this approach are ultraviolet and X-ray PHOTOELECTRON SPECTROS-COPY. Energies are measured for electrons that are ejected from the sample. The energies of the ejected electrons reflect the difference between the energy of the incident radiation and that required to remove an electron from a molecular energy level. Photoelectron spectroscopy ejects electrons from valence energy levels, and X-ray photoelectron spectroscopy ejects electrons from core energy levels (electrons that are more tightly held by the atom). Another spectroscopy that is nonresonant is Raman. In this experiment intense light with a frequency ν_0 shines on a sample. Light scattered by the sample is measured. If during the scattering process a vibration with energy ν_v is excited, the scattered light is observed at $\nu_0 - \nu_v$. If $h\nu_0$ matches the energy of an electronic transition of the molecule (a resonance condition), the experiment becomes resonance Raman spectroscopy (see RAMAN SPECTROMETRY for a discussion of nonresonant and resonant types).

ENERGIES OTHER THAN ELECTROMAGNETIC RADIATION

Some spectroscopies are based on energies other than those of electromagnetic radiation. An important example is MASS SPECTROMETRY, or mass spectroscopy. In mass spectroscopy, a molecule in the gas phase is converted to a charged ion. Fragmentation may occur during the ionization process. Electric and magnetic fields are used to separate ions according to the ratio of mass to charge. A mass spectrum consists of a plot of number of ions as a function of mass. The pattern of masses observed characterizes the molecular weight of the compound and the fragments into which the ionized molecule tends to rearrange or decompose.

Timescales of Observation

Molecules in the gas or liquid phase are constantly vibrating, rotating, and translating. In the solid phase, barriers to motion are higher, but motions persist. These motions interconvert states of molecules. Whenever there are two or more states of the molecular species distinguishable by a spectroscopic technique, there exists the potential for studying the dynamic interconversion between these states. If the molecule stays in one form for a time that is long relative to the timescale of the measurement (Table 1), separate signals are observed for the two forms. At the other extreme, if the two forms interconvert in a time that is short compared with the timescale of the measurement, an averaged signal is observed. However, when the time required for interconversion is of the same order of magnitude as the timescale of the experiment, broadened signals are observed. In this regime analysis of the shape of the signal provides detailed information on the KINETICS of the interconversion process. The lineshapes are sensitive to rates of the order of the separation between the states considered, not the energy of the radiation used in the spectroscopic technique. For example, modern proton nuclear magnetic resonance (NMR) (^{1}H NMR is commonly called proton NMR, and even PNMR or PMR) is usually performed with radiofrequency (RF) radiation in the 200 to 600 MHz range, but individual transitions detectable by NMR may differ by only a few Hz. Thus, dynamic processes, such as intramolecular rearrangements or intermolecular interconversions as slow as a few times per second, can affect the observable NMR lineshape. Similarly, in mass spectroscopy ions that interconvert on the timescale required for an ion to move from the source to the detector result in broadened peaks called metastables. Dynamic effects are less frequently observed in spectroscopies such as infrared and visible in which the timescales are very short relative to the time required for most molecular interconversions.

Oscillating energy at such a low frequency that it is called mechanical vibration can be used to monitor motions in solids. The dielectric constant is not a single value, but rather a frequency-dependent response to radiation from static through mechanical vibrational frequencies to microwave and visible frequencies, and so on. This type of spectroscopy, which is very important for determining nonresonant interaction of radiation with matter, can be used to reveal the dynamics of the molecular species.

The lifetime of the excited state of the molecule (the state achieved by absorbing one or more photons) can also have an impact on the width of lines in a spectrum. A sharp line is observed when the energy separation between the ground state and the excited state is well defined. This occurs when the excited state has a relatively long lifetime. As the lifetime of the excited state decreases, the uncertainty in the energy of the transition increases and the width of the line in the spectrum increases. The lineshape for these transitions is Lorentzian. However it should also be noted that spectral linewidths often are not determined by the lifetime of the excited state or by dynamic processes. The apparent linewidth may be determined by the ability of the spectrometer to distinguish between different energies or by the presence of many closely spaced and overlapping lines. Lineshapes in these cases are closer to Gaussian than to Lorentzian.

How Chemists Use Spectroscopy

Spectroscopic techniques are among the most powerful and generally useful in science. The normal characterization of a chemical compound includes at least the infrared (IR) spectrum, because the IR spectrum reveals the vibrations and hence structure of the species, and the ultraviolet (UV) and visible (if colored) spectrum, because the UV-vis spectrum reveals the electronic energy levels of the species. A mass spectrum can be more definitive than elemental analysis for revealing the elemental composition of the species. However, one has to be aware of the possibility of observing only the most volatile or easily charged species in a mixture, and that these might be minor (impurity) species. Nuclear magnetic resonance spectroscopy, since its discovery in 1945, has become one of the most powerful tools for examining chemical substances, from small organic or inorganic species to polymers and proteins. It is possible to observe the NMR spectrum for every nucleus with a nuclear spin, but the nuclei most commonly observed are ^1H, ^{13}C, ^{19}F, ^{31}P, and ^{11}B. NMR is part of the standard spectroscopic characterization of diamagnetic molecules containing these nuclei.

Most spectroscopies can be performed on samples in any of the states of matter. Species to be studied may be solids, liquids, or gases or may be dissolved or dispersed in solids, liquids, or gases. IR, UV-vis, and NMR spectroscopy have been conducted on all types of samples (including living human beings), but some kinds of spectroscopy require specific conditions. Mass spectroscopy is performed on ions in very low-pressure gas phase. A wide variety of techniques has been developed to transform samples into gaseous phase ions for mass separation. MÖSSBAUER SPECTROSCOPY, which depends on the resonant absorption of gamma rays, has such a narrow line width that motion of the absorber must be controlled accurately; so the species to be studied by Mössbauer must be rigidly held in either a crystalline lattice or a glassy matrix, and this is often done at liquid-helium temperatures.

Beyond the initial characterization of a molecule, detailed spectroscopic studies provide insight into the details of electronic and molecular structure. This information can then be compared with calculational models that predict properties for yet unknown substances. The spectrum that is used by one scientist as a "fingerprint" of the molecule may be the basis for extended study for another scientist. Finally, note that each of the spectroscopic techniques provides a different view of the molecular species. The parable of the blind men and the elephant applies to chemical spectroscopy. To use only one spectroscopic method may blind the researcher to the full nature of the species being investigated.

BIBLIOGRAPHY

COLTHRUP, N. B.; DALY, L. H.; WIBERLEY, S. E. *Introduction to Infrared and Raman Spectroscopy*, 3rd ed.; New York: Academic Press, 1990.

DRAGO, R. S. *Physical Methods for Chemists*, 2nd ed.; Fort Worth: Saunders College Publishing, 1992.

LEVER, A. B. P. *Inorganic Electronic Spectra*, 2nd ed.; New York: Elsevier, 1984.

SANDERS, J. K. M.; HUNTER, B. K. *Modern NMR Spectroscopy*, 2nd ed.; New York: Oxford University Press, 1993.

SANDRA EATON
GARETH EATON

Spin

See ANGULAR MOMENTUM.

Spin Quantum Number

See ATOM, ELECTRONIC; QUANTUM NUMBERS.

Standard State

See STATES OF MATTER.

Standard Temperature and Pressure

See STP.

Stars, Chemistry of

See ORIGIN OF THE CHEMICAL ELEMENTS.

States of Matter

Solid, liquid, gas (or vapor), and plasma are the usual states of aggregation considered for pure substances and mixtures. The transition between these states and their characteristics is illustrated by considering the changes in a pure substance as a function of temperature at constant pressure. At temperatures below the melting temperature the substance is a solid characterized by a well-defined lattice or crystal structure with a molecular level order that extends over a long range. The atoms or molecules in this lattice are immobile except for vibrational and rotational motions around their average position. Consequently, solids have extremely low fluidity and diffusivity. There are many different types of lattices, and, even for a single substance, different crystal structures may occur at different temperatures and pressures. For instance, pure water has eight different solid phases, though many of these exist only at very high pressures. As the temperature increases, most commonly the solid melts and a liquid phase is formed, though in some instances sublimation to a vapor occurs. The molecules in a liquid have great mobility, and consequently, there is only imperfect intermolecular ordering, which is of short range. Since molecules are mobile, the fluidity of and diffusivity in a liquid are high, though the distance between intermolecular collisions is only of the order of one molecular diameter, or less than 1 nanometer (nm).

When the temperature is increased further, the liquid converts to a vapor, which is characterized by an almost complete lack of molecular order and a distance traveled between intermolecular collisions ranging from very long in near perfect vacuum to 70 nm for nitrogen at ambient conditions to less than 1 nm for gases at high densities. Other characteristics of a gas are that it will fill the volume in which it is placed, it is very compressible, its viscosity is low, and its diffusivity is high. The critical temperature of a pure substance is the highest temperature at which a liquid will exist; at higher temperatures only a gas exists. It is common to refer to a substance above its critical temperature as a gas, since it cannot be made to condense by increasing the pressure; if the temperature is less than the critical temperature, the state is referred to as a vapor.

On further increasing the temperature, the substance may begin to decompose (typically, first the molecules into atoms). On increasing the temperature still further, the atoms decompose into ions and electrons as the electrons in the outermost atomic orbitals are expelled, resulting in an ionized gas or plasma, which macroscopically is electrically neutral. As a result of their occurrence in the cosmos, plasmas are believed to be the most prevalent form of matter. Plasmas may also be formed at moderate temperatures by electrical discharge or radiation.

While the states described above are stable, substances can also form metastable states and also states whose characteristics are intermediate to the four states described above. A glass is an example of an amorphous solid that is metastable in which the molecules are immobile in a structure that has short-range but not long-range order. If left undisturbed,

the transition from metastable amorphous solid to stable crystalline structures can occur, but generally only over a very long (geologic) period of time as in the case of the transition from diamond to graphite. Liquid crystallinity is an example of a stable state that is intermediate between a liquid and a crystalline solid. Liquid crystals, which occur in some polymeric systems composed of highly asymmetric molecules, have the fluidity typical of liquids, but the molecules are to some degree aligned, resulting in anisotropic behavior; that is, the fluid behaves differently to shear in directions along and perpendicular to the axis of the molecules. The main classes of liquid crystals are the nematic, smectic, and cholesteric forms, in all of which there is orientational order between the molecules. In addition smectic liquid crystals self-organize into layers, while in cholesteric liquid crystals the molecules organize into twisted layers.

The different degrees of local translational ordering in the solid, liquid, and gas states is made quantitative by use of the radial distribution function (RDF). The radial distribution function at a distance r from a central molecule is defined as the ratio of the probability of finding another molecule at that distance r to the corresponding probability if all the molecules were uniformly distributed at the same density. Since gases at very low densities behave as ideal gases with molecules that are uniformly distributed in space, the RDF in this limit is constant and equal to unity. In a liquid, as shown in Figure 1, the RDF has an alternating series of smooth peaks corresponding to the nearest neighbor distances followed by valleys, resulting from steric hindrance, which de-

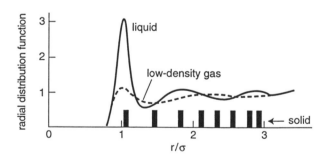

Fig. 1. A schematic representation of the radial distribution function for a liquid (unbroken line) and a low-density gas (dashed line) as a function of position away from a central molecule. For comparison, the locations of surrounding molecules in a lattice are indicated. The radial distribution function for a solid would consist of Gaussian functions of large magnitude and small width centered at these points.

crease in magnitude and gradually approach unity at five or more molecular diameters from the central molecule. In contrast, the RDF of a solid has sharp peaks resulting from the presence of other molecules vibrating around specific lattice sites at well-defined distances from the central molecule. The resulting RDF is periodic and continues out to large distances. If not for defects in the lattice, this perfect ordering would extend to the boundaries of the crystal. The distance between peaks and the size of the peaks in the RDF of a solid give information on its underlying crystal structure, whereas the smoothness and short range of the peaks of the RDF in liquids are the result of the competing effects of interaction potential between the molecules and their translational motion. Experimental information on the radial distribution function in solids and liquids can be obtained from X-ray and neutron-diffraction measurements.

Far from the critical point, the liquid and vapor phases have distinctly different properties. As the critical point is approached along the liquid-vapor coexistence curve, the physical properties of the two phases become increasingly similar, and they are identical at the critical point, which is the endpoint of the coexistence curve. Several models have been proposed for the calculation of the thermodynamic properties of fluids. The simplest thermodynamic model is the IDEAL GAS LAW, which can be derived from statistical mechanics by considering the fluid to consist of noninteracting particles of negligible volume. This leads to the following volumetric and calorimetric relationships for a pure ideal gas:

$$\frac{Pv}{RT} = 1 \quad \text{and} \quad c_v = c_v(T)$$

where P is the absolute pressure, T is the temperature in an absolute scale, R is the universal gas constant, v is the molar volume, and c_v is the heat capacity at constant molar volume, which is a function only of temperature for a pure ideal gas (but also of mole fraction for mixtures of ideal gases). Many volumetric equations of state have been developed to describe real gas behavior. One example, derivable from statistical mechanics, is the virial equation of state in which the pressure is given as a power series expansion in the inverse molar volume:

$$\frac{Pv}{RT} = 1 + \frac{B}{v} + \frac{C}{v^2} + \frac{D}{v^3} + \ldots$$

where $B, C, D, \ldots$ are the so-called virial coefficients that depend on temperature (and on mole fractions for mixtures). In the limit of very large values of the molar volume, that is, very low densities, this equa-

tion reduces to the volumetric ideal gas law. Cubic equations are another class of equations of state of which the first was the VAN DER WAALS EQUATION:

$$P = \frac{RT}{v - b} - \frac{a}{v^2}$$

where $v - b$ is the free volume, that is, the volume available to a molecule as the result of the presence of other molecules of finite size, and the a term accounts for the attraction between molecules. This equation was the first to qualitatively describe the behavior of both vapor and liquid states, the transition between them, and the occurrence of a vapor-liquid critical point. Modern cubic equations of state that are modifications of this classic equation are now commonly used in chemical engineering design.

BIBLIOGRAPHY

GOODSTEIN, D. L. *States of Matter;* New York: Dover Publications, 1985.

TABOR, D. *Gases, Liquids, and Solids: And Other States of Matter;* New York: Cambridge University Press, 1991.

WALTON, A. J. *Three Phases of Matter,* 2nd ed.; New York: Oxford University Press, 1983.

MARCELO CASTIER
STANLEY I. SANDLER

Stationary State

A stationary state in quantum mechanics is one whose physical properties do not vary with time. The concept was introduced by Niels BOHR in connection with his model of the atom (see BOHR MODEL OF THE HYDROGEN ATOM). Classically, an electron in a curved orbit around a positively charged nucleus should continuously radiate energy; in Bohr's quantum-mechanical model, this does not occur, or occurs only in discrete steps between specific energy levels. Bohr postulated that

> An atomic system can exist in certain stationary states, each one corresponding to a definite value of the energy E of the system (Pauling and Wilson, 1935, p. 27).

In the usual representation of quantum mechanics, a stationary state is described by a time-independent wave function $\Psi_k(r)$, which is in turn a solution of the time-independent SCHRÖDINGER equation with energy E_k. The time dependence is given by a pure imaginary exponential, so that the full time-dependent wave function is

$$\Psi(r, t) = e^{-2\pi i E_k t/h} \Psi_k(r)$$

Since probability amplitudes and expectation values of physical observables depend on the product $\Psi^*\Psi$ and an imaginary exponential times its complex conjugate is equal to one, there is no explicit time dependence in the resulting product.

In reality, a quantum state is never strictly stationary, because it eventually absorbs or emits energy to make a transition to another quantum state. Even the GROUND STATE (the lowest-energy state of the system) is subject to external perturbations in any other than a completely isolated system. The concept of a stationary state, however, provides a useful and accurate approximation for calculating the properties of quantum mechanical systems.

BIBLIOGRAPHY

DIRAC, P. A. M. *Principles of Quantum Mechanics;* Oxford: Clarendon Press, 1958.

FEYNMAN, R. P.; LEIGHTON, R. B.; SANDS, M. *The Feynman Lectures on Physics;* Reading, MA: Addison-Wesley, 1963; Vol. 3, pp. 7–2 and 11–22.

PAULING, L.; WILSON, E. B., JR. *Introduction to Quantum Mechanics;* New York: McGraw-Hill, 1935.

RICHTMYER, F. K.; KENNARD, E. H.; LAURITSEN, T. *Introduction to Modern Physics;* New York: McGraw-Hill, 1955; p. 145.

JEFFREY STEINFELD

Statistical Mechanics

One wishes to describe the macroscopic properties of matter, given an understanding of the behavior of individual molecules. The latter is understood from quantum mechanics, but the connection to macroscopic properties needs an average over the properties of the molecules in rather specific ways. In part, the process is described in the entry STATISTICAL THERMODYNAMICS. Here, the basis for the averaging will be described. Some of the calculations that can be carried out with computers to do the averaging in a practical way will also be described.

General Principles

PHASE SPACE

When we speak of the averaging that must be done, we mean primarily over the modes of energy of the individual molecules and the interaction energy among the molecules. To do this, we must first have a coordinate system in which to describe the motion. There are three main types of motion (called degrees of freedom) of individual molecules: translation (movement of the entire molecule), rotation, and vi-

bration. In vibration, the atoms of the molecule move with respect to each other. There are three directions of motion for translations in three-dimensional space, there are three axes about which to rotate (unless the molecule is linear, in which case only two axes exist), and all the rest of the motions are vibrations. Since three numbers are needed to describe the position of each atom, we need all together 3N numbers (degrees of freedom) to describe the entire molecule, where N is the number of atoms in the molecule. Therefore there are 3N-6 vibrations (3N-5 for a linear molecule in which only two of the motions are taken up by rotation).

In describing the energy of a molecule, one must know both how fast the molecule is moving in the particular mode of motion and the position. Therefore we need both a position and a momentum (or velocity, although momentum, mass x velocity, turns out to be more convenient) coordinate for each degree of freedom. To illustrate, we will consider not a three-dimensional molecule, but a one-dimensional oscillator with one degree of freedom, like a pendulum. This will have only one position and one momentum and can be represented in two dimensions. When the pendulum is at an extreme position, it is not moving—velocity equals zero, but potential energy equals a maximum. When the pendulum is at the minimum, or center, of its arc, its position is zero, but its velocity and thus its momentum and kinetic energy are maximum. Kinetic plus potential energy is constant. This is true in both directions of motion and is illustrated in Figure 1.

Figure 1 shows a kind of phase space. In general, phase space would have a position and momentum coordinate for each degree of freedom. In three dimensions, 6N total coordinates are needed, where N is the number of atoms in the molecule. A molecule with three atoms, such as water, would have eighteen coordinates, three vibrations, three translations, and three rotations, with a position and a momentum for

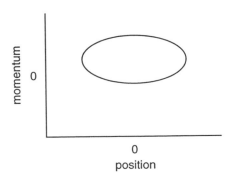

Fig. 1. Phase space.

each. One can also define a phase space with $6NN$ coordinates, where N is the number of molecules in the entire system. Then one point in this space would represent an entire gas (or condensed phase, for that matter). The space would have $6N \times 6 \times 10^{23}$ coordinates for one mole. It is a very useful kind of space, conceptually.

ENSEMBLES

To find a means of translating the information on individual molecules into information on the macroscopic sample of molecules, it is necessary to consider large groups of molecules. To begin with, we define an ensemble. This is a group of systems of molecules, not merely a group of molecules. Imagine a large number—say, a mole—of molecules as a system. Then imagine repeating the system a very large number of times, not physically, but conceptually. This can be an ensemble. To continue defining an ensemble, we need to constrain three variables. Suppose we choose to make the number of molecules, the volume, and the energy constant. We can illustrate the ensemble by repeating the first figure, with a slight variation, as shown in Figure 2.

The inner ellipse has energy E; the outer (corresponding to bigger swings of the pendulum, and so more energy) has energy E + dE (dE implies there is only an infinitesimal difference between the two energies). The shaded area contains all the systems, here pendulums with all possible momenta and positions for energy between E and E + dE; the area is supposed to be small, and all the systems have nearly the same energy.

This is a complementary description to that given in the entry on statistical thermodynamics, where the number of states, W, was connected to the entropy. Here, W is the number of states in the shaded ring, between E and E + dE, for a given N and V. Thus $W = W(N, V, E)$.

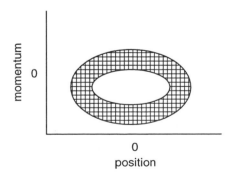

Fig. 2. Ensembles.

We could average any property (e.g., entropy) other than the three that are held constant over these systems. This kind of set of systems is called a microcanonical ensemble. A second possibility is to have the systems in equilibrium with a heat bath. The energy is no longer constant because in the course of exchanging energy with the heat bath, the energy can fluctuate. However, the temperature, T, is now defined by the heat bath. This kind of ensemble is called a canonical ensemble. Finally, among the common ensembles, we could allow exchange with a reservoir of particles. Then N is no longer constant, but the equilibrium with the particle reservoir requires μ, the chemical potential, to be constant (see the entry on statistical thermodynamics for the definition of chemical potential). This is the grand canonical ensemble.

In each case the ensemble is the assembly of all possible states of the system consistent with the constraints that characterize the system (e.g., for the microcanonical ensemble, the energy, number of molecules, and volume).

AVERAGES We must take averages over the ensemble. One could also imagine averages over time for a single system. It is close enough to being true that one can imagine time averages over a single system instead of ensemble averages. The time average supposes that the system visits all possible states (i.e., states consistent with the constraints) given enough time. The ergodic theorem states that the averages will be the same as long as the same states are counted an appropriate number of times and that this in fact happens when the time and ensemble averages are taken. The question will be important when molecular dynamics is discussed. For now, return to ensemble averages.

Are the averages over the different ensembles equivalent? In the thermodynamic limit, they are. The thermodynamic limit means that we consider the system to have so many particles that it is as good as infinite. Under these conditions there is no difference between allowing the temperature (constant E) or the energy (constant T) to fluctuate. The fluctuations are so much smaller than the average value of the properties that they do not matter. When we count the states, we observe that the number rises rapidly with energy because there are many ways to arrange the systems as the energy rises (at zero energy there is only one, the minimum energy configuration, e.g., a perfect crystal; see the third law in the entry on statistical thermodynamics). On the other hand, the probability of reaching a state drops sharply as the energy exceeds the average. Thus only states close to the average have much chance of being populated. For a very large number of systems it makes no difference whether all or nearly all the systems have the average energy. The properties of the system will be the same. Averages involving the grand canonical ensemble also will be the same. The different ensembles allow computations that make different physical assumptions about the system. The form of the distribution law is different for the different ensembles, although the averages are the same.

The entry on statistical thermodynamics describes the transition from microscopic (molecular) to macroscopic systems. There is an assumption underlying the description in that entry, the assumption of equal a priori probabilities. The relation between the number of states, W, and the entropy was discussed. Above, it was seen that W = W(N, V, E). The assumption of equal a priori probabilities states that all these states are, absent constraints, equally probable at thermodynamic equilibrium. This assumption underlies the apparatus of statistical thermodynamics and makes possible the calculations described in the entry on statistical thermodynamics. Because W is the number of states per unit of energy, it is known as the density of states. It is a density in energy space, not ordinary volume. This assumption does not mean that all states are equally occupied; it leads instead to the BOLTZMANN distribution when the appropriate constraints are applied—e.g., constant N and E (see the entry on statistical thermodynamics).

Application of Statistical Mechanics to Real Systems

To compute the thermodynamic properties of real systems, it is necessary to compute the partition function (see the entry on statistical thermodynamics) and other properties of the system. For a perfect gas, this is easy, as it is only necessary to compute the properties of individual molecules. However, for any real system it is necessary to take the interactions between molecules into account. If the interactions are very weak, they can be well described by two or three terms of an approximate equation, called the virial equation. (The equation can in principle be extended to any degree of accuracy but normally is not, for practical reasons.) However, as computers have improved, a technique called simulation has become popular. There are two types of simulation commonly used. Both require that the potential energy of interaction between pairs of molecules be known. Two possible means of obtaining the potentials are from *ab initio* calculations or from doing actual simulations and fitting the results to some of the known thermodynamic data. In the first case, SCHRÖDINGER's WAVE EQUATION is solved for the

molecule's wave function, thus getting the charge distribution and other properties that concern molecular interactions. Alternatively, one may guess as to reasonable parameters and do a simulation, change the parameters to get a better fit, and continue until the parameters are optimized. One might start with *ab initio* values, then optimize further. We are more concerned with the use of the parameters than with how they are obtained.

MONTE CARLO SIMULATIONS

Here, a group of molecules, with their interactions, determined by the parameters described above, are chosen to represent an ensemble. The starting positions being arbitrary, the molecules have to be allowed to equilibrate, by being moved sequentially. After this, additional moves allow the collection of data on the equilibrated ensemble. The data may include thermodynamic values, such as energy, as well as the position and orientation of the molecules. A number of factors must be considered in setting up a simulation. The number of molecules is typically in the hundreds with computers available in the 1990s. It is interesting that this is usually sufficient to get an adequate representation of thermodynamic quantities, except sometimes for heat capacities, which are the derivative (rate of change) of energy with temperature and which therefore have larger errors than the energy itself.

CHOICE OF ENSEMBLE The conditions of the simulation may matter. One possible choice of variables corresponds to a microcanonical ensemble, picking constant N, V, E. Also, the canonical ensemble in which the molecules are taken to be in equilibrium with a thermal reservoir, leading to N, V, T constant, can be used. Constant μ, V, T (grand canonical) also is possible.

BOUNDARY CONDITIONS If the sample of molecules is small, there must be some way to account for the interactions with more distant molecules. Typically, one uses periodic boundary conditions. These allow a repeat of the simulation volume across each boundary, so that each molecule interacts with fictional molecules with real potentials, rather than with hard boundaries.

MOVES The molecules are moved in random directions (and rotated if necessary, or moved in any other degree of freedom that is being simulated); after each move, the interaction energy with the other molecules is calculated. Then it is necessary to decide whether to accept the move as a new configuration or to reject it, making the new configuration identical to the old (either way, the configuration must be averaged in or the sample will be biased). Most commonly, one uses the Metropolis criterion, suggested by Metropolis and coworkers in 1953. If the energy is lower after the move, it is accepted; if it rises, the new move is accepted with probability $\exp(-\Delta E/k_BT)$, with ΔE the increase in energy. The bigger ΔE, the smaller the chance it is accepted. The net effect is to lower energy until equilibrium is reached. By that time, very few moves remain that lower energy further, so only a small number of moves can go down in energy. Many moves raise energy, and a fraction are accepted. The overall effect is to keep energy within a small range and allow the system to be sampled at equilibrium.

Sometimes the Boltzmann factors contain more than one part, and certain sampling tricks can be used to obtain results over a wider range of parameters. These sampling methods include shifts in the energy parameter with separated sampling in different ranges, a scheme called umbrella sampling.

Typically there are several thousand moves per molecule, or of the order of a million moves total, in a simulation.

Before considering the results, let us discuss the main alternative method of simulation.

Molecular Dynamics Simulation

Here the molecules are not moved at random. Instead, they follow the standard Newtonian laws of motion, essentially $F = ma$, where F = force on the molecules, m = mass, and a = acceleration. The force is found from the same molecular parameters discussed previously. When the energy is known for all positions, the rate at which it changes in a given direction is the force; this rule is entirely general when energy is not added to or taken away from the system.

If one knows the law of motion, one can find how the molecule moves in response to the forces exerted by the other molecules. The acceleration is the rate of change of the velocity, so from the original law, one can get the rate at which velocity changes by multiplying acceleration x times. Given the velocities, the new positions (change in position = velocity × time) of the molecules can be found. Then the new forces can be determined and used to get the next change in the velocities, and so on. To do this, one has to take one time step after another. However, if the steps are too long, then the other molecules also will have moved appreciably and the calculation will be inaccurate. Therefore very short time steps are needed, and the usual choices are 0.5 to 2×10^{-15} second (0.5 to 2 femtoseconds, abbreviated fs). Since 1 fs is a very short time, a great many steps are needed to reach equilibrium. Fortunately, most systems will

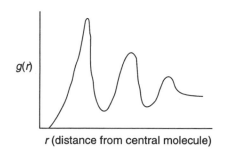

Fig. 3. Radial distribution function.

reach equilibrium in 10^{-9} s or less, so of the order of 10^6 steps will do. However, the method is just as limited as a Monte Carlo simulation for purposes of simulating most time-dependent phenomena for which interesting events generally take more than 10^{-9} s. Shorter events are different in that molecular dynamics has an inherent time scale, unlike Monte Carlo. We assume the validity of the ergodic theorem on the equivalence of ensemble and time averages in doing this type of simulation. Starting the calculation requires initial velocities and positions.

Results: Both forms of simulation can provide the thermodynamic functions of the system. In addition, the radial distribution function, $g(r)$, gives the number of molecules as a function of distance from a central molecule. Of course, the simulation can only give an average. This is particularly valuable for a liquid and can be compared with X-ray or neutron diffraction data. A typical radial distribution function might appear as shown in Figure 3.

There usually are two or three peaks of $g(r)$, measured in number of molecules. These define average intermolecular distances and get broader the farther out they are from the central molecule until eventually they disappear at the average density.

Other important properties that can be found by simulation include phase changes and critical points. These are harder to compute, but they can be found to better accuracy than was possible with approximations available prior to computers.

Summary

The apparatus of statistical mechanics makes it possible to compute the average properties of thermodynamic systems. In the entry on statistical thermodynamics, relations are described that connect the various quantities. In this entry we have described the means by which one can describe a system conceptually and the means by which these ideas can be transformed into calculations of the thermodynamic properties and some other properties of the system, such as the radial distribution function.

BIBLIOGRAPHY

CHANDLER, D. *Introduction to Modern Statistical Mechanics;* Oxford: Oxford University Press, 1987.

HILL, T. L. *An Introduction to Statistical Thermodynamics;* New York: Dover, 1960.

ISIHARA, A. *Statistical Physics;* New York: Academic Press, 1971.

PLISCHKE, M.; BERGERSON, B. *Equilibrium Statistical Physics,* 2nd ed.; Singapore: World Scientific, 1994.

STANLEY, H. E. *Introduction to Phase Transitions and Critical Phenomena;* Oxford: Oxford University Press, 1971.

TOLMAN, R. C. *The Principles of Statistical Mechanics;* Oxford: Oxford University Press, 1938.

MICHAEL E. GREEN

Statistical Thermodynamics

Atoms and molecules have their motions described by quantum mechanics (see QUANTUM MECHANICAL COMPUTATIONS). Large bodies—that is, large compared with the size of atoms—can be adequately described by classical mechanics as understood since the time of Isaac Newton. Besides overall motion there are internal motions that appear as the heat and other energy quantities, which in turn tell us how chemical equilibrium is arranged. These properties also belong to large (macroscopic) bodies, as there is no chemical equilibrium for systems of just a few atoms. This leads to the question of how the connection can be made between microscopic systems, for which quantum mechanics is correct, and macroscopic systems, for which quantum mechanics or even classical mechanics would be unable to keep track of all the atoms. No computer is big enough to follow the motions of all the atoms in a system composed of a million atoms, let alone a million million million, which would be about the minimum size of a solid particle visible to the naked eye. For this reason, statistical techniques were introduced.

There is a second reason for using statistical techniques: The equations of classical and quantum mechanics are time-reversible, but real life is not. If we see a film of someone eating a banana, we see the peel coming off the banana and the banana disappearing little by little into the person's mouth. If the film showed the banana growing as it was bitten, we would know the film was running backward. Nothing in mechanics indicates why there is a difference between forward and backward. However, by looking at the relative probability of certain sets of interactions among molecules, we can get a good idea of what is happening.

Therefore we must look at the interactions among particles (molecules) that lead to the properties of macroscopic matter, in particular, the way the effects of the interactions are averaged to give the thermal and chemical equilibrium properties. A second consequence will be a limited understanding of how the reversible equations describing individual interactions lead to irreversible effects for large numbers of molecules.

Energy

In classical mechanics energies can have any value. Think of a pendulum. This is large in the sense mentioned above. It is therefore classical. The energy of the swinging pendulum depends continuously on the amplitude of the swing. Molecules also oscillate, but they follow quantum mechanical rules; when the atoms in a molecule oscillate, they can have only certain energies. If the minimum energy of the oscillation is $1/2E_o$ (see the entry on quantum mechanical computations), then the only possible energies are $(n + 1/2)E_o$, where n is a whole number, 0, 1, 2, and so on. The quantum oscillator cannot have an energy of, say, $2.34E_o$. This is the difference between the energies of quantum and classical oscillators. When we consider a macroscopic system of a huge number of such oscillators, we average them in particular ways to get the total energy and the other related quantities we need to understand chemical equilibrium as well as the behavior of pure systems.

Boltzmann Distribution

To get an average energy or an energy total for all the molecules in the system, we have to know how many molecules have a particular energy. For example, for our quantum oscillator, we need to add up all the energy of the molecules. For this case, there will be N_o molecules with energy E_o, N_1 molecules with energy E_1, and so on. The total energy is therefore the sum of the molecular energies, $N_oE_o + N_1 E_1 + \ldots$ What we must then know is the number of molecules in each level, N_o, N_1, and so on. This distribution is known as the Boltzmann distribution after Ludwig BOLTZMANN, one of the pioneers of statistical mechanics. The distribution tells us that the relative number of molecules with energy E is proportional to $\exp(-E/k_BT)$, where T is the absolute temperature in kelvins (where K = °C + 273), and k_B is Boltzmann's constant, equal to 1.38×10^{-23} joule K^{-1}. The distribution can be written as

$$N_i = N_o \exp(-E_i/k_B T).$$

N_i is the number of molecules in level i, E_i is the difference in energy of molecules in that level with molecules in the lowest level, and N_o is the number of molecules in the lowest level. Then the total energy of the molecules is

$$E_T = N_o(E_o \exp(-E_o/k_B T) + E_1 \exp(-E_1/k_B T) + E_o \exp(-E_2/k_B T) + \ldots)$$

Partition Function

For the Boltzmann distribution we can add the relative populations of the individual states to get a quantity q:

$$q = \exp(-E_o/k_B T) + \exp(-E_1/k_B T) + \ldots$$

If we take this sum, and operate on the sum by taking the derivative (an operation from the calculus) with respect to $1/k_B T$, we can get the average energy.

$$E_{avg} = (E_o \exp(-E_o/k_B T) + E_1 \exp(-E_1/k_B T) + \ldots)/q$$

In other words, the average energy can be found if we just know q. This makes q a very important quantity, and we give it the name *partition function* (it is sometimes called the "sum over states," from its German name, and this is perhaps a more descriptive title). Knowing q is equivalent to knowing all the thermal properties of the system. Since we get q from the energy levels of the system, this connects the quantized energy levels to the macroscopic quantities, as we sought to do. The partition function acts in a sense as a normalizing factor for the distribution, making the total probability equal to one.

Other Thermodynamic Quantities

Entropy

With the average energy and the partition function, we can next get the entropy. This is a little more complicated, and we will not go through the mathematics to obtain the entropy. However, the entropy is particularly important in understanding the second of the two reasons for applying statistical techniques—that is, understanding irreversibility.

Entropy can be understood as a thermodynamic quantity—that is, as a property of a macroscopic part of matter. It also can be understood directly as a statistical property. In either form, it will have the same value. Therefore, as a macroscopic thermal property, it can be found from the partition function q, but the statistical form provides a more direct way to understand what entropy signifies. This is also the key to understanding irreversibility.

Consider a highly ordered physical system, and then subject it to a natural process. For example, think of a room with a neatly ordered set of children's blocks. Then allow a couple of three-year-olds into the room and come back in an hour. There is an excellent probability that the blocks will be ordered in a rather different and less predictable manner. Similarly, a solid, which is defined by the high degree of order of the atoms in the crystal, will, if heat is applied, change to a less ordered and hence less predictable form, a liquid. Still more heat changes the substance into a still less ordered form, a gas. In the highly ordered form, the original set of blocks or the solid has a low entropy, and applying the three-year-olds or the heat (which moves the molecules randomly) produces a higher-entropy form. If one counts the number of ways in which the molecules arrange themselves, one can make this quantitative. Suppose the blocks, or the solid at absolute zero temperature, to be perfectly ordered; there is then only one way to arrange the molecules (or blocks, which are indistinguishable from each other). Call the number of arrangements W, so at absolute zero temperature we can make W = 1. Entropy must apply to macroscopic systems, and it is the kind of property that depends on the amount of substance: For given conditions, twice as much matter has twice as much entropy if the matter is of the same type. There is only one way, then, to relate W to the entropy (which is always represented by the symbol S):

$$S = k_B \ln W$$

That is, the entropy is proportional to the (natural) logarithm of the number of ways of arranging the system. If you put together two pieces of matter, then the number of arrangements is the number of arrangements of the first piece times the number of arrangements of the second (if you can arrange the first piece 100 ways and the second 200 ways, the combination has 100×200 arrangements because each of the first 100 can be matched to any of the 200 others). However, the total entropy must be the sum of the two entropies. This is possible with the formula above because $\ln (W_1 \times W_2) = \ln W_1 + \ln W_2$, so the total entropy is

$$S = k_B \ln W = k_B \ln (W_1 \times W_2) = k_B \ln W_1 + k_B \ln W_2 = S_1 + S_2$$

(The formula $S = k_B \ln W$ is due to Boltzmann and was his proudest accomplishment; it also serves as his epitaph.)

As one example, consider a system in which a gas is confined in a volume V, which is half of a box of total volume 2V, the other half being a vacuum (see Figure 1). Suppose that the barrier separating the two halves is removed and the gas let free. It will move into the other half, as each molecule now has twice as many states available to it as before. If there are N molecules and each has twice as many states, there will be $2 \times 2 \times 2 \times 2 \times \ldots$ as many states, for a total of 2^N times as many states. This is the increase in W, so the new entropy is therefore $k_B \ln 2^N$ greater than the old, or $N k_B \ln 2$ greater. If N equals AVOGADRO'S NUMBER, one gets an increase of $N_A k_B \ln 2$. Since the gas constant R equals $N_A k_B$, this can be written as $R \ln 2$. The classical thermodynamic result for the increase in entropy on the expansion of an ideal gas is precisely this. This is another example of the way classical and statistical concepts match.

We also observe that we would expect to see the gas expand. At the same time, the entropy increases. In fact, this is quite general; if a process is irreversible, as the expansion of a gas into a vacuum is, entropy increases. This is a way of stating the famous Second Law of Thermodynamics: In any closed system (like our box, a system does not exchange heat or matter with its surroundings), entropy increases until it reaches a maximum. At that point the system is in equilibrium and can change no more unless the system ceases to be closed.

See also STATISTICAL MECHANICS on the law of equal a priori probabilities.

There is also a Third Law of Thermodynamics: It was observed above that a crystal would be perfectly ordered at absolute zero temperature (T = 0), meaning that only one configuration, W, exists, and entropy, proportional to the log of W, becomes zero. This is one way to get an absolute zero of entropy (for which we normally measure only differences). The third law takes the entropy of every pure element in its most stable state at T = 0 to be zero. This makes the entropy of every substance a positive tem-

perature. At T = 0, crystalline compounds may also have entropy = 0.

FREE ENERGY

The free energy of a system can be thought of in several ways. It is, to begin with, the sum of the energy and the temperature times the entropy ($T \times S$ is another kind of energy): $A = U - TS$ (A is free energy). It can be found directly for a gas from the partition function per mole as written:

$$A = -N_A k_B T \ln q$$

(This is approximate; it assumes no additional energy in the interactions of the molecules because q is the partition function of one molecule as we have defined it, and the formula makes another kind of simplification, which we will ignore.) For condensed phases, the formula needs modification mainly by inclusion of molecular interactions. Entropy is the change in heat divided by the temperature, T, for a reversible process. Suppose there is a change in the system (say, a chemical reaction), and the system can exchange heat with its surroundings. The heat change is reversible because the surroundings are so large that the change occurs very slowly, hence reversibly. The surroundings are very large, so this will not change the temperature of the surroundings. Then there will be a change in energy, which will result in the heat generated being dumped into the surroundings, or else the heat absorbed being taken from the surroundings. Then the heat divided by T becomes the entropy change of the surroundings. There will in addition be an entropy change of the system itself. Therefore the free energy change divided by T is the total entropy change of the system plus surroundings, or the total entropy change of the whole reaction, or other change. Since the direction of spontaneous change is increasing entropy, if we can determine the free energy change of the reaction, we know the position of chemical equilibrium. A reaction will proceed until the concentration of the reactants and products is such that the free energy is at a minimum (entropy is at a maximum—the signs are chosen so that the free energy goes down when total entropy goes up). This makes the free energy a peculiarly important quantity, and its direct relation to the partition function means that an equilibrium constant for any chemical reaction can be determined if the partition function of the reactants and products is known. (see the end of this entry for a more detailed discussion).

Finally, the free energy as we have defined it, like the energy, is used at constant volume. Most chemical reactions are run at constant pressure. There is another form of energy, called enthalpy, and of free

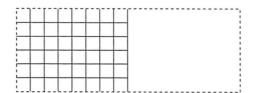

Fig. 1.

energy, called the Gibbs function, which is used at constant pressure. Using these quantities, all the rest of the discussion carries through at constant pressure.

HEAT CAPACITY

The HEAT CAPACITY is the last thermodynamic quantity we will discuss. The heat capacity measures how much heat a substance can absorb as the temperature changes. Some substances absorb a lot—water, for example. Others—say, a metal such as copper—absorb only about a tenth as much per kilogram for each degree of temperature rise. Heat is a form of energy of motion (kinetic energy) and is absorbed by motion of the molecules that make up the substance: by translations, that is, motion of the whole molecule; by rotations of the whole molecule; and by vibrations, like those we have discussed before. The energy distributes itself among these degrees of freedom at equilibrium. The heat capacity is the rate of change of energy (at constant volume) or enthalpy (at constant pressure) with temperature.

Since the degrees of freedom of any monatomic substance are the same if the temperature is high enough to excite them all, all monatomic substances have approximately the same molar heat capacity (molar because we have to compare the same number of atoms, not the same mass). Understanding this point was one of the triumphs of classical physics; its failure at low temperature was a catastrophic failure of classical physics and was one of the first significant phenomena to be explained by quantum mechanics. In 1900 Max PLANCK, a professor in Berlin, explained blackbody radiation by assuming the oscillators in a hot solid could have only certain energies ε, 2ε, 3ε, but nothing in between. In 1905 Albert EINSTEIN, in addition to providing the special theory of relativity and two other major discoveries, pointed out that the vibrations in a cold solid could be thought of similarly. This had the consequence that at very low temperatures there would not be enough thermal energy to excite the oscillators—that is, to move them from the lowest energy state, ε, to the second state, 2ε. As there was no possible energy between these in the quantum theory, the oscillator could absorb no energy at all, and the heat capacity would fall all the way to zero at zero temperature, where none of the oscillators could be excited. This is unlike the classical theory in which the states could not be treated separately, and any energy, no matter how small, would be absorbed, so the heat capacity would remain the same all the way to absolute zero temperature. Einstein's explanation of the experiments on heat capacity at low temperature was a significant step in gaining acceptance for QUANTUM THEORY.

Not only is this of historical interest and of practical importance, but also it illustrates the relationship between the motions of the atoms of which matter is composed and the macroscopic thermodynamic quantities that describe the behavior of ordinary matter.

Chemical Equilibrium

A chemical reaction occurs when atoms rearrange into new molecules. The new molecules in turn can revert to form the original molecules again. The rates at which these two rearrangements occur balance eventually, with as many of the new molecules being formed per unit time as are destroyed. If the formation of new molecules is very fast relative to the reverse, there will be many of the new molecules and few of the original molecules to make the overall rates of formation and destruction balance. If the reverse reaction is fast, there will be little product (new molecules) and much reactant to make the rates equal. That is, the probability of the reverse reaction for each molecule of product is high, and therefore only a few are enough to make as many reactant molecules as the many reactant molecules make of product molecules. If the inherent rates are just equal, there will be equal numbers of reactant and product molecules.

The relative rates are determined by the thermodynamic quantities we have been discussing. Recall that the free energy (the negative of the total entropy change) must be a minimum at equilibrium (entropy a maximum) in any closed system for any process. The change of free energy in reaction is the difference of the total free energy of the products and that of the reactants. In principle we know how to get the free energy from the partition function. Alternatively, we can measure changes in free energy. Once we know the free energy change, we can calculate the position of equilibrium. Usually the position of equilibrium is expressed in terms of an equilibrium constant, which is a kind of ratio of products to reactants. However, this contains no information not present in the free energy change, and is in fact quite simply related to this quantity.

BIBLIOGRAPHY

ATKINS, P. W. *Physical Chemistry*, 5th ed.; New York: Freeman, 1994.

———. *The Second Law*; New York: Freeman, 1984.

ROCK, P. A. *Chemical Thermodynamics*; Mill Valley, CA: University Science Books, 1983.

MICHAEL E. GREEN

Stellar Chemistry

See ORIGIN OF THE CHEMICAL ELEMENTS;
SPACE CHEMISTRY.

Stereochemistry

Stereochemistry (from the Greek *stereos,* meaning solid) is the study of isomeric substances that have the same connectivity but differ in their three-dimensional geometrical arrangements. Two categories exist for comparing different three-dimensional geometries: *configuration* and *conformation*. Because there is an empirical component to these terms, a universally unambiguous distinction between configurational stereoisomers (sometimes simply referred to as stereoisomers) and conformational stereoisomers (sometimes simply referred to as conformations or conformers) has not emerged. According to Eliel and Wilen (1994, p. 102), configurational stereoisomers result from "arrangements of atoms in space of a molecule with a defined constitution without regard to arrangements that differ only by rotation about one or more single bonds, providing that such a rotation is so fast as not to allow isolation of the species so differing." Conformational isomers are then taken to be the result of such bond rotations. Representative examples of conformational and configurational isomers are presented in Figures 1–3.

3-Methyl-1-cyclohexanol has four configurational stereoisomers, as shown in Figure 1. The first of the 3-methyl-1-cyclohexanol stereoisomers shown in Figure 1 is often expressed in terms of an equilibrium between the two "chair" conformational stereoisomers shown in Figure 2. 4-Methyl-1-cyclohexanol is more symmetrical than the 3-methyl isomer and only has two configurational stereoisomers (see Figure 3). The overall molecular symmetry for simple alkenes is analogous to 4-methyl-1-cyclohexanol in that two configurational stereoisomers are possible.

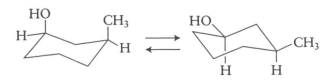

Fig. 2. "Chair" conformational stereoisomers of 3-methyl-1-cyclohexanol.

In the mid-nineteenth century, scientists combined discoveries from crystallography with the developing picture of molecular structural isomerism to move chemistry into the third dimension. Jean Baptiste Biot (1774–1862), a physicist and a crystallographer, observed optical activity in samples of certain liquids such as turpentine, various essential oils, and also solutions of substances such as sugar and camphor. Before then, optical activity had been observed with mineralogical samples such as quartz, where an asymmetrical crystalline form was observable. Because both pure liquids and solutions or organic compounds exhibit optical activity, the property could be attributed not to a characteristic of the solid state but rather to the molecular structure. In 1848, Louis PASTEUR separated an optically inactive sample of a tartaric acid salt into optically active dextrorotatory and levorotatory components by physically segregating the enantiomorphous crystalline forms. He established that what had been designated as a distinctive substance (racemic acid) was a mixture of two compounds: the natural, dextrorotatory tartaric acid and a substance that, although identical in all of its other chemical properties, was yet opposite in its solid-state structure and in its observed rotation of polarized light. In 1860, Pasteur proposed that molecular asymmetry was the basis for these observations, although at that time the idea of using molecular structure to understand reactivity differences was still uncommon.

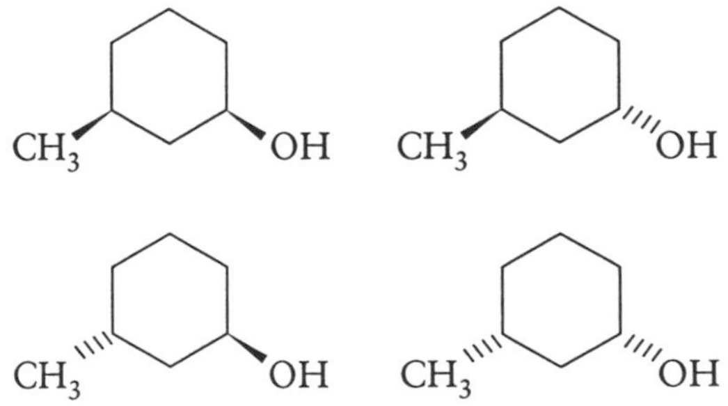

Fig. 1. Configurational stereoisomers of 3-methyl-1-cyclohexanol.

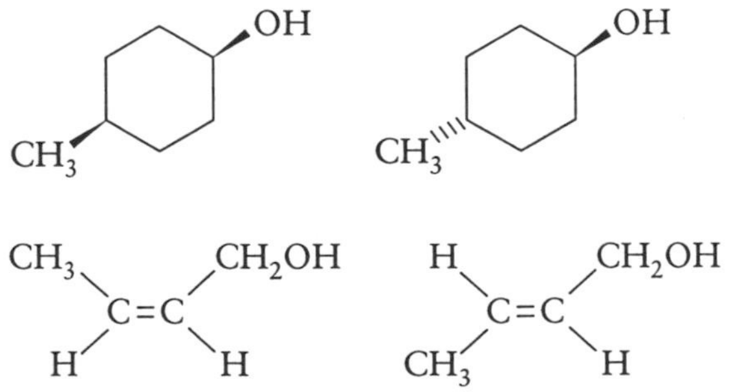

Fig. 3. Configurational stereoisomers of 4-methyl-1-cyclohexanol.

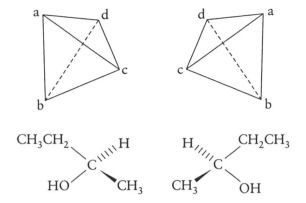

Fig. 4. Nonsuperimposable mirror image (enantiomorphic) relationship in a stereoisomeric pair.

As molecular structural theory matured throughout the late 1800s, the number of substances whose structures were identified increased dramatically. Into this environment, Jacobus VAN'T HOFF (1852–1911) and Joseph Achille LE BEL (1847–1930) independently proposed, in 1874, that molecular asymmetry and its consequences on isomerization could be explained if the arrangement (configuration) of the groups on a tetravalent atom was tetrahedral. Macroscopically or microscopically, a tetrahedral array of four different things gives rise to two and only two different arrangements that have a nonsuperimposable mirror image (enantiomorphoric) relationship (Figure 4). The term *stereoisomers* was introduced by

Victor Meyer in 1888 (Ihde, 1964, p. 328) to describe molecules that differ only in their three-dimensional arrangements. For many years the term *optical isomers* was used synonymously with the term *geometrical isomers* in situations where issues of optical activity were not involved. Today, *stereoisomer* is emerging as the preferred generic term for any isomeric relationship based on three-dimensionality.

Taken in combination, the proposition for tetrahedral atoms enabled chemists to resolve the problems in isomerization that had emerged. In the case of the $C_4H_4O_4$ fumaric and maleic acid isomers, not only could the existence of two geometrical isomers (today: stereoisomers) be confirmed, but also the easy transformation of maleic acid to maleic anhydride could be accounted for structurally based on the spatial proximity of the carboxylic acid functional groups (Figure 5).

The original problem of understanding isomerism in the tartaric acids was reconciled by examining the consequences of two singly bonded and asymmetrical tetrahedral carbon centers. Four isomers had been named: two optically active forms, dextrotartaric and levotartaric acids; and two optically inactive forms, mesotartaric and racemic acids. One of the two optically inactive forms could be understood to be a single symmetrical substance where the sense of the asymmetry in one-half of the molecule was internally compensated by its opposite sense in the other (mesotartaric acid), while racemic acid, as demonstrated by Pasteur, was a mixture of the *dextro* and *levo* forms and not a separate isomer (Figures 6 and

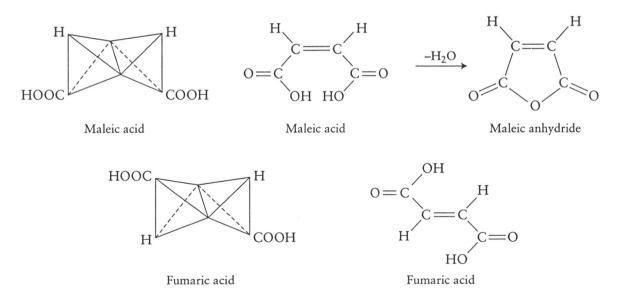

Maleic acid   Maleic acid   Maleic anhydride

Fumaric acid   Fumaric acid

Fig. 5. Stereoisomeric relations in maleic acid and fumaric acid and structures of maleic acid, maleic anhydride, and fumaric acid.

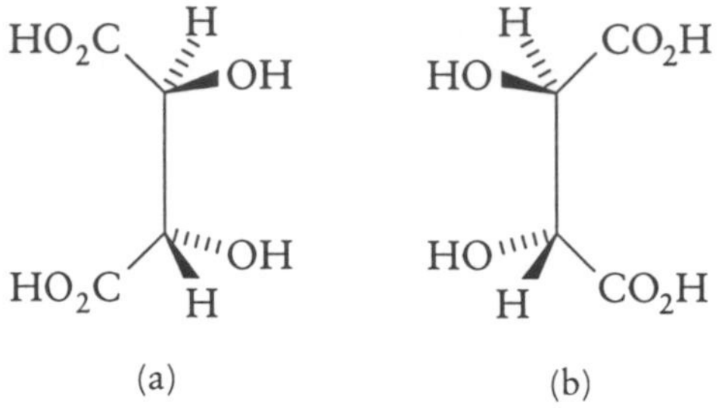

Fig. 6. (a) Levo- and (b) dextrotartaric acids.

7). The term *meso* is still used to classify stereoisomers with such internal compensation, and the term *racemic* evolved to represent a sample made up of equal numbers of molecules with the opposite sense of asymmetry.

In 1875, van't Hoff also predicted that the extended tetrahedron formed by even numbers of cumulative double bonds should result in two (optical) stereoisomers (Figure 8), a phenomenon not observed until 1935 (Eliel and Wilen, 1994, p. 1122).

The term *chiral,* from the Greek for hand and applied to molecular systems whose asymmetry results in handedness, was coined in 1884 by Lord Kelvin (Eliel and Wilen, 1994, p. 4) but did not come into common usage until the 1960s. Asymmetrical atomic centers giving rise to stereoisomers have been known as chiral centers, although the term *stereocenter* (derived from the more generically correct term, stereogenic center) has come into common usage since the 1980s. *Chirality* is a term that can be applied to mixtures in addition to individual molecular species. Mixtures of chiral molecules can range from having 100 percent of the sample representing the same sense of asymmetry (in which case the sample is a collection of homochiral molecules) to equal representation by molecules and their mirror image iso-

mers (in which case the sample is heterochiral, racemic), or any distribution in between heterochiral and nonracemic. Extended tetrahedra (Figure 8) are representative of molecules with an axis of chirality rather than a center of chirality. Molecular chirality results from one degree or another of twisting within a molecular structure, where a turn to the left can be distinguished from a turn to the right. Even a simple stereocenter (Figure 4) or an allene (Figure 8), when viewed from another perspective, presents a molecular twist that emerges as a common theme in the three-dimensional structure of chiral geometries (Figure 9).

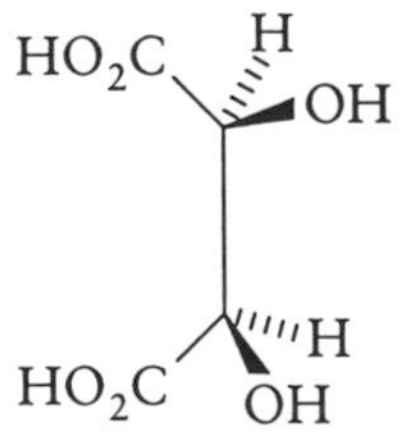

In most cases, an actual conformational (bond twist) change results in the formation of different three-dimensional geometry. In the event that these different molecular geometries represent substances that can be easily isolated from one another, they are generally called configurational isomers. Restricted rotation in biphenyls (Figure 10) is an example of this kind of isomerism, which is called atropisomerism (literally, "not turning," a reference to the restricted rotation). In other cases, the overall molecular architecture causes one sense or another of a twist to form. In helicenes, the simple interconversion of having one end of the molecule move past the other is restricted and results in isolable substances (Figure 11). For many bond rotations or twists, the resulting geometries do not represent easily isolable substances, and they are generally called conformation isomers. Different molecular geometries resulting from the bond rotations in butane, for example, interconvert on a fast timescale at extremely low temperatures. The three staggered geometrical forms for rotation about the C2–C3 bond in butane, represented by the Newman projections shown in Figure 12 (Newman, 1955), are classified as conformational isomers. Like the configurational isomers of tartaric acid, two of the conformational isomers of butane are chiral, while the third is not.

The use of the terms *configurational* and *conformational* as classificational adjectives to describe the empirical or observational distinctiveness of stereoisomers has emerged since the 1970s and is preferred to the noun forms *configuration* and *conformation,* which are both historically significant and in common usage. The two comparative categories into which stereoisomers can be placed are absolutely distinctive and orthogonal in definition. Two stereoisomers that have a nonsuperimposable mirror image relationship are called ENANTIOMERS (reflecting the historical and crystallographic heritage of enantiomorphic forms; *enantiomorphic* literally means "in opposite shape"). The only other category is de-

Fig. 7. Mesotartaric acid.

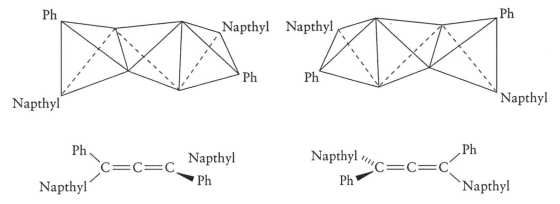

Fig. 8.   Optical stereoisomers resulting from the extended tetrahedron formed by even numbers of cumulative double bonds in a representative allene.

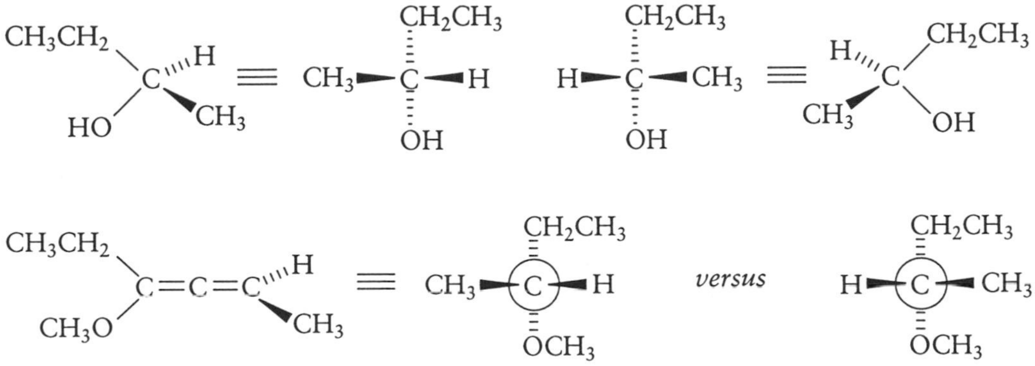

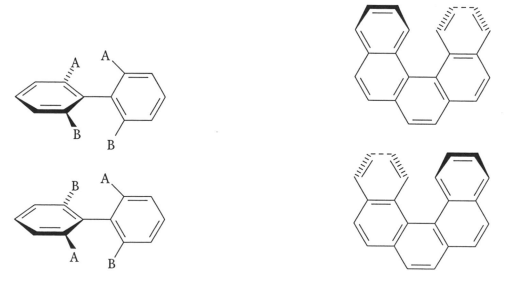

Fig. 9.   Some representative three-dimensional chemical structures exhibiting chiral geometries.

Fig. 10.   An example of restricted rotation in a biphenyl.

Fig. 11.   An example of restricted rotation in a helicene.

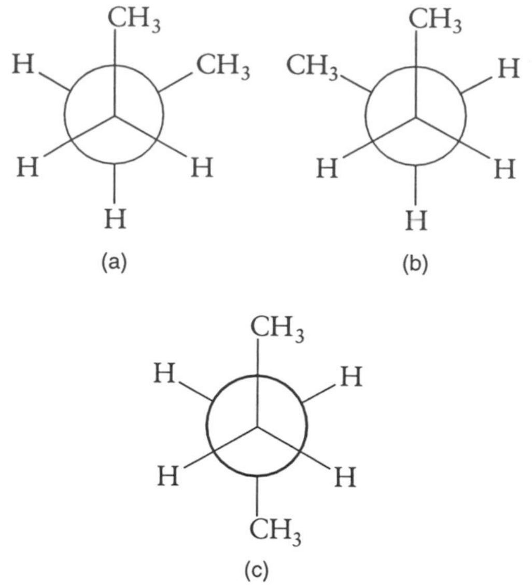

Fig. 12. Newman projections of the conformational isomers of butane: (a) and (b) are chiral, whereas (c) is achiral.

fined negatively. Stereoisomers that are not enantiomers are called DIASTEREOMERS. The molecules in Figures 9, 10, and 11 would also be called configurational enantiomers. The top two drawings in Figure 12 represent conformational enantiomers, while the relationship between either of the top two drawings in Figure 12 and the lower drawing is as conformational diastereomers. Levotartaric and dextrotartaric acid (Figure 6) are configurational enantiomers, and the relationship between either of these and mesotartaric acid (Figure 7) is as configurational diastereomers. Maleic and fumaric acids (Figure 5) are also configurational diastereomers.

The need to represent and to label three-dimensional geometry was another consequence of van't Hoff and Le Bel's proposition. The representations used by van't Hoff derived from solid geometry and were comparable to the tetrahedral shapes shown in Figures 4, 5, and 8. The "lines, dashes and flying wedges" system used throughout these examples

came into common usage in the 1960s. The point of view represented in Figure 9 is implied in the system proposed by Emil FISCHER in 1891 and is still referred to as the Fischer projections (Figure 13).

From 1874 to 1951, there was no experimental method that could be used to distinguish one enantiomeric form from the other. The physical property of optical activity formed the first basis on which labels were made. Dextrotartaric and levotartaric acid came to be known d- and l- or (+)- and (−)-tartaric acids, respectively, where d-(+)- and l-(−)- simply stand for the empirical observation of dextrorotatory and levorotatory optical rotations at a specified wavelength. The use of the letters d- and l- is considered obsolete, and the use of the (+) or (−) is preferred. Racemic mixtures may be labeled dl- or (±) or rac-, although the latter two are preferred. The label $meso$ can be used as a descriptor for an achiral diastereomer that also includes at least one source of chirality in its structure. The problem of not knowing the absolute configuration of a stereocenter was resolved by using relative identification. Naturally occurring (+)-glyceraldehyde [$HOCH_2CH(OH)CHO$] was usually used as the point of reference. Whatever its actual three-dimensional geometry, natural (+)-glyceraldehyde was assigned the designation (D), which stood for the configuration of dextrorotatory glyceraldehyde, not its optical rotation, which was designated as d- or (+). Optically active molecules with single stereocenters of unknown configuration were, by precise chemical transformations, derived from or transformed into (D)-(+)-glyceraldehyde. By monitoring the transformations at the unknown center, it could then be assigned to have been the same as natural glyceraldehyde, hence (D), or to have been that of its enantiomer, hence (L). In 1951, J. M. Bijvoet, A. F. Peerdeman, and A. J. Van Bommel showed, using X-ray crystallography, that the absolute arrangement of atoms in space for sodium rubidium tartrate could be determined. Overnight, 75 years of relative configurational assignments could be shown as actual three-dimensional shapes.

As absolute configurations came to be known, a different system of labels arose. The use of (D) and (L) has become obsolete except where strongly adhered

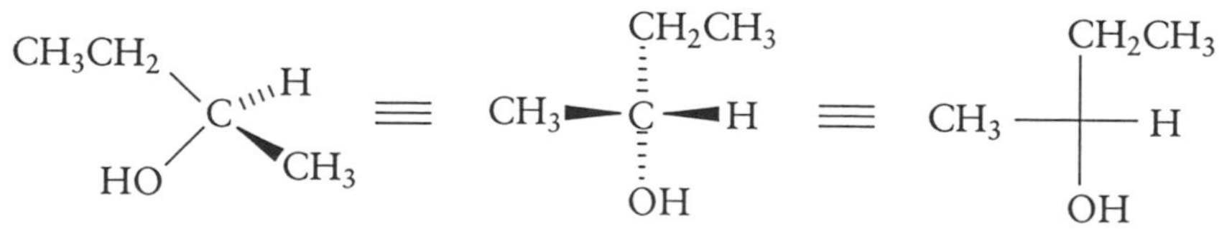

Fig. 13. Fischer projections.

to by historical usage in identifying families of the most common naturally occurring carbohydrates, called the (D)-series, and the common naturally occurring amino acids, called the (L)-series. Using a system of assigning priorities known as the Cahn-Ingold-Prelog rules (Ege, 1994, pp. 211–214), the groups of atoms attached to a stereocenter (as well as other sources of stereoisomers) can be arranged and viewed so as to result in a unique label. For configurational enantiomers that arise through stereocenters of tetrahedral arrays, the configurational descriptors are (R) and (S). The molecular structure shown in Figure 13 represents (R)-2-butanol. Its configurational enantiomer (Figure 4) is (S)-2-butanol. It is not possible to predict, *a priori*, whether the (R)-isomer will be dextrorotatory or levorotatory. However, the specific rotations of (R)- and (S)-isomers will be equal in magnitude and opposite in sign. Helices and other identifiably twisted architectures are assigned descriptors based on the sense of turning of the twist. The label (P), for plus, is used for right-handed twists, while (M), for minus, is used for left-handed ones. Molecular geometries that arise from an axis of chirality rather than a center of chirality, such as allenes and biphenyls, may be labeled according to a modification of the rules for helicity (P and M), or by rules for extended tetrahedra and the labels aR and aS. Sometimes, as in the case of the maleic and fumaric acids (Figure 5), configurational diastereomers result from completely achiral systems. Historically, the labels *cis* (for same-sided groups, as in maleic) and *trans* (for opposite-sided groups, as in fumaric) were used to distinguish the configurational isomers possible for some alkenes. Except for fairly symmetrical cases, these labels cannot be used unambiguously. Instead, the preferred descriptors are based on assigning Cahn-Ingold-Prelog priorities (high or low) to the pairs of groups on each end of the double bond. If the two high-priority groups are *cis*, the stereochemical descriptor is (Z), from the German *zusammen*,

meaning "together," if the two high-priority groups are *trans*, the stereochemical descriptor is (E), from the German *entgegen*, meaning "apart."

The majority of naturally occurring (plant and animal) substances for which stereoisomers are possible exist in nature as single enantiomers. The biochemical pathway results in a single stereoisomer of cholesterol even though its connectivity represents a total of 128 possible stereoisomers. One of the important characteristics in chemical and biochemical synthesis is understanding the stereochemical outcome of reactions that involve the transformation or preparation of stereoisomeric compounds. The epoxidation reaction of the norbornyl derivatives shown in Figure 14 illustrates the use of two terms that are used to describe the stereochemical course of a reaction: *stereospecificity* and *stereoselectivity*. The aspect of these reactions wherein the stereochemical outcome depends on the configurations of the two starting materials is called the stereospecificity. The mechanism of this reaction results in little to no loss of the stereochemical information from the starting materials, so the reaction is highly stereospecific. The aspect of these reactions wherein the stereochemical outcome depends on the difference in energy between the *exo* and *endo* pathways is called the stereoselectivity. The stereoselectivity in these reactions is lower than the stereospecificity in that obvious amounts of products from both reaction pathways are formed. Because the selection represented in this reaction is for one diastereomer over another, the outcome is further described as diastereoselectivity. The relative amounts of each diastereomer would be given in terms of a diastereomeric ratio such as 80 : 20. Reactions in which one enantiomer is formed selectively are expressed in terms of enantiomeric excess. An 80 : 20 ratio would correspond to a 40 percent enantiomeric excess, or 40 percent ee. In reactions where two diastereomeric products are ultimately transformed into a corresponding set of enantiomers, the diastereomeric ratio

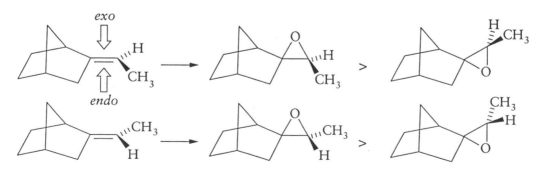

Fig. 14.   Epoxidation reaction of norbornyl derivatives.

can also be expressed in terms of a diastereomeric excess, or de.

Stereochemistry plays a significant role in inorganic chemistry and biochemical and materials science. Examples of stereoisomerism extend from the main group elements, especially carbon, sulfur, phosphorous, and silicon, to transition metal complexes where the increased number of valences creates an even wider variety of stereochemical alternatives. The structure and function of enzymes rely on the interplay of three-dimensional interactions. The primary structure of proteins, which is the linear sequence of amino acids that compose the macromolecule, is a polymeric array of chiral compounds. In water, large fragments of the molecule fold on themselves in predictable ways. The secondary structure of proteins is the set of locally regular arrangements that result, in part, from the series of homochiral molecules bonded together. These stereochemical organizations include helices and pleated sheets (a portion of the polymer chain that folds back on itself with multiple, spaced hairpin turns in order to make a relatively flat molecular surface). A great deal of effort in computational chemistry has been devoted to the protein folding problem. The tertiary structure of globular proteins corresponds to the overall shape, or dynamic family of shapes, of the macromolecule. The shapes of the exterior and interior surfaces and the position and directionality of the involved functional groups all contribute to the biological activity of these chiral molecules. Associations between enzymes and their substrates rely on many different diastereomeric interactions, resulting in highly selective reactions. Stereochemistry also mediates some of the properties in nonbiological polymers. Stereocenters along a polyethylene backbone can end up regularly, alternatingly, and irregularly arranged, each diastereomeric arrangement giving the corresponding polymers different properties.

Macroscopic physical and physiological effects that can be correlated with stereochemical features in key compounds extend beyond enzymatic stereoselectivity. Intermolecular interactions and supramolecular assemblies always rely, in part, on the spatial complementarity of the associated molecules. The biochemical signals for taste and odor are often sensitive to stereochemical differences. The chief molecular component in the essential oil of spearmint, (*S*)-(−)-carvone, is the enantiomer of the molecule found in caraway oil, (*R*)-(+)-carvone (Figure 15). Most natural amino acids have the (*S*)-configuration and they taste bitter, while many of the unnatural (*R*)-isomers taste sweet. By leaving the carboxylic acid side-chain from aspartic acid underivatized in methyl (*R*)-aspartyl-(*R*)-phenylalaninate (aspartame, an artificial sweetener), a portion of the molecule mimics

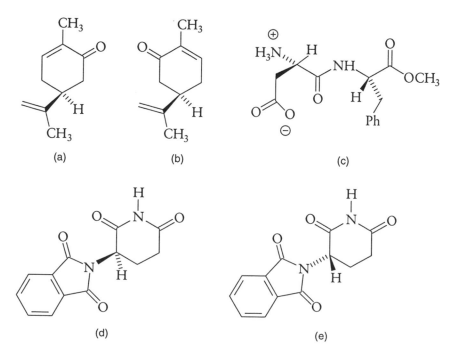

Fig. 15.   (a) (S)-(−)-carvone (spearmint); (b) (R)-(+)-carvone (caraway); (c) aspartame; (d) (R)-(+)-thalidomide; (e) (S)-(−)-thalidomide.

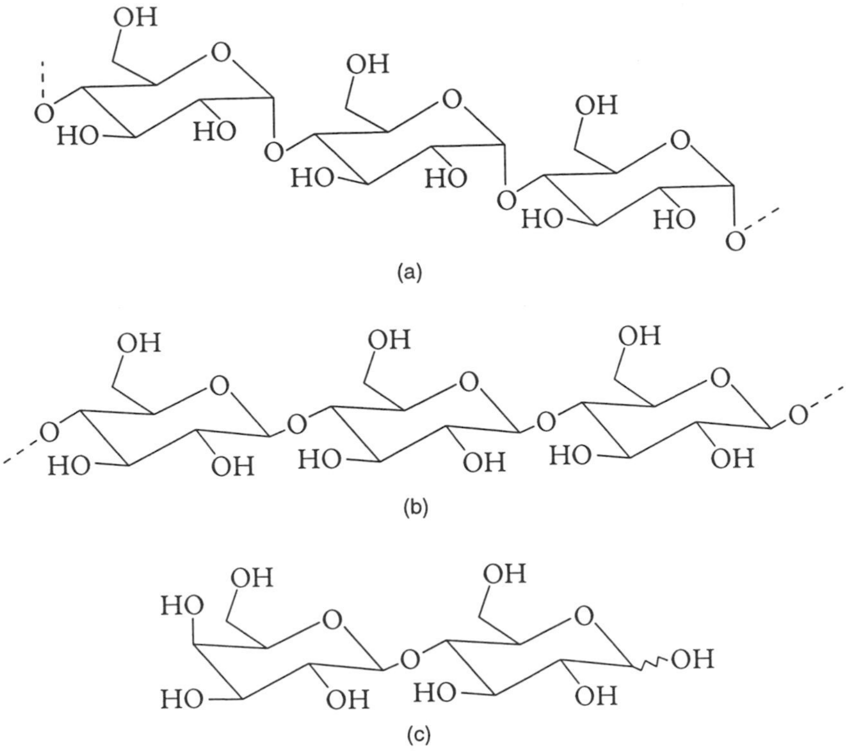

Fig. 16.   (a) A three-piece subunit of starch; (b) a
three-piece subunit of cellulose; (c) natural lactose.

the functional groups in the geometry of an (S)-configuration. The use of thalidomide to treat nausea in pregnant women during the 1960s is a significant incident that heightened stereochemical awareness in the biomedical and pharmaceutical communities. The use of the racemic thalidomide in Europe was fairly quickly correlated with a dramatic increase in babies born with limb deformities. The (R)-(+)-isomer was claimed not to cause deformities in animals even in high concentrations, while the (S)-(−)-isomer was a potent cause of fetal abnormalities as well as an antiabortive that enabled these babies to be carried to term when the mothers might have otherwise miscarried (Figure 15).

The hydrolysis of a polysaccharidelike starch is mediatated by hydrolase enzymes that can be found in the digestive process as early as the mouth. The shape of the axially oriented polyglucose chain is easily accepted by the active site of the hydrolase. In cellulose, the diastereomeric and equatorially oriented polyglucose is neither accepted nor easily hydrolyzed. This selectivity even extends to the hydrolysis of the equatorially oriented linkage found in lactose, which is not hydrolyzed by the starch hydrolases and requires its own enzyme, lactase, in order to

be digested (Figure 16). Lactase deficiency results in unhydrolyzed lactose and consequent gastrointestinal distress in many individuals. Commercial products are available that can artificially boost the concentration of lactase in individuals with lactose intolerance.

BIBLIOGRAPHY

ALLINGER, N. L. "Conformational Analysis in the Elementary Organic Course." *J. Chem. Educ.* 1964, *41*, 70–72.

BAXTER, J. F. "Introduction to the Symposium on Three-Dimensional Chemistry." *J. Chem. Educ.* 1964, *41*, 65.

BIJVOET, J. M.; PEERDEMAN, A. F.; VAN BOMMEL, A. J. "Determination of Absolute Configuration of Optically Active Compounds by Means of X-Rays." *Nature* 1951, *168*, 271–272.

BUSCH, D. H. "The Stereochemistry of Complex Inorganic Compounds." *J. Chem. Educ.* 1964, *41*, 77–85.

EGE, S. N. *Organic Chemistry*, 3rd ed.; Boston: D. C. Heath, 1994.

ELIEL, E. L. "Teaching Organic Stereochemistry." *J. Chem. Educ.* 1964, *41*, 73–76.

ELIEL, E. L.; WILEN, S. H. *Stereochemistry of Organic Compounds*; New York: Wiley, 1994.

IHDE, A. J. *The Development of Modern Chemistry*; New York: Harper, 1964.

INGRAHAM, L. L. "Three-Dimensional Effects in Biochemistry." *J. Chem. Educ.* 1964, *41*, 66–69.

NEWMAN, M. S. "A Notation for the Study of Certain Stereochemical Problems." *J. Chem. Educ.* 1955, *32*, 344–347.

BRIAN P. COPPOLA

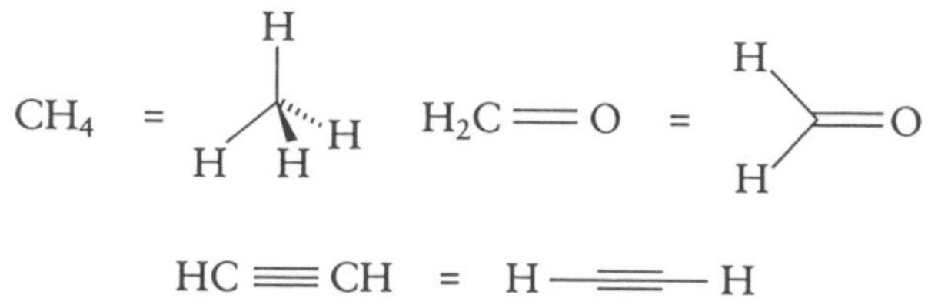

Fig. 1. Example of repulsive interaction between bonding electrons.

Stereoelectronics

The *stereo* part of the word stereoelectronics comes from the Greek *stereos*, meaning "solid" and tells us that we are dealing with objects in three dimensions. In this case the objects are molecules, and stereoelectronics is concerned with interactions of bound electrons, through space between molecules and through and between parts of the same molecule. The strengths of such interactions depend on their geometry and can result in preferences for particular molecular arrangements, interpreted as stereoelectronic effects on structure. Related effects operate on TRANSITION STATES for reactions and may give rise to stereoelectronic control of reactivity. The effects have been identified and interpreted most convincingly in ORGANIC CHEMISTRY, and the examples below are mostly organic. But the underlying interactions are equally pertinent to inorganic systems.

The properties and behavior of molecules—in other words, their chemistry—are determined primarily by the electrons that make up their bonds or are available to form bonds—like the LONE PAIRs of electrons on amines or the unpaired electrons of radicals. These electrons are best thought of as concentrations of electron density, known as orbitals, with rather well-defined geometries, or shapes. These shapes can be calculated by the methods of quantum mechanics and can be visualized experimentally by special crystallographic techniques. The typical interaction between them is repulsive—as would be expected between interacting negative-charge clouds. These repulsive interactions account for the vast numbers of nonproductive collisions between molecules in the gas and liquid phases. To a large extent they account also for the geometrical arrangements around atoms in molecules. Thus the four groups at a saturated carbon center are arranged tetrahedrally; if there are three, the arrangement is trigonal; if only two, a linear arrangement is preferred. In each case the groups end up as far from each other as possible, as expected if the dominant interaction is repulsion between the bonding electrons—see Figure 1.

Similar repulsive interactions between bonding electrons operate across one or more bonds and are the basis of conformational preferences. The normal preference across a single bond is for the staggered arrangement (Figure 2, a), again minimizing interactions, this time between the groups at the two ends of the bond. A series of preferences of this sort results in the observed chair conformation (Figure 2, b) of a cyclohexane. In this structure the largest remaining interactions (indicated by the arrow in b) are those between the vertical (axial) CH bonding electrons, so any larger substituent, such as the methyl group of (b), is forced into the alternative equatorial position as shown. Effects of this sort, depending on the relative sizes of groups, are called steric effects.

The geometries of reactions between molecules can in many cases be explained in similar terms: thus the inversion of configuration that accompanies concerted nucleophilic substitution at a saturated carbon involves (Figure 3, c) approach of the incoming nucleophile Nu from the side remote from the leaving group L. This minimizes repulsive interactions between Nu and L and between all five groups attached

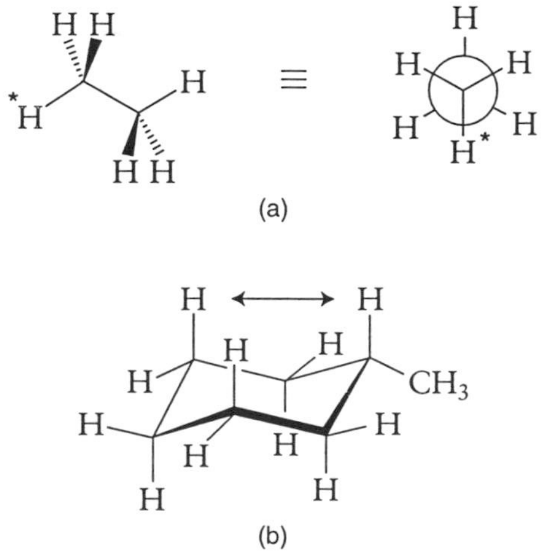

Fig. 2. (a) Staggered arrangement across a single bond; (b) observed chair conformation of a cyclohexane.

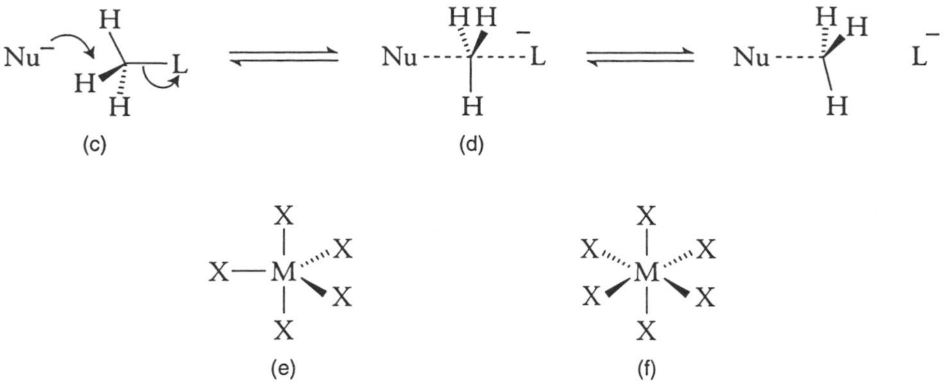

Fig. 3. Inversion of configuration accompanying nucleophilic substitution.

to the central carbon atom in the transition state (d). This has the trigonal bipyramidal arrangement (e) typical of stable inorganic molecules MX_5 with five groups attached to a central atom. Note that the five groups once again end up as far away from each other as possible. (Molecules MX_6, with six groups attached to a central atom, similarly adopt the octahedral geometry, f).

This simple approach can go a long way toward explaining both the shapes of molecules and the geometry of reactions between them, but it is not the whole story. In particular it does not take into account vacant orbitals. This may sound like a contradiction in terms, since we defined an orbital above as a concentration of electron density; but the idea is strictly logical. Take the reaction called S_N1 (see Figure 4) in which a good leaving group L (good because it gives a very stable anion L^-) departs from a suitable carbon center able to accommodate the positive charge left behind. The carbon center of the cation (Figure 4, g) is short two electrons; with the remaining three groups at the central carbon atom it adopts the expected trigonal geometry, leaving a vacant p-orbital available to accept a pair of electrons from a suitable donor.

We know (from both experiment and calculation) all about the properties of p-orbitals—their sizes, shapes, and energies being the most important. In particular, we know that the interaction with an electron-pair donor is most efficient when the two orbitals involved—the p-orbital and the orbital containing the electrons of the donor—can overlap properly. This depends on their shape and symmetry, and for the p-orbital of (g) means that the donor must approach along its axis. The same applies to the donor orbital. So for a productive interaction, leading to bond formation, a trajectory is preferred by which the two interacting orbitals—one filled (+) and one vacant—approach along a common axis (as shown in Figure 5, h).

The situation is only a little more complicated when the acceptor molecule has a full complement of bonds and no obvious vacant orbital. In fact every electron-pair bond is described by a bonding orbital (σ or π), which is occupied by the bonding electrons, and an antibonding orbital (σ^* or π^*). This antibonding orbital has a similar basic geometry but is vacant (if it were occupied, no bond would be formed). The C=O group of a carbonyl compound, for example, has a vacant π^*-orbital (dashed in Figure 6, i), which can accept an electron pair from a suitable donor in much the same way as described for (h). The most accessible (lowest-energy) vacant orbital in a molecule and the orbital containing the most reactive (highest-energy) electrons are called the FRONTIER ORBITALS.

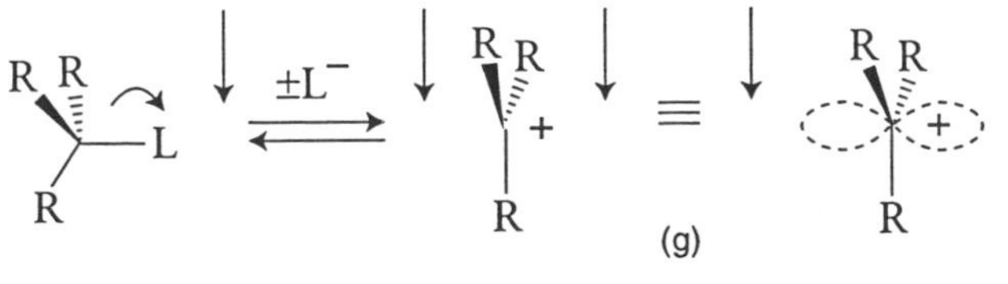

(g)

Fig. 4. S_N1 reaction.

(h)

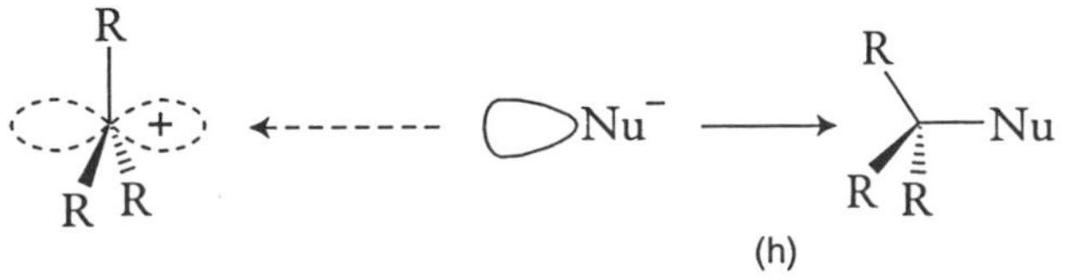

Fig. 5. Interaction leading to bond formation along a common orbital axis.

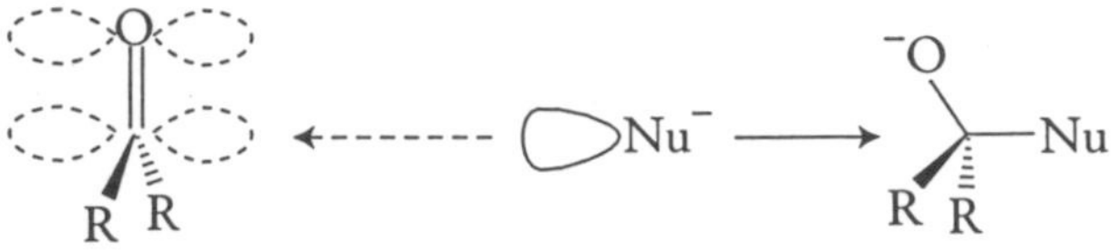

Fig. 6. Frontier orbitals.

The direction of approach of a nucleophile adding to a carbonyl group is thus controlled both by its repulsive interactions with the three substituent groups on the carbonyl carbon atom and by the preferred geometry of attractive, bonding overlap between the two filled and vacant orbitals making the new bond. The same is true of concerted nucleophilic substitution at a tetrahedral carbon, described above (Figure 3, c and d). The bonding interaction in this case is between the donor orbital of the nucleophile and the vacant σ^*-orbital of the C—L σ-bond. This has axial symmetry like the σ-bonding orbital, and the preferred interaction is again along the common axis of the two orbitals directly involved (Figure 7, j).

In these cases the repulsions between filled orbitals and the attractive interactions between the filled and vacant orbitals at the reaction center reinforce each other. Qualitatively, either explanation is successful, and interpretation in terms of interactions between filled and vacant orbitals is a refinement. But other, important classes of interaction have been identified in which the effects are opposed. The evidence is typically that a simple explanation in terms of repulsions between filled orbitals cannot account for the observed preference for a particular geometry or the failure to observe a particular reaction. It is these unexpected observations, requiring a special explanation, that are usually referred to as stereoelectronic effects.

Stereoelectronic Effects on Conformation

The geometries of saturated compounds are determined primarily by steric effects. The conformation of *n*-butane, for example, is as shown by Figure 8 (k) because the two largest substituents, the two terminal methyl groups, end up as far away from each other as possible in this arrangement. Thus all the steric repulsions between the groups at the two ends of the central C—C bond are minimized. However, the preferred conformation of 1,2-difluoroethane is as shown by Figure 8 (l), with the two fluorine atoms as close to each other as possible in the staggered conformation. Fluorine is the smallest atom or group after hydrogen, but it is nonetheless larger than H, and on steric grounds an extended conformation corresponding to (k) would be expected. Because the two fluorine atoms in the preferred conformation (l) are gauche to each other (torsion angle of 60°), this is called the *gauche effect*.

This is a stereoelectronic effect. The C—F bond is strongly polarized by the electronegative fluorine atom, and as a result the antibonding σ^*-orbital is particularly low in energy and thus particularly accessible. The best available donor electrons are those in the σ-bonds attached to the other carbon atom and specifically those in the C—H bonds rather than the C—F bond. Donor-acceptor interactions of this sort are well known to be most efficient when the orbitals concerned are parallel, which in this situation means coplanar and *trans* to each other. In this case the donor-acceptor interaction is dominant, so the molecule adopts the conformation shown by Figure 9 (m), with both C—F bonds *trans* to C—H.

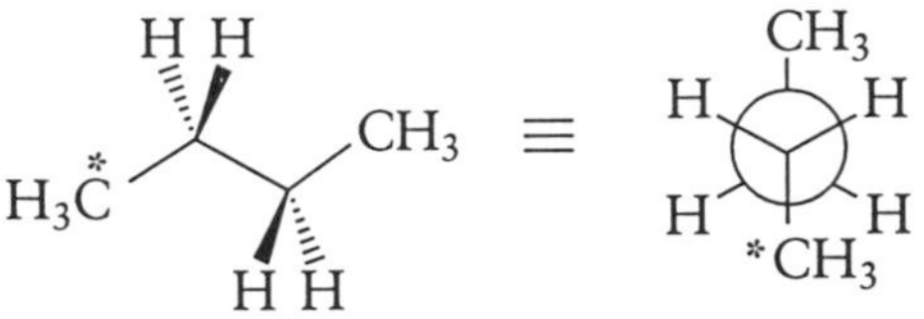

(k)

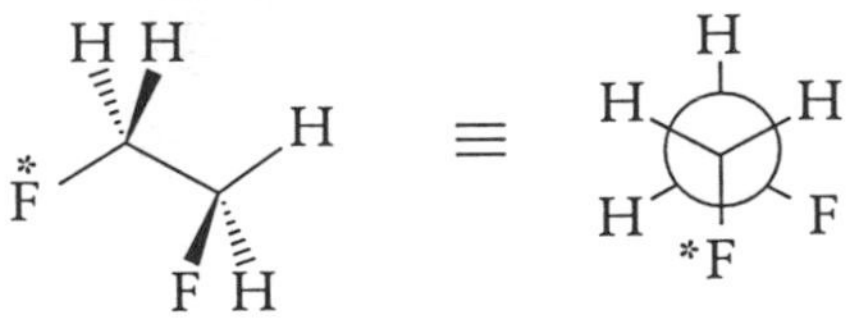

(l)

Fig. 8. Conformation of *n*-butane (k) and 1,2-difluoroethane (l), with examples of the *gauche* effect.

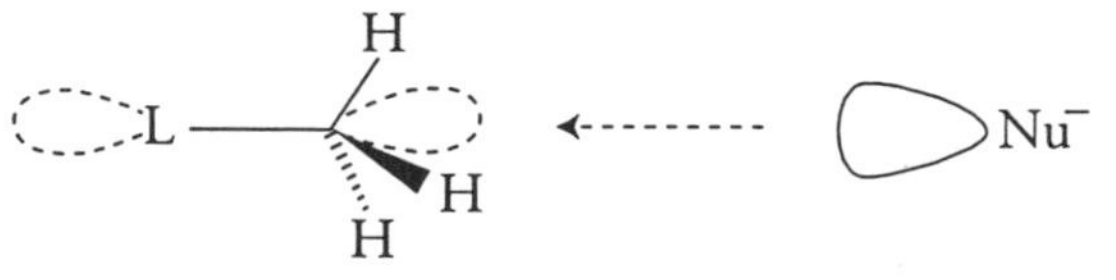

(j)

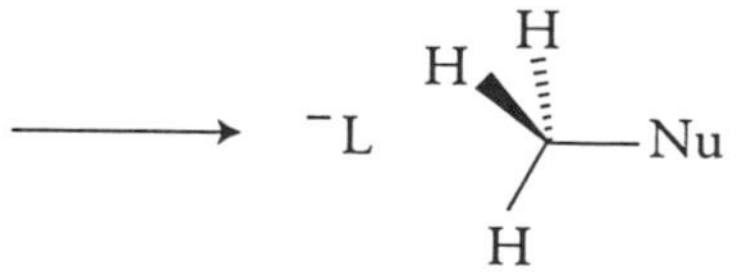

Fig. 7. Bonding interaction along the common axis of two orbitals.

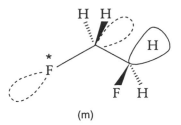

Fig. 9.   Conformation of
1,2-difluoroethane
molecule.

contain the basic structural unit O—C—C—O many times over and because carbohydrates are polysaccharide acetals with their structure in three dimensions controlled by the effect illustrated in Figure 9 (m). In fact, the effect was first identified in SUGAR chemistry, as an unexpected preference for axial conformations in the usual pyranoside structure, shown in Figure 11 (o) for a glucoside.

This is the same effect that was illustrated in (n). The axial form of a six-membered cyclic acetal (Figure 11, p) can adopt the same basic conformation as Figure 10 (n) about both C—O bonds, but the conformation of the equatorial isomer (q) is constrained by the ring in a less favorable conformation about the (in-ring) C—O bond and so is of higher energy. This preference for the sterically less favorable axial isomer is called the *anomeric effect*.

To summarize, there is a stereoelectronic preference for conformations in which the best donor lone pair or bond is *trans* coplanar with respect to the best acceptor bond.

The size of such effects depends on how good the match between the donor and acceptor orbitals is and especially on their relative energies. The effect described for (m) is greatest for two fluorine atoms but smaller—though still observed—for two methoxy (OMe) groups. This is because OMe groups are larger than fluorine atoms and because oxygen is less electronegative. But, given an electron-donor orbital better than a C—H bond, a methoxy group can also show a large effect. The best electron-donor orbital is a nonbonded electron pair (lone pair) of the sort found on amine nitrogen or bicovalent oxygen. So it is no surprise that the conformations of acetals and related compounds with two hetero-atoms on the same carbon (or other) center should show similar effects. Thus the preferred conformation of a simple acetal like $(MeO)_2CH_2$ has the same geometry (Figure 10, n) as 1,2-difluoroethylene (Figure 9, m) with O and OMe as close as possible (*gauche* to each other) across the C—OMe bonds. (The compound is symmetrical, so this applies to both C—OMe bonds.) The reason is the same as for (m); the interaction of the σ^*-antibonding orbital of the bond to the electronegative atom (in this case oxygen) with the best donor orbital available—here one of the lone pairs on oxygen—is dominant.

These effects have far-reaching consequences in structural biology because sugars and nucleic acids

Stereoelectronic Effects on Reactivity

Bond making and bond breaking involve interactions between filled and vacant orbitals that are more drastic than those that affect only conformation, and efficient overlap between the orbitals concerned is if anything more important. The same geometrical requirements are involved, and the most important ones have already been discussed, at least implicitly. For example, the donor-acceptor interaction between a C—H and a C—F bond leads naturally to a polarization in the sense H(+) . . . F(−) and predisposes the system to an elimination reaction. The elimina-

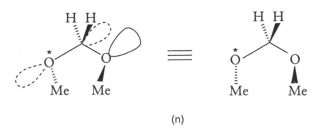

(n)

Fig. 10.   Preferred conformation of a simple acetal, $(MeO)_2CH_2$.

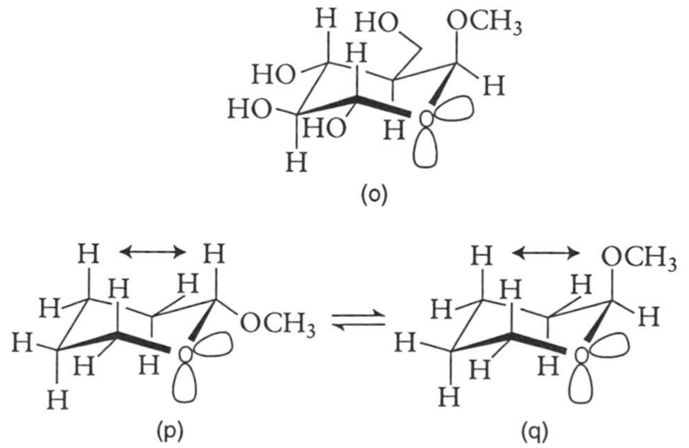

Fig. 11.   (o) Pyranoside structure of a glucoside; (p) the axial isomer of a six-membered cyclic acetal; (q) the equatorial isomer of the same cyclic acetal.

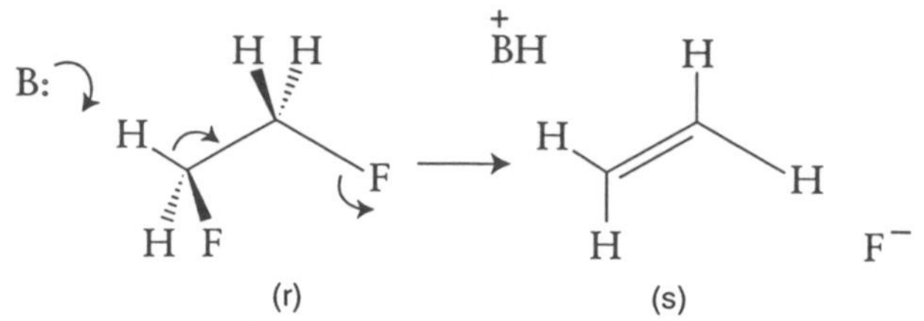

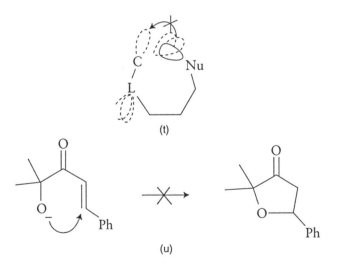

Fig. 12. Elimination reaction under favorable conformational conditions.

tion thus goes best in the conformation that maximizes this interaction (Figure 12, r), and often not at all where it is prevented by the geometry in a rigid system.

The same generalization applies to many other important classes of reaction, such as rearrangements and fragmentations, in which major structural reorganizations may occur. In each case the reaction involves one or more key stereoelectronic interactions, which have their own geometrical requirements. When these are denied, the reactions concerned face higher activation energies and may not be observed at all under the usual conditions.

Some of the clearest examples are seen in cyclization reactions, where the size and shape of the ring being formed impose well-defined geometrical restrictions. Concerted nucleophilic substitution, for example, requires a linear transition state (Figure 7, j); this clearly cannot be accommodated in a six-membered or smaller ring (see, for example, Figure 13, t), so such reactions can be safely ruled out on

Fig. 13. Hypothetical reactions that cannot take place because of stereoelectronic limitations.

stereoelectronic grounds. Similarly, the addition of a nucleophile to a C=O groups takes place from above the plane (9, above), so that apparently simple reactions like the example shown in Figure 13 (u), are not observed—again, because the ring is too small to accommodate the geometry imposed by the stereoelectronic requirements of the reaction. The practical consequences for such cyclizations are summarized as Baldwin's rules.

BIBLIOGRAPHY

DESLONGCHAMPS, P. Stereoelectronic Effects in Organic Chemistry; Oxford: Pergamon Press, 1983.

KIRBY, A. J. The Anomeric Effect and Related Stereoelectronic Effects at Oxygen; New York: Springer-Verlag, 1983.

KIRBY, A. J. Stereoelectronic Effects; New York: Oxford University Press, 1996.

ANTHONY J. KIRBY

Steroids

Steroids are widely found in nature, occurring in both plants and animals, from single-celled organisms to the most complicated species. They function as HORMONES that regulate various growth and developmental processes, as structural elements in cell membranes, and as toxins with a defensive role in some plants and animals. Humans have long appreciated the latter property. The stimulatory effects on the heart of extracts from leaves of the common foxglove plant, Digitalis purpurea, were first described in 1785, and the steroid responsible for this activity is still in use for the treatment of heart failure. To aid in capturing game animals, certain of the indigenous peoples of South America have historically used extracts of toad skins as arrow-tip poisons, and steroids are among the active ingredients in these poisons. In more recent times, the important roles that steroids play in human physiology have been recognized, and this has led to the development of synthetic steroids to be used as therapeutic agents.

Steroids form a special class of lipids with structures that are difficult to precisely define because so many variations and special cases exist. In general, however, steroids are recognizable by their cyclopentanoperhydrophenanthrene ring system, as seen, for example, in CHOLESTEROL, a celebrity steroid of profound biological importance. Cholesterol has achieved notoriety because of its role in the genesis of atherosclerosis, but in its other roles as a constituent of mammalian cell membranes and as a precursor in the biosynthesis of the bile salts and the steroid hormones, it is essential to life (see Figure 1).

Cholesterol in humans is derived not only from the diet but also from endogenous synthesis that occurs in virtually all cells, though the greatest quantities are made in the liver. The biosynthesis of cholesterol is closely regulated by a negative feedback mechanism; that is, an increase in plasma cholesterol concentration is a signal for decreased biosynthesis, and a decrease in plasma concentration is a signal for increased biosynthesis.

The mechanism of this regulation has been found through investigation of the cholesterol biosynthetic pathway. The process can be considered to begin with acetate, for every carbon atom in the cholesterol molecule is derived from acetate. The biosynthesis proceeds through a number of enzymatically catalyzed steps including the formation at one stage of the intermediate, mevalonic acid. This reaction, catalyzed by 3-hydroxy-3-methylglutaryl coenzyme A reductase (HMG CoA reductase), is the rate-limiting step in the pathway and is the site of the negative feedback regulation. A high concentration of cholesterol in the plasma resulting from the sum of the dietary and biosynthetic inputs inhibits the activity of the HMG CoA reductase, resulting in a decrease in the amount of cholesterol synthesized. The effect of inhibiting this enzyme, then, is to lower the plasma cholesterol concentration. Because high plasma cholesterol is a risk factor for cardiovascular disease, this enzyme is a target for drug therapy, and drugs such as lovastatin have been developed that pharmacologically inhibit its activity.

Pure cholesterol is virtually insoluble in blood. In the circulation, therefore, it is associated with proteins as soluble complexes. These complexes are formed in the liver and released to the circulation, where after several conversions the complexes become low-density lipoproteins (LDL). Cells have receptors that specifically recognize LDL, and it is the

LDL form that delivers cholesterol into the cell via these receptors, where it becomes available for important biosynthetic processes. The risk of atherosclerosis arises, however, when excessive amounts of LDL remain in the circulation.

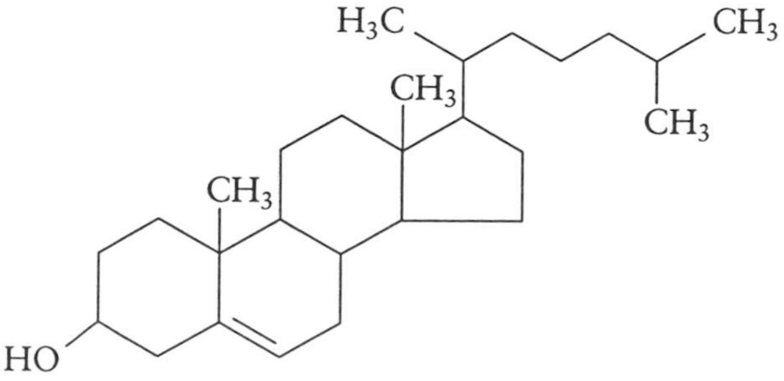

The development of atherosclerotic plaques begins with the deposition of cholesterol crystals on the inside surfaces of arterial walls, a process that involves LDL. If cholesterol concentration remains high, the crystals grow larger, the surrounding tissue begins to proliferate, and a complex series of incompletely understood events leads to still larger plaques and a loss of flexibility of the blood vessel. Heart attacks occur when the plaques grow large enough to occlude the vessels that supply blood to the heart (coronary artery disease), or strokes occur when the occlusion occurs in vessels supplying the brain. The HMG CoA reductase inhibitors are effective and widely used drugs that lower plasma cholesterol concentrations and consequently inhibit plaque formation. Clinical studies have shown that the use of these drugs by individuals with coronary artery disease significantly reduces the risk of having a coronary event, and mortality is reduced. Further studies are now being carried out to determine if these drugs can be used prophylactically to prevent the onset of the disease (Witztum, 1996).

But cholesterol has more than just a pathological role. As a component of mammalian cell membranes, it helps to preserve the structural integrity of all cells. As a precursor to the bile acids, it is necessary for the digestion and absorption of fats. And as the common starting point for the biosynthesis of the steroid hormones, it is necessary for the regulation of many bodily processes. The most important of the steroid hormones in humans are (1) the hormones of the adrenal cortex, which include cortisol and aldosterone; (2) the sex hormones, especially estradiol, progesterone, and testosterone; and (3) calcitriol, the active metabolite of vitamin D.

Adrenocortical Hormones

Aldosterone is a mineralocorticoid, so named because its effects are on mineral (primarily sodium and potassium) balance. Acting in the kidney, aldosterone causes sodium ions to be reabsorbed from the tubular fluid, which in turn drives the reabsorption of water, thereby protecting the body from excess loss of sodium and water. Cortisol is a glucocorticoid, so named for its effects on glucose metabolism, although cortisol also has mineralocorticoid effects, influences cells of the immune system, and is fundamentally involved in the biological response to stress

Fig. 1. Cholesterol.

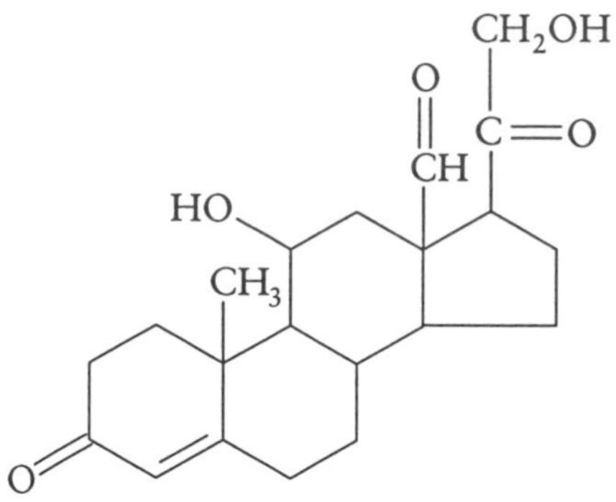

Fig. 2. Adrenocortical hormones: (a) aldosterone; (b) cortisol.

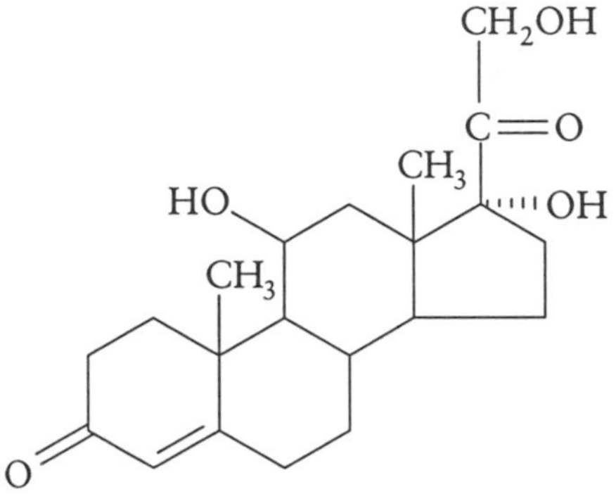

(Figure 2). These hormones, synthesized in the cortical area of the adrenal glands, are essential for life. They help to maintain normal function in the kidneys; the skeletal muscles; and the immune, cardiovascular, endocrine and nervous systems. Removal of the adrenal cortex leads to death within a few days unless a strictly regulated environment is provided to compensate for the loss. These profound effects are due to the influence of these hormones on metabolism, on water and electrolyte balance, and on the synthesis of a wide variety of cellular mediators (Guyton and Hall, 1996).

The glucocorticoids and their synthetic analogs are important to the practice of medicine. The dramatic effect of glucocorticoids in relieving the symptoms of rheumatoid arthritis, first demonstrated in 1949, has led to the development of synthetic drugs such as prednisone, which have fewer side effects and which are used in the treatment of a wide variety of diseases. The therapeutic efficacy of the glucocorticoids stems largely from their strong anti-inflammatory properties, which mimic certain of their physiological actions. Generation of an inflammatory response involves the production by the immune system of cellular mediators such as interleukins and eicosanoids. Glucocorticoids in their physiological role as regula-

tors of the immune system slow the synthesis of these mediators, and administration of glucocorticoid drugs results in a further inhibition of their synthesis, thereby attenuating the inflammatory response and providing significant relief of disease symptoms (Schimmer and Parker, 1996).

Sex Hormones

The steroids with the most immediately recognizable effects are the sex hormones. The female hormones are collectively known as estrogens and progestins and the male hormones as androgens. The most important female estrogen is estradiol, which is synthesized in the ovary. The significant progestin is progesterone, which is synthesized in the corpus luteum (Figure 3). In the male, testosterone, synthesized in the testes, is the most important androgen. Regulators of sex hormone synthesis are themselves hormones, peptides rather than steroids, and they are synthesized in the hypothalamus and the anterior pituitary. These peptide hormones are identical in males and females, and their actions on the gonads are to increase the synthesis of the sex hormones. The physiological effects of the sex hormones are varied and complex, but broadly, they promote the growth and development of the reproductive tissues, leading to sexual maturity. Thereafter, their actions maintain

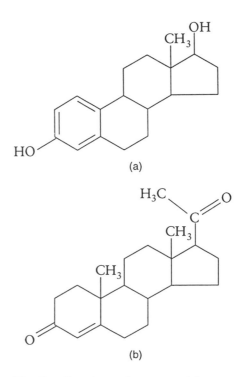

Fig. 3. Female sex hormones: (a) estradiol; (b) progesterone.

reproductive function in adults, though with recent research, there is an increasing realization that they have significant effects on nonreproductive tissues as well.

Estradiol and progesterone are regulators of the female reproductive cycle. In the first week to ten days following the cessation of menstruation, the hypothalamic and pituitary peptide hormones stimulate a rising concentration of estradiol, which helps to promote the release of a mature ovum and initiates changes in the lining of the uterus to prepare it for implantation of a fertilized egg. At the time of ovulation, the ovum is released from the ovarian follicle (which becomes the corpus luteum), and progesterone acts to maintain the uterine lining in this receptive state for several more days. If fertilization does not occur, plasma progesterone concentration begins to fall, the uterine lining can no longer be maintained, and it begins to slough off, signaling the onset of the menstruation.

Both estradiol and progesterone have negative feedback effects on their own synthesis, and this phenomenon is the basis for the most important mechanism in the actions of the oral contraceptives. High plasma concentrations of estrogens and/or progestins inhibit the synthesis of the hypothalamic and pituitary peptide hormones. Since ovulation depends upon the presence of these peptide hormones, it does not occur in their absence. The use therefore of pills containing both an estrogen and a progestin (combination oral contraceptives) or of pills containing a progestin only results in the suppression of ovulation, and pregnancy cannot occur.

Taken orally, estradiol and progesterone are rapidly inactivated after absorption from the gut. Synthetic analogs of both hormones have been developed, however, that are resistant to this inactivation and are therefore suitable for use as oral contraceptives. The most common synthetic estrogen in these preparations is ethinyl estradiol. A number of synthetic progestins have been developed, including norgestrel, which is used in pill form as well as in capsules implanted under the skin. In the latter form, norgestrel is slowly released from the capsules, providing a contraceptive action for at least five years. This eliminates the burden of taking a pill every day (or daily for three weeks out of every four in the case of the combination pills) and the possibility of accidental pregnancy.

With an estimated 60 million young, healthy people around the world regularly taking oral contraceptives, the incidence and severity of adverse effects are of no small concern. Early reports associated an increased risk of death from cardiovascular disease with the use of pills containing high doses of estrogen, but since the 1980s the estrogen dose has been progressively reduced. More recent studies do not show such a relationship, except in women over thirty-five who are heavy smokers. Many studies on cancer risk have also been carried out, and the data indicate that the incidence of endometrial and ovarian cancer is actually decreased among oral contraceptive users. The situation with breast cancer is less clear. There are reports that women who began taking the high-estrogen-dose contraceptives before the age of twenty, and who used them for a long period of time, have a 50 percent increase in their risk for breast cancer by age forty-five. Other studies have not confirmed this, however, and for the large majority of women who do not belong in this category, there appears to be no increased risk (Baird and Glasier, 1993).

Oral contraceptives confer certain health benefits not related to their contraceptive actions. In addition to the protection against endometrial and ovarian cancers mentioned above, there is a decreased incidence of fibrocystic breast disease and pelvic inflammatory disorders among users. It is also interesting to note that from a statistical point of view the risks of adverse consequences associated with pregnancy and childbirth are greater than those associated with the use of oral contraceptives.

At approximately the age of fifty, the production of estrogen in the ovaries begins to progressively decline, and menopause begins. In some women this results in the onset of symptoms that are serious enough to warrant the administration of estrogen, a treatment known as estrogen replacement therapy (ERT). The most significant condition treated with ERT is osteoporosis, a condition that affects nearly one-third of all postmenopausal women. Osteoporosis is a disease characterized by a loss of bone mass and an increased susceptibility to fractures. Estrogen acts to slow the loss of bone, and hence it slows the development of osteoporosis, though it does not reverse it.

In the 1960s and 1970s ERT was used extensively to treat postmenopausal symptoms, but with time it became apparent that this treatment was associated with an increase in endometrial cancer. This led to the use of combination therapy, where both estrogen and progestin were administered together. This procedure, called hormone replacement therapy (HRT), removed the association with endometrial cancer and has even had a protective effect against this cancer (Belchetz, 1994).

More recently, the possibility that ERT or HRT may lead to an increased susceptibility to breast cancer has been a serious concern (Marshall, 1993). Numerous studies in many countries on various populations have provided conflicting results, but the consensus of opinion now is that no convincing evidence exists for such a risk. At most, there may exist a 30 percent increase in breast cancer after fifteen years of continuous use, but that risk disappears if the therapy is discontinued (Steinberg et al., 1991). An exception is a patient with a family history of breast cancer. In this case, ERT or HRT is not advised.

Another consideration in the use of HRT is the protection it gives against cardiovascular disease. The leading cause of death in women over sixty-five years of age in the United States is heart disease, especially myocardial infarction, or heart attack. Studies have shown that HRT protects against heart disease in women, and this is a significant benefit of the therapy (Gura, 1995).

The use of ERT or IRT presents a dilemma. The possible risk of breast cancer that may arise with long-term, continuous use must be weighed against the relief from postmenopausal symptoms and the protection afforded against the risk of osteoporosis and heart disease. The risk factors for these diseases vary from person to person, and any individual who may be contemplating this therapy should personally evaluate them.

Of the androgenic steroids in men, testosterone is the physiologically most significant (Figure 4). Increasing testosterone production at the time of puberty promotes the growth and development of the male reproductive organs, and it maintains their function throughout life, as in, for example, the stimulation of sperm cell production in the testes. On the negative side, testosterone also stimulates the growth of some tumors of the prostate gland, and drugs have been developed to treat this kind of cancer either by inhibiting the synthesis of testosterone or by inhibiting its activity.

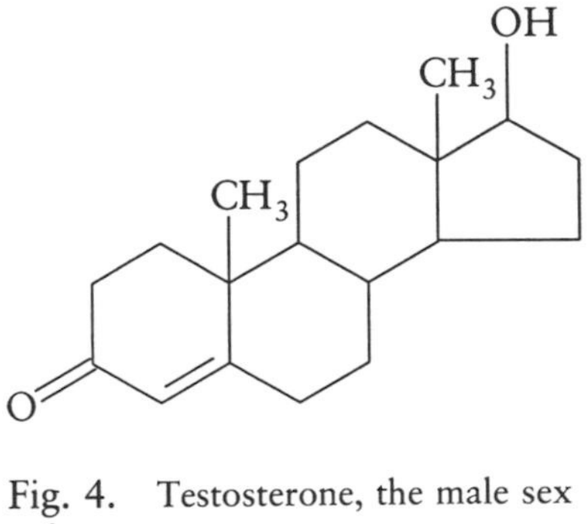

Fig. 4.   Testosterone, the male sex hormone.

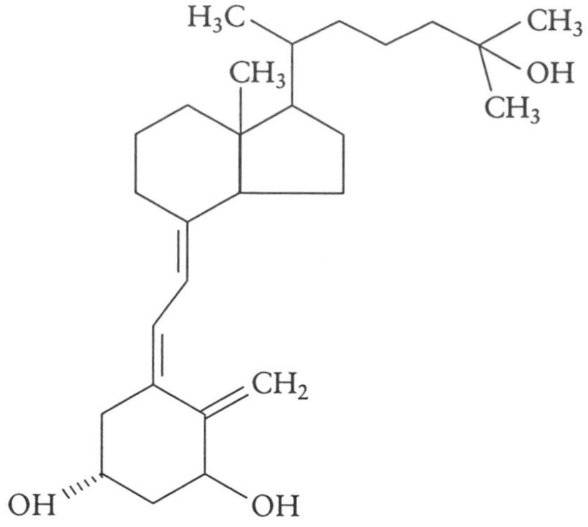

Fig. 5.   Calcitriol.

Anabolic steroids are synthetic analogs of testosterone. The androgens, unlike the estrogens, have actions that stimulate an increase in lean muscle mass. They are said to increase strength, speed, and stamina, and some athletes have used them in attempting to increase their performance capability. There has been controversy over whether these drugs actually accomplish the gains attributed to them, and clinical studies have failed to consistently establish any benefit. These studies, however, have not used the dosages that are used by high-performance athletes and have often not been accompanied by regular training and dietary control. After an extensive review of the literature, the American College of Sports Medicine concluded that the use by some individuals of anabolic steroids does indeed result in a small but statistically significant increase in lean body mass and muscular strength, though only when accompanied by an adequate diet and regular, high-intensity training (American College of Sports Medicine, 1987).

It is unlikely, however, that the effects of anabolic steroids on athletic performance will be further investigated in any kind of scientific way in the foreseeable future. Reports of adverse effects associated with their use such as liver problems, possible stroke or heart attack, and violent and aggressive behavior, though rare, will preclude rigorous and systematic testing in humans (Yesalis and Bahrke, 1995).

Vitamin D

The active form of vitamin D is 1,25-dihydroxy vitamin D_3, or calcitriol (Figure 5). Like the other steroid hormones, calcitriol binds to an intracellular receptor and initiates the production of new proteins (see below). Calcium-binding proteins are among those induced by calcitriol, and they help to carry out one of the major functions of the hormone, a facilitation of calcium absorption from the intestine. Calcium is an important mineral for the formation of bones and teeth, and it is also vital for many other cellular functions. Calcitriol is an important component of the regulatory system that maintains calcium concentrations in the body at an optimal level.

Discussions of vitamin D usually consider only the effects on calcium homeostasis, but recent discoveries have shown that it has other important functions as well. Positive effects on the immune system have been noted, and calcitriol has been shown to effectively treat certain skin diseases, notably psoriasis. Calcitriol also has actions to inhibit the growth of malignant cells and is being studied as a possible treatment for several cancers including leukemia.

Mechanisms of Steroid Hormone Action

The steroid hormones affect cells and tissues in a way unique among the hormones. Other hormones, such as the peptide hormones catecholamines or eicosanoids, initiate their actions by binding to a hormone-specific molecule on the surface of the target cell called a receptor. The receptor-hormone complex then activates one or more intracellular signaling pathways that ultimately change the activities of the cell to those specified by the hormone—e.g., bronchial smooth muscle relaxation in response to epinephrine, or increased glucose uptake from the circulation in response to insulin. The steroid hormones also bind to hormone-specific receptors, but the steroid receptors are located inside the cell, not on the outside surface of the membrane. Because the steroids are lipid-soluble, they pass easily across the cell membrane and gain access to the cell interior, unlike the other hormones. When the steroid receptor-hormone complex forms, it binds to DNA and regulates the expression of proteins. The synthesis of certain proteins will be turned on, the synthesis of others will be turned off, and this new population of proteins is responsible for the new, hormone-specific activities of the cell—e.g., increased sodium retention in the kidney after aldosterone or increased calcium absorption from the gut in response to calcitriol.

While steroids are studied far less in plant cells than in animal cells, it is now clear that the plant steroids can also act at the DNA level. Brassinolide, a steroid found in many types of plants, has been shown to be essential for normal light-stimulated growth and development in *Arabidopsis thaliana*, a flowering plant with well-characterized genetics (Russell, 1996). This finding implies that brassinolide is required for the expression of genes involved in plant development in a way that is analogous to steroid action in animal cells.

Steroid hormones have recently been shown to have effects other than the direct DNA effects. Akin to the actions of the nonsteroid hormones, receptors located on certain cell membrane surfaces activate intracellular signals in response to steroid hormone binding. Aldosterone, for example, rapidly activates the sodium-proton exchanger in vascular smooth muscle cells in a way that does not depend on changes in DNA transcription. Steroid signaling from the membrane is a relatively new concept, and discovering the physiological relevance of this phenomenon and how it relates to the genomic effects has led to many exciting new studies (Wehling, 1995).

From arrow-tip poisons to oral contraceptives, steroids have been important agents in human society. Continuing research into the actions of the steroid hormones is leading to a still greater appreciation of their many physiological roles, and new therapeutic strategies for the treatment of various cancers, immune disorders, and inflammatory diseases may form around this new knowledge.

BIBLIOGRAPHY

American College of Sports Medicine. "Position Stand on the Use of Anabolic-Androgenic Steroids in Sports." *Med. Sci. Sports and Exercise* 1987, *19*, 534–539.

Baird, D. T.; Glasier, A. F. "Hormonal Contraception." *N. Engl. J. Med.* 1993, *328*, 1543–1549.

Belchetz, P. E. "Hormonal Treatment of Postmenopausal Women." *N. Engl. J. Med.* 1994, *330*, 1062–1071.

Gura, T. "Estrogen: Key Player in Heart Disease among Women." *Science* 1995, *269*, 771–773.

Guyton, A. C.; Hall, J. E. "The Adrenocortical Hormones." In *Textbook of Medical Physiology*, 9th ed.; Philadelphia: Saunders, 1996; pp. 957–970.

Marshall, E. "Search for a Killer: Focus Shifts from Fat to Hormones." *Science* 1993, *259*, 618–621.

Russell, D. W. "Green Light for Steroid Hormones." *Science* 1996, *272*, 370–371.

Schimmer, B. P.; Parker, K. L. "Adrenocorticotropic Hormone; Adrenocortical Steroids and Their Synthetic Analogs; Inhibitors of the Synthesis and Actions of Adrenocortical Hormones." In *Goodman & Gilman's The Pharmacological Basis of Therapeutics*, 9th ed.; J. G. Hardman, L. E. Limbird, eds.; New York: McGraw-Hill, 1996; pp. 1459–1485.

Steinberg, K. K.; Thacker, S. B.; Smith, J.; Stroup, D. F.; Zack, M. M.; Flanders, D.; Berkelman, R. L. "A Meta-Analysis of the Effect of Estrogen Replacement Therapy on the Risk of Breast Cancer." *JAMA* 1991, *265*, 1985–1990.

WEHLING, M. "Looking beyond the Dogma of Genomic Steroid Action: Insights and Facts of the 1990s." *J. Mol. Med.* 1995, *73*, 439–447.

WITZTUM, J. L. "Drugs Used in the Treatment of Hyperlipoproteinemias." In *Goodman & Gilman's The Pharmacological Basis of Therapeutics,* 9 ed.; J. G. Hardman, L. E. Limbird, eds.; New York: McGraw-Hill: 1996, pp. 875–897.

YESALIS, C. E. BAHRKE, M. S. "Anabolic-Androgenic Steroids. Current Issues." *Sports Med.* 1995, *19*, 326–340.

DONALD H. MILLER

Stoichiometric Laws

See CLASSICAL STOICHIOMETRIC LAWS.

STP

STP is the abbreviation of standard temperature and pressure, a temperature of 0°C = 273.15 K and a pressure of 1 atm = 760 mmHg = 1.01325 bar. Under these conditions, 1 mole of an ideal gas occupies a volume of 22.414 L, known as the STP molar volume. (See also GASES; IDEAL GAS LAW.)

ROBERT K. WISMER

Strain

In 1876 Victor Meyer noted that reactions that might have led to strained rings of three, four, or five members invariably resulted in the formation of unsaturated open-chain isomers. He concluded that the synthesis of cycloalkanes was not to be expected. Despite the less-than-optimistic prediction of Victor Meyer, W. H. Perkin, Jr., working in the laboratory of Adolf von Baeyer in Munich, soon described the synthesis of the diesters labeled (a) and (b) in Figure 1 by what would now be regarded as "classical" malonic ester syntheses. In 1882, Freund reported that treatment of 1,3-dibromopropane with sodium metal yielded a gas that was later shown to contain cyclopropane. This and the subsequent work of Perkin attracted considerable attention and helped to provide the foundation of modern chemistry. An earlier report in 1881 by the Russian chemists Markownikoff and Krestownikoff describing the synthesis of the cyclobutane dicarboxylic acid (Figure 1, item c) seems to have aroused considerably less attention.

After the pioneering work of Perkin, Adolf von Baeyer began to examine the bonding of tetravalent

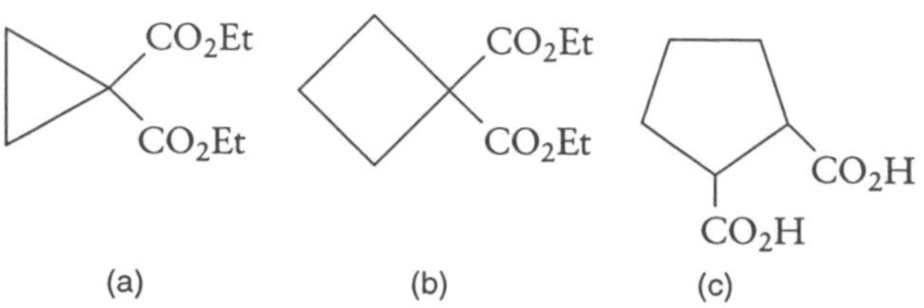

Fig. 1. Two diesters (a and b) synthesized by malonic ester synthesis and the cyclobutane dicarboxylic acid (c).

carbon in cyclic systems (see Huisgen, 1986). In 1885, with the aid of specially prepared models, he gradually worked out the now famous theory of ring-closure, or strain theory, which says that steric factors explain the observation that five- and six-membered chains can be formed readily, while a shorter or longer chain undergoes such closure with difficulty, or not at all. He correctly deduced that three- and four-membered carbocycles would be less stable than their five- and six-membered homologues because of the essential deviation in bond angles from the normal tetrahedral value of 109°28'.

One consequence of the Baeyer strain theory was that the five-carbon ring with an internal angle of 108° should be the most stable and most easily formed of all rings. The synthesis of C by Perkin in 1885 as a well-defined mixture of *cis*- and *trans*-isomers was accepted by Baeyer as confirmation of his theory. It was in this way that the foundations for understanding strain within organic molecules were laid.

Baeyer's idea of rings existing as planar entities was seriously questioned in 1890 when Sachse reported that two nonplanar models of cyclohexane can be formed without distorting the tetrahedral bond angle. It soon became apparent that larger rings are also nonplanar and that strain, although present in the cycloalkanes, is not simply caused by bond angle deformation (Baeyer strain). These effects are generally discussed today in terms of bond angle and bond length distortions as well as torsional effects and nonbonded interactions, although the twisting and bending of double bonds as well as other factors also need to be considered (see Greenberg and Liebman, 1978; Halton, 1991).

The strain energy of a molecule is the additional energy compared with a suitable strain-free model. To evaluate the strain energy of a molecule, it is first necessary to know the heat of formation. The heat of formation is defined as the heat content change accompanying the formation of a compound from its elements. Abbreviated ΔH_f, it is the heat content of the compound minus the heat content of the precursor elements. Most compounds have a negatively signed heat.

$$\Delta H_f = [\text{heat content of compound}] - [\text{heat content of precursor elements}]$$

Cyclohexane, which is considered to be strain free, has a heat of formation (ΔH_f) of -29.5 kcal/mole or -4.92 kcal/mole per methylene unit. The fact that a methylene unit in a strain-free straight chain alkane has a value of -4.926 kcal/mole demonstrates the nearly strain-free nature of cyclohexane.

Cyclopropane is the simplest unavoidably strained organic compound—unavoidably because there is no means to remove the strain of this simple cyclic hydrocarbon. For cyclopropane with three methylene units the strain-free energy is therefore $3 \times (-4.92) = -14.76$ kcal/mole, and this compares with a heat of formation of $+12.73$ kcal/mole. The difference $[12.73 - (-14.76)] = 27.5$ kcal/mole is the strain energy (SE) of the molecule. The strain energies of the homologous cycloalkanes are 26.5 6.2, 0.0, and 6.3 kcal/mole for cyclobutane to cycloheptane, respectively.

The strain in other three-membered rings such as ethylene oxide is about the same as that in cyclopropane. The high strain energy of ethylene oxide and its derivatives leads to high chemical reactivity including ring-opening reactions by a wide variety of organic reagents. Strain-sensitive thermally induced reactions of ethylene oxide include the rearrangement to acetaldehyde at about the same rate that cyclopropane rearranges to propene.

It is sometimes useful to express strain energies either in terms of the C—C bond (SE_b) or per carbon atom (SE_c). For the cycloalkanes, SE_b and SE_c are the same. For cyclopropane and cyclobutane these values are 9.2 and 6.6 kcal/mole, respectively.

One of the most easily achieved estimates of the energy of a relevant strain-free model is by use of the Franklin group equivalents (see Franklin, 1949). As was noted above, the energy per CH_2 unit from unconstrained paraffins is available, as are data for the other units used to build hydrocarbon assemblies. Ethene is assumed to be strain free, but values for substituted π-bonds are available.

For compounds whose heats of formation have not been determined and for molecules yet to be synthesized, estimates of strain energies can be computed with a high degree of reliability. The heat of reaction (ΔH_r) for the hypothetical reaction illustrated in Figure 2 has been calculated at various levels of theory. The ΔH_r values would represent the increase in strain of the product bicyclo[4.1.0]hept-1,6-ene, shown in

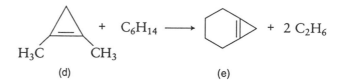

Fig. 2. (d) Dimethylcyclopropene and (e) bicyclo[4.1.0]hept-1,6-ene ($\Delta H_r = 17.97$ kcal/mole).

Figure 2, example e, over that for dimethylcyclopropene (Figure 2, example d) which had been determined earlier by experiment.

Of more interest than the strain energy is the olefinic strain (OS), which is defined as the difference in strain between the alkene and the corresponding alkane. The geometry of the three-membered ring requires that the angle deformation for a trigonal sp^2 center will be notably greater than that for the sp^3 equivalent. The strain energy of cyclopropene (55 kcal/mole) is almost twice that of cyclopropane. By comparison, the smaller angle deformation for the sp^2 sites in cyclobutene (120° vs. 90°) results in a strain-energy increase of only 2 kcal/mole compared with cyclobutane. In cyclopentene the olefinic strain is negative, since the unsaturation reduces the number of H—H eclipsing interactions.

Compounds also exist in which classical π-orbital overlap is affected. In one type of distorted alkene, the framework exerts a twisting action of the π-bond, which results in rotation of the π-orbitals in opposite directions. Distortion of the first type is exemplified by the incorporation of a trans-double bond into a ring. trans-Cyclooctene (SE ~16.5 kcal mol^{-1}) is some 11 kcal/mole less stable than its cis-isomer. As the ring size is decreased, the twist angle increases and the strain energies of the lower trans-cycloalkenes are such that trans-cycloheptene and trans-cyclohexene exist only as reaction intermediates at ambient temperatures. By comparison, the larger ring cycloalkenes can have the trans-isomer as the thermodynamically stable form.

Another way of twisting a π-bond is to locate the double bond at the bridgehead of a bicyclic molecule. One such compound is the bicyclic alkene, shown in Figure 3, item (f). The inability of the p-orbitals on bridgehead positions to overlap with p-orbitals adjacent to them is a statement of Bredt's rule, which states that double bonds cannot be placed at a bridgehead. Bredt's rule is violated for very large ring systems, in which p-orbitals can adopt positions of maximum π-type overlap. The recently reported synthesis of the ring system illustrated in Figure 3, item (g),

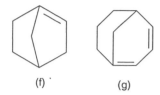

(f)   (g)

Fig. 3.   (f) Bicyclic alkene and (g) a recently synthesized ring system.

(i)   (j)

Fig. 5.   (i) Spiropentane and (j) spiropentadiene.

suggests that the limits of Bredt's rule may not be as broad as had previously been believed.

Strained Compounds Exemplified

Strained compounds have played a major role in the development of important concepts such as chemical reactivity and aromaticity. Since their high energy content often results in unexpected reactions, they continue to serve as a playground for the organic chemist. Selected strained compounds that have been either synthesized or discussed in the literature are presented in this section.

Cubane (see Figure 4, item h) is a member of the $(CH)_n$ family of Platonic hydrocarbons. This remarkable compound was synthesized in 1964 at the University of Chicago by Philip E. Eaton and his graduate student T. W. Cole, Jr. The experimentally derived strain energy (157 kcal/mole) may be considered to be the sum of the strain energies of the six cyclobutane faces.

In contrast to cubane, the spirocyclic compounds display a higher strain than the sum of their independent components. Spiropentane (Figure 5, item i) has 8 kcal/mole of additional strain compared with two cyclopropane units. This additional destabilization is thought to arise from the rigid spiro center. Spiropentadiene (Figure 5, item j) is one of the most strained hydrocarbons to have been reported. Theoretical calculations predict a standard heat of formation of 157.4 kcal/mole, which is more than twice the experi-

mental heat of formation of 66.2 kcal/mole for cyclopropene, and is consistent with the high energy content expected for spiropentadiene. As a consequence of the high reactivity of this extremely energetic compound, it exists only fleetingly at $-105°C$.

Many organic compounds have BOND ANGLES that deviate significantly from that of a regular tetrahedron. It is not surprising therefore that bond angle and bond twist deformations at saturated carbon have aroused special interest with regard to the synthesis of compounds with planarized carbon. Although the idealized structure [4.4.4.4]fenestrane (windowpane), Figure 6, item (k), remains to be prepared, the all cis-[5.5.5.5] compound (l) is estimated to have SE $\sim$182 kcal mol^{-1}.

Deformation of an idealized geometry by twisting is best exemplified by constrained bicyclics that are $trans$-fused and especially the $trans$-bicyclo[n.1.0] alkanes. The overlap in such a bond is particularly poor, and molecules having this feature are more reactive than their cis-fused isomers. The fused bond is distorted in two planes and has been described as "twist-bent" by Paul Gassman of the University of Minnesota. See Figure 7.

The compound $trans$-bicyclo[4.1.0[heptane (Figure 8, item o) is about 42 kcal mol^{-1} more strained than the cis-fused isomer, which exhibits typical cyclopropane ring strain. The compound undergoes easy $trans$ to cis isomerization. As the bridging ring size increases, the strain energy decreases to that of cyclopropane, and this is achieved in 19 ($n = 6$). Not surprisingly, this also corresponds to the SE of the cis-isomer.

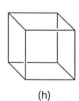

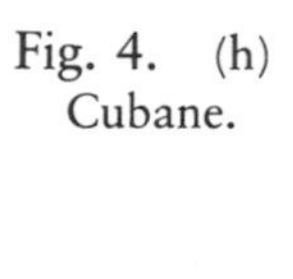

(h)

Fig. 4.   (h) Cubane.

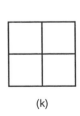

(k)   (l)

Fig. 6.   (k) Hypothetical structure [4.4.4.4] fenestrane and (l) all cis-[5.5.5.5].

(m)   (n)

Fig. 7.   (m) Symmetrical bond and (n) unsymmetrical bond.

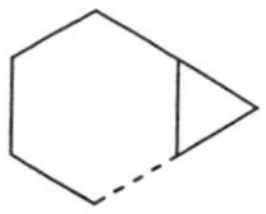

(o)

Fig. 8.   (o) The compound *trans*-bicyclo[4.1.0] heptane.

It is clear that the concept of strain in organic chemistry has had a profound impact upon the development of the field. Molecules so unstable that Baeyer and Perkin would not have dreamed of their existence are available today. Advances in computing technology continue to allow the theoretician to predict the properties of unknown compounds, and new developments in the handling of highly reactive molecules serve to stimulate the practitioner at the bench.

BIBLIOGRAPHY

FRANKLIN, J. L. "Prediction of Heat and Free Energies of Organic Compounds." *Ind. Chem. Eng.* 1949, *41*, 1070.

GREENBERG, A.; LIEBMAN, J. F. *Strained Organic Molecules,* New York: Academic Press, 1978.

HALTON, B., ed. *Advances in Strain in Organic Chemistry,* London: JAI Press, 1991; Vol. 1.

HUISGEN, R. "Adolf von Baeyer's Scientific Achievements: A Legacy." *Angew. Chem. Int. Ed. Engl.* 1986, *25*, 297.

W. E. BILLUPS

Substitution Reaction

The nucleophilic substitution reaction in organic chemistry involves replacement of a leaving group (Lg) by a NUCLEOPHILE:

$$Nu: + A—Lg \rightarrow Nu—A + Lg:$$

The scope of this reaction includes centers (A), which can be aliphatic, aromatic, cabonyl, and vinyl, as shown in Figure 1.

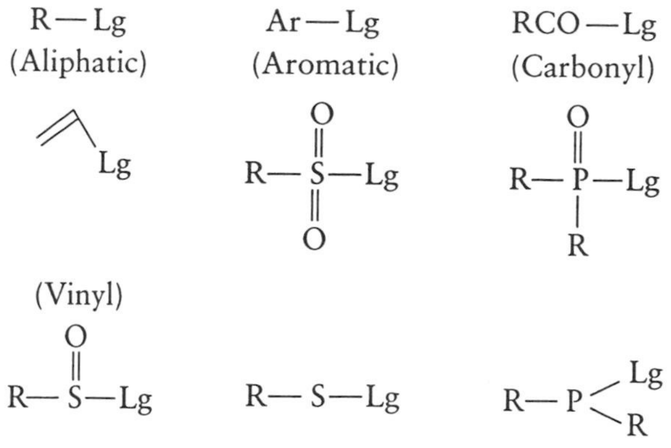

R—Lg   Ar—Lg   RCO—Lg
(Aliphatic)   (Aromatic)   (Carbonyl)

(Vinyl)

Fig. 1.   Detailed schematic of aliphatic, aromatic, carbonyl, and vinyl centers of nucleophilic substitution reactions.

Substitution at sulfur or phosphorus atoms, as in hydrolysis or aminolysis of phosphate or sulfate esters, is usually considered to be within the province of ORGANIC CHEMISTRY.

Nucleophilic substitution occupies an important place in the historical development of mechanistic organic chemistry. Lively discussions centering on the Hughes and Ingold groups at University College, London, resulted in the formulation of the general mechanistic scheme explaining aliphatic substitution. Hughes's 1937 observation of identical rate constants for inversion of configuration and incorporation of label in the identity reaction of [128]iodide ion with (+)2-iodooctane indicated the geometry of an aliphatic process; it explained the observations of inversion in aliphatic substitution resulting from Walden's initial experiments in the nineteenth century. Bender's experiment (in 1951) showed [18]oxygen exchange in carbonyl oxygen during ester hydrolysis and provided the first unequivocal evidence for addition intermediates in carbonyl substitution. Heterolytic mechanisms of substitution depend on the relative energies of nucleophile, Lg, and putative intermediates and are summarized in Figure 2. The mechanisms include the $D_N + A_N$ process, where the nucleophile attacks *after* the Lg has departed; this mechanism is also known as the S_N1 process if a cation is an intermediate, and an El_{cb} mechanism if the cation is formally stabilized by ionization of a proton (e.g., RNHCO-Lg $\rightarrow$ RN=C=O $\rightarrow$ RNHCO-Nu). The borderline process (A_ND_N) involves a concerted displacement of the Lg by nucleophile and is often known as the S_N2 mechanism. The path ($A_N + D_N$) involves formation of an addition intermediate and is often denoted $B_{Ac}2$ and S_NAr for displacements at carbonyl and aromatic centers, respectively.

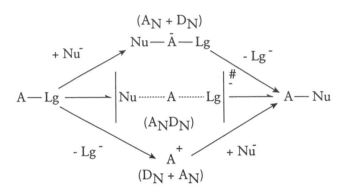

Fig. 2.   Heterolytic mechanisms of substitution.

Homolytic mechanisms have been demonstrated involving chain reactions (Figure 3) and have been observed with both aromatic and aliphatic centers (A). An analogous single-electron transfer mechanism (SET) has been postulated, which involves no expression of free radicals:

$$[Nu: A\text{-}X] \rightarrow [Nu \cdot A \cdot \cdot X] \rightarrow$$
$$[Nu \cdot \cdot A :X] \rightarrow [Nu\text{-}A :X]$$

This mechanism is analogous to the $A_N + A_N$ or $A_N D_N$ processes in Figure 2 and develops into a FREE RADICAL pathway if A· and X· survive to diffuse from their reaction complexes. Heteroaromatic substitution sometimes involves the ANRORC process (addition nucleophilic ring opening and ring closure; Figure 4).

Nucleophilic substitution also includes attack at metal centers in coordination complexes; while this is usually a separate topic, the mechanisms fit the same general schemes as for organic substitutions.

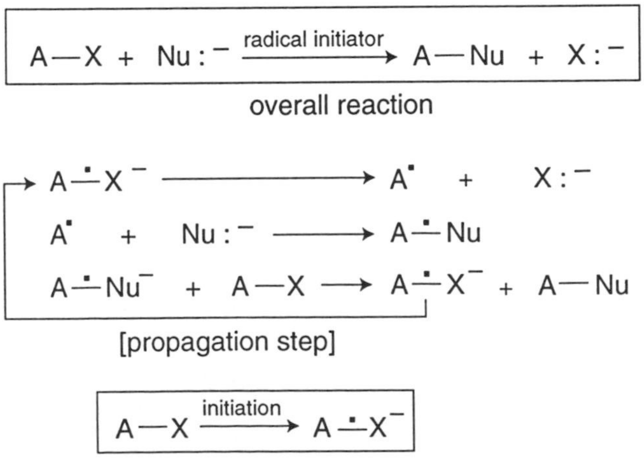

A—X + Nu : ⁻ $\xrightarrow{\text{radical initiator}}$ A—Nu + X : ⁻

overall reaction

A $\overset{\bullet}{-}$ X ⁻ ⟶ A· + X : ⁻

A· + Nu : ⁻ ⟶ A $\overset{\bullet}{-}$ Nu

A $\overset{\bullet}{-}$ Nu⁻ + A—X ⟶ A $\overset{\bullet}{-}$ X⁻ + A—Nu

[propagation step]

A—X $\xrightarrow{\text{initiation}}$ A $\overset{\bullet}{-}$ X⁻

Fig. 3.   Homolytic mechanisms involving chain reactions.

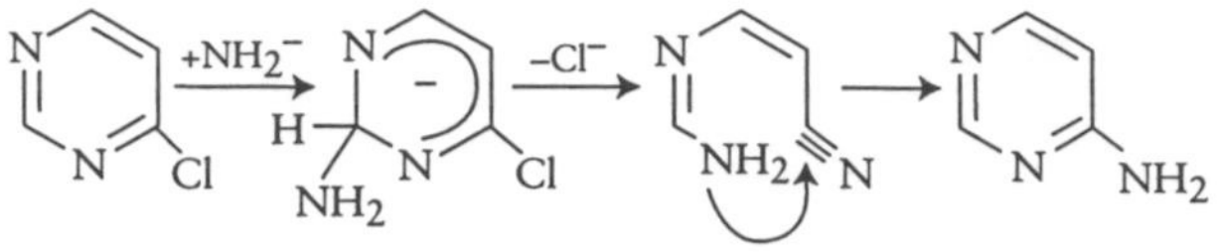

Fig. 4.   The addition nucleophilic ring opening and ring closure (ANRORC) process.

Reactivity

The speed of a heterolytic substitution reaction increases with the electron-releasing power of the entering nucleophile, with the stability of the Lg, and with increasing electropositivity of the central atom. A general order of reactivity for a given Lg and nucleophile is

Vinyl << Aromatic << Heteroaromatic
                << Aliphatic << RSO_2- << RCO-

(least reactive)   (most reactive)

Substitutions at the vinyl and aromatic centers are usually very slow because the increase in negative charge on the system resulting from interaction of the center with the nucleophile (a or b in Figure 5), cannot be easily accommodated by carbon; in the aromatic case the loss of resonance is a further disincentive.   Incorporating   electron-withdrawing substituents (X, Y, or Z) into the vinyl or aromatic systems enables substitution to proceed readily as a result of the ability of these substituents to stabilize the increased negative charge of the intermediate.

Nucleophilic substitution can be catalyzed by conversion of the Lg; thus, the iodide ion is a well-known catalyst for the nucleophilic substitution of an alkyl chloride because it is an exceptional nucleophile and substitutes for the chloride ion to yield the alkyl iodide, which reacts faster (since the iodide ion is also an exceptional Lg) than the chloride. The catalysis of hydrolysis of an alkyl chloride by iodide is

$$I^- + R\text{-}Cl \rightarrow R\text{-}I$$
$$R\text{-}I + OH^- \rightarrow ROH + I^-$$

Imidazole (Im) plays a role similar to that of iodide ion in the catalyzed hydrolysis of phenyl esters.

$$Im + RCO\text{-}OAr \rightarrow Im\text{-}CO\text{-}R$$
$$Im\text{-}CO\text{-}R + H_2O \rightarrow RCOOH + Im$$

Stereochemistry

The relative geometry of the trajectories taken by ligands as their bonds to the central atom form and break is either "in-line" or "adjacent" (Figure 5, c

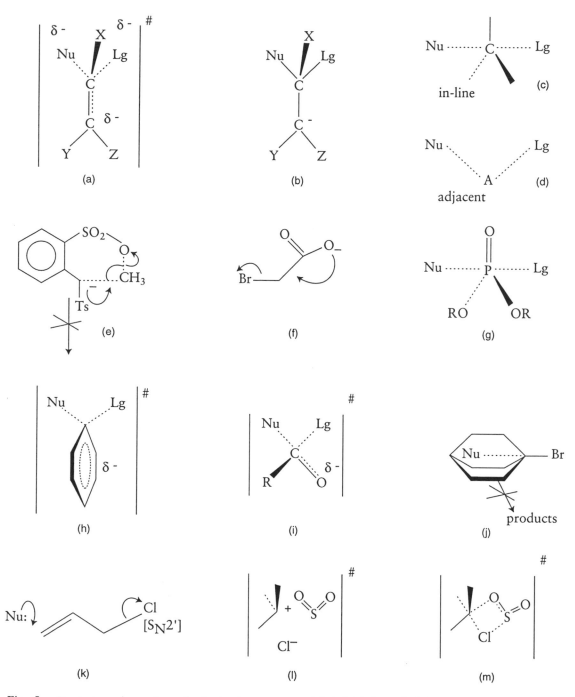

Fig. 5. Depiction of a variety of substitution processes, labeled (a) through (m).

and d). Substitution at tetrahedral carbon has in-line geometry whereas at trigonal carbon it is adjacent. Eschenmoser's classical experiment showed that the methyl group (e) does not transfer directly from the oxygen to the carbon because the preferred geometry is in-line, a situation that can only be supported by an intermolecular mechanism. An intramolecular nucleophilic displacement reaction can occur (f), since the Lg fission is exocyclic and places little constraint on the stereochemistry. Stereochemical studies of concerted substitution reactions at the phosphyl (g) and analogous sulfyl centers indicate that these proceed with the inversion of configuration expected for in-line geometry. Displacements at aromatic (h) and carbonyl (i) centers are expected to have adjacent geometries.

The geometry of the S_N2 displacement reaction is neatly demonstrated in (j), where the nucleophile is constrained to attack from an impossible direction for an in-line mechanism. An interesting substitution process is the S_N2' process (k), which is a special case of a cyclization reaction. Baldwin formulated empirical rules that predict the cyclization reactivity as a function of ring size, the COORDINATION NUMBER of the central atom, and whether ligand departure is exocyclic or endocyclic. The predictions have exceptions but are useful for the evaluation of synthetic routes.

Inversion of configuration is not diagnostic of a concerted displacement mechanism but does indicate in-line geometry. Many aliphatic substitution reactions exhibit inversion of configuration and yet proceed by a mechanism involving ion pair intermediates. For example, the retention of configuration in chlorination of hydroxyl groups with thionyl chloride is due to the formation of a tight ion pair intermediate (l) and is not evidence for an "adjacent" path (m).

Symbolic notation is used as a shorthand way of representing mechanism both in speech and in documentation and was introduced to facilitate the early discussions of substitution; current practice is to evaluate the International Union of Pure and Applied Chemistry (IUPAC) code as a replacement for that originally used by Ingold (S_N2, S_N1, etc.), which is based on experimental observation. The IUPAC notation is based on mechanism, and although it can appear complicated, it is not subject to confusion.

See also STEREOCHEMISTRY.

BIBLIOGRAPHY

BUNNETT, J. F. "Aromatic Substitution by the $S_{RN}1$ Mechanism." *Acc. Chem. Res.* 1978, *11*, p. 413.

GUTHRIE, R. D.; JENCKS, W. P. "IUPAC Recommendations for the Representation of Reaction Mechanisms." *Acc. Chem. Res.* 1989, *22*, 343.

INGOLD, C. K. *Structure and Mechanism in Organic Chemistry,* 2nd ed.; Ithaca, NY: Cornell University Press, 1969.

LOWRY, T. H.; RICHARDSON, K. S. *Mechanism and Theory in Organic Chemistry,* 3rd ed.; New York: Harper & Row, 1987; pp. 327, 640, 661.

MARCH, J. *Advanced Organic Chemistry,* 4th ed.; New York: Wiley, 1992; p. 293.

RAPPOPORT, Z. "Nucleophilic Vinylic Substitution. A Single- or a Multi-Step Process?" *Acc. Chem. Res.* 1981, *14*, 7.

SNEEN, R. A. "Organic Ion Pairs as Intermediates in Nucleophilic Substitution and Elimination Reactions." *Acc. Chem. Res.* 1973, *6*, 46.

SYKES, P. *A Guidebook to Mechanism in Organic Chemistry,* 6th ed.; New York: Wiley, 1987; pp. 77, 130, 203.

WILLIAMS, A. "Concerted Mechanisms of Acyl Group Transfer Reactions in Solution." *Acc. Chem. Res.* 1989, *22*, 387.

ANDREW WILLIAMS

Sugars

Sugars—or more generally carbohydrates—are the world's greatest renewable source of biomass and potential energy. They are widely distributed in plants, representing around three-quarters of the plant world as, for example, the cellulose and hemicelluloses of WOOD and cotton, the starches of plant seeds and potatoes, the gums and mucilages of trees and seaweed, and the outer coats of many bacteria and fungi. Approximately 70 percent by weight of seeds is carbohydrate, present primarily in the endosperm. The carbohydrates are principally present in nature as polysaccharides, which are polymers of the simple sugars called monosaccharides.

Structure

MONOSACCHARIDES

The monosaccharides, of which glucose is the most common and will be used as our example, are composed of carbon, hydrogen, and oxygen in the ratio $C(H_2O)$, the so-called hydrates of carbon. The number of carbons may be three or greater (although no more than eight carbons are common in nature), arranged as a straight chain to which are attached hydrogen and hydroxyl groups. One of the carbons is a CARBONYL group, so that glucose with six carbons, which is called a hexose, can be represented as in Figure 1. The carbonyl group is either an ALDEHYDE, as in glucose, or a KETONE, as in fructose (Figure 2).

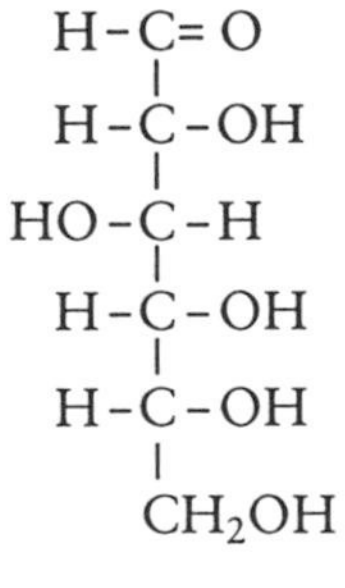

$$
\begin{array}{c}
H-C=O \\
| \\
H-C-OH \\
| \\
HO-C-H \\
| \\
H-C-OH \\
| \\
H-C-OH \\
| \\
CH_2OH
\end{array}
$$

Fig. 1.   D-glucose, an aldohexose.

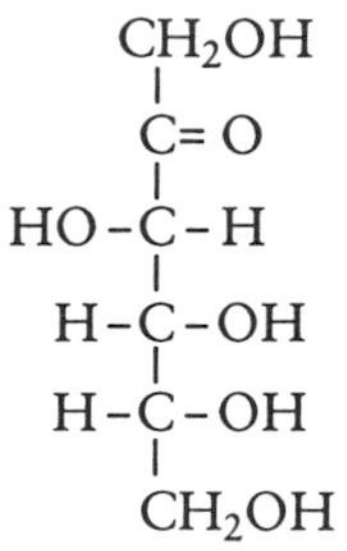

$$CH_2OH$$
$$|$$
$$C = O$$
$$|$$
$$HO - C - H$$
$$|$$
$$H - C - OH$$
$$|$$
$$H - C - OH$$
$$|$$
$$CH_2OH$$

Fig. 2.   D-fructose, a ketohexose.

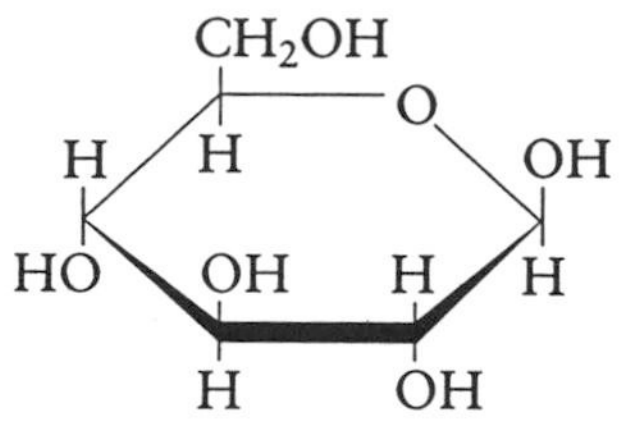

D-glucose can assume two six-membered ring forms, as in Figures 4 and 5.

Fig. 4.   β-D-glucopyranose.

It is seen that four of the carbons in glucose have different groups on each of their four valences, and by the rearrangement of these groupings 4^2, or 16, optical ISOMERS can be identified. Similarly, for monosaccharides with different numbers of carbon isomers, there are families of isomers—the aldopentoses, for example, constituting 3^2, or 9, optical isomers. All of these families of sugars can be considered to be structurally derived from the two optical isomers of the three-carbon aldotriose, glyceraldehyde (Figure 3). This relationship was recognized by Emil FISCHER, who called the glyceraldehyde with the hydroxyl group at carbon two on the right side, as shown in Figure 3, the D-isomer; its mirror image is L-glyceraldehyde. This designation has been shown by X-RAY CRYSTALLOGRAPHY to be correct for a related sugar acid, tartaric acid. The D- and L-designations bear no ready relationsihp to the dextro- or levo-optical rotatory property of these isomers.

D-glucose is one of the eight aldohexoses in the D-series; some of the other naturally occurring monosaccharides are D-mannose and D-galactose (aldohexoses); D-fructose and L-sorbose (ketohexoses); and D-ribose, D-xylose and L-arabinose (aldopentoses).

The polyhydroxy keto- or aldehydo-sugars are very reactive in the open-chain form (Figures 1 and 2), and the C=O group condenses spontaneously with an accessible intramolecular hydroxyl group to form a ring at which time the C=O oxygen becomes a hemiacetal hydroxyl group, which may be on one side or the other of the plane of the sugar ring. Thus

In this nomenclature -ose indicates a sugar, *pyran*- shows the six-membered ring, *gluco*- determines the arrangement of the OH and H groups above or below the plane of the pyran ring, and α-, β- are the anomeric forms of the sugars formed when hemiacetal ring closure occurs. The alternative stable ring is five-membered, relating to furan, so that we have α-D-glucofuranose (Figure 6).

In aqueous solution the monosaccharides exist in an equilibrium between the open-chain *aldehydo*- or *keto*-form and the α- and β-anomers of the pyranose and furanose ring forms. For most monosaccharides the open-chain structure is very minor, and the pyranose anomers are the principal constituents. The sugars crystallize most easily in one of these forms, crystalline glucose being the α-D-anomer. When all forms are dissolved in water, their equilibrium is established by a process called mutarotation.

The representation of the sugars in two dimensions, as in Figures 1 and 2, gives a false picture of the shape of the molecules, which are better represented in the three-dimensional formulas of Haworth, in which the pyranose and furanose rings are presented as flat planes. In fact, the rings assume a boat or chair shape in space, with the substituents either being parallel to the axis of the ring (axial groups) or projecting from the ring (equatorial groups). Steric effects principally determine the most stable ring shape, with as many as possible of the bulky groups, such as —OH and —CH_2OH, being equatorial. Thus, α-D-glucose is most stable, as shown in Figure 7, although

$$CHO \qquad\qquad CHO$$
$$| \qquad\qquad\qquad |$$
$$H - C - OH \qquad HO - C - H$$
$$| \qquad\qquad\qquad |$$
$$CH_2OH \qquad\quad CH_2OH$$

D-glyceraldehyde   L-glyceraldehyde

Fig. 3.   (a) D-glyceride and (b) L-glyceride.

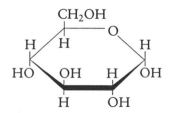

Fig. 5.   α-D-glucopyranose.

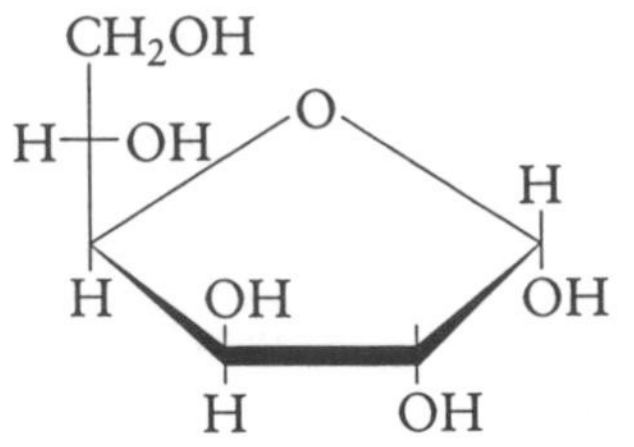

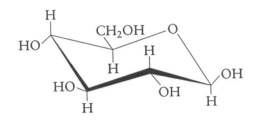

Fig. 6.   α-D-glucofuranose.

Fig. 8.   β-D-anomer of α-D-glucose.

the energy difference between it and its β-D-anomer (Figure 8) is small, accounting for the mutarotation to equilibrium representing nearly equal amounts of the two anomers.

Derivatives of the monosaccharides are found in nature in which a hydroxyl group is replaced by —NH₂ to give the amino sugars, such as glucosamine, or by —H to give the deoxy sugars, such as the 2-deoxy-ribose in RNA (see NUCLEIC ACIDS). Another common derivative is formed by oxidation of the primary —CH₂OH group to —COOH, giving the uronic acids, such as glucuronic acid.

Oligosaccharides and Polysaccharides

Monosaccharides link by condensing the anomeric hydroxyl group and a hydroxyl group in another sugar with the elimination of water. Thus lactose, or milk sugar, is the disaccharide β-D-galactosyl D-glucose (Figure 9), shown as the α-D-anomer.

The anomeric carbon, by the equilibrium of all of the sugar forms, has reducing properties such that the sugars will produce metallic silver from ammoniacal silver nitrate and cuprous oxide from copper sulfate. Such sugars are called reducing sugars, but the reducing group can be derived to give nonreducing compounds. The reducing sugar may be reduced to an alcohol called, in general, a gly*citol*; or oxidized to an acid, a gly*conic* acid; or reacted by aldol condensation with an alcohol to give a gly*coside*. The disaccharide sucrose, common cane sugar,

is a nonreducing compound in which the α-D-glucopyranose condenses with the anomeric hydroxyl group of β-D-dructofuranose to give α-D-glucopyranosyl β-D-fructofuranoside (Figure 10), a "double" glycoside. As these aldol condensations continue, higher-molecular-weight molecules are formed, called oligosaccharides, proceeding further to polysaccharides such as starch (polymer of α-D-glucose) and cellulose (polymer of β-D-glucose), each with a molecular weight in the millions of daltons. Given the many sugar building blocks and the several different hydroxyl groups in each whereby the condensations can occur, it is not surprising that thousands of polysaccharides are found in nature with a great variety of physical properties that have directed their use in food, biotechnology, and industry.

Nutrition

All higher forms of animal life are totally dependent upon the photosynthetic process (see PHOTOSYNTHESIS), whereby plants use the sun's energy to convert CO_2 and H_2O first to carbohydrates, from which amino acids and fatty acids are formed. The principal carbohydrates of human nutrition are starch, lactose, and sucrose; vitamin C (L-*xylo*-ascorbic acid) is the only carbohydrate essential for human (and guinea pig) health.

The reactions that occur in biological systems are catalyzed by ENZYMES, without which the reactions proceed too slowly to be physiologically effective. The food of animals is digested by enzymes in the mouth and the small intestine so that starch, sucrose, and lactose are hydrolyzed to their component monosaccharides. Cellulose cannot be digested by humans and so is excreted in the feces, but it is used as the main carbohydrate source by cows, horses, sheep, and other polygastric animals.

Considering most particularly human biology, the salivary α-amylase during the mastication of food partially digests starch to give oligosaccharide fragments. This digestion is temporarily halted by the acidic conditions in the stomach. Passage from the

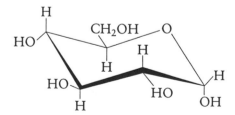

Fig. 7.   α-D-glucose.

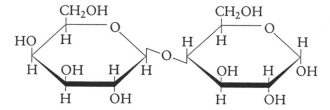

Fig. 9.   Lactose.

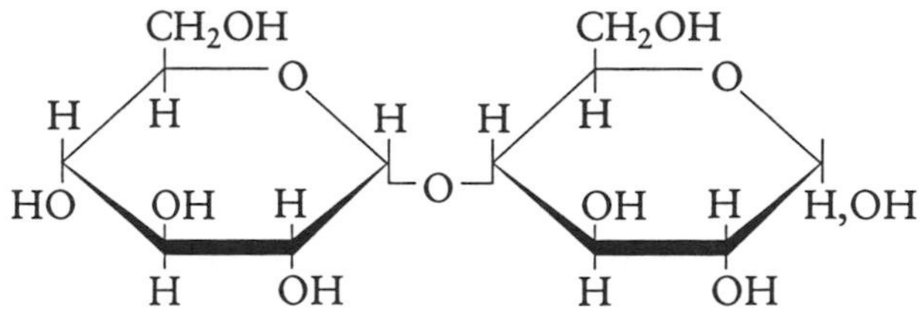

Fig. 11.   Maltose.

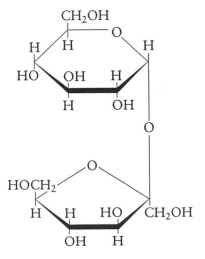

Fig. 12.   Isomaltose.

stomach to the small intestine neutralizes the stomach acid and under alkaline conditions the pancreatic amylase continues the digestion of starch to produce glucose, maltose (Figure 11), and isomaltose (Figure 12). These disaccharides together with lactose (Figure 9) and sucrose (Figure 10) are hydrolyzed by enzymes in the brush border cells of the small intestine to give a mixture of glucose, fructose, and galactose. These sugars are carried in the portal vein to the liver where the galactose and fructose are converted to glucose.

It is not essential that humans include carbohydrates in their diet, since all the necessary sugars can be biosynthesized from certain AMINO ACIDS and fatty acids when needed (see also FOOD METABOLISM). It is advisable, however, to derive at least 20 percent of the body's energy needs from sugars in order to metabolize fat without causing metabolic imbalances such as acidosis and ketosis. Carbohydrates are a principal energy source, 4 cals/g being converted to adenosine triphosphate (ATP), which is the "energy currency" of biological systems that use it for muscle contraction, nerve conduction, movement of molecules across cell membranes, and biosynthesis.

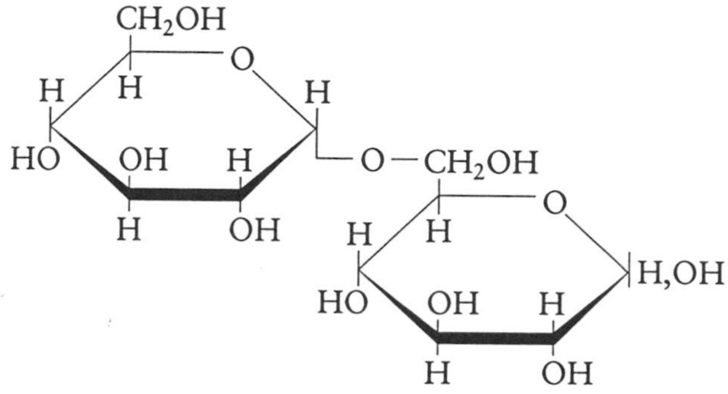

Biochemistry

The biochemistry of carbohydrates in mammals centers upon D-glucose, which is the normal blood sugar. The concentration of the latter is under the hormonal control of insulin, epinephrine, and glucagon. After a carbohydrate meal, the blood sugar concentration increases, insulin is secreted from the pancreas, and the glucose is stored in tissues as the polysaccharide glycogen, which is sometimes called animal starch. Between meals the glycogen is broken down to maintain the glucose supply to the body, which is so critically important for brain function. The glucose is used by cells in several ways:

1. The glucose is metabolized by a series of steps to CO_2 and H_2O, the energy being 40 percent conserved as ATP. Three metabolic sequences are involved. Glucose, as the 6-phosphate, enters glycolysis in which it is converted to pyruvate in the cytosol of the cell with the generation of 8 mols of ATP. The pyruvate enters the mitochondria and is converted to acetyl coenzyme A, which by the Krebs cycle is oxidized to CO_2 by nicotine adenine dinucleotide (NAD^+) and flavin adenine dinucleotide (FAD) with the generation of 2 mols of ATP equivalents, 6 mols of $NADH,H^+$, and 2 mols of $FADH_2$. These reduced coenzymes, $NADH,H^+$ and $FADH_2$, are oxidized by oxidative phosphorylation in the respiratory chain to 6 mols

Fig. 10.   Sucrose.

of water with the formation of 24 mols of ATP. The ATP equivalent of the metabolic oxidation of NADH,H[+], FADH, and glucose is given as the maximum value. Depending upon the cellular efficiency it may be less.

2. Most biosynthetic pathways have reduction steps, whereas the catabolic pathways are principally oxidative. The reducing agent is nicotinamide adenine denucleotide phosphate, NADPH, which is produced in large measure by an alternative oxidative pathway of glucose catabolism, the pentose phosphate pathway. In this pathway the oxidation to CO_2 and H_2O is accomplished by NADP[+] in contrast to O_{2-} in glycolysis and the Krebs cycle. By this complex pathway glucose is also converted to D-ribose used in RNA biosynthesis.

3. The glucose is converted to derivatives, such as glucuronic acid and 2-deoxy-2 amino-D-glucose (D-glucosamine), and to other naturally occurring monosaccharides that are incorporated into biopolymers containing carbohydrates, such as glycogen, glycoproteins (as in ANTIBODIES), glycoconjugates (as in chondroitin sulfate of tendons), and glycolipids (as in cell membrane receptors). These polymerizations are not directed by a "template" such as found in protein biosynthesis, so that the resulting polymers are molecularly dispersed.

Commercial Uses

The principal uses of carbohydrates are in animal and human foodstuffs—animals consume corn, wheat, and oats; humans consume cereals, baked goods, and soft drinks. Biotechnology has commercialized the enzymatic isomerization of D-glucose to D-fructose for inclusion into soft drinks, fructose being a much sweeter sugar than glucose. One of the oldest applications has been FERMENTATION, and many products other than ETHANOL are produced by fermentation, such as CITRIC ACID, essential amino acids (lysine and tryptophan), and monosodium glutamate. The move to use ethanol as a motor fuel from renewable sources has received much impetus to reduce environmental and balance-of-trade concerns.

The unique physical properties of the polysaccharides or their derivatives have lent them to wide applications, including food stabilizers (carrageenin in ice cream), drilling muds, pectins in jams and jellies, paper coatings (starch derivatives), adhesives, pharmaceuticals, cosmetics, metallurgy, and textile finishes.

See also ORGANIC CHEMISTRY.

BIBLIOGRAPHY

ASPINALL, G. O. *The Polysaccharides;* New York: Academic Press, 1982; Vols. 1 and 2.

EL KHADEM, A. S. *Carbohydrate Chemistry: Monosaccharides and Their Oliogmers;* New York: Academic Press, 1988.

HORTON, D., ed. *Advances in Carbohydrate Chemistry and Biochemistry;* New York: Academic Press, 1945–1995; Vols. 1–51.

KENNEDY, J. F., ed. *Carbohydrate Chemistry;* New York: Oxford University Press, 1988.

PIGMAN, W.; HORTON, D., eds. *The Carbohydrates, Chemistry and Biochemistry,* 2nd ed.; New York: Academic Press, 1970; Vols. 1, 2, and 3.

REX MONTGOMERY

Sulfur

16
S
32.06

Melting Point: 386 K (rhombic); 392 K (monoclinic); **Boiling Point:** 718 K; **Density:** 2.07 g/cm[3] (rhombic); 1.96 (monoclinic); **Most Common Ion:** −2

Sulfur is the sixteenth most abundant element in the Earth's crust. The element's name is derived from the Latin word for sulfur. The element occurs free in nature and is one of the two nonmetals known to early civilizations, with carbon being the other. Mentioned in the Bible as brimstone, sulfur was burned in ancient rituals. The poet Homer described the power of burning sulfur in fumigation. The Egyptians used SO_2, obtained by burning sulfur, as a bleach for cotton as early as 1600 B.C., and about 500 B.C. the Chinese created gunpowder, using sulfur as one of the components. The identification of the substance as an element is credited to GAY-LUSSAC and Thenard in 1809, although LAVOISIER proposed the elemental nature of sulfur in 1777.

Sulfur occurs naturally in large deposits, frequently near volcanoes. Large natural deposits of elemental sulfur in Texas, Louisiana, and Mexico have been mined for many years. In addition, sulfur is found in crude oil and natural gas as hydrogen sulfide, and combined in several minerals, for example, cinnabar, Epsom salts, galena, gypsum, pyrite, and stibnite.

Sulfur, an element whose yellow appearance is quite descriptive, has several allotropic forms. At least three of these, α-orthorhombic, β-monoclinic, and γ-monoclinic, possess an S_8 ring. The relatively low melting point of sulfur is used to great advantage by the Frasch process. This process supplanted the mining of sulfur from volcanic surface deposits. Developed about 1891 by Herman Frasch, the process uses superheated water to melt sulfur contained in subsurface deposits. Compressed air forces the molten liquid to the surface.

The manufacture of SULFURIC ACID is an important industrial use for sulfur. Sulfuric acid is made by the contact process, and the extent of its production has been used to measure a nation's economic health. About 90 percent of recovered sulfur is transformed into the acid, with the remainder being used in the pulp and paper industry, in the production of tires (vulcanization), in insecticides and fungicides, and in the chemical industry to produce organic compounds.

Sulfur is an essential element to both plant and animal life. The AMINO ACIDS cysteine, cystine, and methionine contain sulfur. NITROGEN FIXATION also involves iron-sulfur and iron-molybdenum-sulfur proteins.

See also SULFUR CYCLE.

BIBLIOGRAPHY

BIXBY, D. W., FIKE, H. L.; SHELTON, J. E.; WIEWIOROWSKI, T. K. "Sulfur." In *Encyclopedia of Chemical Technology*, 3rd ed.; R. E. Kirk, D. F. Othmer, M. Grayson, eds.; New York: Wiley-Interscience, 1981; Vol. 22, pp. 78–106.

GREENWOOD, N. N.; EARNSHAW, A. *Chemistry of the Elements;* New York: Pergamon, 1984; pp. 757–782.

LIDE, D. R., ed. *Handbook of Chemistry & Physics*, 71st ed.; Cleveland: Chemical Rubber Co., 1990–1991.

ALTON J. BANKS

Sulfur Cycle

The volatility of sulfur is the probable reason why it is deficient in the Earth relative to its cosmic abundance. Sulfur is the sixteenth most abundant element in the Earth's crust, with a mean concentration of 260 ppm, whereas in the whole Earth it is the sixth most abundant element, with a concentration of 1.84 percent (18,400 ppm).

Atmosphere

The steady state amount of sulfur in the atmosphere is about 4.3 Tg (Tg [teragram] = 10^{12} g), only 2×10^{-5} percent of the total amount in the crust, hydrosphere, biosphere, and atmosphere (Figure 1). The annual inputs to and losses of sulfur from the atmosphere are, however, large (Figure 1), indicating a rapid turnover. Some of the main sulfur compounds in the atmosphere along with their concentrations,

their total amounts, their lifetimes and the distances they are transported are listed in Table 1.

Carbonyl sulfide, COS, is the least reactive sulfur species in the atmosphere. Its longer lifetime (>1 year) means it gets into the stratosphere, where it is photochemically decomposed:

$$COS \xrightarrow{h\nu} S + CO \tag{1}$$

Sulfur dioxide is then produced in further reactions, and this may be the origin of its elevated levels in the upper troposphere:

$$S + O_2 \rightarrow SO + O \tag{2}$$

$$SO + O_2 \rightarrow SO_2 + O \tag{3}$$

The lifetimes of other sulfur compounds in the atmosphere are short. This is, in part, because of the tendency of the sulfur compounds to become oxidized. A simplified reaction sequence is shown in Figure 2. All reactions are homogeneous except for the conversion of SO_2 to SO_3 that occurs in aerosol droplets catalyzed by metal ions such as Fe^{2+} and Mn^{2+}. Some or all of the reactions occur before sulfur emitted to the atmosphere is returned to the Earth's surface. One product, H_2SO_4, is deposited in wet precipitation, (acid rain) and produces rainwater with a pH of 4–6 (and at times as low as 2). European and North American areas are mostly affected by this type of pollution.

Nearly equal amounts of natural and artificially generated sulfur species occur in the atmosphere. Of the 120 Tg of sulfur extracted per year (fossil fuels and sulfide ores), about 70 Tg goes into the atmosphere from combustion and ore processing, 25 Tg flows into rivers and sewers, and 25 Tg is used as fertilizer (see FERTILIZERS). Common sources of atmospheric sulfur are volcanoes (H_2S, CS_2, COS, SO_2, H_2SO_4), animal wastes, wood pulping, microbes, decay of organic material, sewage treatment (H_2S, R_2S, RSH), ocean spray (SO_4^{2-}, $(CH_3)_2S$), petroleum manufacture (H_2S, R_2S, RSH, CS_2, COS), fuel combustion (SO_2, SO_3), algae (($CH_3)_2S$)), and forest fires (COS, SO_2). The chemical form depends on the process—i.e., whether it is oxidizing or reducing. Sulfur dioxide emissions have increased at about 4 percent per year over the past few decades.

Biosphere

Sulfur, an essential element for living tissue, exists in a variety of compounds in plants, animals, and microorganisms. These include AMINO ACIDS (cysteine,

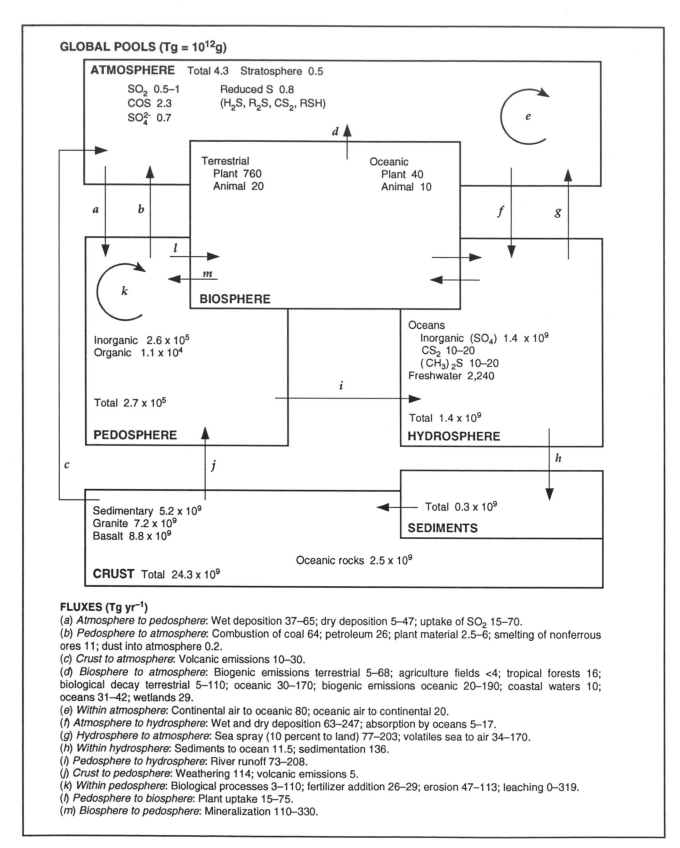

Fig. 1.   Global pools and fluxes for sulfur.

FLUXES (Tg yr⁻¹)

(a) *Atmosphere to pedosphere*: Wet deposition 37–65; dry deposition 5–47; uptake of SO_2 15–70.

(b) *Pedosphere to atmosphere*: Combustion of coal 64; petroleum 26; plant material 2.5–6; smelting of nonferrous ores 11; dust into atmosphere 0.2.

(c) *Crust to atmosphere*: Volcanic emissions 10–30.

(d) *Biosphere to atmosphere*: Biogenic emissions terrestrial 5–68; agriculture fields <4; tropical forests 16; biological decay terrestrial 5–110; oceanic 30–170; biogenic emissions oceanic 20–190; coastal waters 10; oceans 31–42; wetlands 29.

(e) *Within atmosphere*: Continental air to oceanic 80; oceanic air to continental 20.

(f) *Atmosphere to hydrosphere*: Wet and dry deposition 63–247; absorption by oceans 5–17.

(g) *Hydrosphere to atmosphere*: Sea spray (10 percent to land) 77–203; volatiles sea to air 34–170.

(h) *Within hydrosphere*: Sediments to ocean 11.5; sedimentation 136.

(i) *Pedosphere to hydrosphere*: River runoff 73–208.

(j) *Crust to pedosphere*: Weathering 114; volcanic emissions 5.

(k) *Within pedosphere*: Biological processes 3–110; fertilizer addition 26–29; erosion 47–113; leaching 0–319.

(l) *Pedosphere to biosphere*: Plant uptake 15–75.

(m) *Biosphere to pedosphere*: Mineralization 110–330.

Table 1. Sulfur Compounds in the Atmosphere

Species	Concentration (μg m^{-3})			Total Amount (Tg)	Lifetime	Transport Distance (km)
	Polar	Rural	Urban			
H$_2$S					1–4 days	100–1,000
(CH$_3$)$_2$S	0.05	0.2–1	1–75	0.8	1–3 days	100–1,000
CH$_3$SH					1–3 days	100–1,000
CS$_2$		<0.1	<0.1		40 days	
COS		0.7	0.7	2.3	>1 year	Global/stratosphere
SO$_2$	0.15	0.2	5–20	0.5–1	1–10 days	10–100
SO$_4^{2-}$	0.5	0.5	2–5	0.7	Few days	100–1,000

methionine), cofactors (thiamine, ferredoxin), sulfate esters (choline sulfate, tyrosine sulfate), and sulfonates (cysteic acid, sulfolipid). About 90 percent of plant sulfur is in amino acids. Plant uptake of sulfur is estimated at around 15–75 Tg yr^{-1}.

Sulfur is absorbed by plants as the SO$_4^{2-}$ ion. This is converted to organic-sulfur compounds intracellularly, generally, but not always, by reduction of the sulfate to sulfide. Excess sulfur remains as sulfate. Also, organic-sulfur can be oxidized to SO$_4^{2-}$ by bacterial processes in plants.

Hydrosphere

The principal sulfur species in the hydrosphere is the sulfate ion, around 1.4×10^9 Tg in the ocean and 2,240 Tg in freshwater. Lesser amounts of (CH$_3$)$_2$S, CS$_2$, and H$_2$S occur in the hydrosphere, their concentrations being greatest in anaerobic areas such as stagnant waters. The oceans are a global source of COS, emitting about 0.6 Tg yr^{-1}. The amount of sulfur transferred to ocean sediments is approximately 132 Tg yr^{-1}, mainly as the reduced species S^{2-} and HS$^-$, which then form iron sulphides.

The average concentration of SO$_4^{2-}$ in rivers is approximately 11.2 mg L^{-1}, giving a flux of 132 Tg yr^{-1}. At least 28 percent and maybe as much as 45 percent derives from pollution sources such as fossil fuel burning (SO$_2$) and fertilizer runoff (SO$_4^{2-}$).

Sulfate is the major species in pore water near the surface, but its concentration decreases with depth, whereas the concentration of sulfide increases. At lower levels still the sulfide concentration also decreases owing to its removal as iron pyrites and by diffusion.

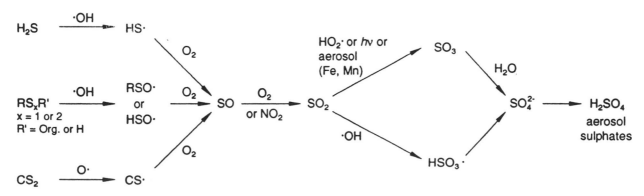

Fig. 2. Reactions of sulfur species in the atmosphere. (Adapted from T. E. Greadel, *Revs. Geophys. Space Phys.*, 1977, *15*, 421–428.)

Table 2. Soil Sulfur Reactions

Chemical Interconversions	Process	Bacteria
SO_2, $SO_4^{2-} \rightarrow$ Org.S	Assimilation	Many bacteria
Org.S $\rightarrow H_2S$, R_2S, RSH	Decay	Many anaerobic and aerobic bacteria
Org.S $\rightarrow SO_4^{2-}$	Oxidation of organic sulfides	Many plant and animal bacteria
SO_4^{2-}	Reduction	*Desulfovibrio, Desulfuromonas*
$H_2S \rightarrow SO_4^{2-}$	Oxidation	*Thiobacillus, Beggiatoa*

Pedosphere

The concentration of sulfur in soils is quite variable, and for most soils lies in the range 0.005–0.5 percent (50–5,000 ppm), median 0.01–0.06 percent (100–600 ppm). The primary source is weathered rock, but significant amounts, as SO_4^{2-} and SO_2, are added from wet and dry deposition and fertilizers.

The major forms of sulfur in soils are organic (as amino acids, sulfonates, organic esters of H_2SO_4, and other organic-sulfur species) and inorganic sulfate (soluble, insoluble, and absorbed). Very little sulfur is present as sulfide or bisulfide, the highest concentration occurring in waterlogged and anaerobic conditions.

Weathering processes bring about the dissolution of gypsum ($CaSO_4 2 H_2O$) and oxidation of pyrites (FeS_2) and other iron sulfides, e.g.:

$$2 \, FeS_2 + 2 \, H_2O + 7 \, O_2 \rightarrow 2 \, FeSO_4 + 2 \, H_2SO_4 \quad (1)$$

$$4 \, FeSO_4 + O_2 + 10 \, H_2O \rightarrow 4 \, Fe(OH)_3 + 4 \, H_2SO_4 \quad (2)$$

Sulfuric acid is produced when FeS_2 is exposed to air and water. This is a significant process in the formation of acid mine water in old coal mine shafts. The preindustrial rate of weathering of sulfide was around 36 Tg yr^{-1}.

Numerous interconversions of sulfur species occur in soils (Table 2), many of which involve bacteria. The oxidation of S^{2-} is rapid if sufficient dioxygen is present. Sulfides such as H_2S, $(CH_3)_2S$, CS_2, and CH_3SH are formed in decay processes.

Crust

The pool of sulfur in Earth's crust is estimated at around 24.3×10^9 Tg, of which approximately 75 percent is sulfide. The main sources of crustal sulfur are rocks, fossil fuels, pyrites, gypsum, volcanic deposits, and free element deposits. About 132 Tg yr^{-1} of sulfur (SO_4^{2-}, S^{2-}, organic-S) transfers from the oceans to sediments, which eventually become part of Earth's crust. Over several million years, significant amounts of dioxygen have been used in the weathering and oxidation of reduced sulfur.

BIBLIOGRAPHY

BRIMBLECOMBE, P.; LEIN, A. Y., eds. *Evolution of the Global Biogeochemical Sulphur Cycle;* SCOPE 39; Chichester, U.K.: Wiley, 1989.

COLE, J. A.; FERGUSON, S. J. *The Nitrogen and Sulphur Cycles;* Cambridge, U.K.: Cambridge University Press, 1988.

HUTZINGER, O., ed. *The Handbook of Environmental Chemistry;* Berlin: Springer-Verlag, 1980; Vol. 1A.

IVANOV, M. V.; FRENEY, J. R., eds. *The Global Biogeochemical Sulphur Cycle;* SCOPE 19; Chichester, U.K.: Wiley, 1983.

LIKENS, G. E., ed. *Some Perspectives of the Major Biogeochemical Cycles;* SCOPE 17; Chichester, U.K.: Wiley, 1981.

SVENSSON, B. H.; SÖDERLUND, R., eds. *Nitrogen, Phosphorus and Sulphur—Global Cycles;* SCOPE 7; Ecol. Bull., 1976, no. 22.

WAYNE, R. P. *Chemistry of Atmospheres;* Oxford: Clarendon Press, 1991.

JACK E. FERGUSSON

Sulfuric Acid

Sulfuric acid, H_2SO_4, is a colorless, odorless, slightly viscous liquid, m.p. 10.49°C, b.p. ~290°C, and $D_4^{15} = 1.8356$, which decomposes at approximately 340°C into sulfur trioxide and water. It is miscible with water in any ratio (with vigorous heat evolution), and aqueous sulfuric acid solutions are available commercially in different concentrations (weight percent H_2SO_4) ranging from 78 percent (60° Baumé [Bé]) to 100 percent, and as oleums (fuming sulfuric acid, $H_2SO_4 + SO_3$).

The major process used for the production of sulfuric acid is the contact process (oxidation of SO_2 to SO_3 over vanadium pentoxide as catalyst), which produces concentrated acid (98 percent). In this process, the initially formed gas stream (SO_2 + some O_2) either is dried prior to oxidation or is oxidized in the presence of water vapor with subsequent acid condensation and removal. Sulfuric acid from the process is used as the drying agent and also to absorb the sulfur trioxide produced. Environmental concerns have led to widespread adoption of the double absorption modification, which drastically reduces SO_2 emissions, and in the United States a sulfur dioxide conversion efficiency of 99.7 percent is required. The sulfur dioxide used in the contact process is obtained from a variety of sulfur-bearing raw materials including elemental sulfur (the principal source), hydrogen sulfide containing waste gases, flue gases from the combustion of sulfurous fossil fuels, contaminated and diluted sulfuric acid, and the roasting of sulfide ores of nonferrous metals or iron pyrites.

Sulfuric acid is by far the most widely used industrial chemical. The largest single consumer is the FERTILIZER industry, mostly for use in the production of PHOSPHORIC ACID and subsequently phosphates. It has innumerable other uses, principal among which are its use in the steel industry for pickling (removal of mill scale from plate and strip steel) and in the metallurgical industry to dissolve minerals such as bauxite (for the manufacture of alumina and aluminum), fluorspar (for hydrogen fluoride), and ilmenite (for titanium dioxide pigment). It is a strong acid that can be used as a dehydrating medium in sulfonations, in nitrations, and in many other synthetic transformations. It is typically less costly than any other acid, is chemically stable, and can be handled easily in steel (or common-alloy) vessels at commercial concentrations (including >100 wt.%).

Water/sulfuric acid mixtures offer an extremely wide and continuous range of acidity. While the pH of such solutions cannot be defined accurately, values of H_0 (acidity function) ranging from +0.84 (1 wt.% H_2SO_4) to −11.94 (100 wt.% H_2SO_4) at 25°C have been determined using basic indicators of overlapping range. These values are not absolutes and somewhat different results have been obtained using different bases or methods of determination.

Pure sulfuric acid and its aqueous solutions are extremely irritating, corrosive, and toxic to tissue, resulting in rapid destruction and severe burns. Repeated contact with dilute aqueous solutions can cause dermatitis and prolonged mist inhalation can lead to chronic bronchitis. Erosion of teeth due to exposure to strong sulfuric acid fumes has been recognized. The TLV for H_2SO_4 mist is 1 mg/m^3 and, in general, exposure to 0.125–0.50 ppm H_2SO_4 is slightly annoying, 1.5–2.5 ppm is definitely unpleasant, and 10–20 ppm is unbearable.

BIBLIOGRAPHY

HAWLEY, G. G., ED. *The Condensed Chemical Dictionary*, 9th ed.; New York: Van Nostrand Reinhold, 1977; p. 827.

SANDER, U. H. F.; FISCHER, H.; ROTHE, U.; KOLA, R. *Sulphur, Sulfur Dioxide, and Sulphuric Acid*; Bonn: Verlag Chemie, 1984.

KEN TURNBULL

Sumner, James Batcheller (1887–1955)

James Sumner was born in Canton, Massachusetts, in 1887. In 1904, an accidental shooting by his hunting companion resulted in the amputation of Sumner's left arm. He had been left-handed, but was able to gain dexterity with his right hand. He became an expert tennis player, skier, and mountain climber and engaged in these activities throughout his life. In 1906, he entered Harvard. He received a bachelor's degree in chemistry in 1910. He became a graduate student in biochemistry at the Harvard Medical School in 1912. His professor, Otto Folin, urged him to transfer to the law school, believing that Sumner's disability rendered him incapable of laboratory research. The young chemist's skill and determination, however, won over Folin, and Sumner received his doctoral degree in 1914. He then taught biochemistry at Cornell University until his retirement in 1955. He died in Buffalo, of cancer, shortly after his retirement.

Sumner offered the first courses in biochemistry at Cornell. His teaching load was exceptionally heavy, and he decided that he must devote his limited research time to something of great import. Because the chemical nature of enzymes was still a mystery, Sumner made this subject a lifelong quest. Two German chemists, Emil FISCHER and Richard Willstätter, dominated the thinking about enzymes. Fischer thought that enzymes might possess a protein-like character; Willstätter believed that enzymes were small reactive molecules adhering to and stabilized by colloidal carriers. Sumner adopted Fischer's view and tried to isolate the enzyme urease from its richest source—jack beans—by means of methods used for purifying proteins. By 1926 he had isolated four crys-

1413

talline globulin proteins and claimed that one of them was urease.

The response was negative. In 1926, Willstätter's conceptions prevailed, and chemists dismissed Sumner's claim to have isolated an enzyme in pure form for the first time. In the mid-1930s, the protein view of enzymes began to overcome opposition, and the key figure in this change was John Northrop at the Rockefeller Institute. Between 1930 and 1935, Northrop obtained pepsin and several other enzymes as crystalline proteins by improving methods of isolation and providing a host of physicochemical analyses to prove that the pure crystals were enzymes.

Sumner also isolated several enzymes during the 1930s. He demonstrated the importance of metals for protein and enzyme activity. In 1936, he proved that one of the four globulins in jack beans was a hemagglutinin. The protein, however, was inactive by itself; it required the presence of divalent metals such as zinc for its activity. In 1937, he isolated catalase. The enzyme's protein was inactive without an iron porphyrin entity as coenzyme. Further research by Sumner, Otto Warburg, and others revealed that many enzymes were metalloproteins.

During 1937 and 1938, Sumner collaborated with Theodor Svedberg in Sweden to determine the molecular weights of his crystalline enzymes by means of Svedberg's ultracentrifuge method. In 1946 he shared the Nobel Prize for chemistry with Northrop and Wendell Stanley, who adopted the methods of Sumner and Northrop in achieving the first isolation of a virus in 1936.

The resolution of the controversy over the nature of enzymes led to proliferation of enzyme studies as a branch of protein chemistry. For biochemistry, the protein nature of enzymes became the basis for exploring catalysis in living systems and the relationships between enzymes and substrates in accomplishing the multitude of reactions taking place in living systems.

See ENZYMES AND CATALYSIS and PROTEINS for discussion of enzymes and proteins.

BIBLIOGRAPHY

CORI, C. "James B. Sumner and the Chemical Nature of Enzymes." *Trends in Biochem. Sci.* 1981, 6, 194–196.

COSTA, ALBERT B. "James Sumner and the Urease Controversy." *Chem. Br.* 1989, 25/8, 788–790.

MAYNARD. L. "James Batcheller Sumner." *Biog. Mem. Nat. Acad. Sci.* 1958, 31, 376–396.

———. "The Story of Urease." *J. Chem. Ed.* 1937, 14, 255–259.

ALBERT B. COSTA

Superacids

Chemists long considered mineral acids such as SULFURIC ACID and NITRIC ACID to be the strongest protic acids. More recently this view has changed considerably with the discovery of extremely strong acid systems that are hundreds of millions, even billions, of times stronger than 100 percent sulfuric acid. Such acid systems are termed *superacids*. This term was first suggested by Hull and Conant (1927) to describe acids such as perchloric acid in glacial acetic acid, which were capable of protonating certain weak bases such as ALDEHYDEs and KETONES.

Superacids encompass both BRØNSTED-LOWRY (proton-donor) and Lewis (electron-acceptor) acids as well as their conjugate pairs. The concept of acidity and acid strength can only be defined in relation to a reference base (see ACID-BASE THEORY). According to an arbitrary but widely accepted suggestion by Gillespie (see Gillespie and Robinson, 1967), all Brønsted (protic) acids stronger than 100 percent sulfuric acid are classified as superacids.

Lewis acids also cover a wide range of acidities extending beyond the strength of such frequently used substances as $AlCl_3$ or BF_3. Olah et al. (1985) suggested the use of anhydrous aluminum trichloride, the most widely used Friedel-Crafts catalyst, as the best available to measure protic superacid strengths using Hammett's acidity constant measurements ($-H_0$ value, a logarithmic scale).

The combination of Brønsted and Lewis superacids leads to conjugate superacids of enormous strengths. There remain, however, many difficulties in measuring the strength of Lewis acids. Various physical, electrochemical, and spectroscopic methods are available to measure protic superacid strengths using Hammett's acidity constant measurements ($-H_0$ value, a logarithmic scale).

Examples of Brønsted superacids include perchloric acid, $HClO_4$ ($-H_0$, 13.8); fluorosulfuric acid ($-H_0$, 15.1); anhydrous hydrogen fluoride, HF ($-H_0$, 15.1); teflic acid, TeF_5OH ($-H_0$, $\cong 15$); trifluoromethanesulfonic acid, CF_3SO_3H ($-H_0$ 14.1); and higher homologous perfluoroalkanesulfonic acids, $C_nF_{2n+1}SO_3H$ ($-H_0$, 12–14).

Some of the well-known Lewis superacids are aluminum trihalides (AlX_3, X = Cl, Br, I), boron trifluoride (BF_3), arsenic pentafluoride (AsF_5), antimony pentafluoride (SbF_5), tantalum and niobium pentafluorides (TaF_5 and NbF_5), boron tristrifluoromethanesulfonate B $(OSO_3CF_3)_3$, and gold, tantalum, platinum, and niobium fluorosulfonates: Au $(OSO_2F)_3$, Ta $(OSO_2F)_5$, Pt $(OSO_2F)_4$, and Nb $(OSO_2F)_5$, respectively.

Conjugate Brønsted-Lewis superacids include polysuluric acids (oleums), which are SO_3-containing sulfuric acids. With 50 mol percent SO_3, a $-H_0$ value of 14.5 is reached. A fluorosulfuric acid–antimony pentafluoride combination is called *Magic Acid* (a trade name). With 90 percent SbF_5 content in the system, the $-H_0$ reaches to 26.5. With a fluoroantimonic acid mixture ($HF : SbF_5$), acidities in excess of 30 ($-H_0$) can be estimated for 1:1 composition. Other conjugate Brønsted acids include $HB(HSO_4)_4$, $CF_3SO_3H : B(OSO_2CF_3)_3$, $FSO_3H : SO_3$, $FSO_3H : AsF_5$, $FSO_3H : HF : SbF_5$, $FSO_3H : SbF_5 : SO_3$, $FSO_3H-Nb(FSO_3)_5$ and $FSO_3H-Ta(FSO_3)_5$, $FSO_3H-Au(FSO_3)_3$, $FSO_3H : Pt(FSO_3)_4$, $CF_3SO_3H : SbF_5$, $HF : TaF_5$, and so-called conjugate Friedel-Crafts acids ($HBr : AlBr_3$, $HCl : AlCl_3$, $HI : AlI_3$, $HF : BF_3$). The latter conjugate acid systems in 1:1 composition are not stable in the absence of proton-acceptor bases.

Solid superacids contain either Brønsted (proton donor, often OH group) or Lewis-type (electron acceptor metal) sites. These include polymeric perfluorinated sulfonic acids (Nafion-H®) as well as many transition metal oxides, sulfates, and silicoaluminates (zeolites). However, the latter substances exhibit superacidic character only at high temperatures. Both Brønsted and Lewis superacids can be immobilized in solid supports such as graphite and fluorinated graphite.

Applications of Superacids

The high acidity and the extremely low nucleophilicity of the counter-ions of superacidic systems are especially useful for the preparation of stable, electron-deficient cations (see ANIONS AND CATIONS), including CARBOCATIONS. Many of these cations, which were formerly suggested to be fleeting metastable intermediates in many organic reactions and were

detected only in the gas phase in mass spectrometric studies, can be conveniently studied in superacid solutions. New chemical transformations and syntheses that are not possible using conventional acids can be achieved with superacids. These include transformation and syntheses of many industrially important HYDROCARBONS. The unique ability of superacids to bring about hydrocarbon transformations and even to activate methane (the principal component of natural gas) for electrophilic reactions has opened up a fascinating new field in chemistry.

Superacids such as Magic Acid or fluoroantimonic acid have made it possible to prepare stable, long-lived carbocations, which are too reactive to exist as stable species in more basic solvents. Stable superacidic solutions of a large variety of carbocations have been obtained. Many structurally diverse carbocations, as well as related acyl cations and acidic carboxonium ions and other heteroatom stabilized carbocations, have been prepared in superacidic solutions and even isolated as stable salts. Some of the examples are shown in Figure 1 (counter-anions are not shown).

The astonishing acidity of Magic Acid and related superacids allows protonation of exceedingly weak bases. Not only all conceivable π-electron donors (such as olefins, acetylenes, and aromatics) and *n*-donors (such as ethers, amines, and sulfides) but also weak σ-electron donors such as saturated hydrocarbons including the parent alkane, methane, are protonated. The ability of superacids to protonate saturated hydrocarbons (alkanes) rests on the ability of the two-electron, two-center covalent bond to share its bonded electron pair with empty *p*- or *s*-oribtals of a strongly electron-deficient reagent such as a protic acid (see Figure 2).

Superacids catalyze the isomerization of linear hydrocarbons to branched hydrocarbons. Straight-chain alkanes with five to eight carbon atoms have

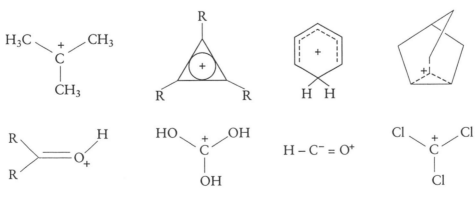

Fig. 1.

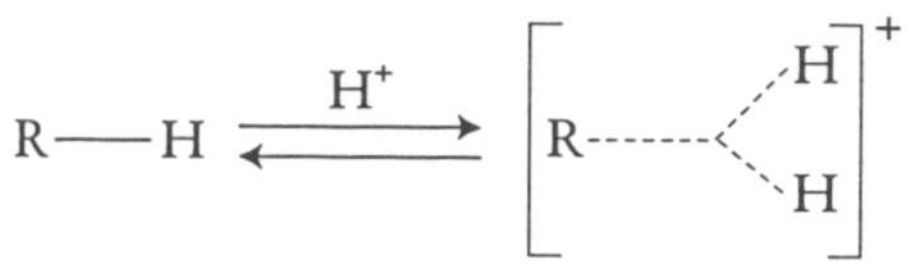

Fig. 2.

considerably lower octane numbers than their branched isomers, and hence there is a need for higher-octane branched isomers. Isomerizations are generally carried out under thermodynamic control leading to equilibrium conditions. The ionic equilibria in superacid systems generally favor increasing amounts of the higher-octane branched isomers at lower temperatures.

Aromatic as well as aliphatic alkylation can be easily carried out under superacid catalysis. Even polymerization of saturated hydrocarbons, including the conversion of methane to hydrocarbons as large as those in gasoline, can be effected under superacid treatment. Many important organic transformations such as ionic hydrogenation, carbonylation, halogenation, nitration, hydroxylation, and amination of aromatics as well as alkanes can be conveniently carried out under superacidic catalysis. In many such reactions superelectrophilic activation of the electrophile occurs, leading to enhanced reactivity. Many inorganic sulfur, selenium, telurium, and halogen cations can also be generated in superacid media.

See also ACID-BASE REACTIONS; ACIDS AND pH.

BIBLIOGRAPHY

GILLESPIE, R. J.; PEEL, T. E. *Adv. Phys. Org. Chem.* 1972, *9*, 1.

GILLESPIE, R. J.; ROBINSON, E. A. *Nonaqueous Solvent Systems;* T. C. Waddington, ed.; New York: Academic Press, 1967, p. 117.

HULL, N. F.; CONANT, J. B. *J. Am. Chem. Soc.* 1927, *49*, 3047.

OLAH, G. A.; PRAKASH, G. K. S.; SOMMER, J. *Superacids;* New York: Wiley Interscience, 1985.

GEORGE A. OLAH

G. K. SURYA PRAKASH

Superconductivity

Superconductivity is the ability of some elements, metallic alloys, or chemical compounds to conduct electricity with essentially zero electrical resistance. The charge carriers form a special state of matter in which the usual influences that give rise to resistance to the passage of electrical current, such as scattering from atomic thermal vibrations or impurity atoms, do not have any effect. Superconducting materials have a normal degree of metallic electrical resistance at room temperature (comparable, for instance, to that of copper), but when cooled to sufficiently low temperatures they undergo a sharp transition to a zero resistance state. This critical temperature, T_c, is generally close to absolute zero, limiting the use of superconducting materials in everyday technology. Modern MAGNETIC RESONANCE IMAGING (MRI) machines are the most common present application, in addition to high-energy particle accelerators, employing strong electromagnets made of superconducting wire cooled in a bath of liquid He at 4.2K. Superconductors make excellent windings for high-field magnets because minimal heat (the energy lost in electrical resistivity becomes heat) is generated by the large electrical currents necessary to produce high fields.

Superconductors display an additional remarkable property associated with zero resistance known as the Meissner effect. When a magnetic field (see MAGNETISM) is applied to any normal material, it penetrates that material to interact with the atoms and the electrons present. That interaction is an important characteristic of all materials and is known as the "magnetic susceptibility." When a superconductor is cooled in a magnetic field, it acts like a normal material above T_c, but below T_c, for small magnetic fields, the penetrating field is completely expelled from the superconductor. This flux expulsion is important in the characterization of superconducting materials, and the manner in which the flux finally penetrates into the superconductor with increasing magnetic field is an important consideration for their technological application.

Superconductivity was first discovered by H. Kammerlingh Onnes in Leiden, Holland, for solid elemental Hg cooled below 4.2K. About thirty of the pure crystalline elements have been found to be superconducting, ranging from the high T_c's of Nb (9.2K) and Pb (7.2K) to the low T_c's of Be (0.026K) and W (0.015K). These elements are Type I superconductors. The application of a magnetic field (such as would be generated by the magnet made of superconducting wire) *abruptly* destroys the superconductivity above a critical field strength, where the field suddenly and completely permeates the superconductor. The critical fields for elemental superconductors are too low for them to be of practical use.

For simple solid solutions made from mixtures of two nearby elements in the transition metal series, there is a universal relationship between the superconducting T_c and the average number of valence electrons per atom, known as the Matthias rule. The superconducting T_c's of such mixtures are strongly peaked just below 5 and just below 7 electrons/atom, with a strong minimum near 5.75 electrons/atom. This can be understood by a simple band-filling argument. The electronic orbitals of constituent atoms in a solid strongly interact, forming extended electronic orbitals that are broader in energy than the individual atomic orbitals and are called bands. The valence electrons from the metal atoms go into a set of energy bands that do not change across the series; only the degree of electronic filling of the bands changes. At filling levels where the density of electronic states (number of electrons per unit energy) is highest, T_c is highest. Elements in their amorphous (noncrystalline) form can also be superconducting. Special synthetic methods involving very fast cooling of melts or low temperature deposition of vapor are often necessary to suppress crystallization. In general, the T_c's of the best crystalline superconductors are decreased for the amorphous forms (crystalline Nb = 9.2 K, amorphous NB = 5.7K) and the T_c's of the worst crystalline superconductors are increased for the amorphous forms (crystalline Mo = 0.9K, amorphous Mo = 8−9K). The long-range disorder of amorphous materials washes out sharp features such as maxima and minima in the electronic density of states, "homogenizing" the majority of the amorphous materials' T_c's into the 5K regime.

The 1950s saw two important advances in superconductivity research. It was found that superconducting metallic alloys did not abruptly become non-superconducting in applied magnetic fields, but rather lost superconductivity only gradually as the magnetic field increased. From these Type II superconductors, high-field magnets could be fabricated. In addition, a successful theoretical explanation for superconductivity was finally put forward by Bardeen, Cooper, and Schrieffer (the BCS theory). This theory explained that although electrons should normally repel each other, under certain special conditions in solids they could "condense" into a sea of loosely bound pairs that could not lose energy to the underlying crystal lattice (see CRYSTAL LATTICE ENERGY). These pairs, therefore, experienced no resistance to their motion. For the superconductors then being studied, the "glue" holding the pairs together was found to be a beneficial coupling between the electron motion and crystal lattice vibrations. This is called electron-phonon coupling: the stronger the electron-phonon coupling, the higher the T_c. For very strong electron-phonon coupling, however, the crystal structure can be altered, destroying superconductivity.

Intermetallic superconductors with the highest critical temperatures are based on Nb alloyed with various elements in a cubic symmetry compound with stoichiometry Nb_3X, where, for example, X = Ga, Ge, Sn, Al, or Si. These compounds have the A15 crystal structure. This crystal structure has three nonintersecting chains of Nb atoms running parallel to the three cube edge directions. The highest T_c is for Nb_3Ge (23.2K), but because of chemical instability and brittleness, alloys of this family are not currently in wide use commercially. Commercial applications employ an NbTi solid solution alloy (Nb and Ti mixed together randomly in the same crystal structure as elemental Nb), which has a T_c comparable to pure Nb but acts as a Type II superconductor, therefore sustaining high magnetic fields. The alloy also has acceptable forming characteristics and mechanical properties.

Many chemical compounds are known to be superconducting, but many thousands of times more are not. The problem is to get normal metallic conductivity in the first place, which is itself not common. The vast majority of simple ionic salts and oxides (such as NaCl and SiO_2), for instance, are insulating: a full charge transfer from cation to anion occurs in the chemical bond; for electronic conductivity delocalization of charge in the chemical bond is needed. There are several important classes of superconducting chemical compounds, some of which have been discovered only in the past decade.

Metallic conductivity is relatively rare in organic compounds, having been observed for the first time in 1973. Superconductivity in organic charge transfer salts was first observed (T_c near 1K) in 1979 under the application of high pressure to materials of the family $(TMTSF)_2 X$ where X = PF_6, AsF_6, SbF_6, etc., and TMTSF stands for tetramethyltetraselenafulvalene. The TMTSF acts as an electron donor, a cation; and the X group as an electron acceptor, an anion (see ANIONS AND CATIONS). Approximately twenty of the organic superconductors presently known are based on compounds of the donor molecule BEDT-TTF (or ET): bis(ethelenedithio)tetrathiafulvalene, with various acceptor molecules, which include inorganic cations. At the present time, the highest known T_c (12.8K) is for the crystallographic kappa form of $(ET)_2Cu[N(CN)_2]Cl$, under an applied pressure of 300 bar.

Considerable strides have been made in the development of new materials in recent years. Structurally, the organic superconductors are highly layered in nature, with the large, flat donor molecules stacked like cards in a deck, with the acceptor molecules in interstitial positions. The manner in which orbitals overlap between one donor molecule and the next is very important to the physical properties, as are the details of the chemistry and structure of the donor molecules themselves. The prevalent thinking is that electron-phonon coupling gives rise to superconductivity in these materials, but that is a matter of current research.

In 1990, a technique was developed by which relatively large amounts of a new form of carbon, C_{60}, could be made and isolated. C_{60} molecules have a structure like that of a soccer ball, with carbon atoms at the sixty vertices resulting in twelve pentagonal and twenty hexagonal carbon rings. These balls form a close-packed structure in an extended solid, with a unit cell of a face-centered cubic (FCC) configuration (one ball at each corner of a cube, and one ball in each cube face). Soon after this announcement, it was found that reaction with alkali atoms, at a stoichiometry A_3C_{60}, resulted in superconductivity for A = K and Rb. The highest T_c in the family (33K) is presently known for a mixture of Rb and Cs, $RbCs_2C_{60}$: the superconducting transition temperature increases as the size of the alkali ion increases. At the stoichiometry A_3C_{60}, the alkalis K, Rb, and Cs occupy all the available octahedrally coordinated and tetrahedrally coordinated (i.e., bonded to six C_{60} or four C_{60} molecules, respectively) interstitial positions in the FCC array of C_{60} molecules in the solid C_{60}. Band structure calculations have shown that three electrons are donated to the C_{60} molecule by the three alkalis at the superconducting stoichiometry and half fill a particular molecular orbital. The increased distance between C_{60} molecules in the A_3C_{60}'s with larger A intercalates changes the overlap between wave functions on neighboring C_{60} balls and changes T_c. The alkaline earth ions Ca and Ba have also been found to result in superconducting compounds when reacted with C_{60}, but the details of their crystal structures, stoichiometries, and electronic states have not yet been established.

High-temperature superconductivity was discovered in copper oxide–based compounds in 1986. To date, approximately thirty-five chemically and structurally unique members of this family have been synthesized and studied. Although some have T_c's in the 30K range (the first discovered, $La_{1.8}Ba_{0.2}CuO_4$, has a T_c of 28K), most are superconducting in the 50 to 100K range, with the highest T_c for any compound so far known reported for $Tl_2Ba_2Ca_2Cu_3O_{10}$: an astounding 127K. These materials have had an impact on science in at least three significant ways. First, it is widely believed that it is impossible for conventional

coupling between electrons and phonons to be the glue that holds the superconducting charge-carrier pairs together at such high temperatures. A whole new theoretical explanation is needed. The possibilities currently under consideration range from electronic states in which the charge and spin of electrons are separated, to pairing by purely electron-electron interaction, to exotic forms of electron-phonon interaction, and magnetism. As yet there is no consensus. Second, with such high T_c's, immersion in relatively inexpensive and easy-to-handle liquid N_2 at 77K would seem to promise the possibility of much more widespread and inexpensive technological application. Considerable effort is now being expended to overcome problems these materials have carrying current in high magnetic fields at 77K. The most promising superconductor for practical applications known at the present time is the 92K superconductor $YBa_2Cu_3O_7$. Finally, there are many new physical phenomena associated with this family of materials that are the continuing subject of intensive basic research.

The copper oxide–based superconductors have several chemical and structural characteristics that make them unique materials. First, they are based on Cu^{2+}, which with the electron configuration $3d^9$ has a single missing electron in the 3d-orbital. Owing to the splitting of the 3d-orbital energies in the solid oxides, the result is a half-filled energy band. At half-filling, electron energy states are strongly correlated. The electrons do not simply act as a "charged sea" filling in the available energy states as they do in conventional metals. The half-filled band is especially stable, causing added or subtracted electrons to interact strongly with the electrons already present. Second, copper and oxygen form an unusual chemical bonding pair. The copper 3d- and oxygen 2p-orbital energies are very similar, making the chemical bond between them essentially covalent. Such bonding occurs generally only between identical or very similar atoms. For unlike atoms, ionic bonding is the general rule. Between nickel and oxygen, for example, the bond is much more ionic. Finally, the crystal structures that form are highly two-dimensional in character, with stacking of infinite sheets of CuO_2 with alternate structural layers. The CuO_2 layers are the electronically important area in the crystal structure where superconductivity occurs. Both the crystal structures and the degree of band filling away from the Cu^{2+} state are known to be critical parameters controlling T_c. Optimal band filling, obtained by chemical manipulation of the elements other than copper present, is about 0.2 electron-holes per Cu.

There are several additional families of superconducting materials that are of interest. Chevrel phases, typified by $PbMo_6S_8$, have modest T_c's (12K for $PbMo_6S_8$) but very high critical fields. Their crystal structures consist of Mo_6S_8 cubic clusters (S at the eight corners of small cubes and Mo at the center of the six cube faces) aligned along the cube diagonals to form chains, with Pb atoms acting as spacers. Heavy fermion superconductors, typified by $CeCu_2Si_2$, have very low T_c's, with the superconducting state owing to the condensation of charge carriers of very heavy mass. For $CeCu_2Si_2$, for example, $T_c = 0.65$ K and the carrier mass $= 1,000$ times the mass of an electron. This is because the carriers are in a very narrow Ce 4f energy band: the narrow electronic band makes the carriers act as if they have an effective mass many times that of the free electron. Finally, another family of oxide superconductors, based on $Ba_{.6}K_{.4}BiO_3$ ($T_c = 31K$), $BaPb_{.75}Bi_{.25}O_3$ ($T_c = 12K$), and $BaPb_{.75}Sb_{.25}O_3$ ($T_c = 3.5K$), is considered unusual because the T_c's are very high when compared with intermetallic superconductors with the same electronic density of states.

Research in the field of superconductivity has undergone a rebirth with the recent advances in high T_c cuprates, intercalated C_{60}, and organic materials. The next decade promises the possibility of some exciting additions to the families described here.

BIBLIOGRAPHY

CAVA, R. J. "Superconductors beyond 1-2-3." *Sci. Am.* 1990, 262, no. 8, 42–49.

EVETTS, JAN, ed. *Concise Encyclopedia of Magnetic and Superconducting Materials;* Oxford: Pergamon Press, 1992.

KRESIN, V. Z.; WOLF, S. A. *Fundamentals of Superconductivity;* New York: Plenum Press, 1990.

R. J. CAVA

Sweeteners

Taste is generally classified as sweet, bitter, sour, and salty. Sometimes a fifth category is added, which is called umami and represents the taste of $C_5H_8O_4NaN$, monosodium glutamate (MSG). Sweetness is one of the most enjoyable of the tastes, and many foods have sugar added to enhance their acceptability to consumers. Soft drinks constitute a huge market in which sweeteners play a central role.

The SUGARS, part of the important food group known as simple carbohydrates, are natural sweeteners and are found in many fruits and berries. It may be that during evolution, human beings who enjoyed the sweet taste were more successful because the fruits, berries, and seeds provided highly caloric sugars, which gave the much needed energy required for survival. Other naturally occurring sweet substances that are not sugars have been discovered, but generally are not important commercially.

Natural Sweeteners

Of the many sugars known to scientists, common table sugar has historically been of the greatest commercial importance, with global annual production approximating 115 million tons per year. Another name for table sugar is sucrose, and it is the only natural material produced in such high purity in large quantities. A more scientific name for sucrose would be β-D-fructofuranosyl α-D-glucopyranoside. Sucrose, also the sugar in honey, is classified as a disaccharide because it is composed of two simpler sugars, the monosaccharides glucose and fructose.

Glucose is the most important energy source utilized in living systems and is therefore of great biological importance. Many glucose molecules, when bonded together to form long chains (with α-linkages), compose the important foodstuff known as starch, a complex carbohydrate. Long chains of glucose molecules bonded in a slightly different way (with β-linkages) give us another important material called CELLULOSE, the main component in cotton and WOOD. (Long chains, called POLYMERS or macromolecules, are important in living systems in the form of starch, cellulose, PROTEINS, DNA, and some HORMONES. Plastics are polymers of great importance made from petroleum, with many applications in our everyday lives.)

Fructose is slightly sweeter than sucrose and, in the form of high-fructose corn syrup, has recently emerged as a strong competitor for sucrose and has captured a sizable portion of the soft drinks market. It is also available in the marketplace as a crystalline sugar resembling sucrose.

The chemical formulas for glucose, fructose, and sucrose appear at the top of the next page.

Although it is not very sweet, lactose, an important sugar in milk, is also produced as a by-product of the dairy industry. Many commercially important carbohydrate products are produced for nonfood purposes.

Sucrose is the standard for taste characteristics. The major taste characteristics can be classified as potency and quality. Potency is a measure of the amount of sweetness for a certain weight. All the

Glucose

Fructose

Sucrose

natural sugars are low-potency sweeteners, some not sweet at all. The quality of the taste includes sensations such as cooling and menthol or licorice-like tastes, metallic tastes, and even bitterness or sourness mixed with the sweetness. More subtle features such as how the substance feels in the mouth are important, but of major importance are the temporal characteristics, which are manifested as delayed reaction times and prolonged sweet sensations, which are called persistence. That is, for most artificial sweeteners, the sweet taste may not be immediate, but once it sets in, it may last for several minutes, resulting in what might be called an aftertaste, especially if it is undesirable.

Artificial Sweeteners

As the fields of chemistry and biochemistry developed, the highly caloric nature of the sugars and their importance in our diets were soon recognized. With this understanding came an appreciation for the potential of a noncaloric sweetener, especially in a society where obesity and health problems such as diabetes and tooth decay are common. As a result, when nonsugar substances were found to be sweet, they were of immediate interest. Artificial sweeteners are used in soft drinks, desserts, chewing gum, candy, iced-tea mixes, breakfast cereals, and toothpastes.

Until recently, all these materials were discovered accidentally. Saccharin was discovered by REMSEN and Fahlberg at Johns Hopkins University in 1879 and is about 300 times sweeter than sucrose. Later came the discovery of cyclamates by Michael Sveda at the University of Illinois in 1937. Cyclamates are about 30 times sweeter than sucrose, and both cyclamates and saccharin bear very little resemblance to

any known foodstuff. Early studies suggested that cyclamates and saccharin were carcinogens because they caused cancerous bladder tumors in test animals fed large quantities of the sweeteners. As a result, cyclamates were banned in the United States, but a moratorium on the banning of saccharin was declared due to public pressure. Since cyclamates are still being used in more than 50 countries, the test results are being reevaluated, and cyclamates are again being considered for use in the United States.

The most successful synthetic sweetener, aspartame, was discovered accidentally by James Schlatter at G. D. Searle & Co. in 1965. The scientific beauty of aspartame is that it is a dipeptide. Protein, one of the important foodstuffs, is a polypeptide. That is, proteins are natural polymers composed of hundreds of amino acid units bonded together to form a very long chain. A dipeptide is simply two amino acid units bonded together, whereas in aspartame one of the amino acids has a trivial modification. So basically, aspartame is like protein. At the time of the discovery of aspartame, all dipeptides were thought to be bitter, so the fact that aspartame was sweet was a surprise indeed. Moreover, the taste characteristics of aspartame were quite good, much like those of sucrose. However, certain people are born with a genetic disease called phenylketonuria (PKU) and must carefully control their intake of the amino acid phenylalanine, one of the amino acids found in aspartame. The safety of aspartame has come under attack for other reasons also, but the Food and Drug Administration (FDA) has considered it acceptably safe when used in moderation, and it is currently the preferred artificial sweetener on the market.

Since the discovery of aspartame, many compounds of similar structure have been synthesized and evaluated. Many were found to be sweet, but none had all the desirable taste characteristics found in aspartame. In recent years, an intense effort has been expended by industry to discover new sweet compounds to compete with aspartame, and some have been introduced in the United States and in other countries.

The structures of several artificial sweeteners are shown at the top of the next page.

Currently there are only three artificial sweeteners that are approved by the FDA for sale in the United States—aspartame, saccharin, and acesulfame K. Acesulfame K is sold in some European countries and has recently been approved for marketing in the United States. A number of other compounds possess desirable characteristics for an artificial sweetener, and approval by the FDA is being sought for several. Of the currently available artificial sweeteners, aspar-

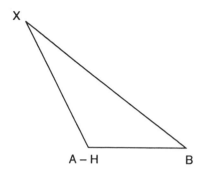

Sodium saccharin

Sodium cyclamate

Aspartame

Acesulfame K

tame seems to have the best taste characteristics and is the preferred sweetener.

Theories of Sweetness

Since the field of chemistry has come of age, there has been much interest in the theory of the various tastes. While the story is still incomplete, some progress has been made. There is a wide variety of chemical structures that elicit a sweet taste, and it is difficult, perhaps impossible, to organize such a large body of data into one comprehensive theory. One of the more productive concepts was that presented by Shallenberger and Acree (1967), and it is usually referred to as the AH-B theory. Based on previous ideas, it involved complementary sites on the sweet molecule and on the taste bud receptor site that would allow the sweet substance to adhere to the taste bud in such a way that the taste sensation could be detected. Figure 1, the A-H, B concept, shows an attractive force between the A-H and B parts of the sweet molecule and the receptor site. In the early 1970s, Kier added a

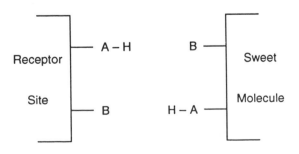

Fig. 1.   The A-H, B concept.

Fig. 2.   The Kier triangle.

third site, X, which now formed a triangle with the AH-B sites (Kier, 1972). The Kier triangle, shown in Figure 2, represents just the sweet molecule. The receptor would have complementary sites to bind with the sweet molecule. This was an important contribution and helped explain several key features of sweetness.

During the 1980s, Tinti and Nofre unveiled a model with a total of eight sites for the molecule. While that seems complex, they were able to use the model to guide them in their efforts to synthesize new sweet compounds of a particular type. Indeed, they have prepared several that are the most highly potent sweeteners known, one being as much as 200,000 times sweeter than sucrose. There are probably other successful models being used by industrial chemists, but they are proprietary (secret) and, even if they exist, have not been disclosed to the public. These models of sweetness seemed to work best on the particular class of substances for which they were designed. However, after all this effort, it is still not possible to predict whether any given substance will elicit a sweet or bitter taste.

The theory of sweetness is a very complex field and spans many disciplines, including chemistry, physics, psychology, physiology, neurophysiology, and others. The evaluation of sweetness is rather involved and requires screened, experienced tasters, who work under very controlled conditions using reference sweeteners and other standardization techniques. One of the greatest problems is discovering the characteristics of the taste-bud receptor sites because those must come from the human tongue. Some animals seem to respond positively to sweet substances, but there is as yet no way of knowing for certain whether it is the same sensation that humans experience. So there is a variation in the sense of taste across many species, and this adds to the complexity. In addition, there is evidence that two or more kinds of taste buds or taste-bud cells are operating, and these seem to be served by more than one nerve group and by nerves serving several different cells. There is even considerable variation among humans; certain compounds such as phenylthiourea are almost tasteless to some people and intensely bitter to others. For many people, but not all, saccharin has an undesirable aftertaste. All these factors taken together make for a very complex system indeed, and a solution to the puzzle of sweetness does not seem to be forthcoming in the near future.

BIBLIOGRAPHY

KIER, L. B. *J. Pharm. Sci.* 1972, *61*, 1394.

SHALLENBERGER, R. S.; ACREE, T. E. "Molecular Theory of Sweet Taste." *Nature* 1967, *216*, 480.

JERRY W. ELLIS

System of Weights and Measures

See INTERNATIONAL SYSTEM OF UNITS.

T

Taste

Taste is called a chemical sense because the stimuli for taste perception are soluble chemicals that enter the mouth. For most animals the chemical senses (taste and smell) play an important role in the acquisition of nutrients. The chemical senses are used not only to detect food sources but also to provide information about the suitability of the food. The importance of the chemical senses is often not appreciated by humans, who rely so heavily on vision and hearing.

There is often some confusion between taste (gustation) and smell (olfaction). In terrestrial animals olfaction is considered the distance sense, while gustation is the contact sense. Odorants waft their way to us in the air from a distance. Tastants, on the other hand, must be placed directly in our oral cavity to be detected. Volatile compounds found in food gain entry to the nasal cavity from the mouth through the nasopharynx, where they stimulate olfactory receptors. While we usually call this sensation taste, it should better be considered flavor. Chewing food releases more of the volatiles and thus enhances flavor. Other contributors to the flavor of food include its temperature, viscosity, texture, and pungency—together called "mouth feel" and mediated primarily by sensory endings of the trigeminal nerve.

Sapid food molecules taken into the oral cavity stimulate taste receptors on the tongue, palate, phar-ynx, and larynx. The taste-receptor cells (TRCs) are grouped together in microscopic taste buds that on the tongue are arranged on papillae, elevations on the tongue surface (see Figure 1). Approximately five thousand taste buds are found on a typical human tongue, but there is a large variation in the number of taste buds from person to person. There are three major types of papillae—fungiform, foliate, and vallate—each having different distributions on the tongue surface, as shown in Figure 1. Three cranial nerves—facial, glossopharyngeal, and vagus—innervate TRCs in different areas of the oral cavity. The nerves convey taste information to the central nervous system.

Sensations directly elicited from stimulating TRCs can be grouped into four distinct qualities: sweet, salty, sour, and bitter. (Some researchers believe there is a fifth taste quality called umami, which is the taste of monosodium glutamate [MSG].) There clearly is a relationship between chemistry and taste. Many salts (e.g., NaCl and LiCl) taste salty, and most acids (e.g., HCl, citric acid, and acetic acid) taste sour. There is a wide variety of sweet molecules (e.g., sucrose, aspartame, and sodium saccharin), and several models for sweetness have been proposed (see SWEETENERS). The chemicals that produce bitter tastes are highly diverse. They range from KCl (which is primarily reported as bitter but also has a salty component) to the organic molecules quinine and caffeine, to denatonium chloride (one of the most bitter compounds known).

1423

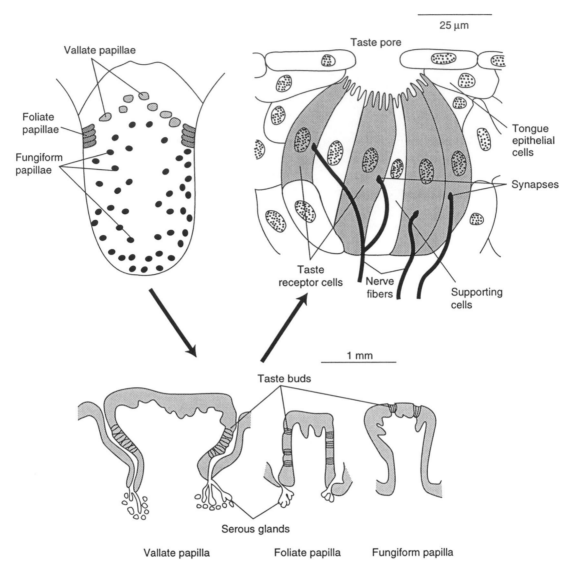

Fig. 1. Schematic diagram of the tongue (top left); the three papillae found on the tongue (bottom); and a taste bud (top right).

Taste molecules reach the TRCs through the taste pore, an opening that penetrates the cornified cell layers in the tongue above the taste bud. Taste molecules are then presumed to interact with the microvillar membranes on the apical end of the TRCs. The interaction of the taste molecule with the membrane results in a depolarizing membrane potential in the TRC (the inside becomes more positive with respect to the outside). This depolarization activates a synapse at the base of the TRC, which elicits action potentials in the taste nerves.

What is the molecular basis of taste quality at the receptor level? In recent years a number of varied and diverse mechanisms involving several cell-signaling elements have been elucidated. Only one mechanism

for each of the four taste qualities is discussed below (see Figure 2).

Salty taste is mediated by passive movement of Na^+ ions through Na^+-selective channels in the TRC microvilli. When you eat something salty, the high concentration of Na^+ outside the TRC causes the Na^+ ions to diffuse down their concentration gradient, through the Na^+ channels and into the cell. The movement of positive ions into the cell depolarizes it, as shown in Figure 2. Like many similar channels in other types of epithelia (e.g., kidney tubule cells), the Na^+ channels in TRCs are blocked by the drug amiloride.

Acid transduction appears to be mediated by a pH-sensitive K^+ channel. In the absence of stimulation

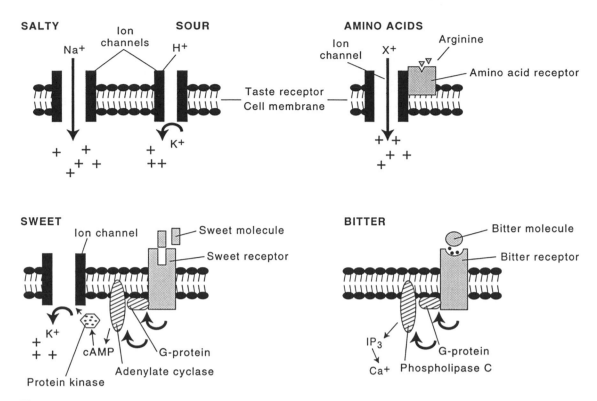

Fig. 2. Schematic diagram of a proposed mechanism by which tastants from each of the four taste qualities (plus amino acids) elicit responses from taste receptor cells.

there is a small, constant, outward K^+ current through the apical membrane of the TRC. Hydronium ions present in sour foods block the K^+ channel, turning off the K^+ current and thus causing the cell to depolarize (see Figure 2).

Sweet taste appears to be mediated in part by a second messenger system in which some sweet molecules first bind to a membrane receptor, which activates a G-protein, which activates adenylate cyclase. This leads to the production of cyclic adenosine monophosphate (cAMP) inside the TRC. The cAMP, in turn, activates an enzyme, protein kinase, which chemically modifies and closes K^+ channels, causing a membrane depolarization by blocking the exit of the positively charged K^+ ions (see Figure 2).

The mechanisms responsible for bitter taste may be the most diverse. One mechanism for denatonium chloride stimulation appears to involve a receptor protein for this compound coupled to a G protein that activates membrane-bound phospholipase C. This, in turn, generates inositol trisphosphate (IP$_3$), which stimulates the release of Ca^{++} ions from internal stores. The Ca^{++} ions can then directly activate the synapse at the base of the TRC (see Figure 2).

Another mechanism has been elucidated for amino acids that are particularly effective taste stimuli for

fish. A receptor for arginine is directly coupled to an ion channel. The presence of arginine causes the channel to open, and cations move into the cell, depolarizing it (see Figure 2).

A number of chemicals are classified as taste modifiers—compounds that alter the taste sensations produced by other compounds. Gymnemic acid, extracted from the leaves of a South American plant, abolishes sweetness. Miraculin, a large protein found in the berries of the African miracle fruit bush, makes sour compounds taste sweet. A chemical in artichokes imparts a sweet taste to subsequent nonsweet substances. Finally, detergents—like those found in toothpaste—alter the membrane properties of TRCs and cause a distortion of the basic tastes. This is why orange juice tastes different after you brush your teeth.

BIBLIOGRAPHY

GILBERTSON, T. A.; KINNAMON, S. C. "Making Sense of Chemicals." *Chem. & Biol.* 1996, *3*, 233–237.

KUANG-CHIC, T.; HUA-ZHONG, H. "Structural Theories Applied to Taste Chemistry." *J. Chem. Educ.* 1987, *64*, 1003–1009.

LINDEMANN, B. "Chemoreception: Tasting the Sweet and the Bitter." *Current Biology* 1996, *6*, 1234–1237.

LINDEMANN, B. "Taste Reception." *Phys. Rev.* 1996, 76, 719–766.

MILLER, I. J., JR. "Anatomy of the Peripheral Taste System." In *Handbook of Olfaction and Gustation*, R. L. Doty ed.; New York: Marcel Dekker, 1995; pp. 521–547.

SCOTT, T. R.; PLATA-SALAMAN, C. R. "Coding of Taste Quality." In *Smell and Taste in Health and Disease*; T. V. Getchell, R. L. Doty, L. M. Bartoshuk, J. B. Snow, Jr., eds.; New York: Raven Press, 1991; pp. 345–368.

SHALLENBERGER, R. S. *Taste Chemistry*; New York: Routledge, Chapman, & Hall, 1993; p. 613.

SIMON, S. A.; ROPER, S. D., eds. *Mechanisms of Taste Transduction*; Boca Raton, FL: CRC Press, 1993.

WAYNE L. SILVER

Tautomerization

Compounds with structures that differ markedly in arrangement of atoms, but that exist in easy and rapid equilibrium, are called tautomers. In most cases of tautomerization, which is also known as tautomerism, a proton shifts from one point of attachment on a molecule to another. For example, as illustrated in Figure 1, in keto-enol tautomerism a KETONE equilibrates with its ENOL by rapid hydrogen shifts (March, 1992; Stoddart, 1979).

Acetone, the simplest ketone, exists in equilibrium with its enol tautomer, but the small equilibrium constant shows that the keto form is dominant (see Figure 2).

Formation of the enol tautomer is much more favorable for 2,4-pentanedione and other beta-dicarbonyl compounds. The positioning of the C=C and C=O double bonds causes the extra stability of the enol. The alternating double-single-double bond structure of the enol is a special, energetically stabilizing feature called a conjugated system. The equilibrium shifts in favor of the enol because the diketo tautomer lacks conjugation (see Figure 3).

The 2,4-pentanedione enol also possesses an intramolecular hydrogen bond that imparts extra stability to this tautomer. The two factors together, conjuga-

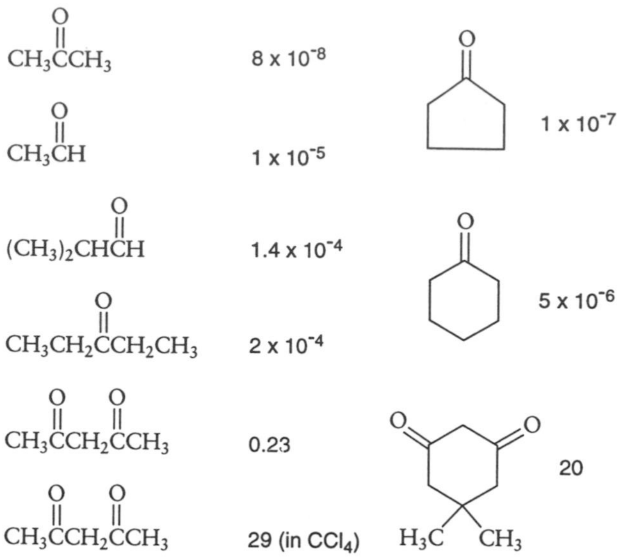

Fig. 2. Tautomeric equilibrium constants in water (adapted from data in Carey, 1990, p. 420).

Compound	Constant
CH_3CCH_3 (O)	8×10^{-8}
CH_3CH (O)	1×10^{-5}
$(CH_3)_2CHCH$ (O)	1.4×10^{-4}
$CH_3CH_2CCH_2CH_3$ (O)	2×10^{-4}
$CH_3CCH_2CCH_3$ (O, O)	0.23
$CH_3CCH_2CCH_3$ (O, O)	29 (in CCl_4)
cyclopentanone	1×10^{-7}
cyclohexanone	5×10^{-6}
dimedone (H_3C, CH_3)	20

tion and intramolecular HYDROGEN BONDING, contribute several kilocalories per mole to stabilizing the enol relative to the diketone.

Both acids and bases (see ACID-BASE REACTIONS; ACID-BASE THEORY) catalyze the interconversion between keto and enol tautomers (Streitweiser, 1992). In the gas phase or in the absence of an acid or base, the rapid tautomeric equilibration does not take place. The catalyst does not change the equilibrium constant; it only increases the rate of the tautomeric process. The rate of tautomerism depends on the strength of the acid or base catalyst.

The solvent strongly influences the keto-enol tautomeric equilibrium. In nonpolar solvents the enol tautomer is more favorable than in polar solvents such as water or METHANOL, which can donate hydrogen bonds. For example, 2,4-pentanedione is a mixture of 81 percent dione and 19 percent enol in aqueous solution, but in CCl_4 solvent the compound is 97 percent in the enolic form. The solvent's ability to disrupt the intramolecular hydrogen bond of the

keto form   enol form

Fig. 1. Keto-enol tautomerism.

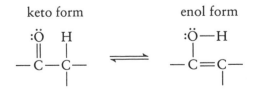

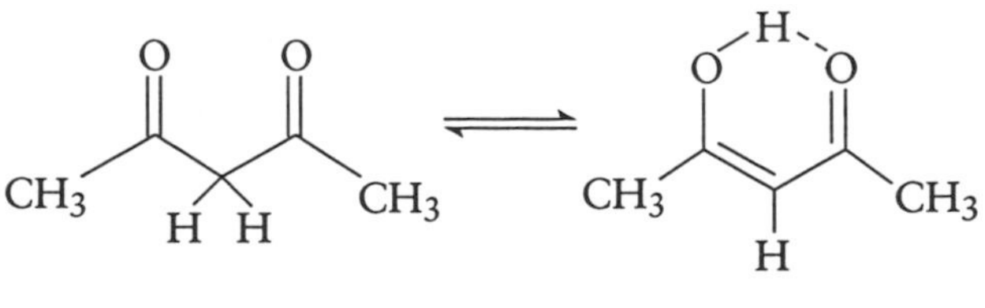

Fig. 3. Formation of the enol tautomer for 2,4-pentanedione compounds.

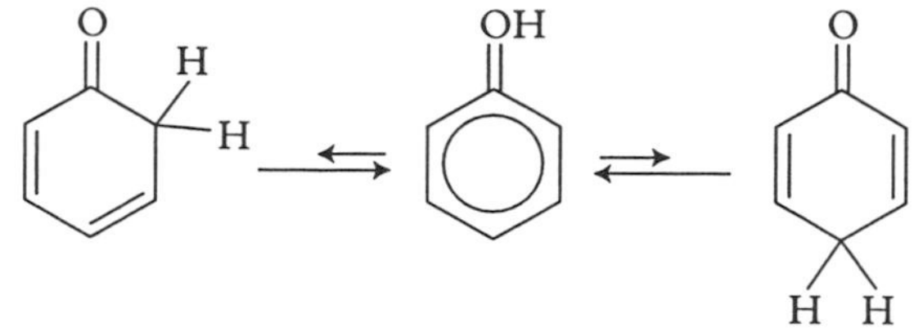

2, 4-Cyclohexadienone   Phenol   2, 5-Cyclohexadienone
                       (a stable enol)

Fig. 4.   The aromatic ring (center) provides
         phenol with added stability.

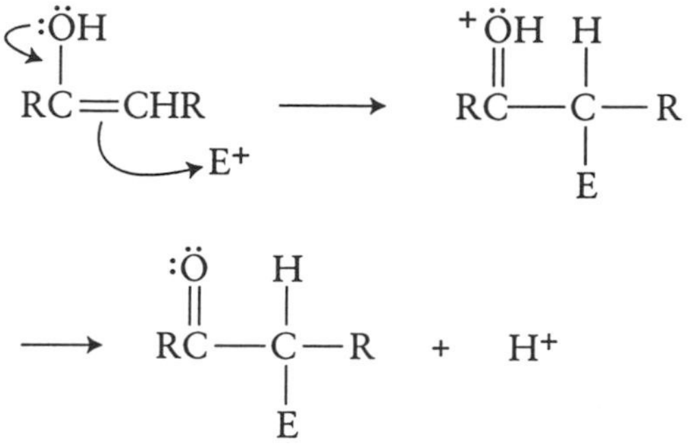

Fig. 6.   Proton loss producing a carbonyl with an
         electrophile attached at the alpha carbon.

enol causes the larger amount of the diketone form in water.

Phenol is formally the enol of both 2,4-cyclohexadienone and 2,5-cyclohexadienone, but actually neither of the dienone forms exists. The reason is that phenol possesses an aromatic ring (see AROMATICITY) that gives added stability. The two dienones are not aromatic, since they do not have six π-electrons that circulate freely around the ring. Tautomerization to the dienones is energetically unfavorable because it disrupts the resonance of the aromatic system. (See Figure 4.)

An enol cannot form from a CARBONYL compound that has no hydrogen attached to the alpha carbon (this is the carbon next to the C=O group). Compounds such as 2,2-dimethylpropanal (pivaldehyde) and 2,2,4,4-tetramethyl-pentan-3-one (di-*tert*-butyl ketone) that do not possess alpha C—H bonds cannot tautomerize. (See Figure 5.)

The chemical reactivity of a ketone is quite different from the reactivity of an enol. If the ketone and enol rapidly interconvert, reactions characteristic of either or both tautomers take place. Nucleophiles attack at the electron-deficient carbonyl carbon of the ketone, and electrophiles attack at the electron-rich double bond of the enol. The enol double bond is more reactive than the double bond of an ordinary ALKENE because the oxygen donates electron density from its LONE PAIR into the double bond.

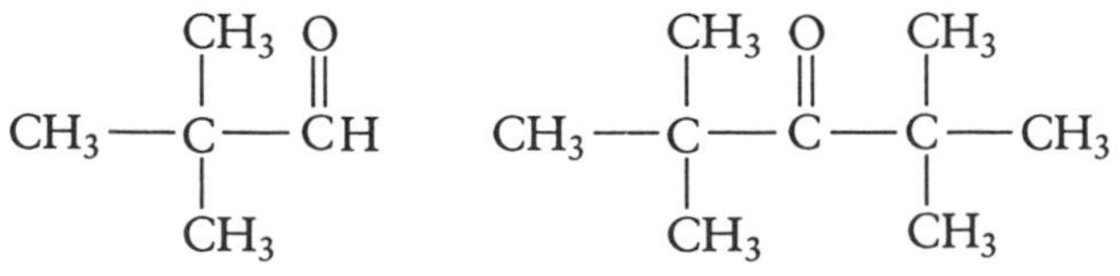

Electrophilic attack on the enol leads to a CARBOCATION intermediate that is protonated on oxygen. Loss of the PROTON gives a carbonyl compound that now has the ELECTROPHILE attached at the alpha carbon. (See Figure 6).

The exchange of the alpha hydrogen of acetone for DEUTERIUM in D_2O solvent illustrates electrophilic addition to an enol. (See Figure 7.) This reaction takes place in the presence of an acidic catalyst. If the reaction continues, all six hydrogens will eventually exchange with the solvent.

The mechanism for acid-catalyzed tautomerization (Loudon, 1995; see Figure 8) of a ketone involves protonation of the weakly basic carbonyl oxygen, followed by loss of a proton from the alpha carbon. The second of these two steps is much slower than the first, when the H_3O^+ ion is the catalyst. Conversion of the enol back to the ketone follows the same pathway in reverse. Weaker acids than H_3O^+, such as ACETIC ACID, also catalyze enolization.

The mechanism for base-catalyzed tautomerization (Loudon, 1995) involves initial loss of a proton from the alpha carbon, followed by protonation of the oxygen anion, as shown in Figure 9. In this case the first step is slower than the second step. Again, weaker bases than OH^-, such as acetate anion, also catalyze the tautomerization.

The tautomeric equilibrium constant K_{eq} is the equilibrium ratio of concentrations of the enol and

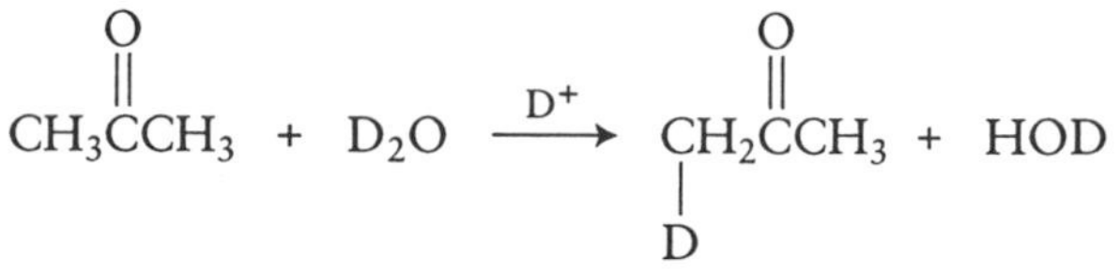

Pivaldehyde   Di-*tert*-butyl ketone

Fig. 5.   The compounds pivaldehyde and di-*tert*-butyl
         ketone cannot tautomerize.

Fig. 7.   An example of electrophilic addition
         to an enol.

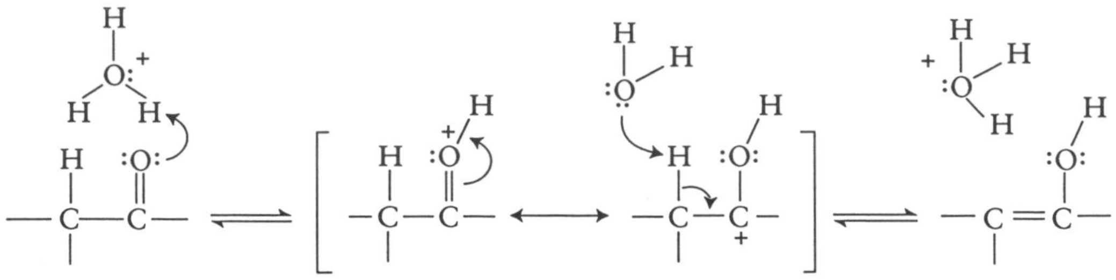

Fig. 8.　An acid-catalyzed tautomerization.

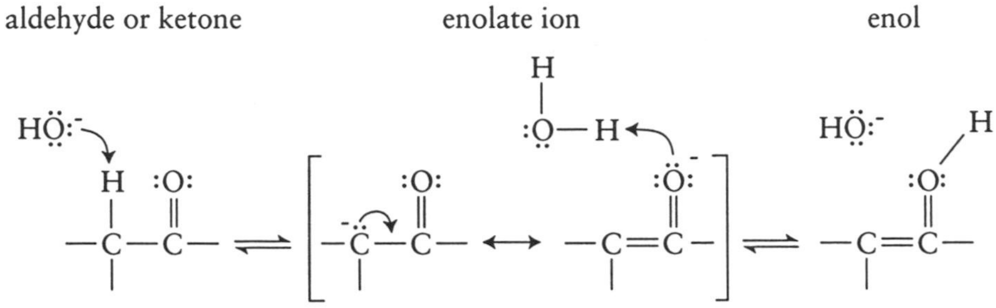

Fig. 9.　A base-catalyzed tautomerization.

ketone. The constant K_{eq} is also given by the ratio k_f/k_r, where k_f is the forward rate constant for conversion of the ketone to the enol, and k_r is the reverse rate constant for conversion of the enol back to the ketone (see Figure 10).

For a given solvent and temperature, K_{eq} is constant, but k_f and k_r increase with increased concentration and strength of the acid or base catalyst. The strongest acid in water is the hydronium ion (H_3O^+), and the strongest base in water is the hydroxide ion (OH^-). For acid-catalyzed tautomerism in water, the fastest possible keto-enol interconversion is with H_3O^+ as catalyst, and for base-catalyzed tautomerism in water, the fastest interconversion is with OH^- as catalyst.

Imine-enamine and lactam-lactim tautomerisms are very similar to keto-enol tautomerism. An enamine has the same relation to an imine that an enol has to a ketone. Just as the keto-enol equilibrium usually favors the keto tautomer, the imine-enamine equilibrium usually favors the imine. A lactam is a cyclic amide, which can interconvert with its less-stable lactim tautomer by 1,3-hydrogen rearrangement. (See Figure 11).

Reaction of an ALDEHYDE or a ketone with a primary amine gives an imine. A primary amine has the general structure RNH_2—one nonhydrogen substituent and two hydrogens are bonded to the nitro-

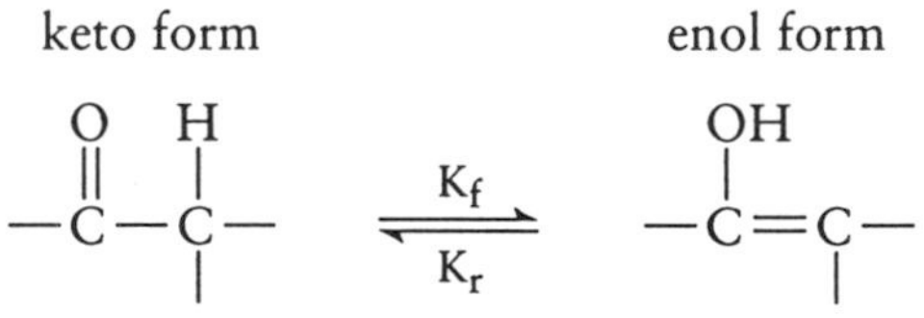

Fig. 10.　Ketone-enol and enol-ketone conversion.

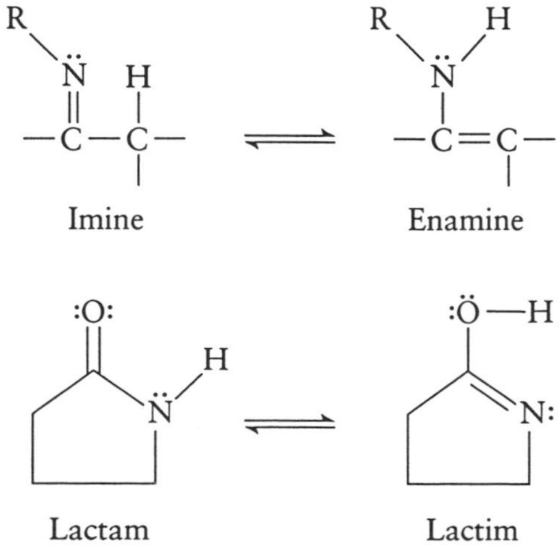

Fig. 11.　Imine-enamine and lactam-lactim tautomerisms.

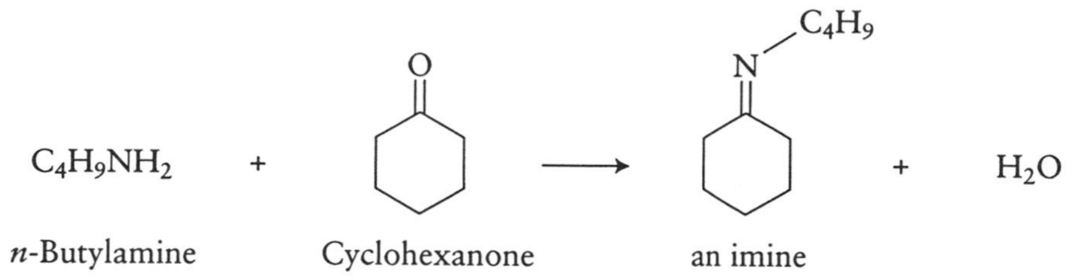

Fig. 12. Reaction of n-butylamine with cyclohexanone.

gen atom. A molecule of water is formed along with the imine. An example, shown in Figure 12, is the reaction of n-butylamine with cyclohexanone, which gives an imine. The water is removed as the reaction proceeds so as to drive the equilibrium toward the products. In dilute aqueous acid, an imine or enamine hydrolyzes to the corresponding carbonyl compound (Streitweiser et al., 1992).

Because secondary amines cannot form imines, they react with ketones or aldehydes to form enamines instead (Streitweiser et al., 1992). A secondary amine has the general structure R_2NH. For

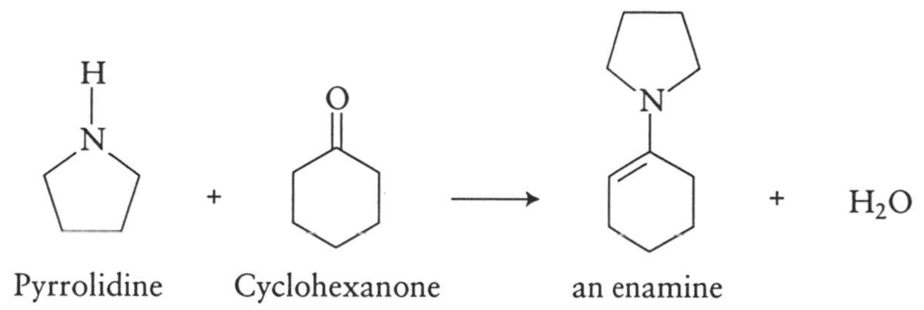

Fig. 13. Reaction of cyclohexanone with pyrrolidine.

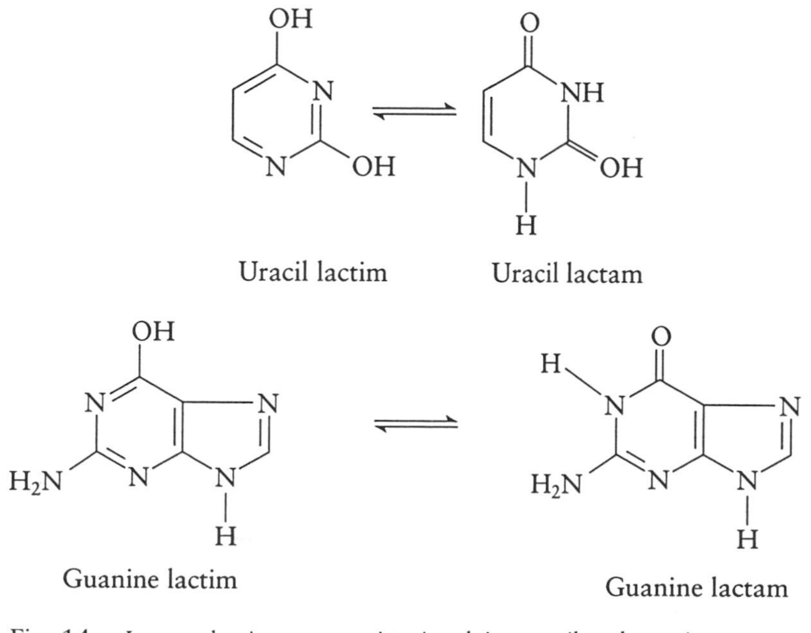

Fig. 14. Lactam-lactim tautomerism involving uracil and guanine found in RNA.

1429

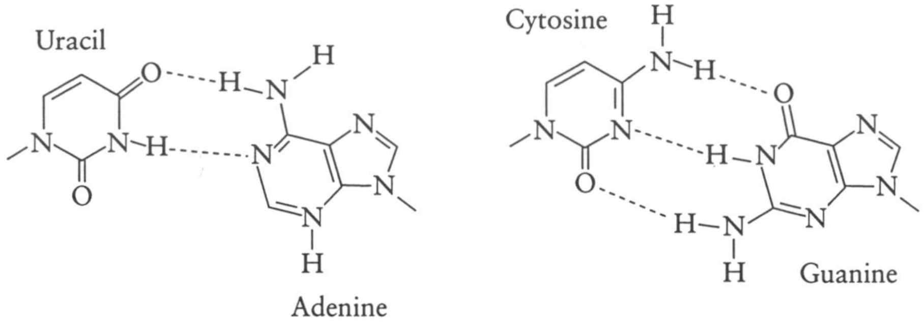

Fig. 15. Base pair formation in RNA—uracil with adenine and guanine with cytosine.

example, reaction of cyclohexanone with pyrrolidine, a secondary amine, gives an enamine along with a water molecule, as shown in Figure 13.

An enamine reacts similarly to an enol, by attacking electrophiles with its nucleophilic double bond. Because the enamine nitrogen is a better electron donor than the enol oxygen, an enamine is more reactive than an enol, and it reacts with weaker electrophiles.

Lactam-lactim tautomerism is important in the purine and pyrimidine bases found in nucleic acids. Uracil, a pyrimidine, and guanine, a purine, are two of the four bases ordinarily found in ribonucleic acid (RNA). Both uracil and guanine exist mainly in the lactam forms in water, even though the lactim forms are aromatic.

In the RNA duplex the uracil lactam forms a hydrogen-bonded base pair with adenine, and the guanine lactam forms a hydrogen-bonded base pair with cytosine (Fessenden and Fessenden, 1994).

The keto-enol, imine-enamine, and lactam-lactim tautomerisms are only a few of the many proton-shift tautomerisms found in nature. Other types of tautomeric equilibria are also known that involve rapid rearrangements of nonhydrogen atoms (March, 1992).

BIBLIOGRAPHY

CAREY, F. A.; SUNDBERG, R. J. *Advanced Organic Chemistry,* 3rd ed.; New York: Plenum Press, 1990; Part A, Chapter 7.

COFFEY, S., ed. *Rodd's Chemistry of Carbon Compounds,* 2nd ed.; Elsevier: Amsterdam, 1976; Vol. IV, Part D, pp. 14–15, 44–45; Part F, pp. 169, 174–175.

FESSENDEN, R. J.; FESSENDEN, J. S. *Organic Chemistry,* 5th ed.; Pacific Grove, CA: Brooks/Cole, 1994; Chapter 26.

LOUDON, G. M. *Organic Chemistry,* 3rd ed.; Redwood City, CA: Benjamin/Cummings, 1995; Chapter 22.

MARCH, J. *Advanced Organic Chemistry,* 4th ed.; New York: Wiley, 1992; pp. 69–74, 1133–1134.

STODDART, J. F., ed. *Comprehensive Organic Chemistry;* Oxford U.K.: Pergamon, 1979; Vol. 1, pp. 1024–1026.

STREITWIESER, A.; HEATHCOCK, C. H.; KOSOWER, E. M. *Introduction to Organic Chemistry,* 4th ed.; New York: Macmillan, 1992; pp. 421–432, 765–768.

KOOP LAMMERTSMA
NATHAN J. HARRIS

T_c

See CRITICAL PHENOMENA.

Temperature

Temperature refers to our sensations of hotness and coldness. Temperature is a property common to all matter, but it is not easily or accurately distinguished by our senses. Temperature is called an intensive property of a substance, meaning that different samples of the same material can have the same temperature regardless of their amounts or sizes. Thus, a spoonful, a cupful, and a potful of boiling water are all said to have the same temperature. Temperature is to be distinguished from heat, which is an extensive property and varies with the amount or size of the sample. In the example above, the potful of water will have more heat than the spoonful.

Consistent with the above distinction, *heat* is interpreted as the total random kinetic energy of the particles (atoms or molecules) in a sample of a substance. Temperature, on the other hand, is related to the average random kinetic energy of the particles.

Objects left in the same environment for long periods of time are said to have the same temperature, even though one object might feel much colder to the touch than another. An example of this would be metal and wood: ice-cold metal feels colder than

Fig. 1.   White phosphorus.

wood at the same temperature. (This is due to other properties of the materials such as thermal conductivity and heat capacity.) For this reason we use devices, called thermometers, that are independent of our subjective sensations to measure temperature. Thermometers work on the principle that many properties of matter vary with the temperature: gases and liquids expand, metals display variations in electrical conductivity, and pairs of different metals exposed to different temperatures generate an electrical voltage (thermocouples). Each of these variations has been used as the basis for a different kind of thermometer, producing a reading that is interpreted in degrees on a thermometric scale.

Two common temperature scales are the Fahrenheit (F) scale, in which water freezes at 32°F and boils at 212°F, and the Celsius (C, formerly known as the centigrade) scale, in which water freezes at 0°C and boils at 100°C. The variation of volume or pressure of a sample of air decreases linearly with temperature over a fairly large range (GAY-LUSSAC's or CHARLES's LAW). If this relationship is graphically extrapolated to "zero volume," the line intersects the temperature axis at a value of −273°C. This is the value of the "absolute zero" of temperature in the Kelvin (K) scale, in which degree intervals are the same as the Celsius scale.

The temperature of very hot objects is also related to the color or wavelength distribution of the light that they emit. This is called "blackbody radiation," and its study gave rise to the concept of the quantum (see QUANTUM THEORY). In everyday language we use "white-hot" and "red-hot" as indicators of extremely hot and very hot temperatures. This concept can be extended to colors or wavelengths that we cannot see, even though we can measure them. This has been done with regard to the "temperature" associated with the background radiation observed in space. Thus, the calculated value of 3°K is consistent with the idea that the universe originated with an explosion 10 to 15 billion years ago and has been cooling ever since.

EZRA SHAHN

Tetratomic

Tetratomic species consist of four atoms and are the smallest molecules that can be nonplanar. Although there are few homonuclear tetratomic species, one important example is an ALLOTROPE of phosphorus called white phosphorus. As shown in Figure 1, white phosphorus is made of discrete tetrahedral molecules and has the molecular formula P_4. The bonds in white phosphorus are made by overlapping the $3p$-orbitals of each atom. The geometry requires that all bonds have 60° angles, so the $3p$-orbitals cannot overlap in the preferred "head-on" fashion. This results in relatively weak and highly reactive bonds; white phosphorus bursts into flames when exposed to air. There are a wide variety of heteronuclear tetratomic species, and many are used in both industrial and household settings. AMMONIA, NH_3, shown in Figure 2, is a pyramidal four-atom species that is used in household cleaners. FORMALDEHYDE, H_2CO, shown in Figure 3, is an example of a planar heteronuclear tetratomic molecule and is commonly used in making plastics. None of the preceding three examples has a linear, or "chain" backbone. Tetratomic molecules are the smallest molecules that can have a nonchain structure.

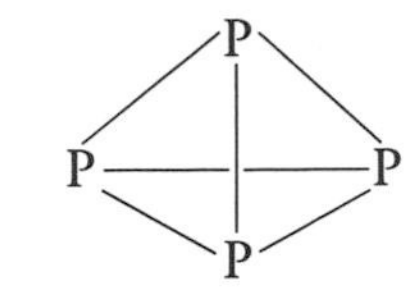

The majority of tetratomic species are asymmetric; thus, they will have a net dipole moment and will orient in an electric field. An example of a symmetric tetratomic with no net dipole moment is boron trifluoride, BF_3, shown in Figure 4. White phosphorus also has no net dipole. Although ammonia is highly symmetric, the pyramidal structure places the more electronegative nitrogen out of the plane defined by the hydrogens, leading to a net dipole moment, with the hydrogen side of the molecule having a partial positive charge. All four-atom molecules will have nine internal motions, or modes, that deform the molecule from its equilibrium geometry. The majority of these molecules will have six vibrational modes and three rotational modes; however, a linear tetratomic species will have seven vibrational modes and two rotational planes. An example of a linear four-atom species is ethyne, commonly called acetylene, having the molecular formula C_2H_2 and shown in Figure 5. Be-

Fig. 2.   Ammonia. Solid triangles indicate that the bond is coming out of the paper; dashed triangles indicate that the bond is going into the plane of the paper.

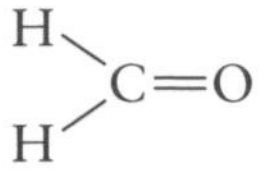

$$ \overset{\displaystyle H}{\underset{\displaystyle H}{}} \!\!\! C = O $$

Fig. 3. Formaldehyde. The double line between oxygen and carbon indicates a double bond.

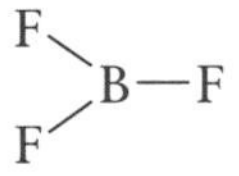

$$ \overset{\displaystyle F}{\underset{\displaystyle F}{}} \!\!\! B - F $$

Fig. 4. Boron trifluoride. Each bond is polar, but the net of the three yields zero dipole for the molecule.

$$ H - C \equiv C - H $$

Fig. 5. Ethyne. The triple line between the carbons indicates a triple bond.

cause of its linear symmetric structure, ethyne has no net dipole moment. The exact nature of the vibrations and rotations will depend upon the connectivity of the atoms within the molecule. As with all molecules, there is a characteristic energy associated with each vibrational and rotational motion; thus, the vibrations and rotations may be studied by a variety of spectroscopic techniques.

BIBLIOGRAPHY

ATKINS, P. W. *Physical Chemistry*, 5th ed.; New York: Freeman, 1994.

McMURRY, J.; FAY, R. C. *Chemistry*; Englewood Cliffs, NJ: Prentice Hall, 1995.

KELLY S. GRIFFITH

Theoretical Chemistry

Theoretical chemistry is concerned with the use of quantum mechanics (see QUANTUM THEORY), classical mechanics, and statistical mechanics to understand the structures and dynamics of chemical systems and to correlate, understand, and predict their thermodynamic and kinetic properties (see BIOENERGETICS; KINETICS; THERMOCHEMISTRY). Modern theoretical chemistry may be divided roughly as follows:

1. Chemical structure
   a. Electronic structure, potential energy surfaces, and force fields
   b. Vibrational-rotational motion
   c. Equilibrium properties of condensed-phase systems and macromolecules
2. Chemical dynamics
   a. Bimolecular kinetics and the collision theory of reactions and energy transfer
   b. Unimolecular rate theory and metastable states
   c. Condensed-phase and macromolecular aspects of dynamics

A critical issue that crosses all boundaries is the interaction of matter and radiation, since spectroscopic experiments are used as both structural and dynamic probes, and they must be understood theoretically. There are also many subfields of theoretical chemistry—e.g., biomedical structure-activity relationships, the molecular theory of NUCLEAR MAGNETIC RESONANCE spectra, and electron-molecule scattering—that fit into two or more of the areas listed.

Another source of overlap among the categories is that some of the techniques of theoretical chemistry are used in more than one area. Thus, statistical mechanics comprises the theory and the set of techniques used to relate macroscopic phenomena to those at the atomic level, and it is used in all six subfields listed. Furthermore, the techniques of quantum mechanics are used in all six subfields, and classical mechanical approximations to quantum mechanics are used profitably in all six subfields as well. Condensed-phase phenomena are often treated with gas-phase theories when solvent or lattice effects are not expected to dominate. Often progress in one subfield is made by importing techniques developed originally for use in another, but there are many specialized theories, models, and approximations as well. Thus, the outline provides a useful starting point for understanding the main areas of current work in theoretical chemistry.

Because quantum and statistical mechanics are also parts of physics, theoretical chemistry is sometimes considered a part of chemical physics. There is no clear border between theoretical physical chemistry and theoretical chemical physics.

Sometimes modern science is said to proceed by three modes—experiment, theory, and computation. This same division may be applied to chemistry. From this point of view, theoretical chemistry is based on analytical theory, whereas computational chemistry is concerned with predicting the properties of a complex system in terms of the laws of quantum mechanics (or classical approximations to quantum mechanics, in the domain of validity of such classical approximations) governing its constituent atoms or its constituent nuclei and electrons without using intermediate levels of analytical chemical theory. Most researchers, however, consider that computational chemistry is a subfield of theoretical chemistry, and indeed computational advances and theoretical understanding are becoming more and more closely linked as the field progresses. Computational chemistry is sometimes called MOLECULAR MODELING or molecular simulation.

Perhaps the single most important concept in theoretical chemistry is the separation of electronic and nuclear motions, often called the Born-Oppenheimer approximation, after the seminal work of Max Born and Robert Oppenheimer (in 1927), although the basic idea is also due to Walter Heitler, Fritz London, and John Slater. The critical facts that form a basis for this approximation are that the electrons are coupled to the nuclei by coulomb forces, but they are much lighter—by a factor of 1,800 to 500,000—and thus, under most circumstances, they may be considered to adjust instantaneously to nuclear motion. Technically we would describe this situation by saying that a chemical system is usually electronically adiabatic. In such cases, the treatment of a chemical system is greatly simplified. For example, the H_2 molecule is reduced from a four-body problem to a pair of two-body problems: one, called the electronic structure problem, for the motion of two electrons moving in the field of fixed nuclei; and another, called the vibration-rotation problem or the dynamics problem, for the two nuclei moving under the influence of a force field set up by the electronic structure. In general, since the energy of the electronic subsystem depends on the nuclear coordinates, the electronic structure problem provides an effective potential energy function for nuclear motion. This is also called the potential energy hypersurface. The atomic force field—i.e., the set of all the forces between the atoms—is the gradient of this potential energy function.

Thus, when the separation of electronic and nuclear motion is valid, topic 1a in the chart becomes a problem in electronic structure, and the results of this step become the input function for work on topics 1b, 1c, and 2. Some important problems where the Born-Oppenheimer separation breaks down are photochemical reactions involving visible and ultraviolet radiation and electrical conductivity. Even for such cases, though, it provides a starting point for more complete treatments of electron-nuclear coupling.

In the subfield of theoretical dynamics, the most important unifying concept is TRANSITION STATE THEORY. A transition state is a fleeting (lifetime on the order of 10 femtoseconds, fs) intermediate that represents the hardest-to-achieve configuration of a molecular system in the process of transforming itself

from reactants to products. A transition state is sometimes called an ACTIVATED COMPLEX or a dynamical bottleneck. In the language of quantum mechanics, it is a resonance or metastable state, and in the language of classical mechanics, it is a hypersurface in phase space. Transition states are often studied by semiclassical methods as well; these represent a hybrid of quantum mechanical and classical equations.

In the early days of theoretical chemistry, the field served mainly as a tool for understanding and correlating data. Now, however, owing to advances in computational science, theory and computation can often provide reliable predictions of unmeasured properties and rates. In other cases where measurements do exist, theoretical results are more accurate than measured ones. Examples are the properties of simple molecules and reactions like $D + H_2 \rightarrow HD + H$. Computational chemistry often provides other advantages over experimentation too. For example, it provides a more detailed view, such as of the structure of transition states or a faster way to screen possibilities. An example of the latter is provided in the field of drug design, where thousands of candidate molecules may be screened for their likely efficiency by approximate calculations, for example, of the electronic structure or free energy of desolvation, and only the most promising candidates synthesized and tested in laboratory studies.

BIBLIOGRAPHY

ALMLÖF, J.; TRUHLAR, D. D.; LYBRAND, T. P. *Interdisciplinary Science Reviews* 1990, *15*, 252–263.

EYRING, H.; WALTER, J.; KIMBALL, G. E. *Quantum Chemistry*; New York: John Wiley, 1944.

LIPKOWITZ, K. B.; BOYD, D. B., eds. *Reviews in Computational Chemistry V*; New York: VCH Publishers, 1994.

DONALD G. TRUHLAR

Thermochemistry

Literally, *thermochemistry* is the study of the temperature changes caused by chemical reactions. Reactions that produce hot products are said to be *exothermic*, as when methane (a fuel, CH_4) burns, producing hot gases, mostly CO_2 and H_2O. Reactions that produce cold products are said to be *endothermic*, as when ammonium nitrate NH_4NO_3 dissolves in water to produce a cold solution in the cold packs applied to sprains.

By predicting a reaction's heat effects, the chemist or engineer can provide for heating or cooling sufficient to control the process. Otherwise, exothermic reactions can overheat, forming undesirable by-products or even exploding, while endothermic reactions may become too slow as the temperature falls.

Energy E is defined as the capacity to do work. Heat *Q* is energy *E* transferred due to a temperature difference. Heat spontaneously "flows" to a region of lower temperature—hot tea cools by losing energy as heat to cooler surroundings, or ice cream melts by gaining energy as heat from warmer surroundings.

Heat transfer ceases only when temperature is everywhere the same, a condition known as *isothermal*. Insulation is used to slow energy (heat) transfer. Perfect insulation, if it existed, would not permit heat to enter or leave. Such (hypothetical) conditions are said to be *adiabatic*, from the Greek for "not through." A thermos bottle with adiabatic walls would keep its contents hot (or cold) forever. Adiabatic conditions are approached in flames and explosions because there is little time for heat transfer.

Heat, work, and energy can be expressed in the same units. Chemists use joules, J, named after James Joule. Other energy units are the calorie (4.184 J), the British thermal unit (1,054 J), and the newton-meter, Nm (1 J).

Heat is measurable by its effects on matter such as a change in volume or temperature. The amount of work needed to produce the same change must be known in order to "calibrate" heat's effect. Work is done when energy is transferred due to action of a force. The unit of force, the newton, N, is approximately that exerted by Earth's gravity on ¼ lb of matter. To lift a ¼-lb hamburger 1 m against gravity requires about one Nm of work, about one-millionth of a hamburger's metabolic energy content.

In the "ice calorimeter," an early device used for quantitative measurement of heat, the reaction of interest occurs in a container surrounded by an ice-water mixture. Calibration experiments using work to produce heat showed that each 6,007 J melt 1 mol (18 g) of ice, decreasing the ice-water volume by 1.626 cm³. Then the heat *Q* can be found from the observed volume change ΔV:

$$Q = (6{,}007 \text{ J mol}^{-1})(\Delta V)/(-1.626 \text{ cm}^3 \text{ mol}^{-1})$$

Heat Capacity

The amount of heat needed to increase by 1 kelvin the temperature of a given amount of matter is called the HEAT CAPACITY. The temperature increase ΔT caused by the transfer of *Q* joules of heat to *m* grams of, say, water is proportional to *Q*:

$$Q = m \times C \times \Delta T \qquad (1)$$

where *C* is water's heat capacity per gram, J K^{-1} g^{-1}, and ΔT is the temperature change in kelvins. *C* varies from substance to substance and increases with rising temperature. Heat capacities expressed in J K^{-1} mol^{-1} are widely used.

In matter, energy manifests itself as particle motion, mainly translation, rotation, and vibration. In most solids, the particles are too tightly packed to permit translation or rotation, leaving vibration as the main form of motion. Vibrations in solids are usually very complicated—as an atom vibrates within a "cage" composed of its neighbors, each neighbor is doing the same. The resulting jellylike motions make it impossible to predict a solid's heat capacity or other thermal properties. The situation in liquids is essentially the same. Therefore, tabulated values of thermal properties of solids and liquids are experimental.

Particles in gases are free to translate and, if molecular, to rotate about their center of mass and to vibrate as one part of the molecule moves relative to the rest. From spectroscopic data it is possible to calculate the vibrational (vib) and rotational (rot) contributions to the gas's heat capacity and other thermal properties. To a good approximation, each form of motion contributes independently; for example,

$$C_{\text{total}} = C_{\text{tr}} + C_{\text{rot}} + C_{\text{vib}} \qquad (2)$$

Statistical Calculations

The energy ε_{vib} associated with a vibration is a multiple of $h \times \nu$, where *h* is Planck's constant, 6.626×10^{-34} Js, and ν is the vibration's frequency in hertz, Hz, a reciprocal second. The energy $h \times \nu$ is called a quantum (plural, quanta). Diatomic molecules have just one mode of vibrating—bond stretching. More complicated molecules have more modes. Although molecules can vibrate in combinations of modes, only modes that are not combinations—that is, are "fundamental" modes—need be considered. Each fundamental vibration contributes to thermal properties independently of other modes of vibration.

Rotational energy ε_{rot} is also quantized, although the quanta are much smaller. Quanta of translational energy ε_{tr} are extremely small and vary with the size of the container. Here, "small" is relative to $k_B T$, where k_B is Boltzmann's constant, 1.38×10^{-23} J K^{-1}, and *T* is the temperature in kelvins.

At equilibrium, energy quanta are distributed in the most probable way. This is described by the Boltzmann equation, which gives the number of particles N_i having energy ε_i:

$$N_i = N \times \exp(-\varepsilon_i/k_B T)/\Sigma \exp(-\varepsilon_i/k_B T) \qquad (3)$$

where Σ is a summation from $i = 0$ to $i = $ infinity. For example, for a vibration of 1×10^{13} Hz at 298.15 K,

$$\varepsilon/k_B T = (1 \times 10^{13} \text{ s}^{-1})(6.626 \times 10^{-34} \text{ Js})/(1.38 \times 10^{-23} \text{ JK}^{-1} \times 298.15\text{K}) = 1.61$$

Then

$$\Sigma = \exp(-0/k_B T) + \exp(-1.61/k_B T) + \exp(-3.22/k_B T) + \cdots = 1.25$$

From equation 3, the fraction of the molecules having, say, one quantum of this vibrational energy is

$$N_1/N = \exp(-1.61)/1.25 = 0.16$$

With modifications, equation 3 can also be used to compute rotational and translational contributions to thermal properties.

As temperature rises, the amount of thermal energy to be distributed among the quantum states increases, going first into translational motion, then rotational, then vibrational.

The First Law of Thermodynamics

The First Law of Thermodynamics states that *the total energy of the universe is fixed, a constant.* Equivalently, by regarding the universe as composed of the system SY under study and its surroundings SU,

$$E_{SY} + E_{SU} = \text{a constant} \qquad (4)$$

Equation 4 requires that energy leaving the system must enter in the surroundings and vice versa. Using Δ to mean "the change in," the First Law of Thermodynamics can also be stated as

$$\Delta E_{SY} + \Delta E_{SU} = 0 \qquad (5)$$

Or, reflecting the system's gain or loss of energy as heat and work, as

$$\Delta E_{SY} = Q + W \qquad (6)$$

where Q and W are positive quantities if they increase E_{SY}.

Effect of Confinement

When a reaction resulting in a net increase in the moles of gas occurs while confined to a rigid, sealed container—that is, at constant volume V—the gases cannot expand so work cannot be done and the reac-

tion's ΔE becomes heat Q_V. For example, $W = 0$ and $Q_V = -99{,}291$ J mol^{-1} when 1 mol of hydrogen peroxide decomposes at 25°C:

$$H_2O_2 \text{ (liquid)} \rightarrow H_2O \text{ (liquid)} + 0.5O_2 \text{ (gas)}$$

Unconfined, the same reaction produces $Q = -98{,}052$ J mol^{-1} because 1,239 J are used to do the work of pushing back the atmosphere to make room for the 12 or so L of O_2 (gas) formed. The work of changing a volume by ΔV against a constant pressure p is

$$W = -p\Delta V \qquad (7)$$

where the $-$ sign is needed because W is negative when ΔV is positive and vice versa. For the H_2O_2 decomposition, $\Delta V = V(0.5 \text{ mol } O_2)$ to a good approximation because the liquids occupy but a few milliliters. By the ideal gas law, $V = nRT/p$, where $n = $ number of moles of gas, $R = $ the gas constant (8.314 J K^{-1} mol^{-1}), and $T = $ absolute temperature in kelvins (298.15 K, 25°C). Then

$$W = -p(nRT/p) = -(0.5 \text{ mol})(8.314 \text{ J K}^{-1} \text{ mol}^{-1}) (298.15 \text{ K}) = -1239 \text{ J}$$

Reactions for which $\Delta V = 0$ transfer the same amount of heat, confined or unconfined, while reactions for which $\Delta V < 0$ transfer more energy as heat when unconfined.

Heat capacity measured at constant pressure C_p exceeds that measured at constant pressure C_V because some of the added energy is consumed in doing work of expanding against the constant pressure. The difference between C_p and C_V is nearly negligible for solids and liquids.

Enthalpy

In constant-volume processes, ΔE is a convenient measure of the energy change. But for processes occurring at constant pressure it is convenient to define another thermochemical quantity, enthalpy H such that ΔH measures the energy transferred as heat under constant pressure conditions, that is, $\Delta H = Q_p$. The First Law of Thermodynamics requires that

$$Q_p = Q_V + \Delta(pV) \qquad (8)$$

and

$$\Delta H = \Delta E + \Delta(pV) \qquad (9)$$

where $\Delta(pV)$ is the work of expansion or contraction accompanying the reaction.

1435

Predicting $\Delta_r H$, the ΔH of a Reaction

Most $\Delta_r H$ data are experimental. However, it is not necessary to experimentally determine every conceivable $\Delta_r H$ because $\Delta_r H$ is independent of all and any intermediates. This principle, known formally as Hess's law, is a consequence of the First Law of Thermodynamics. Where this not the case it would be possible to create energy by, say, alternately decomposing and forming water.

Hess's law may be illustrated by using it to predict $\Delta_r H$ for a reaction for which direct calorimetric determination would be extremely difficult:

$$C \text{ (graphite)} \rightarrow C \text{ (diamond)}$$

If the sum or difference in two reactions is the desired reaction, the sum or difference in their $\Delta_r H$ values will be the desired $\Delta_r H$. Because they are suitable reactions and their $\Delta_r H$ values for 25°C and 1 atm,

$$C(\text{graphite}) + O_2(\text{gas}) \rightarrow CO_2(\text{gas}),$$
$$\Delta_r H = -393,522 \text{ J mol}^{-1}$$

minus

$$C(\text{diamond}) + O_2(\text{gas}) \rightarrow CO_2(\text{gas}),$$
$$\Delta_r H = -395,388 \text{ J mol}^{-1}$$

leaves

$$C(\text{graphite}) \rightarrow C(\text{diamond}), \Delta_r H = +1,866 \text{ J mol}^{-1}$$

In order to avoid ambiguity when stating a $\Delta_r H$ value, it is necessary to specify

1. *Temperature*, because all $\Delta_r H$ vary with temperature.
2. *Pressure*, because it determines the amount of pressure-volume work, if any, and which phase is most stable. By international agreement, "standard" pressure, formerly 1 atm (101,325 Pa), is 100,000 Pa, that is, 0.1 MPa. Pa stands for the pascal, 1 Pa = 1 N m^{-2}.
3. *Ideal behavior*, because $\Delta_r H$ for reactions involving gases and solutes is somewhat concentration-dependent. Observed values are corrected to what they would be if the substances behaved "ideally."

Standard States

The above specifications are embodied in what are called standard states:

- Solid or liquid: the pure substance at 0.1 MPa
- Gas: the hypothetical ideally behaving gas at 0.1 MPa

- Solute: the hypothetical solution of unit solute concentration in which there are either (1) only solute-solute interactions or (2) no solute-solute interactions and at 0.1 MPa, whichever is more convenient

Standard Enthalpy of Formation $\Delta_f H^0$

The $\Delta_r H$ found in most tabulations refers to isothermally forming a compound from its elements, each in its most stable standard state. The most stable standard state of carbon is graphite, some 1,866 J mol^{-1} lower in energy than diamond. For oxygen, it is gaseous O_2, not O_3 or O. The superscript ^{0} designates standard state. For example, $\Delta_f H^0$ (CO_2, gas) refers to ΔH for the hypothetical reaction

$$1 \text{ mol C(graphite)} + 1 \text{ mol } O_2(\text{ideal gas})$$
$$\rightarrow 1 \text{ mol } CO_2(\text{ideal gas})$$

occurring at 0.1 MPa and at any temperature such that $O_2(\text{gas})$ and graphite are the most stable forms of oxygen and carbon. At $T = 298.15$ K, $\Delta_f H^0(CO_2$ gas$) = -393,522$ J mol^{-1}. As a corollary, $\Delta_f H^0 = 0$ for an element in its most stable standard state. Extensive tables of $\Delta_f H^0$ are available (Wagman et al., 1982).

Calculating $\Delta_r H^0$

To calculate a $\Delta_r H^0$ for temperature T, the $\Delta_f H^0$ values for the products and reactants at T are each multiplied by their respective stoichiometric coefficients. Then the total obtained for reactants is subtracted from the total for the products. For example, for

$$3 H_2(\text{gas}) + O_3(\text{gas}) \rightarrow 3 H_2O \text{ (liquid)},$$
$$\Delta_r H^0 = 3\Delta_f H^0(H_2O, \text{liquid}) - 3\Delta_f H^0(H_2, \text{gas}) - 1\Delta_f H^0(O_3, \text{gas})$$

To compute this reaction's $\Delta_r H^0$ at 298.15 K, tabulated $\Delta_f H^0$ values for 298.15 K are available (Wagman et al., 1982).

$$\Delta_f H^0(H_2O, \text{liquid}) = -285,830 \text{ J mol}^{-1}$$
$$\Delta_f H^0(H_2, \text{gas}) = \text{zero, an element in its standard state}$$
$$\Delta_f H^0(O_3, \text{gas}) = +142,674 \text{ J mol}^{-1}$$

Then, $\Delta_r H^0$ at 291.15 K

$$= 3(-285,830) - (\text{zero}) - 142,674$$
$$= -1,000,164 \text{ J mol}^{-1}$$

The enthalpy of mixing, ΔH_{mix}, is the ΔH accompanying the mixing of substances A and B to form a solution or the dilution of a solution. The major component is usually called the solvent; the minor component(s), the solute(s). The ΔH_{mix} can be either exothermic or endothermic as solute-solvent interactions replace solute-solute and solvent-solvent interactions and as solvation becomes complete.

The integral heat of solution ΔH_{IS} is the ΔH for the isothermal mixing process:

$$1 \text{ mol B} + n_A \text{ mol A} \rightarrow \text{solution of B in } n_A \text{ mol A} \tag{10}$$

The ΔH_{IS} reaches a constant value as the solution becomes infinitely dilute ($n_A = \infty$) because solvation reaches completion, solute-solute separations become infinite, and disruption of solvent-solvent attractions is reduced to zero.

The $\Delta_f H^0$ of a solute B in a solution in A is obtained by adding the $\Delta_f H^0$ (B) to ΔH_{IS}(B in ∞ mol of A):

$$\Delta H_{IS}(\text{B in } n_A \text{ A}) = \Delta_f H^0 \text{ (B)} + \Delta_f H(\text{B in } \infty \text{ A}). \tag{11}$$

The concentration dependence of $\Delta_f H^0$ (B in n_A mol of A) is that of $\Delta H_{IS}(B$ in n_A mol of A). Both approach a constant value as $n_A \rightarrow \infty$.

Heat of Ionic Reaction in Solution

A solute that is dissociated in solution—for example, HCl in water—can be regarded as consisting of ions that are nearly independent. Then $\Delta_f H^0$ (HCl in ∞ H_2O), $-167,159$ J mol^{-1}, can be regarded as $\Delta_f H^0$ (H^+ in ∞ H_2O) + $\Delta_f H^0$ (Cl^- in ∞ H_2O). By taking $\Delta_f H^0$ (H^+ in ∞ H_2O) as zero, it follows that $\Delta_f H^0$ (Cl^- in ∞ H_2O) = $-167,159$ J mol^{-1}. With this arbitrary but conventional choice, it is possible to establish values of $\Delta_f H^0$ for other ions in solution. For example, $\Delta_f H^0$ (NaCl in ∞ H_2O) = $-407,111$ J mol^{-1}. Subtracting $\Delta_f H^0$ (Cl^- in ∞ H_2O) leaves $\Delta_f H^0$ (Na^+ in ∞ H_2O), $-239,952$ J mol^{-1}.

Wagman et al. (1982) denote data for compounds fully dissociated in water with "ai." Weak electrolytes may have a second entry, labeled "ao," for the undissociated compound. For example, $\Delta_f H^0{}_{298}$ in kJ mol^{-1} for acetic acid is -486.01 in the "ai" state, -485.76 in the "ao" state, and -484.5 for the pure liquid. As with strong electrolytes, the standard state of a weak electrolyte solute is the hypothetical solution of unit concentration (one molal in Wagman et al., 1982) in which the $\Delta_f H^0$ is that in the infinitely dilute solution.

Spontaneity

Certain chemical systems are capable of changing spontaneously, meaning without outside influence.

Helium released into air spontaneously mixes; alcohol poured into water spontaneously dissolves. Yet these processes never spontaneously reverse themselves; the trace of helium in air and the alcohol in water never spontaneously separate.

A driving force for spontaneous change is a natural preference for mixtures over pure substances. The probability of growing a million-atom crystal of, say, pure silicon from a melt that is 99.999 percent pure silicon involves a type of calculation familiar to gamblers: the chance of repeating a certain outcome, the random selection of a silicon atom from the melt to become part of the crystal. The probability of selecting a silicon atom in any one of a million selections is (0.99999) to the 1,000,000th power, 0.0000454 or 1/22,026—one pure crystal out of 22,026 attempts.

Nature's preference for mixtures applies to liquid and gaseous mixtures as well. However, a substance's pure crystalline state is less probable and more ordered than its pure liquid or pure gaseous states because its atoms have specific locations; those in the gas or liquid do not. Allowing particles to move about increases disorder.

In the absence of thermal energy—that is, at absolute zero—only physical state and purity determine disorder.

The thermochemical quantity that is a measure of disorder is entropy, S. Its relation to probability is logarithmic:

$$S = (\text{a constant}) \times \text{Ln(relative probability)} \tag{12}$$

Relative probability is the numerator of the absolute probability fraction—the 1 in 1/22,026. But Ln(1) = 0. Then the entropy of a pure crystal is zero at absolute zero. This is the Third Law of Thermodynamics.

Entropy of Mixing

The entropy of an impure crystal at 0 K exceeds zero by the entropy increase associated with mixing, ΔS_{mix}. If n_A mol of substance A randomly mix with n_B mol of substance B to form a solution,

$$\Delta S_{mix} = n_B R \text{Ln}[n_B/(n_B + n_A)] - n_A R \text{Ln}(n_A/[n_B + n_A]) \tag{13}$$

where R is the gas constant, 8.314 J K^{-1} mol^{-1}. Random mixing occurs in solids, liquids, and gases.

Thermal Entropy

Thermal energy is distributed among a substance's quantum states in the most probable way, specified by the Boltzmann equation. For gases, equations exist for accurately predicting the entropy contribution of each kind of quantum state. In most cases, the total entropy is made up of independent rotational, vibrational, and translational contributions:

$$S_{total} = S_{rot} + S_{vib} + S_{tr} \qquad (14)$$

S_{rot} depends on the molecule's bond lengths, bond angles, and symmetry. For N_2 at 298.15 K, $S_{rot} = 41.13$ J K^{-1} mol^{-1}.

S_{tr} depends on the pressure p (Pa) and the mass per molecule m (kg).

$$S_{tr} = R[2.5 + 2.5 \, Ln(T) - Ln(p) \\ +100.355 + 1.5 \, Ln(m)] \quad (15)$$

For N_2 at 100,000 Pa and 298.15 K, equation 15 gives $S_{tr} = 150.4$ J K^{-1} mol^{-1}.

S_{vib} depends on the vibrational characteristics of the molecule. The entropy due to a vibration of frequency ν is given by

$$S_{vib} = R \times x[\exp(x) - 1]^{-1} - R[1 - \exp(-x)] \quad (16)$$

where $x = (4.80 \times 10^{-11})(\nu/T)$. For N_2, $\nu = 7.07 \times 10^{13}$ Hz. Then at 298.15 K, $S_{vib} = 0.003$ J K^{-1} mol^{-1}. This high frequency, caused by N_2's triple bond, makes the vibrational quantum large, which in turn makes this vibration hard to excite: according to equation 3, only 0.001 percent of N_2 molecules are in this vibrational state. Equation 16 is applied to each vibration to obtain the total vibrational entropy.

The complicated multibody interactions in solids and liquids have, to date, prevented writing a general equation for predicting thermal entropies of solids or liquids; calorimetric data are used instead.

For very small changes in temperature, the change in entropy with temperature is given by

$$S_{T_2} - S_{T_1} = (C/T)\Delta T \qquad (17)$$

where C is the substance's heat capacity. When ΔT is large, equation 17 is inaccurate because C/T varies with temperature. Integration is then required. In graphical integration, $S_{T_2} - S_{T_1}$ is the area under a plot of C/T versus T from T_1 to T_2. Integration may be used provided no phase transition occurs between T_1 and T_2.

Entropy Change Due to a Phase Transition ΔS_{trs}

The gain or loss of energy, ΔH_{trs}, accompanying a phase transition produces an entropy change:

$$\Delta S_{trs} = \Delta H_{trs}/T_{trs} \qquad (18)$$

where T_{trs} is the absolute temperature at which the phases are in equilibrium. The ice $\rightarrow$ water transition at 0°C and 0.1 MPa requires an energy input of 6,007 J mol^{-1}. From equation 18,

$$\Sigma S_{trs} = (6{,}007 \text{ J mol}^{-1})/273.15 \\ \text{K} = 22.0 \text{ J K}^{-1} \text{ mol}^{-1}.$$

A phase change can occur when the phases are not in equilibrium, but the change is irreversible under the conditions. For example, at 0.1 MPa and −10°C, water spontaneously freezes but ice does not spontaneously melt.

The Second Law of Thermodynamics

By dividing the universe into the system under study and its surroundings,

$$S_{univ} = S_{SY} + S_{SU} \qquad (19)$$

and

$$\Delta S_{univ} = \Delta S_{SY} + \Delta S_{SU} \qquad (20)$$

When both the system and its surroundings are considered, all processes cause an irreversible increase in the entropy of the universe. To make ice cubes, a refrigerator uses coolant at −10°C to remove heat from the water (the system) at 0°C. For water $\rightarrow$ ice at 0°C,

$$\Delta S_{SY} = -6007/273.15 = -22.00 \text{ J K}^{-1} \text{ mol}^{-1}$$

according to equation 18. On entering the coolant, this 6,007 J increases the coolant's entropy by

$$\Delta S_{SU} = 6{,}007/263.15 = 22.83 \text{ J K}^{-1} \text{ mol}^{-1}$$

The net effect on this bit of the universe is an entropy increase of

$$\Delta S_{univ} = 22.83 - 22.00 = 0.83 \text{ J K}^{-1} \text{ mol}^{-1}$$

At the same time, the irreversible effects of friction in the refrigerator motor and compressor, electrical resistance in all wires from the power plant to the refrigerator, and so on convert work into heat, adding to ΔS_{univ}. Although some power plants can convert heat into work, inevitable losses in the conversion further increase S_{univ}.

Planck's version of the Second Law of Thermodynamics generalizes the situation succinctly:

Every physical or chemical process in nature takes place in such a way as to increase the sum of the entropies of all the bodies taking part in the process.

In the limit—i.e., for reversible processes—the sum of the entropies remains unchanged.

There are few statements more thought-provoking. To think that the universe is growing increasingly and irreversibly disordered, that yesterday the universe was a bit less disordered, that the universe when new was, presumably, much more orderly—this leads one finally to wonder if, in the beginning, the entropy of the universe was zero!

Absolute Entropy

The entropy of a substance at T kelvins is the sum of the entropy changes required to bring it to T from 0 K. To find water's $S_{298.15}$, heat capacity data for ice from 0 K to the melting point at 273.15 K are integrated. To this are added the ΔS for melting (fusion) ice and the ΔS for warming water from 273.15 to 298.15 K:

$$S_{298} = (S_{273} - S_0)_{ice} + \Delta S_{fus} + (S_{298} - S_{273})_{water}$$

Tables of absolute entropies of substances in their standard states, S^0 at 298.15 K, are available (see Wagman et al., 1982). Values for ionic solutes are relative to that of the aqueous H^+ ion.

Absolute entropies may be used to predict the $\Delta_r S^0$ at a specific temperature for reactions of interest. Each product's absolute entropy S_p^0 is multiplied by its stoichiometric coefficient ν_p, and the resulting quantities are summed. From this is substracted the sum obtained when each reactant's absolute entropy S_r^0 is multiplied by the reactant's stoichiometric coefficient ν_r and the resulting quantities are summed:

$$\Delta_r S^0 = \Sigma(\nu_p \times S_p^0) - \Sigma(\nu_r \times S_r^0) \qquad (21)$$

Free Energy

A reaction's $\Delta_r S$, with its $\Delta_r H$, can be used to predict the reaction's equilibrium constant and its capacity for spontaneous change. For spontaneous change $\Delta S_{univ} > 0$, then

$$\Delta_r S_{SY} + \Delta S_{SU} > 0 \qquad (22)$$

where, by the First Law of Thermodynamics,

$$\Delta S_{SU} = -\Delta_r H/T \qquad (23)$$

Combining equations 22 and 23 yields the condition for spontaneity:

$$\Delta_r H - T\Delta S_{SY} < 0 \qquad (24)$$

By defining the Gibbs free energy G such that $\Delta_r G = \Delta_r H - T\Delta_r S$, the criterion for spontaneous change is simply

$$\Delta_r G < 0 \qquad (25)$$

and the criterion of chemical equilibrium is

$$\Delta_r G = 0 \qquad (26)$$

where the now-superflous SY subscript is omitted.

Standard Free Energy of Formation $\Delta_f G^0$

The free energy change attending the formation of a compound in a standard state from its elements in their standard states has been calculated for many substances (Wagman et al., 1982). These values, usually for 298.15 K, are in fact

$$\Delta_f H^0 - T \times \Delta_f S^0 \qquad (27)$$

where $\Delta_f S^0$ is obtained by applying equation 21 to the formation reaction.

Equilibrium

$\Delta_r G$ becomes zero when the concentrations and pressures reach their equilibrium values, but $\Delta_r G^0$ remains constant because it is defined in terms of the fixed standard-state pressures and concentrations. For a general reaction at a given temperature T

A (gas at pressure p_A) $\rightarrow$
$\qquad\qquad$ B (gas at pressure p_B) (28)

the relation between these quantities is

$$\Delta_r G = \Delta_r G^0 + R \times T \times Ln(p_B/p_A) \qquad (29)$$

where p_A and p_B have arbitrary values. At equilibrium, however, (p_B/p_A) has a unique value called the equilibrium constant K_{eq}:

$$K_{eq} = (p_{B,eq}/p_{A,eq}) \qquad (30)$$

When A and B are in solution, p_B and p_A are replaced by concentrations B and A in equations 29 and 30.

Temperature Dependence of K_{eq}

A K_{eq,T_2} can be computed from knowledge of its value at some other temperature, K_{eq,T_1}, using

$$\ln(K_{T_2}/K_{T_1}) = -(\Delta_r H/R)(1/T_2 - 1/T_1) \qquad (31)$$

BIBLIOGRAPHY

COCHRAN, W. *Dynamics of Atoms in Crystals;* Forest Grove, OR: International Scholarly Book Services, 1973.

HIRSCHFELDER, J.; CURTISS, C.; BIRD, R. *Molecular Theory of Gases and Liquids;* New York: Wiley, 1965. A classic. A comprehensive treatment of the subject.

KAUZMANN, W. *Kinetic Theory of Gases;* White Plains, NY: W. A. Benjamin Co., 1966.

KESTIN, J.; DORFMAN, J. *A Course in Statistical Thermodynamics;* New York, Academic Press, 1971.

NASH, L. *Elements of Classical and Statistical Thermodynamics;* Reading, MA: Addison-Welesy, 1968. A very lucid discussion of the statistical and classical views of entropy.

STULL, D.; PROPHET, H. *JANAF Thermochemical Tables,* 2nd ed., NSRDS-NBS 37; Washington, DC: Supt. of Documents, U.S. Govt. Printing Office, 1971. Tables of C_p, S, $H_T - H_{298}$, ΔH_f^0, ΔG_f^0 and log K_f at 100 K intervals for inorganic compounds and some elements. Supplements: *J. Phys. Chem. Ref. Data* 1974, *3*, 311; 1975, *4*, 1; and 1978, *7*, 739.

WAGMAN, D., et al. "NBS Tables of Chemical Thermodynamic Properties." *J. Phys. Chem. Ref. Data* 1982, *11* (Suppl. 2). Similar to JANAF tables but limited to 0 K and 298 K values and including solutions.

WALL, F. T. *Chemical Thermodynamics;* New York: W. H. Freeman, 1974; pp. 290ff.

T. C. EHLERT

Thermodynamics

See THERMOCHEMISTRY.

Thermodynamics, Statistical

See STATISTICAL THERMODYNAMICS.

Thermoluminescence

The phenomenon of visible photons released from certain solids upon heating is called thermoluminescence. The word *thermoluminescence* includes the fact that heat is needed (thermo-) to start the emission of light (-luminescence).

The simplified mechanism of this phenomenon can be described by analyzing the energy-bond model for an insulating crystal, which consists of a valence band where all the energy levels are filled with electrons. The next highest energy region in a pure substance is called a forbidden band because electrons cannot occupy this band. However, if impurities and/or defects are introduced into the crystal structure, local energy levels in this forbidden band, called "electron traps" or "hole traps," may form (see Figure 1). Above the forbidden band there is a conduction band with available energy levels for electrons. When such a thermoluminescent material is exposed to high-energy radiation, an electron from the valence band can be promoted to the conducting band (transition 1 in Figure 1), leaving a positive hole in the valence band. Both the positive hole in the valence band and the electron in the conduction band are free to move through the crystal (transitions 2 and 3 in Figure 1) until they either recombine or are trapped (transitions 4 and 5 in Figure 1) in the metastable electron or hole traps shown. In most cases, the hole trap is a luminescence center, a center which, with the capture of an electron, gives rise to the emission of a photon. In the case presented here, the trapped electron cannot recombine with the hole in the luminescence center until is raised from its electron trap in the forbidden band back up to the conduction band. This process of promoting the electron from its electron trap to the conduction band is accomplished by heating the crystal (transition 1 in Figure 2). Now, the electron is free to move and can recombine with a hole in the luminescence center (transitions 2 and 3 in Figure 2). The energy gain by the luminescence center is released in the form of photons (light). It is important to note that if the energy required to release an electron from its trap is larger than the thermal energy received due to heating, electrons will remain trapped and no luminescence will occur. The simple mechanism described here is only an example of thermoluminescence process, and other possibilities exist.

Thermoluminescence is widely used in radiation dosimetry. Thermoluminescent dosimeters are based on the fact that thermoluminescent materials can store energy from radiation for long periods of time

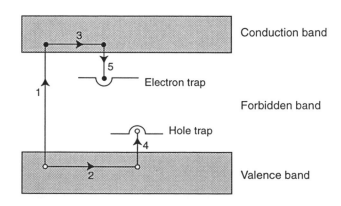

Fig. 1.   Energy-band model of an insulating crystal exposed to ionizing radiation.

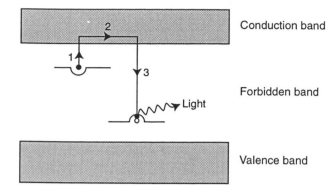

Fig. 2. Energy-band model for luminescence process initiated by heating an insulating crystal previously exposed to ionizing radiation.

and release it in the form of luminescence when heated. Such dosimeters are used in radiotherapy, where it is important to be able to measure small doses of radiation. In this case, a small thermoluminescence dosimeter is placed at the center of the patient therapy treatment field during each radiotherapy exposure. If the same capsule is used for each successive treatment, the total radiotherapy exposure can be determined. The intensity of the light emitted during the thermoluminescence process is a measure of the original radiation dose.

BIBLIOGRAPHY

CAMERON, J. R.; SUNTHARALINGAM, N.; KENNEY, G. N. *Thermoluminescent Dosimetry;* Madison, WI: University of Wisconsin Press, 1968.

DIANE M. BUNCE

EVA LABUDA

Thin-Layer Chromatography

Thin-layer chromatography (TLC) is a planar chromatographic method in which the stationary phase is coated (ca. 0.2 mm thickness) on a flat surface, usually a glass or plastic plate. Although all types of stationary phases for liquid column chromatography are available for TLC work, the most common layer material is pure silica gel, SiO_2.

The sample mixtures are solutions with concentrations of 0.01 to 0.1 percent. Very small volumes (1–5 μL in conventional TLC) of the mixtures are applied to the stationary phase as spots about 1–2 cm from one edge of the plate. For best separation the spots should be very small (5 mm diameter or less) and

uniform. Although this crucial step can be performed manually with a capillary tube or syringe, commercially available mechanical devices give more reproducible results with minimal damage to the stationary phase surface. Usually ten or more mixtures can be applied to a single 20 × 20 cm plate.

After the samples have dried, the plate is placed, sample edge down, in a closed tank containing a shallow depth (a few millimeters) of the mobile phase. As the solvent flows upward by capillary action, it passes over the sample spots. During this *development* process the components move in the same direction, but less rapidly than the solvent front. The migration is retarded by the relative attraction of the solute to the stationary phase.

When silica gel is used, the interaction with the stationary phase is primarily via ADSORPTION to the solid surface, and retention increases roughly as the polarity of the solute increases. Likewise, a polar mobile phase (e.g., ethyl acetate, methanol) competes more effectively than a nonpolar liquid (e.g., hexane, carbon tetrachloride) for adsorption sites and moves the solutes more rapidly. If there is a proper balance of interactions, the components of the sample are retarded to varying degrees and are separated into a line of small spots, also called *zones*.

The plate is allowed to develop until the solvent has moved about three-quarters of the distance to the top edge (several minutes to several hours). It is then removed, the solvent front is marked, and the plate is dried. The zones are made visible by spraying with a suitable solution. Often this contains iodine, which condenses preferentially on the solute zones, or H_2SO_4, which chars organic compounds when heated gently. Alternatively, a stationary phase containing a fluorescent compound can be used. When viewed with a UV lamp, the spots appear black against a lighted background.

A sketch of a TLC plate after development and visualization is shown in Figure 1. Each component is typified by its *retardation factor*, R_f, equal to d_Z/d_M, the distance of zone travel divided by the distance moved by the solvent front. Although not sufficiently precise or reproducible to be used for QUALITATIVE ANALYSIS of a total unknown, R_f values serve to identify the components suspected of being present in the sample, if a similar mixture has previously been chromatographed by the same procedure. For a more definitive identification, a spot can be scraped from the plate. After the compound is extracted, it is identified by various spectroscopic and/or chemical tests. With suitable calibration the extract can also be analyzed to determine the amount of component present, but the procedure is cumbersome. Quantitative anal-

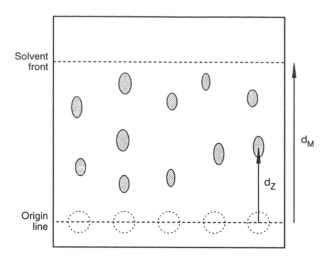

Fig. 1. Typical TLC plate after development and visualization.

preliminary screening (e.g., in forensic work) prior to analysis by HPLC or other methods. Because the stationary phases for HIGH-PERFORMANCE LIQUID CHROMATOGRAPHY (HPLC) and TLC are similar, TLC is a convenient means used to choose optimum conditions for HPLC work, with minimal waste of valuable samples and easy change of solvent.

BIBLIOGRAPHY

HAMILTON, R.; HAMILTON, S. *Thin Layer Chromatography.* Analytical Chemistry by Open Learning Series, D. Kealey, ed.; Chichester, U.K.: Wiley, 1987.

TOUCHSTONE, J. C.; DOBBINS, M. F. *Practice of Thin Layer Chromatography,* 2nd ed.; New York: Wiley, 1982.

KATHRYN R. WILLIAMS

yses are automated by scanning the plate with a device called a densitometer, which measures the intensity of transmitted or reflected light at various locations.

Two-dimensional TLC is often used for separations of complex mixtures. A spot of the mixture is applied to a corner of the plate and developed vertically, so that a series of incompletely separated blobs is aligned along one edge. After drying, the plate is rotated 90° to place the blob edge at the bottom of the tank, and development is continued with another solvent. The separated zones are distributed in both directions on the fully developed plate.

As in all areas of CHROMATOGRAPHY, the practice of TLC has benefited in recent decades from technological advances, especially the development of stationary phases with small (ca. 5 μm) uniform particles. High-performance plates can provide sharper separations in shorter development times, but to realize this capability it is usually necessary to develop the plate with a constant mobile phase flow rate, made possible by commercially available instrumentation for forced flow operation.

One of the major advantages of TLC relative to liquid-column chromatography is the capability for simultaneous separation of a large number of mixtures. Because the plates are discarded after use, sample pretreatment is also less critical. Thus, TLC is very popular as a separation method for a variety of mixtures, including pharmaceuticals, amino acids, carbohydrates, vitamins, and inorganic materials, and has been utilized heavily in biological laboratories. It is widely used for routine product purity checks (e.g., in the pharmaceutical industry) and for

Thiol Functional Group

Sulfur occurs in organic molecules in many different forms, ranging from thiols (Figure 1, item 1), the sulfur equivalent of an alcohol; through disulfides (item 2) and thioethers (3); and on to more oxidized forms such as sulfoxides (4), sulfones (5), sulfonic acids (6), and the related sulfonamides (7). These different functionalities give organosulfur compounds many different chemical and physical properties, which lead to extensive use both in the industrial world and in biological systems.

A particularly important characteristic of thiols, disulfides, and thioethers is their smell. The odors are usually very strong and unpleasant. The simplest thiol, hydrogen sulfide (H_2S), gives rotten eggs their characteristic odor, and 3-methylbutane-1-thiol (Figure 2, item 8) is used by skunks to repel predators. More pleasantly, the aroma of coffee can be traced to the molecule 2-furyl methanethiol (item 9). Diallyl

$$R-S-H \qquad R-S-S-R \qquad R-S-R$$
$$(1) \qquad\qquad (2) \qquad\qquad (3)$$

$$\overset{O}{\underset{}{\overset{\|}{R-S-R}}} \quad \overset{O}{\underset{\overset{\|}{O}}{\overset{\|}{R-S-R}}} \quad \overset{O}{\underset{\overset{\|}{O}}{\overset{\|}{R-S-OH}}} \quad \overset{O}{\underset{\overset{\|}{O}}{\overset{\|}{R-S-NR'_2}}}$$
$$(4) \qquad (5) \qquad\quad (6) \qquad\quad (7)$$

Fig. 1. Basic forms of (1) thiols; (2) disulfides; (3) thioethers; (4) sulfoxides; (5) sulfones; (6) sulfonic acids; and (7) sulfonamides.

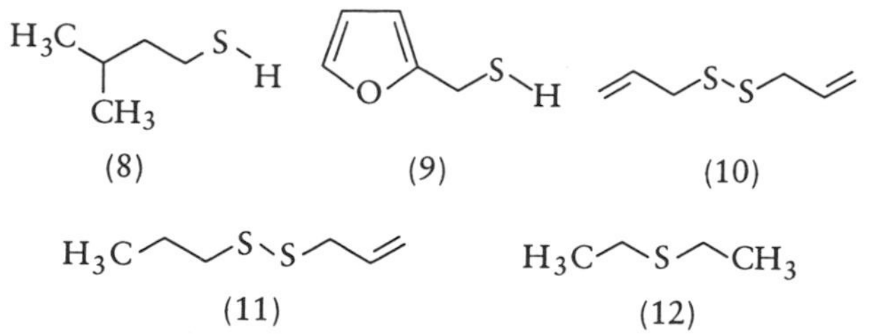

Fig. 2.   Structures of (8) 3-methylbutane-1-thiol;
(9) 2-furyl methanethiol; (10) diallyl
disulfide; (11) allyl propyl disulfide; and
(12) diethyl sulfide.

disulfide (10) gives garlic its characteristic odor, whereas the closely related compound allyl propyl disulfide (11) is responsible for the smell of onions. Diethyl sulfide (12) is commonly used as a trace component of gas which allows leak detection because it is detectable by the human nose at very low concentrations.

Sulfur is a crucial element for life. The human body typically contains around 150 g of sulfur, and most of this is present in the amino acids found predominantly in the proteins of hair, skin, and nails. Cysteine (Figure 3, item 13) is a naturally occurring amino acid that contains a thiol group. Two cysteine molecules can be oxidized to form cystine (item 14), where the thiol groups have been joined by a disulfide bond. Such bonds play an important role in controlling the structure and mechanical strength of proteins. For example, insulin consists of two chains of amino acids, one with 21 residues (the A chain), the other containing 30 residues (the B chain)—see Figure 4. The two chains are joined by two disulfide bonds between four specific cysteine residues of the six present in insulin. The remaining two residues are in the same chain and form a disulfide bond that is important for controlling the overall shape of the molecule.

Hair and wool are made up of a protein called α-keratin, which utilizes disulfide bonds between cystine residues that are part of a long peptide chain to provide extra strength. Claws, nails, and hooves are

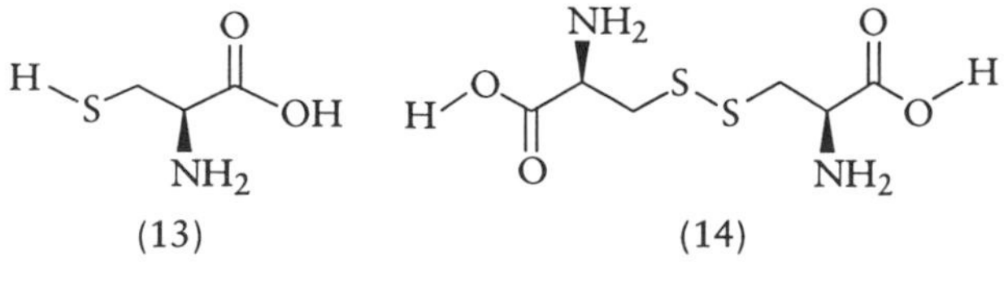

Fig. 3.   Structures forms of (13) cysteine and
(14) cystine.

also made from α-keratin but contain many more disulfide bridges, leading to more rigid structures. Rubber can be strengthened by heating with sulfur in a process known as vulcanization. The extra strength results from the formation of disulfide bonds between neighboring molecules, similar to those naturally occurring in proteins.

Disulfide bonds can be broken and reformed relatively easily, and this is the basis of the common hairdressing technique of permanent waving. First, the disulfide bonds between cystine residues are broken using sodium mercaptoacetate. This makes the hair very flexible, and it can be shaped as desired. The hair is then treated with an oxidizing agent, usually hydrogen peroxide, which reforms the disulfide bonds between nearby cysteine residues, thus "freezing" the hair in its new form.

There are many other sulfur compounds that occur naturally. Thioethers (also known as sulfides) usually have two alkyl substituents directly bonded to the sulfur atom—see Figure 5. The simplest is dimethyl sulfide (Figure 5, 15), which is a volatile foul-smelling liquid at room temperature. The amino acid methionine (16) is a naturally occurring thioether. One of its most important roles in biology is to form S-adenosyl methionine (SAM, 18) by reacting with adenosine triphosphate (ATP, 17). SAM is a biochemical methylating agent (the body's equivalent of an alkyl halide) crucial to many biological processes (see Figure 5), typically reacting with alcohols and amines to form the methylated species (which often have very different biological activities from their nonmethylated precursors), along with S-adenosyl homocysteine (19).

Penicillins (see Figure 6, item 20) are also very important naturally occurring sulfur-containing molecules and are part of a group of compounds referred to as the β-lactam antibiotics. They were discovered in 1928 by Sir Alexander Fleming, and following work by Sir Howard Florey and Ernst Chain, they were introduced as antibiotic medicines in 1943. The revolutionary ability of these new drugs to treat bacterial infections led to the award of the Nobel Prize to the discoverers. β-Lactams work by inactivating an enzyme that bacteria use to form cells walls. When this occurs, the high internal pressure of the cell can no longer be sustained because of the weakened walls, and the bacteria burst and are killed.

Soon after penicillins were introduced, strains of bacteria which were resistant to treatment began to appear. Newer, more effective penicillins have since been prepared by modifying the basic structure, and this, along with the discovery of a new class of β-lactam antibiotics, the cephalosporins (Figure 6,

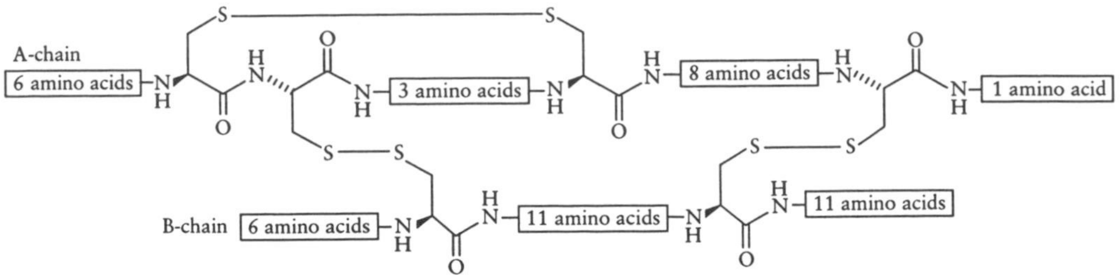

Fig. 4. Two chains (A and B) of amino acids contained in insulin, linked by two disulfide bonds.

item 21), have helped to overcome problems of resistance. However, it is only a matter of time before strains of bacteria resistant to these newer drugs appear, and this remains one of the major threats to human health in the future.

Not all thioethers are lifesaving medicines; they have also been put to more sinister uses. Mustard gas, or *bis*(2-chloroethyl)thioether (Figure 7, 22), was first used in 1917 during World War I. Unusual for a volatile thioether, it does not have a strong odor and therefore is not immediately detectable by smell, which has obvious advantages when it is used as a weapon. It induces formation of blisters when it comes into contact with skin or is inhaled, and it can lead to a relatively rapid death when encountered in high concentrations, or more slowly, from subsequent infections of blisters and general poisoning. Significantly, the autopsies of soldiers killed in World

War I indicated that (22) has a profound effect on rapidly dividing cells and suggested that related compounds may be effective as antitumor agents. Unfortunately, (22) turned out to be too toxic for clinical use, but it led to the discovery of less toxic related antitumor compounds (e.g., 23). This work marks the beginning of modern cancer chemotherapy.

One of the most important reactions of sulfides is their oxidation. This initially leads to sulfoxides (4), where the sulfur atom now has the formal +4 oxidation state. The simplest sulfoxide is dimethyl sulfoxide (DMSO, 24), which is widely used as a polar solvent. The very much higher boiling point of DMSO (b.p. 189°C) relative to dimethyl sulfide (b.p. 38°C) is a reflection of its much greater polarity, resulting in stronger electrostatic interactions between the molecules. Interestingly, many sulfoxides are also chiral, as they can have four different groups (including the

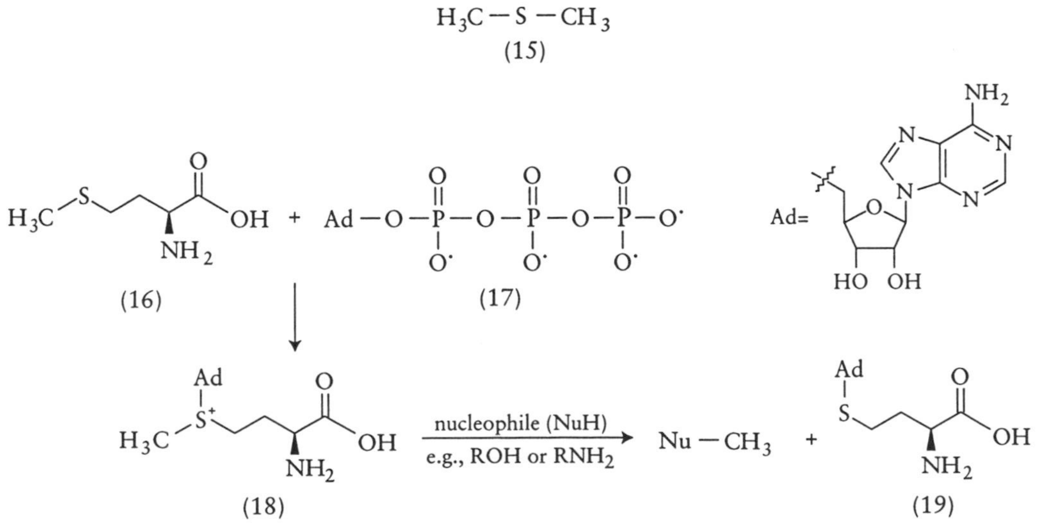

Fig. 5. Representative thioethers: (15) dimethyl sulfide and (16) the amino acid methionine. Nucleophiles are methylated by S-adenosyl methionine (18), also forming S-adenosyl homocysteine (19). This reaction plays an important role in many biological processes.

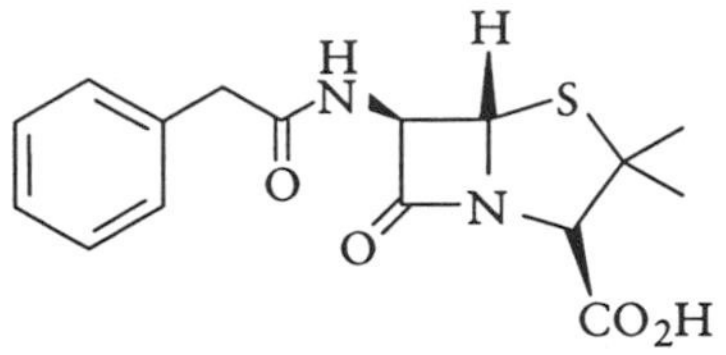

(20)

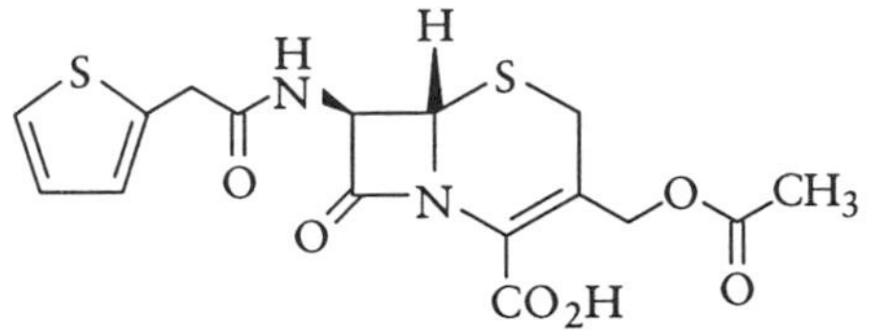

(21)

Fig. 6. A representative penicillin (20) and a representative cephalosporin (21).

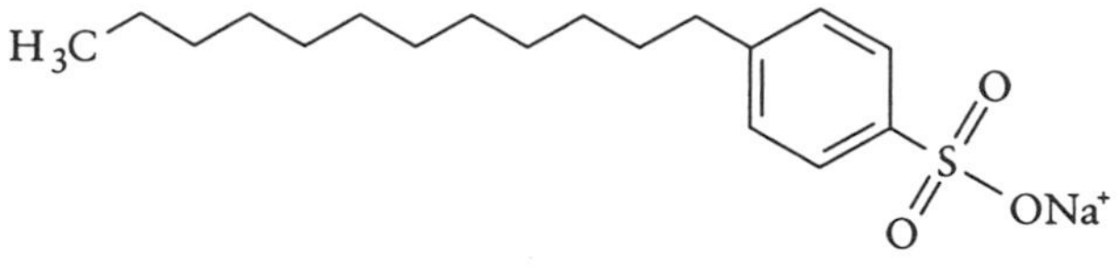

(26)

Fig. 8. The sodium salt of *para*-dodecylbenzene sulfonic acid (26).

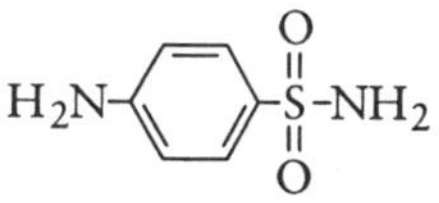

lone pair of electrons) around the tetrahedral sulfur center (e.g., 25). This means that they often occur as racemic mixtures, but they can also be prepared in an optically pure form. This can be useful for the synthesis of single enantiomers of other chiral molecules such as drugs, where this is particularly important, as the two enantiomers often have very different biological activities.

If sulfoxides are further oxidized, then sulfones (5) are formed. These have the sulfur in its highest oxidation state (+6) and are no longer chiral. Sulfones, however, are still used extensively as reagents in synthetic organic chemistry.

Sulfonic acids (6) again have the sulfur in its +6 oxidation state. They are used extensively as chemi-

cal reagents and consumer products. For example, the sodium salt (Figure 8, 26) of *para*-dodecylbenzene sulfonic acid is used as a detergent and has the advantages that, unlike alternative carboxylic acid–derived detergents, it tends not to form a scum with hard water and is also biodegradable.

Just as carboxylic acids can form amides, sulfonic acids can form sulfonamides (7). One of their most important uses has been as antibacterial agents, which were first used in the 1940s and are credited with saving thousands of lives during World War II by preventing infection of the wounded. The parent compound is referred to as sulfanilamide (Figure 9, 27, *para*-aminobenzenesulfonamide). Its activity was discovered by accident when prontosil (28), a sulfonamide dye used to stain bacteria, was found to be an antibacterial agent. It was later discovered that protonsil was not actually the effective agent but was degraded in the cell to form sulfanilamide. Sulfanilamide works by preventing the biosynthesis of folic acid, a compound essential for bacterial growth. It is able to do this because it resembles *para*-aminobenzoic acid (PABA, 29), one of the building blocks used by cells to synthesize folic acid, and it binds to the

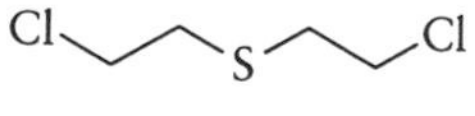

(22)

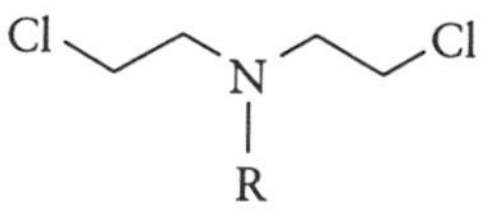

(23)

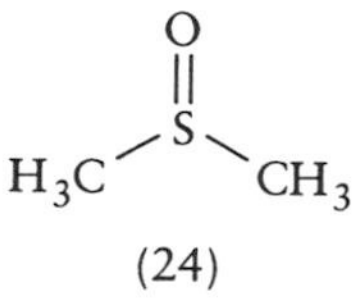

(24)

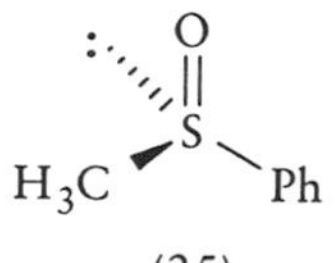

(25)

Fig. 7. More representative thioethers: (22) the highly toxic mustard gas *bis*(2-chloroethyl) thioether; (23) a less toxic amine-derived antitumor compound; (24) dimethyl sulfoxide; and (25) a chiral sulfoxide.

(27)

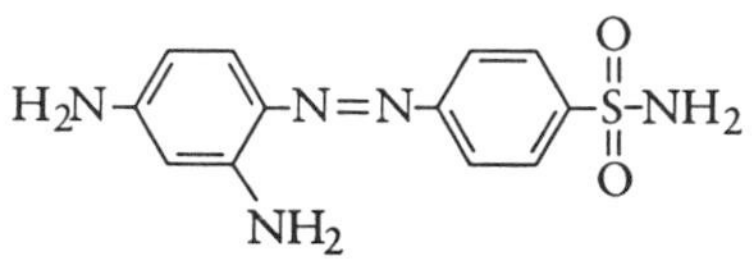

(28)

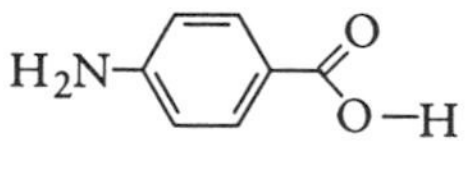

(29)

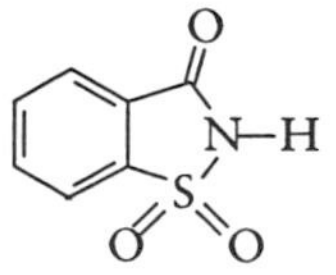

(30)

Fig. 9. More representative sulfonamides; (27) *para*-aminobenzenesulfonamide; (28) prontosil; (29) the related *para*-aminobenzoic acid; and (30) saccharin.

folic acid–synthesizing enzyme in place of PABA. Many different sulfonamide drugs have now been synthesized; however, more recently, more effective families of antibacterial agents have been discovered, and nowadays sulfonamides are not usually used alone to treat infections in humans.

One other very common sulfonamide is saccharin (30). It supposedly was discovered in 1879 by accident by a dirty, careless chemist who failed to wash his hands after a session in the laboratory and noticed a sweet taste to his food. It is approximately 300 times sweeter than sugar; 0.03 g of saccharin is equivalent to a heaped teaspoon (10 g) of normal sugar (sucrose). Saccharin remains unchanged as it passes through the body and so is useful as a sugar substitute for those who need to restrict their sugar intake for medical reasons (e.g., diabetics) or to control their weight.

BIBLIOGRAPHY

ATKINS, P. W. *Molecules;* New York: Scientific American Library Series, 1987.

BLOCK, E. *Reactions of Organosulfur Compounds;* Organic Chemistry, a Series of Monographs; New York: Academic Press, 1978; Vol. 37.

BROWN, W. H. *Organic Chemistry;* Fort Worth, TX: Saunders College Publishing, Harcourt Brace College Publishers, 1995.

COULSON, C. J. *Molecular Mechanisms of Drug Action;* London: Taylor and Francis, Ltd., 1988.

HART, H. *Organic Chemistry: A Short Course,* 8th ed., Boston: Houghton Mifflin Company, 1991.

SILVERMAN, R. B. *The Organic Chemistry of Drug Design and Drug Action;* San Diego: Academic Press, 1992.

SZMANT, H. H. *Organic Building Blocks of the Chemical Industry;* New York: Wiley, 1989.

WEATHERALL, M. *In Search of a Cure;* Oxford: Oxford University Press, 1990.

CHRISTOPHER M. RAYNER

Thomson, Joseph John (1856–1940)

Joseph John (J. J.) Thomson was born in Cheetham Hill near Manchester, England, in 1856. Thomson had originally intended to become an engineer; however, financial constraints forced him to enter Owen's College (now the University of Manchester) in Manchester at the age of fourteen. There he studied chemistry, mathematics, and physics. In 1876 he won a scholarship to Trinity College, Cambridge, where he majored in mathematics. He graduated with a B.A. degree in 1880 and was awarded a fellowship a year later, allowing him to remain at Trinity. In 1884 Lord RAYLEIGH recommended that Thomson replace him as the Cavendish Professor of Experimental Physics and director of the Cavendish Laboratory. Despite an uproar from the scientific community, the twenty-eight-year-old Thomson was chosen by the selection committee as an appropriate replacement, and he would hold the post until 1919. Thomson wrote a winning essay in 1882 for the Adams Prize Competition on the theory of vortex rings, which preceded his experiments on gaseous discharges, eventually leading to his 1897 discovery of the ELECTRON.

Thomson was awarded the Nobel Prize for physics in 1906 for his theoretical and experimental investigations on the behavior of electricity in gases (see GASES, BEHAVIOR OF). Two years later he was knighted for his accomplishments in science. At this time, he was studying "canal" rays using instrumentation that eventually, with refinement by one of Thomson's students, Francis ASTON, became the mass spectrometer (see MASS SPECTROMETRY). After 1912 Thomson focused mainly on teaching and administration. He was elected president of the Royal Society in 1915 and became the first scientist to be appointed master of Trinity College in 1918. He was responsible for the substantial growth the Cavendish Laboratories experienced during his thirty-five years as director. His talents as an instructor can be demonstrated by the fact that seven of his students were each awarded the Nobel Prize. Thomson remained active in research, studying optics, electricity and magnetism, photoelectricity, thermionics, and radioactivity. During World War I, Thomson was adviser to various government departments and a member of the Bureau of Inventions and Research.

J. J. Thomson is most remembered for his discovery of the first subatomic particle—the electron. Thomson's early theroetical work was concerned with electromagnetic theories, which led to the study of gaseous conductors of electricity, particularly the nature of cathode rays. Thomson noticed that cathode rays could be deflected by both electric and magnetic fields, meaning that the rays consisted of charged particles, which he called corpuscles. Investigations of the charge-to-mass ratio showed that corpuscles had a definite charge and mass. Further research showed that these particles were more than one thousand times smaller in mass than the lightest known atomic particle, the hydrogen ion. Thus, Thomson concluded that the atom was no longer the smallest entity, but instead the corpuscle, later called the electron, was the smallest. J. J. Thomson was a genius who enjoyed reading, following sports, and conversing in a variety of areas, especially politics. He was a charismatic person who inspired and nurtured his students, and this made him popular with both students and faculty, who routinely honored him at annual banquets. Thomson died in Cambridge in 1940 and is buried in Westminster Abbey.

BIBLIOGRAPHY

IHDE, A. J. *The Development of Modern Chemistry;* New York: Harper and Row, 1964; pp. 480–483.

JAFFE, B. *Crucibles: The Story of Chemistry: From Ancient Alchemy to Nuclear Fission;* New York: Dover Publications, 1976; pp. 197–217.

JULIE BREDAEL

Thomson, William

See KELVIN, LORD.

Tin

50
Sn
118.69

Melting Point: 505 K; **Boiling Point:** 2,543 K; **Density:** 5.75 g/cm^3 (gray); 7.28 (white); **Most Common Ions:** +2, +4

Tin is the forty-eighth most abundant element in the Earth's crust, with an average abundance of approximately 2.1 ppm. The element's symbol is derived from the Latin word *stannum*. The element does not occur free in nature, and it is most frequently found as the oxide in the mineral known as cassiterite, SnO_2. The ease of reduction of this oxide with carbon (coal) accounts for the fact that tin was known and used early in human history. Production of tin dates to about 3500 B.C., about the beginning of that period known as the Bronze Age—an age characterized by the use of bronze tools. While several alloys of metals with copper are known as bronzes, the tin-copper bronze has been the most widely used of these. The alloy solder (Sn-Pb) was used by the Romans in 79 A.D.

Tin has three polymorphs: α-tin (gray tin), β-tin (white tine), and γ-tin. The gray form is stable below 14°C. White tin is the stable form at room temperature, and γ-tin is the most stable above 161°C.

The metallurgy of tin is almost totally based upon recovery of the metal from cassiterite. Major production of tin occurs in Bolivia, China, CIS (former USSR), Indonesia, Malaysia, and Thailand. Reverberatory furnaces reduce the oxide with carbon at temperatures about 1,300°C. The cost of tin is relatively high compared with the common metals. As a consequence, tin is rarely used as the element, but is used instead as an alloy or a thin layer of tin applied to a lower-cost metal. Tinplate, a low-carbon sheet steel coated with a thin layer of pure tin, was developed to address the need for packaging materials for food products—the famous tin cans. Tinplate containers have been used for aerosol containers—for example, paints, hair spray, and shaving cans.

Other than tinplate, tin alloys form a large class of useful substances. Tin-lead alloys (solders) have been used for joining metals, but the toxicity of lead has reduced the widespread use of these solders. Copper-tin alloys (bronzes) have been used for manufacturing pumps, bushings, gears, and propellers for ships. Bell metal, owing to its tonal quality, has been used to make bells. Babbitt metal is a tin-lead-copper alloy and has excellent wear resistance for bushings. Pewter, a tin-antimony-copper alloy, has been used to make items ranging from coffee and tea servers to pipes for pipe organs. While it is declining in use owing to modern printing techniques, type metal, a lead-antimony-tin alloy, has been valuable in the printing industry. Titanium-aluminum-tin alloys are used for aircraft and spacecraft construction, and silver-tin-mercury alloys have been used as dental amalgams. Tin containers have also been used for the tubes in which pharmaceutical ointments and artists' paints are sold. The tin-niobium alloy, Nb_3Sn, is a superconducting material at low temperatures and is finding use in high-field superconducting magnets.

Deposition techniques have been used to deposit a layer of SnO_2 on glasses to strengthen the glass. Most recently, thick layers of SnO_2 form electrically conducting layers to assist in defogging windshields in automobiles. Organotin compounds have been used as stabilizers for polyvinyl chloride (PVC) plastics and as additives to marine antifouling paints.

BIBLIOGRAPHY

GREENWOOD, N. N.; EARNSHAW, A. *Chemistry of the Elements;* New York: Pergamon, 1984; pp. 427–465.

LIDE, D. R., ed. *Handbook of Chemistry & Physics,* 71st ed.; Cleveland: Chemical Rubber Co., 1990–1991.

MAYKUTH, D. J. "Tin and Tin Alloys." In *Encyclopedia of Chemical Technology,* 3rd ed.; R. E. Kirk, D. F. Othmer, M. Grayson, eds.; New York: Wiley-Interscience, 1981; Vol. 23, pp. 18–35.

ALTON J. BANKS

Titanium

22
Ti
47.88

Melting Point: 1933 K; **Boiling Point:** 3,560 K; **Density:** 4.54 g/cm³; **Most Common Ions:** +2, +3, +4

Titanium is the ninth most abundant element in the Earth's crust. The metal was discovered by William Gregor in 1791 in the magnetic ore ilmenite, $FeTiO_3$. Gregor, a clergyman, proposed the name menaccanite in honor of his local congregation. In 1795 the German chemist M. H. Klaproth discovered the ore rutile, TiO_2, in Hungary. Klaproth determined that rutile was the oxide of a new element and named it after the mythological Greek Titans. Despite Gregor's initial claim, Klaproth's name for the element was accepted.

Rutile and ilmenite are the most important sources of titanium. Deposits of ilmenite are mined in Australia, Canada, Russia, Malaysia, and Scandinavia, with the principal deposits of rutile occurring in Australia. An important deposit of ilmenite was located in the Ilmen Mountains in Russia, after which the ore was named.

Recovery of the metal from the oxides was initially difficult, but the metal was obtained in 1910 by Hunter. The metal, unlike many, cannot be recovered by reduction with carbon, owing to the formation of carbides. In 1932 Wilhelm Kroll of Luxembourg produced the metal by reducing $TiCl_4$ initially with calcium. In 1940 he succeeded in using magnesium and sodium for the reduction. The Kroll method produces $TiCl_4$ from either ilmenite or rutile with chlorine and carbon. The tetrachloride is subsequently distilled and reduced with molten magnesium under an inert atmosphere.

Titanium was first employed in the United States as an additive to iron and steel. It is resistant to many environments, owing to the formation of a protective oxide layer. Titanium is used to handle process liquors in the chemical industry. It also finds service as a building material for heat exchangers and coolers. Titanium is used chemically in the production of acetaldehyde from ethylene. The tetrachloride when mixed with triethylaluminum is used as a Ziegler-Natta catalyst in the production of polyethylene from ethylene. Propeller shafts and other parts of boats exposed to seawater take advantage of the element's chemical resistance. The aircraft and spacecraft industry use the metal's low density, temperature resistance, and good mechanical strength to advantage in the production of airframes and spacecraft. The acceptance of titanium by body tissues has made the metal useful in hip joint replacements, heart valves, screws, and external braces. Annual world production of titanium is currently about 10^5 tons.

Titanium dioxide is perhaps the most important compound of titanium. A major portion of it is used in paints, with lesser fractions being used as surface coatings on paper and fillers in rubbers and plastics. Titanium carbide is one of the hardest carbides known and is used in cutting tools. Titanium dioxide provides the asterism of star sapphires.

BIBLIOGRAPHY

GREENWOOD, N. N.; EARNSHAW, A. *Chemistry of the Elements;* New York: Pergamon, 1984; pp. 1111–1118.

KNITTEL D. "Titanium and Titanium Alloys." In *Encyclopedia of Chemical Technology,* 3rd ed.; R. E. Kirk, D. F. Othmer, M. Grayson, eds.; New York: Wiley-Interscience, 1981; Vol. 23, pp. 98–130.

LIDE, D. R., ed. *Handbook of Chemistry & Physics,* 71st ed.; Cleveland: Chemical Rubber Co., 1990–1991.

ALTON J. BANKS

Titration

Titration is the process by which a substance is determined by measuring the quantity of a reagent required to react quantitatively or stoichiometrically with it. In titrimetry or volumetric analysis, a solution of precisely known concentration is added to the sample solution, its volume being precisely measured by means of a buret, until the reaction is just complete. The titration point at which there is chemical equivalency between the titrant (reagent) and titrand (analyte in the sample solution) is called the EQUIVALENCE POINT, whereas the point at which the addition of titrant is actually terminated (and the reaction is observed to be complete) is called the ENDPOINT. The endpoint may be detected in a variety of ways, depending on the reaction. For example, the pH may be measured and plotted as a function of titrant volume in an acid-base titration, to obtain a titration curve. More often, a visual indicator is used that changes color at or near the equivalence point (see below).

Since the concentration of the titrant is known (e.g., moles/liter) and the volume to reach the endpoint is measured, then the number of moles reacted is known (moles / liter × liter = moles). From the stoichiometry of the reaction (see CLASSICAL STOICHIOMETRIC LAWS), the number of moles of analyte titrated is calculated. Thus, in the titration of sodium carbonate with hydrochloric acid to produce sodium chloride and carbonic acid:

$$NaCO_3 + 2\,HCl \rightarrow 2\,NaCl + H_2CO_3$$

the moles of sodium carbonate are one-half the moles of hydrochloric acid used in the titration. The grams of analyte are then found by multiplying the number of moles by its gram formula weight; e.g.,

$$Na_2CO_3 = moles\ Na_2CO_3 \times 105.99\ grams/mole$$

Standard Solutions

The solution used to titrate the sample is known as a standard solution. Its concentration is known, usually to four figures, e.g., 0.1067 moles/liter (volumetric analyses are capable of precision of a few tenths percent). A *primary standard solution* is prepared by dissolving an accurately weighed quantity (to 0.1 milligram, four or more figures) of a highly pure material called a *primary standard* and diluting to an accurately known volume in a volumetric flask. The primary standard is usually dried in an oven at 110–120° for 1 to 2 hours and cooled in a desiccator before weighing. If a primary standard titrant is not available, a *secondary standard solution* is prepared approximately, and the concentration is accurately determined by standardization (titrating a known amount of primary standard or a primary standard solution). Approximately prepared tenth molar hydrochloric acid solution, for example, is standardized by titrating a known weight of primary standard sodium carbonate.

Back Titrations

A reaction may be slow to go to completion, rendering a sharp endpoint difficult to achieve. In such cases, a back-titration procedure may be used. An excess of the standard solution is added by pipet, and the reaction is allowed to go to completion. Then the amount of unreacted solution is determined by titration with a second standard solution. The difference gives the amount reacted with the sample, from which the amount of analyte is calculated. The moles standard reacted (MSR) is calculated as follows:

moles taken − moles back-titrated

The moles analyte is equal to

MSR × factor (moles analyte/mole standard)

Titration Curves

A titration curve is a plot of a measured property of the sample solution that changes with reaction (e.g., pH), as a function of volume of titrant. The plotted property changes markedly at or near the equivalence point, and this change is taken as the endpoint. Figure 1 illustrates titration curves for the titration of a strong acid (hydrochloric acid) with a strong base (sodium hydroxide), demonstrating the dependence of the magnitude of the endpoint break on concentration. The larger the endpoint break, the more precisely the endpoint can be determined. At the beginning of the titration, the pH is determined by the concentration of hydrochloric acid (e.g., 0.01 moles/liter hydrochloric acid has pH 2.0). The pH changes gradually throughout the titration as part of the acid is neutralized, with the pH determined by the remaining acid. At the equivalence point, a solution sodium chloride exists, and so the pH is 7.0. Beyond the equivalence point, the pH is determined by the concentration of excess (unreacted) sodium hydroxide.

A similar set of titration curves for the titration of a weak acid (acetic acid, represented here as HOAc) is shown in Figure 2. Here, the acid at the beginning of the titration is only a few percent ionized, and so the initial pH is higher (pH = −log [H$^+$]). As the titration proceeds, part of the acid is neutralized to its conjugate base, the salt sodium acetate. This mixture constitutes a BUFFER. At the midpoint of the titration, the concentrations of acid and salt are the same

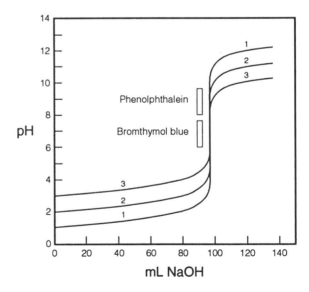

Fig. 1. Titration curves for the titration of hydrochloric acid (HCl) with sodium hydroxide (NaOH), illustrating the dependence of the magnitude of the endpoint break on concentration. Curve 1: 100 mL 0.1 *M* HCl versus 0.1 *M* NaOH. Curve 2: 100 mL 0.01 *M* HCl versus 0.01 *M* NaOH. Curve 3: 100 mL 0.001 *M* HCl versus 0.001 *M* NaOH. (From Christian, G. D. *Analytical Chemistry*, 4th ed.; New York: Wiley, 1986; with permission.)

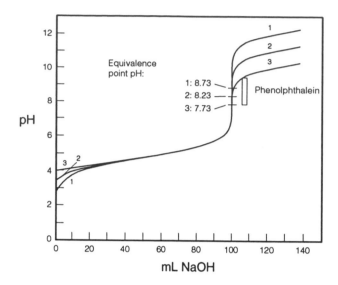

Fig. 2. Titration curves of a weak acid (acetic acid, represented here as HOAc) at different concentrations. Curve 1: 100 mL 0.1 M HOAc versus 0.1 M NaOH. Curve 2: 0.01 M HOAc versus 0.01 M NaOH. Curve 3: 0.001 M HOAc versus 0.001 M NaOH. (From Christian, G. D. *Analytical Chemistry*, 4th ed.; New York: Wiley, 1986; with permission.)

and the pH is equal to pK_a of acetic acid (see the entry on pK_a); the pH of a buffer is given by the Henderson-Hasselbalch equation:

$$pH = pK_a + \log [salt] / [acid]$$

At the equivalence point, a solution of sodium acetate exists. (Acetate is represented here as Ac.) The acetate ion (OAc^-) is a weak Brønsted base, owing to its partial hydrolysis (since acetic acid is a weak acid, it prefers to extract a proton from water):

$$OAc^- + H_2O \rightleftharpoons HOAc + OH^-$$

Hence, the solution is slightly alkaline at the equivalence point. The more concentrated the acetate, the more alkaline the solution. The equivalence point pH values for solutions of different concentrations are indicated in the figure. The pH beyond the equivalence point is the same as for Figure 1, since it is again determined by the concentration of the excess sodium hydroxide. The weaker an acid is, the greater is the hydrolysis of its salt, and the more alkaline is the endpoint; the pK_a is larger and the midpoint of the titration curve is more alkaline, resulting in a smaller endpoint break.

Similar behavior is found for the titration of strong or weak bases with a strong acid. Titration curves for sodium hydroxide titrated with hydrochloric acid

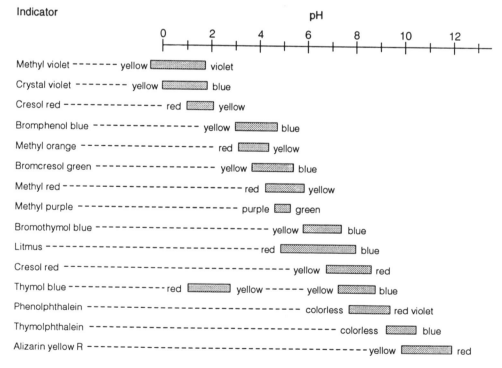

Fig. 3. pH transition ranges and colors of some common acid-base indicators. (From Christian, G. D. *Analytical Chemistry*, 4th ed.; New York: Wiley, 1986; with permission.)

would be the mirror images of those in Figure 1. The titration of a weak base (e.g., ammonium hydroxide, NH_4OH) would create an acidic endpoint due to the hydrolysis of its conjugate acid, the ammonium ion:

$$NH_4^+ + H_2O \rightleftharpoons NH_4OH + H^+$$

Indicators

An acid-base indicator is a dye that is a weak acid or a weak base. (See INDICATORS, ACID-BASE.) The acid form is one color, and the base form a different color. A small amount of the indicator is added to the solution being titrated. The indicator should change color at or near the equivalence point. If we represent the indicator by HIn, its ionization is given by

$$HIn \rightleftharpoons H^+ + In^-$$

The pH of the solution determines the ratio of HIn and In^- as given by the Henderson-Hasselbalch equation:

$$pH = pK_{In} + \log [In^-] / [HIn]$$

pK_a is the acidity constant of the indicator. The eye discerns only one of the two colors when the ratio is about $1:10$ or $10:1$. Hence, the *transition range* of the indicator occurs over about 2 pH units. The midrange of the transition will occur at $pH = pK_{In}$ (when $[In^-] = [HIn]$). So the indicator pK_{In} should be near the equivalence point pH.

Figure 3 gives the transition ranges for several indicators. Phenolphthalein can be used for the titration of acids as strong as or stronger than acetic acid if the solution is not too dilute (Figures 1 and 2). Bromthymol blue is better for dilute solutions.

BIBLIOGRAPHY

CHRISTIAN, G. D. *Analytical Chemistry*, 5th ed.; New York: Wiley, 1994; Chapter 3.

GARY D. CHRISTIAN

Torr

The *torr* is the name of the unit of pressure equal to exactly 1 mmHg. It is named after Evangelista Torricelli (1608–1647), who produced the first mercury barometer. One atmosphere of pressure supports a mercury column 760 mm or 760 torr high. In comparison with other pressure units, 1.0000 torr equals 1.452 mbar and 0.0193 psi (lb/in.^{2}) (See also BAR; PRESSURE, GASEOUS.)

ROBERT K. WISMER

Touch, Sense of

See NEUROTRANSMITTERS.

Tracers, Radioisotope

Tracers are very useful in studying the mechanisms of chemical and biochemical processes. In the tracer, one atom of a molecule is replaced by one of its isotopes. For example, hydrogen (H) may be replaced by deuterium (D) or tritium (T); carbon-12 may be replaced by carbon-13; and so on. When the tracer is detected, the fate of the "normal" molecule without the tracer can be inferred. Either stable or radioactive isotopes may be used as tracers, although the availability of counting equipment, suitable isotopes, or instrumentation to quantitatively detect the isotope limits their usefulness. Radioactive isotopes are often employed because these are detectable at extremely low levels. Nonradioactive labels are detected by a variety of methods. For example, compounds containing ^{18}O may be determined by MASS SPECTROMETRY, ^2D by infrared or NUCLEAR MAGNETIC RESONANCE SPECTROSCOPY (NMR spectroscopy), and ^{13}C by ^{13}C NMR.

When one isotope is substituted for another, the only change is in the mass of this particular nucleus. The electronic structures of the unlabeled and labeled molecules are essentially the same. Thus, the reactivities are similar, but the label provides a method of tracing where certain atoms end up in the final product. This provides information about the relationship between the structure of the product and the structure of the reactant. It can even increase the number of observable products in a reaction by exposing rearrangements that might normally be indistinguishable. In this way we can get information about the symmetry (or lack of symmetry) of intermediates or transition states of a reaction. For example, monitoring the reaction of radioactive ^{128}I incorporated into $(+)2$-iodooctane in the formation of $(-)2$-iodooctane (see Figure 1) established the inversion of configuration during a biomolecular (S_n2) displacement reaction.

Isotopic labeling studies are analogous to substituent labeling studies. However, varying a substituent (e.g., substituting an ethyl group for a methyl group) may change the nature of the reaction to such an extent that it is no longer the same as the reaction of the original molecule. Such changes can be the result of differences in steric and electronic effects between the two substituents. These types of changes are negligible in isotopic labeling studies, but there can be measurable differences in the rates of reaction.

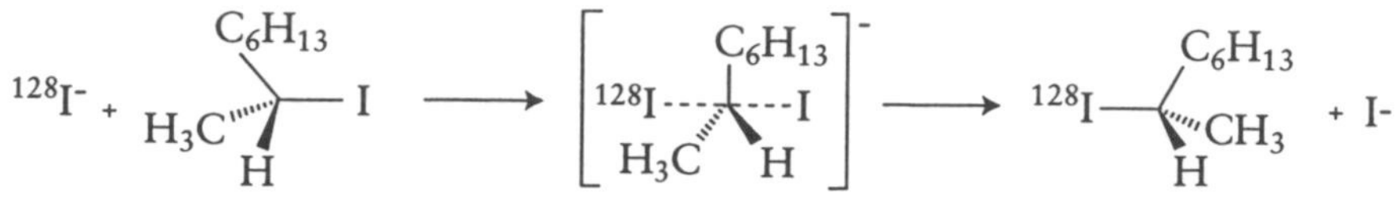

Fig. 1.

These small changes in rates, known as KINETIC ISOTOPE EFFECTS, also give information about the mechanism of the reaction.

Common isotopes used in organic chemistry include: ^{14}C, ^{18}O, ^{15}N, ^{36}Cl, and ^{131}I. It is not usually necessary to use completely labeled compounds. The ^{13}C isotope has been particularly useful since the development of carbon nuclear magnetic resonance (NMR) spectroscopy. Proper instrumentation allows the identification of each carbon atom in a molecule by a distinguishably different signal. Comparing a ^{13}C spectrum of a sample containing ^{13}C in natural abundance (about 1 percent) with the spectrum from experiments involving enriched ^{13}C samples allows the identification of the exact location of a ^{13}C label in a product.

One important application of tracers is to study reactions in which a particular atom of the product could have originated in any one of several reactants. For example, in the hydrolysis of esters (see Figure 2), it is not clear from traditional experiments exactly which carbon-oxygen bond is broken. The use of ^{18}O-labeled water allows one to trace the path of the oxygen atom by analyzing the mass of each product with a mass spectrometer.

Isotopic labeling studies can also give information about the interaction of reactant molecules with the solvent (exchange reactions) as well as internal rearrangement reactions. For the same reasons, isotopic labeling studies are just as useful in studying mechanisms of inorganic reactions.

There have been many applications of tracers in biochemistry. Metabolic pathways can be studied using isotopically labeled molecules. This is particularly important for metabolic studies of compounds that are normally found in the body. If the introduced molecules are not labeled, they will be indistinguishable from the same molecules already in the body—thus impossible to trace. One major consideration in this type of study is the form in which to administer the isotope. The isotope may be in a simple molecule such as $^{14}CO_2$, ^2H_2O, ^3H_2O, $^{24}NaCl$, $^{32}PO_4$, or $K^{131}I$, or may be incorporated into larger molecules through synthesis. Radioactive isotopes are very useful in determining anatomical distributions, since it is easy to detect very low levels. Radioisotopes are also useful for tagging specific or active sites of large molecules such as enzymes. Sometimes it is useful to include more than one label in the same substance. However, this does not necessarily mean that each molecule of the compound needs to contain both labels. For example, mixing a ^{14}C-labeled sample of glycine with a ^{15}N-labeled sample of glycine yields a sample that is effectively double-labeled.

Radioisotopic labeling played an important role in sequencing DNA. This involved labeling a DNA strand at one end with a phosphate group containing phosphorous-32, allowing detection by autoradiography. Various improvements have been introduced, including the use of ^{32}S as a label to increase the sharpness of the radiographic image.

The advantage of tracer methods is that the experiments are generally very simple, yet the results are definitive. The difficulties of synthesis and the number of isotopes that are available for a given element limit the use of tracers.

BIBLIOGRAPHY

KATAKIS, D.; GORDON, G. Mechanisms of Inorganic Reactions; New York: John Wiley, 1987; p. 71.

OLD, R. W.; PRIMROSE. S. B. Principles of Gene Manipulation: An Introduction to Genetic Engineering, 5th ed.; Boston: Blackwell Scientific, 1994; Chapter 9.

SYKES, P. A. Guidebook to Mechanism in Organic Chemistry, 6th ed.; New York: Longman, 1986; pp. 46–49.

WHITE, A.; HANDLER, P.; SMITH, E. L.; HILL, R. L.; LEHMAN, I. R. Principles of Biochemistry, 6th ed.; New York: McGraw-Hill, 1978; pp. 319–320.

PATRICK L. HUSTON

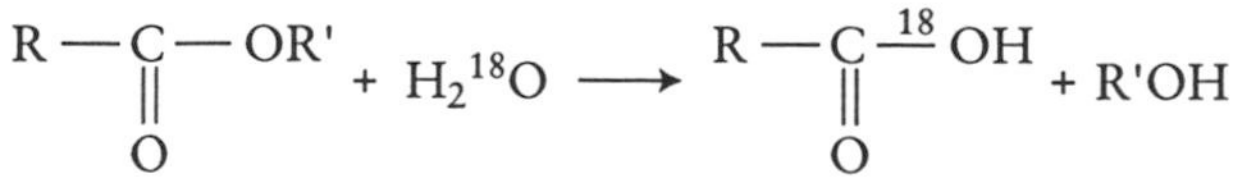

Fig. 2.

Transitional Elements

The transitional elements, also known as transition elements or transition metals, are the collection of thirty elements of the periodic table in Groups 3 through 12 arranged in three rows of ten. The first of the transition metals begins with atomic number 21, scandium (Sc), and continues through number 80, mercury (Hg), at the end of the third row. The seven most recent synthetic elements are assigned to the presently incomplete fourth row beginning with atomic number 103, lawrencium (Lr), through number 109, unnilennium (Une). The elements occur naturally in the compounds known as minerals and ores in the Earth's crust. Iron (Fe), in Group 8, is the fourth most abundant element, while ruthenium (Ru) and osmium (Os) in the same group are less abundant than gold (Au). The symbols of the elements are usually found in the first letters of the names, except for tungsten (W), named from wolframite, its ore. Also exceptions are the five elements known in the metallic form since antiquity, which take symbols from their Latin (L.) names: copper (Cu), L. *cuprum*; gold (Au), L. *aurum*; iron (Fe), L. *ferrum*; mercury (Hg), L. *hydrargyrum*; and silver (Ag), L. *argentum*. Compounds of most of the remaining elements were isolated and identified in the two centuries between 1700 and 1900, although the pure metals for some were not prepared until after 1900. The element scandium was identified in 1876 with the chemical and physical properties predicted by MENDELEEV when the period law and periodic table were developed in the late 1860s. The final element needed to complete the first three rows was produced in 1937 and named technetium (Tc), from the Greek *technetos*, "artificial," the first artificially prepared element.

The history of the concept of a collection of transition elements between groups of regular elements began with Mendeleev's periodic table, which he derived from the chemical and physical properties of the elements. The term *transition elements* was originally applied to the set of elements identified as Group VIII, which were the iron, cobalt, and nickel families of early periodic tables. The concept was continued after the development of Niels BOHR's model of the atom (see also BOHR MODEL OF THE HYDROGEN ATOM; RUTHERFORD-BOHR-SOMMERFELD MODEL OF THE ATOM) when Charles Bury in 1921 applied the concept of layered electronic arrangements in atoms to the structure of the periodic table. Bury proposed that a transition series of elements occurred during the change of an inner layer from a stable group of eight electrons to one of eighteen. Using this theory, the elements from titanium to copper form the first transition series, and other series follow in later periods. This arrangement of the elements was not commonly accepted. It was almost twenty years later that it was introduced again, when Laurence Foster proposed the periodic

chart in its current form. He applied the term *transitional elements*, however, only to the eight families of elements from scandium to nickel. The present concept of the ten families from scandium to zinc is derived from the quantum model of electronic configurations (see QUANTUM THEORY), which generally assigns electrons in these elements to five d-orbitals that have a capacity of ten electrons. In addition to their place in the table, the special qualities of the metallic forms of these elements and the unique features of their compounds are related to their use of the d-orbital space for outer electrons.

The metallic forms of individual transition elements and alloys are the basic structural materials for tools, machines, hardware, and buildings. The majority of these metals are silvery-white, hard solids with high melting points; they are obtained from the ores by chemical and electrochemical reduction reactions. Tungsten has the second highest melting point of all elements, while the Group 12 elements have exceptionally low melting points, decreasing down the family to mercury, which is a liquid at room temperature. The coinage metals copper, silver, and gold exhibit the highest electrical conductivity and are very malleable and ductile. Metals with unique properties are selected for specific application: examples are tantalum (Ta), a nonirritating, noncorrosive metal used in surgical appliances; and titanium (Ti), which has the best strength-to-weight ratio and is therefore utilized in aircraft and mountain bikes. High-strength steels contain molybdenum (Mo), while iron is alloyed with cobalt (Co) and nickel (Ni) to make permanent magnets.

The compounds of transition metals provide the most distinctive features of this collection of elements. The observed variety of compounds for these elements is accounted for by assigning a wide range of oxidation states to each element, as with the eleven oxidation states assigned to manganese (Mn), from -3 in $[Mn(CO)_4]^{-3}$ to $+7$ in MnO_4. Many of the compounds exhibit a variety of colors—from the blood-red iron (III) thiocyanate, $Fe(SCN)_3$, to the deep violet of anhydrous chromium (III) chloride, $CrCl_3$. A color change indicates a compound's chemical reaction, as when dry cobalt (II) chloride reacts with water vapor in the air and changes from blue to pink. Another distinctive feature of some transition metal compounds is PARAMAGNETISM, a weak attraction to magnetic fields, including coordination compounds like $[Ni(H_2O)_6]Cl_2$, which is unusual because it contains an even number of electrons.

Compounds of transition metals are classified as ionic, coordinate covalent, or covalent by the type of bonding of atoms to the metal. Some compounds exhibit primarily ionic bonding by electrostatic attraction between the metal cation and simple or complex anions, as in palladium bromide, $PdBr_2$, or zirconium silicate, $ZrSiO_4$, the zirconium diamond. In coordination compounds, the metal, using unoccupied orbitals, forms from two to eight coordinate covalent bonds by accepting electron-pairs from the surrounding ligand atoms. In tetraamminecopper (II), $[Cu(NH_3)_4]^{2+}$, the unoccupied orbitals of the copper(II) ion accept electron pairs from the nitrogen atoms of four ammonia molecules to form this coordination complex. ORGANOMETALLIC compounds usually contain covalent metal-to-carbon bonds such as in dimethylmercury, $Hg(CH_3)_2$, or as in metal cluster compounds like iridium carbonyl, $Ir_4(CO)_{12}$. Organometallic compounds are used extensively as catalysts in the chemical industry—for example, the production of ALDEHYDES uses cobalt octacarbonyl, $Co_2(CO)_8$, as the catalyst.

All of the transition elements in the first row, from titanium to zinc plus molybdenum and cadmium, also occur in biological systems as essential elements, primarily in the form of coordination compounds. They occur in enzymes, as in nitrogen-fixing bacteria that contain a complex of iron and molybdenum. Complexes of iron, copper, or vanadium are used by different organisms to carry oxygen in their circulatory systems. While at some level all transition metals are toxic, they also are widely used for chemotherapeutic agents and drugs, as observed for the anticancer agent $[Pt(NH_3)_2Cl_2]$ named cis-diamminedichloroplatinum (II), which is shortened to cisplatin.

BIBLIOGRAPHY

FOSTER, L. S. "Why Not Modernize the Textbooks Also? I. The Periodic Table." *J. Chem. Educ.* 1939, *16*, 409–412.

HUHEEY, J. E.; KEITER, E. A.; KEITER, R. L. *Inorganic Chemistry: Principles of Structure and Reactivity;* New York: HarperCollins, 1993; Chapter 19.

LIEBL, M. "KC? Discoverer: Exploring the Properties of the Chemical Elements." *J. Chem. Educ.: Software* 1991, *IVA*, 2.

SORRELL, C. A. *Minerals of the World;* New York: Golden Press, 1973.

GORDON EGGLETON

Transition State Theory

The lack of success of the COLLISION RATE THEORY called for a more satisfactory treatment of reaction rates. A very successful theory was transition-state theory, developed independently in 1935 by Henry Eyring and by Meredith Gwynne Evans and Michael Polanyi. Transition state theory focuses attention on the ACTIVATED COMPLEX and postulates that there is a special type of equilibrium between an activated complex and the reacting substances, in the sense that the activated complex is formed in a state of equilibrium. It is thus possible to calculate the concentration of an activated complex in terms of the concentrations of reactants. The rate of reaction is then expressed as the concentration of the activated complex multiplied by the rate with which the reactants are transformed into products.

The equilibrium constant between the reactants and activated complexes can be treated in two different but equivalent ways: by the methods of statistical mechanics, or on the basis of thermodynamics. Both methods require estimates of the structure and properties of the activated complex; if a potential-energy surface can be calculated, this can be done satisfactorily, but otherwise only approximate estimates can be made. Nevertheless, transition state theory has provided a working tool for those who are not so much concerned with making accurate calculations as with gaining some insight into how chemical reactions occur. The theory is particularly useful in dealing with such matters as isotope effects, solvent effects, and relative rates of similar processes. (See also KINETICS.)

BIBLIOGRAPHY

LAIDLER, K. J. *Chemical Kinetics;* New York: Harper & Row, 1987; pp. 89–127.

KEITH J. LAIDLER

Transmutation

The dream of the ancient alchemist was to transmute base metals into gold. While modern nuclear chemists have transmuted lead or bismuth into gold, the cost is enormous, and the process is of no practical significance. The modern definition of *transmutation* is the synthesis of new nuclei or new chemical elements.

Transmutation has occurred and continues to occur in nature. The chemical elements have been and are now being synthesized by nuclear reactions (nucleosynthesis) in stars and supernovas (see ORIGIN OF THE CHEMICAL ELEMENTS). We believe the universe started as hydrogen, and all the heavier elements have been made from this primordial material in the nuclear furnaces of the stars. The observation of spectral lines of elements such as technetium, whose HALF-LIFE is short ($\sim 10^6$ y) compared with the age of the universe, in the light from nearby stars leads us to conclude that element synthesis has continued.

There are various methods used in the laboratory today to synthesize new nuclear species and elements. The lighter TRANSURANIC ELEMENTS are primarily made in NUCLEAR REACTORS by neutron capture reactions. For example, element 94 (plutonium) is made by the following sequence of nuclear reactions:

$$^{238}U(n, \gamma)\ ^{239}U \xrightarrow{\beta^-} \ ^{239}Np \xrightarrow{\beta^-} \ ^{239}Pu$$

The transuranic elements of greatest practical importance (Np, Pu, Am, Cm) are made primarily by neutron capture reactions in nuclear reactors, utilizing uranium as a starting material. The high neutron fluxes present in nuclear weapons have also been used in element synthesis, most notably the synthesis of elements 100 and 101: fermium and mendelevium. (These elements were discovered in the debris of a large thermonuclear bomb test conducted in the Pacific in the 1950s.)

Another important method of transmutation involves nuclear reactions induced by charged particles from a particle accelerator. Most of the transuranic elements were first synthesized by charged-particle induced reactions such as

$$^{208}Pb(^{58}Fe, n)\ ^{265}Ns$$

Many of the nucleosynthetic nuclear reactions involve radioactive projectile and target nuclei. Modern nuclear scientists are currently seeking to emulate nature by using particle accelerators that produce beams of radioactive nuclei.

One of the most important parts of any nuclear synthesis reaction is the ioslation and identification of the product nuclei. When the product half-lives are long ($t_{1/2} \gtrsim 1$ sec), chemical separations can be used. For shorter half-lives, physical techniques utilizing mass separators, for example, are used.

Transmutation reactions are of great practical significance. The heavy elements Np, Pu, Am, and Cm are important by-products of nuclear reactors and weapons. Their safe disposal is of paramount importance. Radiopharmaceuticals containing nuclides like $^{99}Tc^m$ and ^{131}I play an important role in imaging and therapy. Many radiotracers such as ^{60}Co are widely used in industry, while other tracers such as ^{14}C and ^3H are part of much ongoing research in chemistry and biochemistry (see also TRACERS, RADIOISOTOPE).

BIBLIOGRAPHY

KERVRAN, C. L. *Biological Transmutations*; Woodstock, NY: Beekman Publishers, 1980.

ZIEGLER, K.; MARTIN, B. *Cyclotrons and Their Application*; River Edge, NJ: World Scientific Publishing, 1991.

WALTER LOVELAND

Transuranic Elements

The transuranic or transuranium elements are those elements whose atomic number (number of protons in the nucleus), Z, is greater than 92 (the atomic number of uranium). There are 20 transuranium elements, all of which are human-made. The elements, their atomic numbers, and chemical symbols are 93, neptunium, Np; 94, plutonium, Pu; 95, americium, 95; 96, curium, Cm; 97, berkelium, Bk; 98, californium, Cf; 99, einsteinium, Es; 100, fermium, Fm; 101, mendelevium, Md; 102, nobelium, No; 103, lawrencium, Lr; 104, rutherfordium, Rf; 105, dubnium, Db; 106, seaborgium, Sg; 107, bohrium, Bh; 108, hassium, Hs; and 109, meitnerium, Mt. After much delay and dispute, IUPAC formally approved the names for elements 104–109 on August 30, 1997, but gave elements 110 and 111 the respective temporary names of ununnilium (Uun) and unununium (Uuu).

The ACTINIDE elements (Z = 89–103) form a transition series, like the lanthanide elements, in which the $5f$ electronic subshell is filled. The transactinide elements (104, 105, etc.) have different chemical properties from the actinide elements, in that the valence electrons are in the $7s$ and $6p$ electronic subshells. The study of the chemistry of these elements is quite challenging because it must be done "one atom at a time"—i.e., because of the difficulty of producing these elements, typical experiments involve only a few atoms of the element of interest. Because of the high atomic numbers of these elements (leading to large electronic binding energies) their chemical properties are expected to show relativistic effects, such as a contraction and increase in stability of the $7s_{1/2}$- and $6p_{1/2}$-orbitals along with a destabilization of the d- and f-orbitals. The chemistry of rutherfordium has been shown to be different from that of the actinides and similar to that of hafnium, a clear demonstration of the end of the actinide series at lawrencium. Hafnium shows a gas phase chemistry like that of Group VB elements.

The transuranic elements are all radioactive, with the principal decay modes being α-decay, spontaneous fission, and electron-capture decay. For many years, it was thought that spontaneous fission would

1455

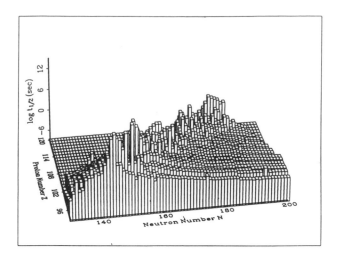

Fig. 1.   A three-dimensional plot of the measured and predicted half-lives of the transuranium nuclei vs. their atomic number Z and their neutron number N.

limit the size of the periodic table and heavy-element stability. It was predicted that at about Z = 106–108, the spontaneous fission half-lives would be so short (<10^{-6} sec) as to preclude detection of these species. That expectation has proven false in that the spontaneous fission half-lives of elements 104–109 are ~10^{-3} seconds, which allows their study in the laboratory. This increased stability against spontaneous fission is due to the influence of nuclear shell structure.

This same nuclear shell structure is predicted to be responsible for the existence of superheavy elements. The superheavy elements are defined as those chemical elements whose stability increases with increasing atomic number (rather than decreasing due to spontaneous fission). Originally it was thought that the nuclei near Z = 114 and N = 182 would form a special island of stable nuclei. Current predictions indicate that the most stable superheavy nuclei have Z~112 (N~182), but do not form a special island. Rather, they are part of a long peninsula of stable nuclei extending out from the region of known nuclei (Figure 1). The prediction maximum values of half-lives of the superheavy nuclei are ~10^5 sec, a value that is so short as to preclude observation in nature. As of this writing, superheavy elements have not been observed despite over two decades of trying to make and observe them.

The actinide transuranic elements were made in a series of experiments beginning in 1939 and continuing until 1961. G. T. SEABORG and A. Ghiorso played important roles in their discovery. The actinide elements have important uses in nuclear reactors

and nuclear weapons. The transactinide elements (Z > 103) are produced in such small quantities as to be of no practical importance.

BIBLIOGRAPHY

PLANCK, M. *Transurane-Transuranium Elements;* New York: Springer-Verlag, 1975.

SEABORG, G. T., ed. *Transuranium Elements: Products of Modern Alchemy;* New York: Van Nostrand Reinhold, 1978.

SEABORG, G. T.; Loveland, W. D. *The Elements beyond Uranium;* New York: Wiley, 1990.

WALTER LOVELAND

Triatomic

Triatomic species consist of three atoms and may be either linear or nonlinear as shown in Figure 1. Homonuclear triatomic molecules consist of three identical atoms and have the generic formula A_3. Heteronuclear triatomic species can have any of the following formulas: ABC, AAB (both of which are asymmetric triatomic molecules), and ABA (a symmetric triatomic molecule). There are very few examples of homonuclear triatomic species, the most famous being ozone (O_3), a nonlinear species. Examples of heteronuclear triatomic molecules include hy-

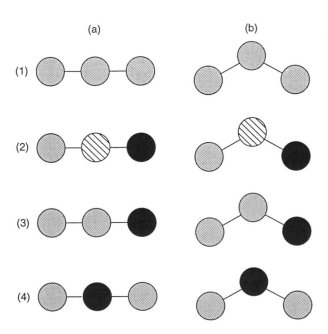

Fig. 1.   Representations of the various types of triatomic molecules: (a) linear and (b) nonlinear; (1) homonuclear, (2) and (3) asymmetric heteronuclear, (4) symmetric heteronuclear.

drogen cyanide (HCN, linear), carbon dioxide (CO_2, which has the structure OCO and is linear), nitrous oxide (N_2O, which has the structure NNO and is linear), sulfur dioxide (SO_2, which has the structure OSO and is nonlinear with a BOND ANGLE of 119.5°), and water (H_2O, which has the structure HOH and is nonlinear with a bond angle of 104.5°). The bonds between the atoms in triatomic molecules are typically polar covalent (since most triatomic species are heteronuclear) and strong. Asymmetric molecules will have a net DIPOLE moment and will orient in an electric field, whereas symmetric triatomic molecules may or may not have a net dipole moment. For example, CO_2, a linear species that has two polar bonds, has no net dipole moment. This is caused by the dipole from one of the carbon-oxygen bonds being canceled out by the dipole of the other carbon-oxygen bond, which is equal in magnitude but exactly opposite in direction. On the other hand, the nonlinear symmetric triatomic molecule H_2O does have a net dipole moment. Here the dipoles from each hydrogen-oxygen bond add together to give an overall dipole moment for the molecule.

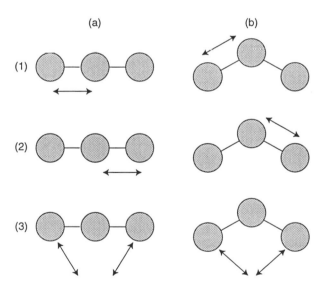

Fig. 2. Diagrams of the types of vibrational motion. The use of a homonuclear symmetric molecule is for symbolic purposes; all triatomic species have these internal modes. For linear species the two rotational planes are in the plane of the paper and perpendicular to the plane of the paper. Nonlinear species add a third rotational motion around one of the bonds. (a) linear triatom vibrations: (1) bond stretch, (2) bond stretch, (3) bending motion in the plane of the paper. There is also a second, independent bending motion that is perpendicular to the plane of the paper. (b) nonlinear triatom vibrations: (1) bond stretch, (2) bond stretch, (3) bend.

All triatomic molecules contain six internal motions (also referred to as modes), or ways to deform from the equilibrium position of each atom as illustrated in Figure 2. Just like diatomic molecules, linear triatomic species contain two perpendicular planes of rotation. The remaining four internal modes are taken up by the two bond stretch motions (which change the distance between atomic centers) and the two bending motions (which are equal in energy, rotated 90° to each other, and change the bond angle from 180°). Nonlinear triatomic molecules have three planes of rotation, the third being rotation around one of the bonds. The remaining three internal modes are the two bond stretch motions and the single bending motion. There are characteristic frequencies of vibration for the stretching and bending motions for both linear and nonlinear triatomic molecules, as well as specific spacing between rotational energy levels. These internal motions can thus be observed by a wide variety of spectroscopic techniques.

BIBLIOGRAPHY

ATKINS, P. W. *Physical Chemistry*, 5th ed.; New York: Freeman, 1994.

BARROW, G. M. *Physical Chemistry*, 4th ed.; New York: McGraw-Hill, 1979.

KELLY S. GRIFFITH

Triboluminescence

Triboluminescence (TL) is defined as the light emitted when a material is stressed to the point of fracture. Triboluminescence is an electrical phenomenon, with charge separation being the primary event (Walton, 1977), although the detailed mechanism has not been fully understood. Many materials (e.g., sucrose, quartz, and various polymers) that exhibit TL are piezoelectric (with no symmetry), a criterion that is necessary and sufficient for the charge separation. Nevertheless, symmetric triboluminescent crystals can achieve charge separation in several ways. For some crystals such as sodium chloride, fracture along certain crystal planes will separate charge. Also, some lattice defects and impurities may result in charge separation, either because of the distortion in the lattice, creating piezoelectricity, or because the impurity molecules have greater or lesser electron affinity than the host. Once fracture separates charge across the opposing faces of the crack or in patches on a single surface, a potential difference is generated. When the potential buildup is sufficient to initiate dielectric breakdown of the surrounding gas, discharge will occur across the gap or along the surface, accompanied by the emission of electrons, positive ions, radio waves, and ultraviolet-visible photons (Sweeting, 1988). If the conductivity of the piezoelectric material is so good that the charge is conducted away before the dielectric breakdown, TL will not be observed.

The essence of TL is to convert mechanical energy into electromagnetic energy. Such conversion is commonly utilized by biological systems. Many biomolecules including amino acids, glycerol, cholesterol, fatty acids, human serum albumin, glucose, nucleic acids, and electrolytes have shown triboluminescent properties. Among these, electrolytes and lipids in the form of oxidized radical products are believed to be the most likely emitters of visible radiation in TL (Baraboi et al., 1987). When these biomolecules are subjected to mechanical influences, their molecular structures are modified to generate separated electric charges between which, in the presence of liquid-containing soluble gases, there emerge weakly luminescent aggregates of ionized particles (Orel, 1993).

Triboluminescence of biomolecules may have applications in medical diagnostics. The circulatory system is one major area of focus. Under conditions of hypercholesterolemia, considerable accumulation of cholesterol ethers and triglycerides in certain white blood cells in the course of their interaction with lipoproteins of very low density (LVLD) has been observed. A decrease of triboluminescent intensity characterizing the surface charge has been observed both in LVLD and in blood plasma in a hypercholesterolemic state. It has therefore been proposed that a TL method may be used for monitoring of atherosclerotic process development (Voziyan et al., 1991). Recent studies on DNA topology have also suggested that mechanical activation of DNA generates an ultraviolet spectrum due to TL which stimulates a reactive phase of tumor cell growth. Carcinogenic factors generate excess numbers of free radicals that affect metabolic processes and consequently change DNA sequence and topology. The pattern of triboluminescence spectral emissions reflects the physiochemical nature of the emission centers, which are determined by the energy gases dissolved in it (Orel, 1993). Studies of triboluminescent intensities at particular wavelengths may facilitate the differentiation between inflammation and cancer (Ganul et al., 1993).

See also BIOLUMINESCENCE; BIOLUMINESCENCE, EXAMPLES OF.

BIBLIOGRAPHY

BARABOI, V. A., et al. "Triboluminescence of Electrolytes and Organic Components of the Blood Serum." *Doklady Akademii Nauk Ukrainskoit SSR. Seriya B: Geologicheskie Khimicheskie I Biologicheskie Nauki* 1987, 2, 66–69 (in Russian).

GANUL, V. L., et al. "Blood Triboluminescence in Patients with Cancer of the Proximal Part of the Stomach and Esophagus." *Voprosy Onkologii* 1993, 38(1–3):55–60 (Leningrad, in Russian).

OREL, V. E. "DNA Triboluminescence and Carcinogenesis." *Medical Hypotheses*, 1993, 40, 267–268.

SWEETING, L. M.; RHEINGOLD, A. L. "Crystal Structure and Triboluminescence, I. 9-Anthryl Carbinols." *J. Phys. Chem.* 1988, 92, 5648–5666.

VOZIYAN, P. A., et al. "Interaction of Very Low Density Lipoproteins with Macrophages and Their Triboluminescence at Hypercholesterolemia." *Ukrainskii Biokhimicheskii Zhurnal* 1991, 63(4):51–56 (in Russian).

WALTON, A. J. "Triboluminescence." *Advances in Physics* 1977, 26, 887–948.

MARY VIRGINIA ORNA, O.S.U.
WEN CHEN

Triple Bonds

See MULTIPLE BONDS.

Tritium

Tritium ($_1^3$H; symbol T; atomic weight = 3.060497 amu) is the heaviest of the three isotopes of hydrogen. It has a mass number of three, with a nucleus that contains one proton and two neutrons. Tritium differs from the other isotopes of hydrogen—PROTIUM and DEUTERIUM—in that it is naturally radioactive. It decays into helium-3 and a beta-particle (β particle):

$$_1^3\text{H} \rightarrow \, _2^3\text{He} + \, _{-1}^0\beta \tag{1}$$

The isotope has a HALF-LIFE of 12.35 years.

Tritium was first prepared in the laboratory in 1934 by Ernest RUTHERFORD by bombarding deuterium with fast-moving deuterium atoms:

$$_1^2\text{H} + \, _1^2\text{H} \rightarrow \, _1^1\text{H} + \, _1^3\text{H} \tag{2}$$

Early attempts to locate naturally occurring tritium were unsuccessful, owing to the very low concentration of atmospheric tritium (approximately one tritium atom per 10^{18} hydrogen atoms in air) and the relative insensitivity of the measuring techniques used. Tritium is produced in nature as a result of the interaction of cosmic rays with molecules in the upper atmosphere. Typical reactions include:

$$^{14}_{7}\text{N} + ^{1}_{0}\text{n} \rightarrow ^{3}_{1}\text{H} + ^{12}_{6}\text{C} \tag{3}$$

$$^{14}_{7}\text{N} + ^{1}_{1}\text{H} \rightarrow ^{3}_{1}\text{H} + \text{fragments} \tag{4}$$

$$^{2}_{1}\text{H} + ^{2}_{1}\text{H} \rightarrow ^{3}_{1}\text{H} + ^{1}_{1}\text{H} \tag{5}$$

Tritium is a by-product of the explosion of hydrogen bombs (see also NUCLEAR ENERGY). The relevant reaction that occurs to produce tritium during detonation is given by equation 1 above. In March 1954, with the advent of atmospheric thermonuclear weapons testing, the concentration of global tritium increased by a factor of more than one hundred. It has begun to decrease as a result of the ban on atmospheric nuclear testing that was imposed during the 1960s. Tritium is also being examined as a possible fuel for nuclear fusion according to the reaction below, which releases approximately 1.7×10^9 kJ per mol of tritium:

$$^{2}_{1}\text{H} + ^{3}_{1}\text{H} \rightarrow ^{4}_{2}\text{He} + ^{1}_{0}\text{n} \tag{6}$$

The preferred large-scale preparation of tritium involves the irradiation of lithium-6 (as an alloy with magnesium or aluminum) in a NUCLEAR REACTOR:

$$^{6}_{3}\text{Li} + ^{1}_{0}\text{n} \rightarrow ^{4}_{2}\text{He} + ^{3}_{1}\text{H} \tag{7}$$

Much of the tritium produced remains trapped in the metallic lattice until it is removed by treatment with acid. The most common method for storing tritium gas is to react it with finely divided uranium to form UT_3. It can be released later by heating the compound above 400°C.

Tritium has been used in a variety of nonnuclear applications owing to the relative ease of its detection, its low expense compared with other radioisotopes, and its relatively low toxicity. It has been used to study the movement of groundwaters and in analyzing the age of bodies of water. It has been used to study hydrogen adsorption phenomena and hydrogenation reactions. One of the more common chemical applications of tritium is in the elucidation of reaction mechanisms and kinetics through hydrogen exchange. In 1956 K. E. Wilzbach discovered that organic compounds could become tritium-labeled simply by being stored under tritium for a few days or weeks. The subsequent introduction of tritium into compounds of biological importance has led to the most important uses of tritium, that is, as a radioactive label in the study of metabolism, enzyme kinetics and assay, pharmacology, radiotherapy, and autoradiography.

BIBLIOGRAPHY

EVANS, E. A. *Tritium and Its Compounds,* 2nd ed.; London: Butterworths, 1974.

GREENWOOD, N. N.; EARNSHAW, A. *Chemistry of the Elements;* Oxford: Pergamon, 1984; pp. 48–50.

MACKAY, K. M.; DOVE, M. F. A. "Deuterium and Tritium." In *Comprehensive Inorganic Chemistry,* 1st ed.; J. C. Bailar, H. J. Emeléus, R. Nyholm, A. F. Trotman-Dickenson, eds.; Oxford: Pergamon, 1973; Vol. 1, Chapter 3.

DOMINICK J. CASADONTE

Tungsten

74
W
183.85

Melting Point: 3683 K; **Boiling Point:** 5933 K; **Density:** 19.35 g/cm^3

Tungsten, a steel-gray metal, is one of three elements whose abundance in the Earth's crust is approximately 1.2 ppm. Molybdenum and terbium are present in equal amounts. The International Union of Pure and Applied Chemistry suggested the name wolfram (German *Wolf,* wolf; *Rahm,* soot or dirt) following the observation by German metallurgists that wolframite ore is a black color. Earlier, Agricola (1556) observed that the tungsten ore frequently occurred with tin minerals and interfered with their smelting much as the wolf consumes animals. The name tungsten comes from the Swedish language (*tung,* heavy; *sten,* stone) in honor of C. W. Scheele. In 1781, Scheele discovered that the mineral, later known as scheelite, yielded lime (CaO) and a yellow acidic powder, "tungstic acid" ($WO_3 \cdot 2H_2O$), upon heating. In 1783 J. J. and F. de Elhuyar proved that scheelite ore ($CaWO_4$) could be reduced to the element by heating with charcoal.

Tungsten has only a few important minerals. Scheelite and wolframite are the most important. Wolframite is a mixture of Fe(II) and Mn(II) tungstates. Other minerals are huebnerite, $MnWO_4$, and ferberite, $FeWO_4$. Large deposits of tungsten are found in China and North Korea, with Bolivia, the former Soviet Union, South Korea, Portugal, and the United States being other major producers.

The metal is recovered from both scheelite and wolframite, both of which form tungstic acid. The acid is roasted to WO_3 and subsequently reduced with hydrogen at 850°C. Annual world production of tungsten in 1980 was 50,000 tons.

Half of the tungsten produced annually is used for the production of tungsten carbide, WC. This carbide is extremely hard and useful in making cutting materials. In 1785 Raspe demonstrated that tungsten hardens steel. The very high melting point of the metal makes tungsten a refractory metal. Tungsten and its alloys are useful in making high-speed and tool steels, heat shields, tool tips, and tire studs. Tungsten is used in the heating elements for electric furnaces as well as contact points for automobile distributors. The TIG welding technique uses tungsten electrodes. Tungsten is also used in the nozzles of solid-fuel rocket engines. Perhaps the most widespread and best known use of the metal is as the lamp filament in incandescent bulbs. Carbon filaments, initially used for this purpose, were replaced with tungsten in 1908.

In 1823 F. WÖHLER reduced Na_2WO_4 and WO_3 with H_2 at red heat to produce tungsten bronzes. These bronzes provide the metallic luster in "bronze paints." Finally, tungsten oxides and sulfides are important in the petroleum refining industry where they form catalysts for hydrocracking, hydrotreating, reforming, isomerizing, and dehydrogenation.

BIBLIOGRAPHY

GREENWOOD, N. N.; EARNSHAW, A. *Chemistry of the Elements;* New York: Pergamon, 1984; pp. 1167–1210.

LIDE, D. R., ed. *Handbook of Chemistry & Physics,* 71st ed.; Cleveland: Chemical Rubber Co., 1990–1991.

MacINNIS, M. B.; KIM, T. K. "Tungsten Compounds." In *Encyclopedia of Chemical Technology,* 3rd ed.; R. E. Kirk, D. F. Othmer, M. Grayson, eds.; New York: Wiley-Interscience, 1981; Vol. 23, pp. 426–438.

MULLENDORE, J. A. "Tungsten and Tungsten Alloys." In *Encyclopedia of Chemical Technology,* 3rd ed.; R. E. Kirk, D. F. Othmer, M. Grayson, eds.; New York: Wiley-Interscience, 1981; Vol. 23, pp. 413–425.

ALTON J. BANKS

U

Ultraviolet Spectroscopy

The ultraviolet (UV) region of the electromagnetic spectrum extends from wavelength $\lambda = 400$ nm to $\lambda = 100$ nm (1 nm, or nanometer, is one-billionth of a meter). The UV region is further divided into the near UV, from 400 nm to 190 nm, and the far UV, from 190 nm to 100 nm. This last is also known as the vacuum UV region because oxygen absorbs in this region; the spectroscope must therefore be evacuated. Data in the UV is also often described in units of wave numbers: 400 nm is 25,000 cm^{-1}, 190 nm is 52,632 cm^{-1}, and 100 nm is 100,000 cm^{-1}. (For units and conversions, see ELECTROMAGNETIC SPECTRUM.)

Quartz sample cells in UV spectrometers serve down to about 210 nm, but fused silica cells can be used to 165 nm. Solvents must be specially purified and their cutoff limits observed. Carbon tetrachloride cuts off at 245 nm; water, methanol, and ethanol at 198 nm; and hexane at 187 nm. Interactions of the solvent with the material being examined must be avoided. Double-beam instruments give best results, with a matching cell containing pure solvent in the reference beam.

Both the position as well as the intensity of absorptions are significant. The latter is given as the molar absorptivity or extinction coefficient ε at the maximum absorption, with $\varepsilon = A/cd,$ where A is the measured absorbance or optical density, c is the concentration in moles L^{-1}, and d is the cell path length in cm. Values of ε can vary from 10 to 10^5.

Most absorptions in the UV derive from electronic transitions in which the molecule, or some part of it, absorbs a band of wavelengths. Fine structure on absorption bands may derive from vibrations and rotations. Information from UV spectroscopy does not usually give a unique identification; this will often result from combining it with chemical tests, other spectroscopic techniques, and synthesis information.

Molecules containing a single chromophore have characteristic absorptions. Some typical examples are given in Table 1, where only the strongest band is listed for each. The huge range of ε values is helpful in identification. Isolated chromophores in a molecule will be approximately additive, but this is not true of chromophores connected by a conjugated system. Both λ and ε increase with the length of a conjugated system of alternating single and double bonds, as in Table 2.

Standard complications of UV-visible absorption spectra can be used for identification of compounds. In the vacuum UV region occur the resonance absorption lines of individual elements, such as H at 121.6 nm, He at 58.4 nm, C at 165.7 nm, and N at 120.0 nm, permitting analysis for inorganic elements.

Table 1. Some UV Chromophore Absorptions

Chromophore	Structure	λ	ε
Nitroso	$-N{=}O$	302	100
Nitrate	$-NO_3$	270	12
Iodide	$-I$	260	400
Anthracene	$C_{14}H_{10}$	252	200,000
Quinoline	C_9H_7N	227	37,000
Toluene	$C_6H_5CH_3$	208	8,000
Bromide	$-Br$	208	300
Ester	$-COOR$	205	50
Amine	$-NH_2$	195	2,800
Ketone	$>C{=}O$	195	1,000
Ether	$-O-$	185	1,000
Benzene	C_6H_6	184	55,000

Table 2. Absorption and the Length of a Conjugated System

Formula	λ	ε
$CH_2{=}CH_2$	165	10,000
$CH_2{=}CH-CH{=}CH_2$	217	21,000
$CH_2{=}CH-CH{=}CH-CH_2$	263	53,000

BIBLIOGRAPHY

COOPER, J. W. *Spectroscopic Techniques for Organic Chemistry;* New York: Wiley, 1980.

LANG, L. *Absorption Spectra in the Ultraviolet and Visible Region;* New York: Academic Press, 1961.

WILLARD, H. H., ed. *Instrumental Methods of Analysis,* 7th ed.; Florence, KY: Wadsworth, 1985.

ZAIDEL, A. N.; SHREIDER, E. Y. *Vacuum Ultra Violet Spectroscopy;* Ann Arbor, MI: Humphrey, 1970.

KURT NASSAU

Ultraviolet-Visible Wavelength

Ultraviolet-visible wavelength, or UV/visible wavelength (λ), is a wave parameter characteristic of UV and visible radiation that ranges from 10 to 780 nm. Electromagnetic radiation is a type of energy transmitted through space and can be viewed as a stream of fast-moving particles that also exhibits properties of waves. *Wavelength* is defined as the distance between crests or troughs of two successive waves and is measured in nanometers (nm). In the electromagnetic spectrum, which is a continuous range of wavelengths from cosmic to radio waves, UV and visible radiation occupy adjoining regions. Visible light, which is the region of the electromagnetic spectrum that we can see with the unaided eye, occupies from 380 to 780 nm. Ultraviolet light, which is a higher-energy radiation, occupies from 10 to 380 nm.

The values of energy associated with wavelengths from the UV/visible region are at the same level as the energies needed to move valence electrons from ground to excited states in materials. To observe these transitions a device called a UV/visible spectrometer is used. In this instrument, light from a continuous source of wavelengths (from 200 to 780 nm) is sent through a wavelength selector (monochromator) and then through the sample. If the energy of this monochromatic light equals the energy of allowed electronic transitions in the atomic or molecular species present in the sample, the light will be absorbed. This results in the reduction of the beam's radiant power, which is measured by a detector. The wavelength absorbed depends upon the nature of species present. The amount of absorbed radiation is related to the concentration of the absorbing species present in the sample. As a result, examination of the intensity of sample absorption at specific UV/visible wavelengths can provide valuable information about the type of species and their concentration.

DIANE M. BUNCE
EVE LABUDA

Uncertainty Principle

See HEISENBERG UNCERTAINTY PRINCIPLE.

Unit Operations

The manufacture of chemicals, pulp and paper, pharmaceuticals, fuels, plastics, foods, metals, semiconductors, and other finished products requires a series of process steps that involves reaction and separation (see SEPARATION SCIENCE). The possible equipment configurations for a process plant are numerous. Given that many such industries are proprietary in nature, there can be significant variation in process flows and equipment required. In order to design and analyze such facilities, it is necessary to have unifying principles that reduce the many possible flowsheets to a set of physical or chemical changes, often embodied in a particular piece of equipment. Since the early 1900s, the field of chemical engineering has adopted the concept of unit operations and a related concept called unit processes in order to have a systematic basis for design of complete processing plants. The use of unit operations greatly reduces the number of individual concepts that must be understood by a chemical engineer. This approach also limits the number of possibilities that must be considered in the synthesis of a processing sequence. In general a unit operation is differentiated from a unit process by the nature of the transformation that takes place in the equipment. If the change is physical, then it is called a unit operation. If chemical reaction also occurs, it is a unit process. However, this distinction has become less important over time, and the term *unit operation* is now generally used to describe the use of equipment in processes where chemical change occurs as well.

One way to classify unit operations is by the transport phenomena that occur within, namely momentum transfer, HEAT TRANSFER, and mass transfer. Momentum transfer refers to fluid dynamics or fluid flow, and thus a pump that simply changes the pressure of a fluid is a unit operation. Other unit operations based on fluid mechanics include mixing, agitation, filtering, sedimentation, classification, and

centrifugation. There are many types of specific equipment that could be utilized in a given application; however, the designer usually selects the least-cost option.

Unit operations that principally involve heat transfer include heat exchangers, condensers, evaporators, furnaces, dryers, and cooling towers. Many equipment options for heat exchangers are available commercially, such as shell and tube, finned, double-pipe, plate, bayonet, falling film, air-cooled, and others. See *Perry's Handbook* for a more detailed list of heat exchangers and their suitability for specific process applications.

The third class of unit operations is that based on mass transfer. Mass transfer operations in chemical plants tend to have a greater influence on product quality than heat and momentum transfer equipment. One taxonomy for classifying unit operations where separations is the main goal has been described by Wankat (1990) and is shown in Table 1.

Historically the main unit operations used in chemical and petrochemical plants and refineries have been distillation (or fractionation) and extraction, and textbooks on unit operations and separations tend to emphasize these separations technologies. However, with improvements in engineered adsorbents and polymeric materials, specialized separations

Table 1.   Classification of Unit Operations for Separations*

Gas-Liquid	Liquid-Solid	Liquid-Liquid	Gas-Solid	Membrane	Chromatography	Mechanical
Adsorption	Adsorption	Extraction	Adsorption	Electrodialysis	Affinity	Centrifuge
Distillation	Molecular sieve	Dual temperature exchange	Freeze drying	Pervaporation	Capillary	Cyclone
• Azeotropic	Clathration	Liquid membrane	Molecular sieve sublimation	Gas permeation	Electrochromatography	Decanter
• Extractive	Crystallization			Reverse osmosis	Gas-liquid	Demister
• Flash	Precipitation			Ultrafiltration	Gas-solid	Electrostatic separator
• Reactive	Zone melting			Gas diffusion	Ion exchange	Emulsion separator
• Steam	Ion exchange					Filtration
• Vacuum	Ion exclusion					Flotation
Evaporation	Leaching					High-gradient magnetic
Foam fractionation	Washing					Impingement separator
	Drying					Magnetic
						Sedimentation
						Scrubber
						Sink-float

*Adapted from Wankat (1990).

such as adsorption and membrane separation have established a niche in several commercial areas. For example, the separation of air into its component gases has been traditionally carried out by cryogenic distillation for larger-scale production. If the plant capacity is smaller and purity requirements for oxygen are not as demanding (such as in steelmaking, enriched combustion, and wastewater treatment), then the use of pressure swing adsorption or membrane separation can become economical.

The advent of materials processing as a major industry in the 1980s is best epitomized by the semiconductor industry. Semiconductors are produced by a series of batch operations, often involving manual transport of a rack of wafers from one piece of equipment to another, usually carried out in the pristine, particle-free environment of a clean room. The manufacture of a finished wafer can involve as many as 200 steps using a set of unit operations that are employed repetitively. These unit operations can include crystal growth, lithography, chemical vapor deposition, physical vapor deposition, etching, ion implantation, and doping of the thin films layered on the wafer. The view of microelectronics manufacturing as a set of unit operations has aided immensely in the improvement in yields from each step as well as in developing a better understanding of how the key operating and design variables influence yield and throughput. As in the traditional chemical industry, there are many processing and equipment variations that can be used in an etching or deposition step.

The unit operations concept has proved to be valuable in the synthesis and analysis of many processes in the chemical and allied industries. Elaborate software packages are available commercially that will perform mass and energy balances, size equipment, and compute capital and operating costs for a given process flowsheet. The software is unit-operations based, so that the engineer can construct and analyze candidate processes in a modular fashion and select the best alternatives using simulation tools. The ability to deconstruct complex processes using the unit-operations concept has proven its worth in modern chemical engineering practice.

BIBLIOGRAPHY

GREEN, D. W., ed. *Perry's Chemical Engineering Handbook,* 6th ed.; New York: McGraw-Hill, 1984.

LEE, H. H. *Fundamentals of Microelectronics Processing;* New York: McGraw-Hill, 1990.

WANKAT, P. C. *Rate-Controlled Separations;* Barking, U.K.: Elsevier, 1990.

THOMAS F. EDGAR

Uranium

See NUCLEAR REACTIONS; NUCLEAR REACTORS; RADIOACTIVITY, HISTORY OF.

Urine Chemistry

Routine chemical analysis of urine is the most important of all screening laboratory tests as it not only provides a correlative biochemical mirror of the serum but also allows evaluation of renal function while assessing for anatomic abnormalities of the kidneys and lower urinary tract. The utility of urinalysis is obvious considering its ease of performance, its low cost, its informational yield, and its noninvasive nature.

This article's purpose is to outline basic renal physiology, the chemical components of urine, their clinical usefulness in terms of diagnosis, and some urinary manifestations of specific metabolic, endocrine, and neoplastic disorders. Microscopic and cytological evaluation of the urine, although of unparalleled diagnostic value in the detection of infections, malignancy, and glomerulonephritis, is not specifically addressed but is covered thoroughly in the referenced publications.

Physiology and Biochemistry of Urine Formation

The primary function of the kidneys is the maintenance of a constant extracellular milieu through regulating the excretion of fluid and ELECTROLYTES. Many molecules are filtered, with the kidney then selectively reabsorbing or secreting those necessary to maintain extracellular homeostasis. Various HORMONES, toxins, pharmaceuticals, and abnormalities of renal perfusion may have profound effects on the ability of the nephron, the primary renal unit microscopically, to perform its function in the formation of urine.

Blood enters the nephron via an afferent arteriole, which branches into a capillary tuft known as the glomerulus. The capillaries then merge to form an efferent arteriole, which subsequently branches into peritubular capillaries. The glomerulus is invaginated into a spherical structure lined by renal tubular epithelium, known as Bowman's capsule. Here tubular epithelial cells are separated from the capillary endothelium by only a thin basement membrane.

Filtration occurs into Bowman's space, with the glomerular ultrafiltrate then entering the proximal convoluted tubule (PCT). Here approximately 70 percent of the utlrafiltrate is reabsorbed as sodium and is cotransported across the tubular epithelium along with glucose, AMINO ACIDS, phosphate, and other organic solutes into the peritubular capillary. Adenosine triphosphate (ATP)-dependent sodium transport also occurs, with chloride and water following passively to maintain electrical and osmotic equilibrium, respectively.

From the PCT, the isosomotic ultrafiltrate enters the loop of Henle (LH), where in its descending segment it becomes much more hypertonic, reaching a concentration of 1,200 mOsm/L, four times the osmolality of the serum, at its turn. This is accomplished by reabsorption of around 75 percent of the water entering the LH into the peritubular fluid and capillaries from the descending limb. By contrast, the descending limb of the LH is only slightly permeable to solutes. The hypertonic environment of the renal medulla allows the movement of water to the interstitium by passive diffusion.

After making a hairpin turn, the ultrafiltrate enters the ascending limb of the LH, where the functional character of the tubular epithelium changes. Here active energy-dependent transport of chloride occurs with sodium following passively to maintain electrochemical equilibrium. Because the ascending limb of the LH is minimally permeable to water, a hypotonic ultrafiltrate is formed and a hypertonic environment is established in the medullary interstitium.

The distal nephron begins at the distal convoluted tubule (DCT), its primary function being the reabsorption of sodium and chloride and the secretion of hydrogen ions and potassium through the action of the hormone aldosterone. As sodium is reabsorbed, water follows passively, expanding the extracellular fluid volume and increasing intravascular pressure. This mechanism of volume expansion is highly adaptive and regulated by the blood flow within the afferent arteriole. At the *juxtamedullary apparatus,* the DCT ascends to contact the afferent and efferent arterioles supplying its glomerulus. Here, a decline in renal blood flow causes production of the hormone renin, which converts the serum protein angiotensinogen to angiotensin 1, which is metabolized to angiotensin 2 in the lung.

Angiotensin 2 leads to direct vasoconstriction in the afferent arteriole and stimulates the secretion of aldolsterone by the adrenal gland, thereby increasing renal perfusion pressure and expanding intravascular volume. Calcium reabsorption also appears to be regulated within the DCT by parathyroid hormone and vitamin D.

Final concentration of the dilute ultrafiltrate leaving the DCT occurs in the collecting duct (CD), where water is reabsorbed passively as it passes through the medullary interstitium. The permeability of the collecting duct and the osmolality of the urine excreted are determined finally by the hormone vasopressin or antidiuretic hormone (ADH). Through altering the CD's permeability to water, a concentrated urine is formed at times of volume depletion while a dilute urine accompanies states of overhydration. The ultrafiltrate leaves the collecting ducts at the renal papillae as urine to enter the renal pelvis, ureter, and finally the bladder.

Physical and Chemical Evaluation of the Urine

Urine for analysis is best collected in a sterile, dry container and should be examined within 1 hour of collection or refrigerated at 5°C. Cellular elements and casts decompose at room temperature and preclude accurate chemical analysis when their intracellular and structural contents enter the urine. General physical parameters of concern include color, appearance, odor, turbidity, pH, specific gravity, and urinary volume produced.

Color

Urine is normally yellow due to the endogenous pigment urochrome, but it darkens on standing as urobilinogen is oxidized to urobilin. Because urochrome is produced and excreted at a constant rate, dilute urine may be colorless while concentrated urine is more yellow, varying only as a function of the accompanying water in the solution. Color may therefore be used to estimate an individual's hydrational status.

Drugs such as phenazopyridine, methylene blue, or metronidazole and foods such as beets, rhubarb, and carrots may affect urine color. Metabolic products such as myoglobin, certain protoporphyrins, and homogenistic acid, abnormally seen in urine during various disease states, may also be discoloring as can bleeding within the urinary tract. With standing, red urine containing whole blood, hemoglobin, or myoglobin appears dark brown or black as the protein moiety is denatured in the acidic urine.

Appearance and Turbidity

Voided urine is normally clear but may appear turbid due to the presence of pus (pyuria), bacteria (bacteriuria), fat (lipiduria), chyle (chyluria), and most commonly crystals (crystalluria). Crystals of calcium oxalate, uric acid, and amorphous urates tend to precipitate in acidic urine while amorphous carbonates and phosphates are seen in alkaline urine. Bacterial overgrowth and/or precipitation of amorphous materials may occur if urine is not examined within 1 hour of collection. The urine of women will commonly appear hazy due to contamination by vaginal epithelial cells or mucus while that of men may appear turbid due to the presence of semen.

Odor

A foul odor is frequently indicative of urinary tract infection although certain bacteria, most notably *Pseudomonas aeroginosa*, may cause a characteristic sweet, fruity odor. The presence of ketone bodies in diabetic acidosis or maple syrup urine disease may also lead to a sweet odor while the odor of individuals with phenylketonuria is described as "mousy." Asparagus causes a very characteristic urine odor in most individuals; however, only those with a particular genetic predisposition can smell the odor.

pH

Normal urinary pH varies from 4.5 to 8, limited primarily by the kidney's ability to excrete hydrogen ions and fixed acids such as $H_2PO_4^-$ and HSO_4^-. The kidney will autoregulate the excretion of acidic metabolic products to maintain a physiological pH in the serum. In most states of metabolic acidosis (diabetic ketoacidosis, starvation ketosis, lactic acidosis, etc.) and chronic respiratory acidosis (chronic obstructive pulmonary disease, pulmonary fibrosis, etc.), an acidic urine will be produced. In alkalotic states, an alkaline urine will be produced. Urinary pH will increase due to infection by urease-producing organisms (i.e., *Proteus mirabilis* and other gram-negative bacteria) as urea is broken down to ammonia.

Renal tubular acidosis (RTA) results in metabolic acidosis due to either continuing loss of HCO_3^- from the proximal tubule (type II RTA) or the inability of the distal tubule to secrete H^+ (type I RTA). In these scenarios, urinary pH will always be alkaline except when type II RTA becomes severe enough that free H^+ bind HCO_3^- to a degree that minimal amounts are actually being filtered at the glomerulus, allowing acidic urine production.

Specific Gravity and Osmolality

Measurement of specific gravity compares the density of the urine specimen to that of an equal volume of distilled water at a comparable temperature. Because it is a ratio of densities, the value of the specific gravity is not only determined by the number of particles present but also by their size.

Measurement of osmolality relates only to the concentration of dissolved particles and is not influenced by their individual molecular sizes. Normal specific gravity varies from 1.003 to 1.035 while urinary osmolality ranges from 50 to 1,200 mOsm/day. Both values will increase with volume depletion due to decreased fluid intake or increased fluid losses through fever, perspiration, diarrhea, or vomiting.

Diabetes mellitus will increase urinary specific gravity as glycosuria increases the particle concentration in solution while the syndrome of inappropriate antidiuretic hormone secretion (SIADH) has the same effect by decreasing the free water available for urine formation.

A decrease in urinary specific gravity and osmolality is seen with increased fluid intake, diuretic use, and diabetes insipidus.

Urinary Volume

Ambient temperature, fluid intake, hormonal milieu, and level of physical activity and renal function affect the volume of urine produced by a given individual. Normally, 600–2,500 mL of urine are produced in a 24-hour period.

High ambient temperatures and strenuous physical activity decrease urinary volume by increasing perspiration, while disproportionate fluid intake will lead to high-volume urine output. Antidiuretic hormone (ADH), produced by the posterior pituitary gland, regulates water absorption in the collecting duct and thereby has a profound influence on urinary output. Diabetes insipidus (DI), due to either the lack of ADH production (central DI) or its inability to act at the kidney (nephrogenic DI), will cause hypotonic urinary excretion and a high serum osmolality. Clinically, it will be manifested as polyuria and excessive thirst (polydipsia) and may be a source of enuresis (bedwetting) in children.

Chemical Composition of the Urine

Urea is the principal end product of protein metabolism in mammals and constitutes about half of the total solids in solution in the urine. It comprises 80–90 percent of the total urinary nitrogen and its excretion is directly related to PROTEIN intake. Urinary urea is decreased by systemic acidosis when metabolic nitrogen is diverted to form the buffer ammonia and also in the case of end-stage liver disease when a decline in protein metabolism leads to decreased urea production.

Creatinine and creatine are by-products of muscle metabolism. Creatinine is the breakdown product of creatine and is excreted at a constant rate regardless of diet (usually 20–26 mg/kg/day). This ratio of creatinine excreted and body weight, known as the creatinine coefficient, serves as a reliable indicator of the adequacy of daily urine collection.

Creatine is present in only very small amounts normally but is increased in pregnancy, starvation, hyperthyroidism, and in states of impaired carbohydrate metabolism, all instances where muscle catabolism is increased. Creatine excretion is decreased in hypothyroidism.

Concentrations of urea and creatinine greater than 600 mg/dL and 50 mg/dL, respectively, are indicative of urine. Analysis for these constituents is particularly useful following urological surgery or renal transplantation when postoperative drainage may be either serous or urinary. The normally low concentrations of glucose and protein in the urine of individuals without diabetes or glomerulonephritis may also be used to distinguish it from spinal, pleural, or peritoneal fluid.

Ammonia, an important metabolic end product in many lower life forms, is present in only minimal amounts in the urine of normal individuals. Nonrenal etiologies of systemic acidosis (diabetes, lactic acidosis, continent urointestinal diversion) lead to increased ammonia formation and excretion by the kidney. When kidney disease is the etiology of systemic acidosis, ammonia excretion may not be increased.

Amino acids are present in very low amounts in the urine of adults. Premature infants and children less than 6 months of age excrete higher amounts, likely due to increased permeability of the proximal convoluted tubule. The renal threshold for amino acids is normally very high, such that only with extremely high-protein diets will their excretion be slightly increased.

The Fanconi syndrome is a hereditary disorder of proximal tubular transport accompanied by marked nonselective aminoaciduria. Cystinuria, on the other hand, is a hereditary proximal tubular disorder in which cystine, ornithine, lysine, and arginine are overexcreted. Individuals having cystinuria are prone to cystine urinary calculi as the solubility of this amino acid is far lower than that of the others at the normally acid pH of urine.

Uric acid is the oxidative end product of all endogenous and ingested purines. In cases of malnutrition, urinary uric acid will be decreased, while pathological states leading to cytolysis and the release of nucleoprotein (i.e., leukemia, other malignancies, and cytotoxic chemotherapy) lead to elevated urinary uric acid levels. Gout and severe liver disease also lead to increased uric acid excretion. Allantoin is an oxidative by-product of uric acid. It is excreted in small quantities in humans but is the chief end product of purine metabolism in some lower mammals.

Uric acid crystals precipitate rapidly in acidic urine (pH < 5.5) and alkalization is commonly used medically to reduce the risk of uric acid stone formation. "Amorphous urate" crystals, the salts of cations and uric acid, are commonly seen in acidic urine.

Oxalates are present in low concentrations in normal urine but are found in large quantities in individuals with the inherited disorder primary hyperoxaluria. Enteric hyperoxaluria is oxalate hyperexcretion due to excessive absorption of oxalates from the gut. It is seen frequently in individuals with Crohn's disease and other malabsorption syndromes in which excessive fat within the gastrointestinal (GI) tract forms soaps with enteric calcium that otherwise would bind enteric oxalates. Because free oxalates are absorbed from the gut in increased amounts, increased excretion occurs at the level of the kidney. Because oxalates form highly insoluble salts with calcium, they are frequent components of urinary calculi.

Sulfates are excreted as either inorganic sulfates, conjugated (organic) sulfates, or neutral sulfur. Urinary sulfates are primarily derived from the metabolism of the sulfur-containing amino acids cystine and methionine and vary in concentration depending on dietary intake.

Phosphates in the urine are influenced by dietary protein content but are also derived from cellular breakdown. They are excreted as sodium, potassium, magnesium, and calcium salts, the latter two precipitating rapidly in alkaline urine. They compose the commonly seen "amorphous phosphate" crystals found in microscopic examination of alkaline urine. Hyperparathyroidism and certain bone diseases increase urinary phosphates while hypoparathyroidism and renal failure lead to decreased phosphate excretion.

Minerals such as sodium, potassium, calcium, and magnesium are present in urine as they are in extracellular fluid. Intake plays a key role, particularly with regard to sodium and potassium, although potassium excretion also increases with excessive cellular catabolism and with systemic alkalosis (where potassium is excreted in favor of hydrogen ions at the DCT, the hydrogen ions thereby buffering the systemic alkalosis).

Chlorides are excreted primarily as the sodium salt and vary in concentration depending on dietary content.

Various enzymes, hormones, and vitamins are also present in small amounts in the normal urine. They may be of diagnostic value in certain disease states.

Abnormal Chemical Constituents of the Urine

Glucose is suggestive of diabetes mellitus, but this diagnosis must be confirmed by blood chemical studies so as to exclude the renal glycosuria that can be seen accompanying generalized proximal tubular dysfunction (Fanconi syndrome). Dipsticks containing the peroxidase enzyme are commonly used to detect the presence of glucose in the urine but may give false-negative results in the presence of bilirubin (excreted in persons with liver disease), homogenistic acid (seen in alkaptonuria), ascorbic acid (seen in persons taking exogenous vitamin C), and epinephrine.

Other sugars such as galactose and lactose may be seen in the urine of individuals unable to metabolize galactose to glucose (hereditary galactosemia) and in nursing infants and mothers, respectively. Congenital fructosuria and pentosuria are rare abnormalities seen in individuals unable to metabolize these sugars.

Ketone bodies include acetoacetic and beta-hydroxybutyric acid. Starvation, pregnancy, certain anesthetics, and particularly impaired carbohydrate metabolism as seen in diabetes cause ketonuria.

Proteins are normally present in only minimal amounts in the urine (30–200 mg/day). Pregnancy is accompanied by proteinuria approximately 33 percent of the time, and other causes of this "physiological proteinuria" include a high-protein diet, strenuous exercise, or orthostatic changes in renal blood flow.

Pathological proteinurias are classified as either prerenal, renal, or postrenal depending on the etiology of the protein present. Factors operating systematically prior to glomerular filtration lead to prerenal proteinuria, while intrinsic renal pathology causes renal proteinuria. Factors operating in the lower urinary tract lead to postrenal proteinuria.

Volume depletion (dehydration) is the chief etiology of prerenal proteinuria. As renal perfusion pressure declines, proteinuria occurs due to tubular dysfunction and may be corrected in many cases with rehydration. Advanced heart failure, liver disease, renal artery stenosis, and certain drugs may also lead to proteinuria of a prerenal origin as well as to a decline in renal function.

Hypertension and diabetes lead to renal insufficiency primarily through vascular phenomena, chiefly obstruction at the level of the medium and small vessels, respectively, thereby leading to renal proteinuria and a decline in creatinine clearance and concentrating ability.

Polycystic kidney disease causes a decline in renal function due to tubular destruction and dysfunction, again resulting in renal proteinuria, while heavy metals such as mercury or arsenic are direct tubular toxins.

Interstitial (allergic) nephritis may cause tubular dysfunction due to immunological mechanisms and is primarily seen after ingestion of ANTIBIOTICS, diuretics, or ANTI-INFLAMMATORY drugs to which an individual is hypersensitive.

Postrenal proteinuria is most commonly due to inflammatory conditions of the lower urinary tract such as cystitis, prostatitis, or urethritis.

Nephrotic-range proteinuria occurs when 3 or more g of protein are lost daily and signifies a marked degree of tubular dysfunction. The nephrotic syndrome encompasses proteinuria as above, hypoalbuminemia, edema, and hyperlipidemia and is most commonly seen in diabetics, patients with glomerulonephritis, and those with autoimmune disorders.

Bence-Jones proteins are immunoglobin lightchains found in the urine of individuals with multiple myeloma and other lymphoreticular malignancies. They may be identified by electrophoresis or by heating the urine to 50–60°C where they will precipitate and then redissolve completely as heating increases to 100°C.

Bilirubin, a by-product of hemoglobin metabolism in the liver, is seen in the urine primarily in cases of obstructive jaundice but must be interpreted in relation to urinary urobilinogen, which is a normal minor constituent of urine. Urobilinogen is a metabolic by-product of bilirubin formed in the gut, a small portion of which is normally absorbed, recirculated, and then excreted renally.

With obstructive jaundice, bilirubinuria is present without urobilinogenuria (since no bilirubin enters the gut for metabolism to urobilinogen).

In cases of hepatitis, urinary urobilinogen increases as the diseased liver fails to reexcrete recirculated urobilinogen into the gut. Urinary bilirubin increases as hyperbilirubinemia presents a larger load of the metabolite to the kidney while fecal urobilinogen decreases.

In the case of hemolytic jaundice, urinary as well as fecal urobilinogen increases due to the larger volume of bilirubin presented to the liver. No bilirubinuria will be present as its serum concentration will not exceed the renal threshold, given normal biliary drainage to the gut.

Blood in the urine (hematuria) may result from glomerulonephritis, calculi, infection, hemoglobinathies, trauma, or neoplasms. A microscopic examination will reveal red blood cells. Red urine in the absence of cellular elements by microscopy may be the result of either hemoglobinuria or myoglobinuria.

Free hemoglobin (hemoglobinuria) may be seen after cases of intravascular hemolysis where the amount of free hemoglobin exceeds the ability of the serum protein haptoglobin to bind it, thereby allowing its filtration by the kidney. A spun blood sample will yield a pink supernatant if hemoglobin is present in the serum. A clear serum supernatant with acellular red urine is usually indicative of myoglobinuria, occurring after muscle trauma in which the protein myoglobin enters the circulation and is subsequently filtered at the glomerulus. Serum protein electrophoresis, haptoglobin levels, and creatine phosphokinase isoenzymes may clarify the diagnosis. Unfortunately, these heme pigments are direct tubular toxins that may lead to acute tubular necrosis and declining renal function.

Urine Chemistry in the Diagnosis of Renal and Systemic Pathology

Glomerulonephritis is a broad term covering sterile inflammatory disorders of the glomerulus that may be acute or chronic and always lead to at least a temporary decline in renal function. Blood, protein, and casts are common findings on urinalysis, while clinically these patients have edema (swelling), oliguria, hypertension, and electrolyte imbalance. Glomerulonephritis may follow skin or respiratory infections by certain strains of group A streptococci, may accompany systemic diseases like systemic lupus erythematosis, or may be precipitated by nephrotoxic chemicals.

Acute renal failure signifies an abrupt decline in renal function and has a number of possible causes ranging from glomerulonephritis, urinary tract obstruction, and renal artery stenosis to radiocontrast administration and interstitial nephritis that was discussed earlier. It is accompanied by an increase in serum creatinine (Cr) and blood urea nitrogen (BUN), clinically known as azotemia, and frequently is reversible. Two causes, declining renal perfusion due to prerenal phenomena and acute tubular necrosis (ATN), account for over two-thirds of all cases, and analysis of the urine will allow their differentiation. A urinary sediment containing numerous casts, urine osmolality < 350 mOsm/kg, and urinary sodium > 30–40 mEq/L are indicative of the tubular dysfunction associated with ATN, while clear sediment, urine osmolality > 500 mEq/kg, and urinary

sodium < 20 mEq/L correspond to prerenal phenomena. A BUN-to-Cr ratio may be calculated using a single blood specimen, with values greater than 20 indicating prerenal azotemia and those less than 15 accompanying ATN. Because urea is filtered and reabsorbed while creatinine is filtered and secreted but not reabsorbed, a disproportionate rise in BUN will occur as the kidney maximizes sodium and water retention with prerenal states.

The fractional excretion of sodium (FE_{Na}) is the percent of filtered sodium that is excreted in the urine and is the most specific method of distinguishing prerenal azotemia from ATN.

$$FE_{Na} = U_{Na} \times P_{Cr}/P_{Na} \times U_{Cr} \times 100$$

In the above formula, U_{Na} and U_{Cr} are urinary sodium and creatinine, respectively, while P_{Na} and P_{Cr} are plasma sodium and creatinine, respectively. Values less than 1 are indicative of prerenal azotemia while those greater than 1 indicate ATN.

Chronic renal failure is the result of an ongoing, chronic insult to the kidneys such as diabetes and/or hypertension, or it may occur after extended acute renal failure where irreversible damage has occurred. Chronic renal failure is generally accompanied by an increase in serum potassium, systemic acidosis, and ongoing proteinuria. The urinary sediment will be "active," containing casts and tubular epithelial cells. Anemia is commonly seen as the diseased kidney no longer produces the hormone erythropoietin, which facilitates red cell production in the bone marrow.

Initial medical therapy includes alkalization with $NaHCO_3$, medication to prevent phosphate absorption from the gut (since phosphates are poorly excreted by the diseased kidney), and frequently vitamin D and calcium supplements since the diseased kidney will no longer produce the active form of vitamin D normally (leading potentially to brittle bones). Recombinant erythropoietin may be administered to these individuals to maintain an adequate red cell mass.

When medical therapy fails to control the metabolic abnormalities associated with chronic renal failure, the blood may be dialyzed against a solute-poor solution either across an artificial membrane (hemodialysis) or across the peritoneal surface (peritoneal dialysis). Many patients with chronic renal failure are candidates for renal transplantation unless they are elderly or suffer from pathological processes that might reoccur in the transplanted kidney.

Renal calculi occur commonly, affecting roughly 10 percent of the population, and are usually due to hypercalciuria, although hyperuricosuria, hyperoxaluria, hypocitraturia, cystinuria, certain medications, and infections may also precipitate stone formation.

Hypercalciuria (>200 mg/day), which accounts for 40 to 75 percent of all renal calculi, may be of three types: absorptive, resorptive, and renal.

Adsorptive hypercalciuria (AH) is due to intestinal hyperabsorption of calcium and may occur irregardless of dietary calcium intake (type 1) or may correct on a low calcium diet (type 2). Type 3 absorptive hypercalciuria occurs due to renal leak of phosphate, which stimulates renal synthesis of 1,25-dihydroxyvitamin D, resulting in enhanced intestinal absorption of calcium and hypophosphatemia.

Therapy of AH Type 1 utilizes a diuretic to lower renal calcium excretion and sodium cellulose phosphate, a calcium-binding resin to lower calcium absorption from the gut. The addition of a low-calcium diet is helpful in AH Type 2 while supplemental dietary phosphate improves the hypophosphatemia of AH Type 3, decreasing vitamin D synthesis.

Renal hypercalciuria results when tubular reabsorption of calcium is impaired, lowering the serum calcium concentration and resulting in enhanced secretion of parathyroid hormone (secondary hyperparathyroidism). The serum calcium will remain in the normal range due to its ongoing renal loss with concomitant mobilization from the bone and enhanced intestinal absorption (both due to the action of parathyroid hormone).

Therapy with trichloromethiazide, a thiazide diuretic, will facilitate calcium reabsorption from the renal tubule, lowering urinary calcium excretion. The addition of potassium citrate will inhibit calcium salt crystallization.

Resorptive hypercalciuria is due to primary hyperparathyroidism, the result of either a parathyroid adenoma or hyperplasia. The enhanced intestinal calcium absorption and bony reabsorption are not accompanied by a renal tubular loss, thereby elevating the serum calcium level. Surgical parathyroidectomy may be curative.

Hyperuricosuria and hyperoxaluria may occur either due to dietary overindulgence or endogenous overproduction.

Liberal intake of purine-rich foods such as meat, poultry, and fish is the most common etiology of elevated urinary uric acid secretion (>800 mg/day) but 30 percent of affected individuals will have endogenous overproduction.

Medical therapy with allopurinol inhibits the action of the enzyme xanthine oxidase and lowers the systemic production of uric acid while alkalization increases uric acid solubility and prevents the formation of uric acid calculi.

Hyperoxaluria, on the other hand, is commonly the result of inflammatory bowel disease, fat malabsorption, and chronic diarrheal states. The intestinal fat binds dietary calcium, which would otherwise complex with oxalate minimizing its reabsorption by the gut. When the available dietary oxalate pool is increased, hyperabsorption occurs, leading to hyperoxaluria. These individuals also experience low urine outputs due to ongoing fluid losses from the gut, making their risk for renal calculi even greater.

Avoidance of high-oxalate foods like nuts, tea, chocolate, and vitamin C, a high fluid intake, and therapy with inhibitors of crystallization like potassium citrate will lower the risk of stone formation in these individuals.

Citrate acts to lower urinary calcium saturation by forming soluble complexes with calcium and directly inhibiting the crystallization of calcium salts. Citrate excretion is impaired by systemic acidosis, hypokalemia, and urinary tract infection and is enhanced by systemic alkalosis, parathyroid hormone, and vitamin D. Normal urinary citrate excretion should be greater than 320 mg/day. Renal tubular acidosis is therefore associated with calcium phosphate calculi due to hypocitraturia-induced crystallization of calcium with the phosphates seen commonly in the alkaline urine of these individuals. Therapy with potassium citrate increases urinary citrate levels, inhibiting stone formation.

Cystinuria is an inherited defect in renal tubular transport resulting in an inability of the proximal convoluted tubule to reabsorb the amino acids cystine, ornithine, lysine, and arginine. Some 65 percent of these individuals will form calculi early in life; these calculi are composed only of cystine because its solubility is much lower than the other amino acids excreted. Cystine crystals are easily identified microscopically by their characteristic hexagonal appearance, and urine containing excess cystine will appear purple after the addition of nitroprusside solution.

Alkalization of the urine increases cystine solubility while medications such as Tiopronin and Penicillamine inhibit the disulfide bonding of cysteine moieties to form cystine. Dosages should be adjusted to keep urinary cystine at less than 200 mg/L.

Urinary tract infection by organisms producing the enzyme urease may be associated with calculi composed of struvite (calcium magnesium ammonium phosphate). The hydrolysis of urea alkalinizes the urine and results in the crystallization of calcium, magnesium, and ammonium with triphosphate ions.

Medical therapy with acetohydroxamic acid, which inhibits bacterial urease, and antibiotic therapy of the infection may prevent further calculi.

Medications such as triamterine, a potassium-sparing diuretic, and certain sulfonamide antibiotics may crystallize in the urine to form calculi when taken in large amounts. Prevention of these calculi is predicated on taking these medications only as directed and at the appropriate dosages.

Disorders of amino acid, porphyrin, and mucopolysaccharide metabolism may have profound effects on the urine chemistry that are of diagnostic importance.

Phenylketonuria, tyrosyluria, alkaptonuria, and melanuria result from the abnormal excretion of phenylpyruvic acid, tyrosine, homogentisic acid, and melanin, respectively. All are the result of enzymatic defects in the phenylalanine-tyrosine metabolic pathway except for melanuria, which may be seen in individuals with systemic melanomas.

Disorders of tryptophan metabolism may lead to abnormal excretion of indican and 5-hydroxyindoleacetic acid (5-HIAA). Exposure to air oxidizes indican to the dye indigo blue, causing the "blue diaper syndrome" seen in babies with Hartnup disease, an inherited metabolic disorder of the intestinal reabsorption of tryptophan. 5-HIAA is produced in elevated amounts by individuals with active carcinoid tumors.

Porphyrias are disorders of the synthetic pathway for heme and lead to porphyrinuria. They result from either inherited metabolic disorders or exposure to certain toxins such as lead. Certain porphyrinurias may cause the urine to fluoresce under ultraviolet light. The detection of elevated urinary aminolevulinic acid, an intermediary in the heme synthetic pathway, is the primary screening test for lead poisoning.

Disorders of mucopolysaccharide (compounds consisting of a proteinaceous core with numerous polysaccharide branches) metabolism lead to connective tissue abnormalities and elevated urinary levels of mucopolysaccharides. Hurler's and Hunter's syndromes are characterized by abnormal skeletal structure and mental retardation. Urinary screening for mucopolysaccharides is a common component in testing panels for infants with retardation or failure to thrive and frequently provides a diagnosis in these very complex cases.

Conclusion

Urine chemical analysis provides an extraordinary amount of important clinical data at a low cost and plays an important role in the diagnosis of many common disorders. The ease of specimen collection and the availability of basic urine testing in most physicians' offices provide rapid answers to sometimes very difficult clinical problems.

BIBLIOGRAPHY

BRADLEY, B.; SCHUMANN, G. B. "Examination of Urine." In *Clinical Diagnosis and Management by Laboratory Methods;* J. B. Henry, ed.; Philadelphia: W. B. Saunders, 1979.

BRENNER, B. M.; RECTOR, F. C. *The Kidney;* Philadelphia: W. B. Saunders, 1986.

KUSUMI, R. K.; GROVER, P. J.; KUNIN, C. M. "Rapid Detection of Pyuria by Leukocyte Esterase Activity." *JAMA* 1981, *245* (16), 1652–1655.

NYAHN, W. L. *Abnormalities in Amino Acid Metabolism in Clinical Medicine;* Norwalk, CT: Appleton-Century-Crofts, 1984.

PREMINGER, G. M.; PAK, C. Y. C. "The Practical Evaluation and Selective Medical Management of Nephrolithiasis." *Seminars in Urology* 1985, *3* (3), 170–183.

RAVEL, R. *Clinical Laboratory Medicine;* Chicago: Year Book Medical Publishers, 1989.

ROSE, B. D. *Pathophysiology of Renal Disease;* New York: McGraw-Hill, 1987.

STRASINGER, S. K. *Urinalysis and Body Fluids;* Philadelphia: F. A. Davis, 1985.

SULLIVAN, L. P.; GRANTHAM, J. J. *Physiology of the Kidney;* Philadelphia: Lea & Febiger, 1982.

C. WILLIAM HINNANT

V

Valence

This concept deals with the combining capacity of atoms and groups of atoms and with the chemical bonds that hold them together. In the nineteenth century, as the formulas of compounds were determined with accuracy, it became obvious that there were similarities. Edward Frankland (1825–1899) in 1852 suggested the concept of combining power, which implied that a definite number of atoms would combine with the original atom.

This suggested that this property could be made quantitative. Initially attempts were made to keep the combining capacity as one constant number for each element. For example, the compound $FeCl_2$ was considered to have the structural formula

$$\begin{matrix} Cl & & Cl \\ & \diagdown & \diagup \\ & Fe{-}Fe & \quad (Fe_2Cl_4) \\ & \diagup & \diagdown \\ Cl & & Cl \end{matrix}$$

so each iron atom had a valence of 3 (i.e., three single bonds). It was soon evident that combining capacity could not be kept constant for many elements. The hydrogen atom was chosen as the standard, since it forms a single bond in its compounds. The atomic weight of one hydrogen atom is 1.0079 amu; the weight of any element that combines with this weight of hydrogen is the EQUIVALENT WEIGHT of that element. For example, 8.000 grams of oxygen combines with 1.0079 grams of hydrogen, and therefore this is the oxygen's equivalent weight, or its equivalent of hydrogen. Valence could be quantized as "the number of atoms of hydrogen (or its equivalent) that combines with any one atom" and Valence = Atomic Weight/Equivalent Weight.

This implies a theoretical definition, where valence is equal to the number of bonds to the atom. The equivalent weight is easy to determine, and no charge is attached to a valence.

The concept of valence was initially most successful in clarifying many formulas in ORGANIC CHEMISTRY. It was also useful for simple inorganic compounds, but problems arose when coordination complexes were found. Valence fitted easily into the periodic table. D. I. MENDELEEV pointed out that with main group elements, the group (or family) number was the same as the maximum valence, so that for elements in group 1A (alkali metals), the maximum valence is 1, and so on. This solution does not work as well in the transition and inner transition elements (see TRANSITIONAL ELEMENTS). However, it works reasonably well in the first half of the transition elements (groups 3B–7B), but not for the second half, and it is also erroneous for some of the lanthanide and ACTINIDE series. The concept was useful in helping to formulate bonding concepts and in the realization that a bond consisted of a pair of electrons.

The problems with coordination complexes, modern bonding theory, and other difficulties led chemists to use more precise definitions, such as oxidation

1473

number. The concept is not widely used in modern chemistry.

See also ATOMIC MASS UNIT.

BIBLIOGRAPHY

BROWN, G. I. *A New Guide to Modern Valence Theory;* 2nd ed; New York: Wiley, 1967.

IHDE, A. J. *The Development of Modern Chemistry;* New York: Dover, 1984.

NORMAN FOGEL

Valence Bond Theory

Classical valence theory harks back more than a century to the work of E. Frankland, A. S. COUPER, Crum Brown, August KEKULÉ, and many others who deduced simple rules for chemical bonding patterns for "well-behaved" molecular species. Valence-bond (or VB) theory is a reinterpretation and extension of these classical ideas in terms of a closely correspondent quantum mechanical view, as advocated by Linus PAULING starting in the 1930s. Thus, VB theory is in principle more quantitative and more widely applicable.

Perhaps the greatest success is in the understanding of the COVALENT BOND. The simplest sort of case has but a single (dominant) classical valence structure, such as for methane:

$$
\begin{array}{c}
\text{H} \\
| \\
\text{H}-\text{C}-\text{H} \\
| \\
\text{H}
\end{array}
$$

where each bond "slash" is identified with an electron pair, as proposed by G. N. LEWIS and I. LANGMUIR in the pre–quantum-mechanical period. In addition, VB theory interprets each such bond in terms of an electron-pair wave function wherein the two electrons each smear out with the greatest concentrations in the bond region. The full many-electron VB wave function is a type of product of pair-wave functions and is obtainable via sophisticated computations based on SCHRÖDINGER'S WAVE EQUATION of quantum mechanics (see also QUANTUM MECHANICAL COMPUTATIONS). Especially for well-behaved cases like methane, reasonable yet simple (lower-order) approximations are available—e.g., in terms of sp^3-hybrid orbitals directed away from the carbon nucleus toward the corners of an imagined tetrahedron centered on the nucleus—and the results merge nicely with classical ideas.

A key point of early VB theory was a well-founded interpretation of the delocalized bonding in species, such as benzene, with two classical valence structures:

Here each structure is mathematically represented by an overall many-electron wave function built from electron pairs much as for methane, but now the full benzene wave function is the sum of the two many-electron wave functions, each associated to one classical structure. There are "wave interference" effects (usually termed RESONANCE effects) between these two simultaneous component wave functions, so that the properties of benzene are not (as correctly recognized before the advent of quantum mechanics) a sum or average of those of the component classical structures—i.e., benzene chemically reacts differently from species with localized double bonds, and many physical properties differ markedly from what is anticipated for a classical mixture of the two structures. One may note that the symmetry equivalence of the two classical structures is essentially irrelevant: the argument applies as well for ortho-xylene or 1,2 dimethylbenzene with two adjacent hydrogens of benzene replaced by two methyls so that the two classical structures are inequivalent in symmetry, though the two corresponding wave functions no longer combine with exactly equal weights.

Historically the first quantum-mechanical treatment of the chemical bond in a stable species coincides with the first VB-theoretic treatment—that for H_2 by Heitler and London in 1927. Thereafter, an alternative molecular-orbital (MO) scheme was developed, involving electron-pair wave functions, in each of which both electrons are identically (and often rather uniformly) spread throughout the molecule. (See also MOLECULAR ORBITAL THEORY.) This MO approach was earlier implemented in a more widely applicable computational format with less dependence on an individual investigator's insight (or lack thereof), so that quantitative computations have more typically been via MO theory, though some less direct correspondences to classical chemical-bonding ideas remain. Now high-quality

VB computations are possible for smaller molecules (say, for those not too much larger than benzene), while cruder VB-theoretic treatments are applicable not only to larger species but also as an intermediate description for smaller molecules. The smaller a molecule (especially in terms of its electron count), the better a wave function that can be computationally realized (by including more terms corresponding to ever less important structures), so that for small atoms or diatomics virtually exact quantitative results are obtainable, not only for GROUND STATES but for EXCITED STATES as well. These higher-order quantitative corrections do not seem so susceptible to simplification, and even for systems so treated the low-order (more qualitative) theories provide a useful understanding.

The quantitative convergence of quantum-mechanical expansions based on either the VB- or MO-theoretic approaches is not in question. Each is a different perspective of an underlying "quantum reality." A preference for one view or another is based on auxiliary criteria: the rapidity of convergence toward quantitative precision; the readiness (or uniqueness) of computational implementation; or the correspondence to a hierarchy of more simplified views merging nicely to classical chemical-bonding ideas.

The standard reference indicating the wide range of applications (in both organic and inorganic realms) of the simplest VB-theoretic ideas is Pauling's masterwork *The Nature of the Chemical Bond*. A varied collection of typically more technical surveys is found in the more recent *Valence-Bond Theory and Chemical Structure*—the first survey is general, short, and less technical; the second is general, longer, and at a level of technicality following a modern undergraduate physical chemistry course. Interesting global discussions are found in the articles by Malrieu, Shaik, Hiberty, et al. in the second book in the bibliography. Currently VB-theoretic research is active with exciting developments in diverse directions.

BIBLIOGRAPHY

KLEIN, D. J.; TRINAJSTIĆ, N., eds. "Valence-Bond Theory and Chemical Structure." *J. Chem. Ed.* 1990, 67, 633–637.

———. *Valence-Bond Theory and Chemical Structure;* Amsterdam: Elsevier, 1990.

MAKSIĆ, Z. B.; ECKERT-MAKSIĆ, M., eds. *Molecules in Natural Science and Medicine;* London: Ellis Horwood, 1991.

PAULING, L. *The Nature of the Chemical Bond;* Ithaca, NY: Cornell University Press; 1st ed., 1939; 3rd ed., 1960.

DOUGLAS J. KLEIN

Valence Electron

Loosely speaking, a valence electron is one that participates in or is available to participate in CHEMICAL BONDING. It is more accurate to refer to valence orbitals because they remain fixed, while the indistinguishable electrons share all available orbitals. The term *valence electron* refers to electrons that occupy valence orbitals. The valence orbitals of atoms are those that are combined in quantum mechanical models to form the molecular orbitals of molecules. Valence electrons (orbitals) are distinguished from others by the properties that make them suitable for bond formation. They have small ionization energies, usually less than 15eV (about 1,500 kJ/mole), while core electrons have ionization energies of 100eV or more. Valence electrons lie in outer atomic regions where they are first to encounter valence orbitals of approaching atoms. Valence orbitals are highly polarizable, i.e., they undergo significant shape distortions in response to the fields surrounding approaching atoms or molecules. These properties allow valence electrons to be easily lost, gained, or shared between atoms during chemical reactions.

The following examples of atomic electron configurations illustrate the valence electrons in groups I and IV of the periodic table. The valence orbitals are italicized and the core orbitals are enclosed in brackets, [ ]. In group I of the periodic table are H with configuration $1s$, Li with $[1s^2]2s$, and Na with $[1s^22s^22p^6]3s$. In group IV are C with $[1s^2]2s^22p^2$, Si with $[1s^22s^22p^6]3s^23p^2$, and Ge with a configuration of $[1s^22s^22p^63s^23p^63d^{10}]4s^24p^2$. Generally, elements in the same group have similar chemistry because they possess similar valence electrons.

RONALD D. POSHUSTA

Valence Shell Electron Pair Repulsion

The valence shell electron pair repulsion, or VSEPR, model is a simple method for predicting the arrangement of the bonds around an atom in a covalent molecule (see also COVALENT BOND). This method is based on the Lewis diagram for a molecule in which valence shell electrons are arranged in pairs that are either bonding (shared) or nonbonding (unshared, or lone) pairs (see also LONE PAIR; LEWIS STRUCTURES). The model was developed by R. J. Gillespie and R. S. Nyholm (1957) from some ideas that were first proposed by N. V. SIDGWICK and

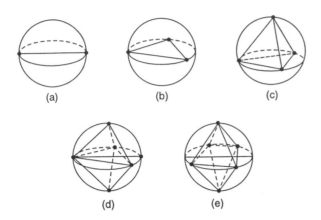

Fig. 1. The points-on-a-sphere model. Arrangements of points that maximize their distance apart: (a) linear arrangement of two points; (b) equilateral triangular arrangement of three points; (c) tetrahedral arrangement of four points; (d) trigonal bipyramidal arrangement of five points; (e) octahedral arrangement of six points.

H. E. Powell (1940), and it was later extended by Gillespie (1972, 1991, 1992).

The basic postulate of the VSEPR model is that a given number of valence shell electron pairs adopt the arrangement that keeps them as far apart as possible; in other words, the electron pairs behave as if they repel one another. These arrangements are shown in Figure 1 for two to six electron pairs, where each electron pair is represented by a point on the surface of a sphere, which represents the valence shell. Alternatively, the region of space occupied by an electron pair—called the electron pair domain—can be approximately represented by a sphere. Therefore, the preferred arrangements of electron pairs are those in which the spheres are packed as closely as possible around a central point representing the core of the atom.

If A is a central atom, X a ligand atom or group, and E a lone (or unshared, or nonbonding) pair, then a molecule with one central atom A is conveniently denoted as AX_nE_m. The shape of the molecule then depends on the numbers of bonding electron pairs n and lone (nonbonding) electron pairs m in the valence shell of A, as shown in Figure 2. This figure illustrates all the possible shapes of molecules with up to six electron pairs in the valence shell of the central atom. An example of each of these molecular shapes is given in Figure 3.

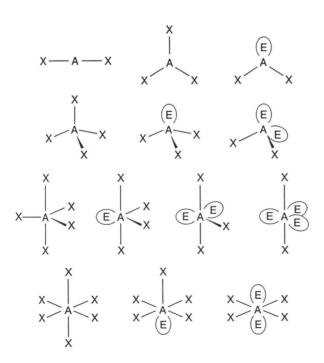

Fig. 2. Predicted shapes for molecules with a central atom A and up to six electrons in their valence shell.

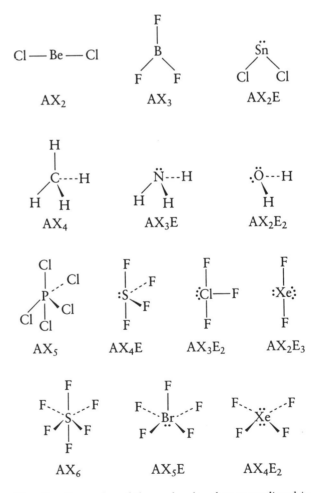

Fig. 3. Examples of the molecular shapes predicted in Figure 2.

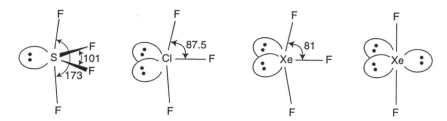

Fig. 4. In molecules with five electron pair domains in the valence shell of the central atom, lone pairs always occupy the equatorial positions and never occupy the axial positions.

An important feature of the VSEPR model is that qualitative predictions about deviations from the ideal shapes given in Figures 3 and 4 can be made very easily on the basis of differences in the sizes of lone pair and bond pair domains.

Nonbonding or Lone Pairs

Lone pair domains are larger than bond pairs because a bond pair is under the influence of two atomic cores, whereas a lone pair is under the influence of only one atomic core. As a consequence, the BOND ANGLES in AX_3E and AX_2E_2 molecules are smaller than the ideal tetrahedral angle, and in AX_5E molecules they are smaller than 90° (Table 1). In a trigonal bipyramidal arrangement of five electron pairs, lone pairs in AX_4E, AX_3E_2, and AX_2E_3 molecules always occupy the less crowded equatorial positions (Figure 4).

Multiple Bonds

The sharing of two electron pair domains between two atoms to give a double bond and the sharing of three electron pair domains to give a triple bond can

be represented as in Figure 5a for the ETHENE and ETHYNE molecules. This domain model corresponds to the classical bent-bond descriptions of these molecules (Figure 5b) and leads directly to the prediction of a planar geometry for ethene and a linear geometry for ethyne. A better, simpler representation is to consider that the two single electron pair domains of a double bond are merged into one double bond (four electron) domain and that the three single electron pair domains of a triple bond are merged into one triple bond (six electron) domain (Figure 5c). Both

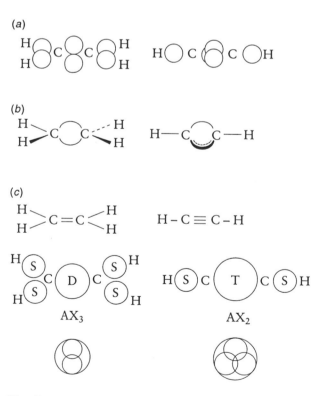

Fig. 5. (a) Electron pair domain models of ethene and ethyne; (b) bent bond models of ethene and ethyne; (c) multiple bond domain models of ethene and ethyne. D is a double bond domain and T a triple bond domain.

Table 1. Bond Angles (°) in AX_3E, AX_2E_2, and AX_5E Molecules

AX₃E		AX₂E₂		AX₃E	
NH₃	107.2	H₂O	104.5	ClF₅	86.0
NF₃	102.3	F₂O	103.1	BrF₅	85.1
PF₃	97.7	SF₂	98.0	IF₅	83.0
PCl₃	100.3	SCl₂	102.0	XeF₅⁺	79.0
PBr₃	101.0	S(CH₃)₂	99.0	SF₅⁻	88.0
AsF₃	95.8	Se(CH₃)₂	96.0	TeF₅	79.0
AsCl₃	98.9	TeBr₂	104.0	SbF₅²⁻	82.0

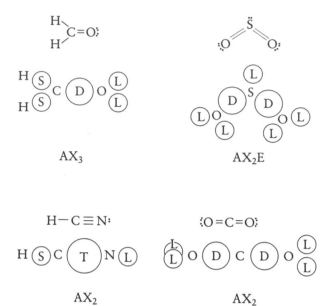

Fig. 6. Domain models of H_2CO, SO_2, HCN, and CO_2. D is a double bond domain, T is a triple bond domain, and L is a lone pair domain.

carbon atoms in ethene then have a planar triangular AX_3 geometry, and the two carbon atoms in ethyne both have a linear AX_3 geometry. Other examples of molecules containing double and triple bonds are given in Figure 6. Because a double bond domain is larger than a single bond domain and a triple bond domain is larger still, these domains take up more

Table 2. Bond Angles (°) in Some Molecules Containing C=C and C=O Double Bonds

$X_2C{=}CY_2$	XCX	YCY	XCC	YCC
$H_2C{=}CH_2$	116.2	116.2	121.9	121.9
$F_2C{=}CH_2$	110.6	119.3	124.7	120.3
$F_2C{=}CF_2$	112.4	112.4	123.8	123.8
$Cl_2C{=}CCl_2$	115.6	115.6	122.2	122.2
$(CH_3)_2C{=}CH_2$	115.6	116.2	122.2	121.9

	XCX	XCO
H_2CO	116.5	121.7
Cl_2CO	111.8	124.1
F_2CO	107.7	124.1
HFCO	110	123

Table 3. Bond Angles (°) in AX_4 Molecules Containing Multiple Bonds

	sb–sb	sb–db(tb)		sb–sb	db–db
POF_3	101.3	117.7	F_2SO_2	96.1	124.0
$POCl_3$	103.3	115.7	Cl_2SO_2	100.3	123.5
$POBr_3$	104.1	115.0	$ClFSO_2$	99.0	123.7
PSF_3	99.6	122.7	$(NH_2)_2SO_2$	112.1	119.4
$PSCl_3$	101.8	117.2	$(CH_3)_2SO_2$	102.6	119.7
$PSBr_3$	101.9	117.1			
NSF_3	94.0	125.0			

space in a valence shell than a single bond domain, so that bond angles involving multiple bonds are larger than those between single bonds in the same valence shell. Examples are given in Tables 2 and 3. For the same reason MULTIPLE BONDS always occupy the equatorial positions of an AX_5 molecule.

Ligand Electronegativity

The space occupied by a bonding domain in the valence shell of the atom A decreases with increasing ELECTRONEGATIVITY of X and decreasing electronegativity of A as electron density moves from the valence shell of A into the valence shell of X. Hence bond angles in an AX_nE_m molecule decrease with increasing electronegativity of X and decreasing electronegativity of A, as can be seen from the data in Table 4. In a trigonal bipyramidal AX_5 molecule, the smaller domains of the bonds to more electronegative LIGANDS always occupy the more crowded axial sites.

The application of the model to molecules with a central atom with a valence shell containing more

Table 4. Effect of Ligand Electronegativity on Bond Angles

Increasing Ligand Electronegativity							
H_2O	104.5	F_2O	103.1				
SF_2	98.0	SCl_2	102.7				
NH_3	107.2	NF_3	102.3				
PI_3	102.0	PBr_3	101.0	PCl_3	100.3	PF_3	97.7
AsI_3	100.2	$AsBr_3$	99.8	$AsCl_3$	98.9	AsF_3	95.8

than six electron pairs and to molecules of the transition metals is discussed in more detailed treatments of the model (Gillespie, 1972 and 1992; Gillespie and Hargittai, 1991).

See also MOLECULAR STRUCTURE.

BIBLIOGRAPHY

GILLESPIE, R. J. *Molecular Geometry;* London: Van Nostrand-Reinhold, 1972.

———. "The VSEPR Model Revisited." *Chem. Soc. Reviews* 1992, *52*, 59.

GILLESPIE, R. J.; HARGITTAI, I. *The VSEPR Model of Molecular Geometry;* Englewood Cliffs, NJ: Prentice-Hall, 1991.

GILLESPIE, R. J.; NYHOLM, R. S. "Inorganic Stereochemistry." *Quart. Rev. Chem. Soc.* 1957, 339.

SIDGWICK, N. V.; POWELL, H. E. "Stereochemical Types and Valency Groups." *Proc. Roy. Soc. A.* 1940, *176*, 153.

RONALD J. GILLESPIE

Vanadium

23
V
50.9415

Melting Point: 2163 K; **Boiling Point:** 3653 K; **Density:** 5.96 g/cm^3

Vanadium is the nineteenth most abundant element in the Earth's crust, with an average abundance of approximately 136 ppm. It was initially discovered in 1801 by Manuel del Rio, who was working on a lead ore. Del Rio could not isolate it and was convinced by others that he had identified PbCrO$_4$. Firm identification of the metal occurred in 1830 when N. G. Sefström isolated the element from an iron ore extracted from a mine in Sweden. Sefström named the element after Vanadis, the Norse goddess of beauty. In 1831 WÖHLER proved that del Rio had found a new element. A very pure sample of the element was not obtained until 1867, when H. E. Roscoe reduced the chloride with hydrogen gas.

The element is widely distributed, but in very low abundance. There are about sixty minerals containing vanadium, with six being of importance. Bravoite, patronite, and sulvanite are sulfides. Davidite is a titanite, and carnotite and roscoelite are oxides. Production of vanadium occurs in Australia, China, CIS (former USSR), Finland, South Africa, and the United States. The metal has also been detected in some meteorites. Vanadium occurs with deposits of uranium—carnotite is a potassium uranyl vanadate—and increased demand for uranium provides for increased production of vanadium from this source.

The metal is currently prepared by one of several techniques: reduction of the chloride with hydrogen or magnesium; reduction of the oxide with aluminum, calcium, or carbon; or electrolysis of the partially refined metal in fused alkali metal halide salts.

The steel industry accounts for over 80 percent of the world's consumption of vanadium as a steel additive. In 1980 the production of vanadium and its alloys exceeded 35,000 tons. Vanadium improves the wear resistance and high-temperature strength of steels by forming carbides in the steels. Thus, vanadium steels are used in rust-resistant steels, spring steels, and high-speed steels, for example, for making taps and dyes, gears and axles for automobile parts, turbine motors, and turbine blades.

Vanadium oxide (V$_2$O$_5$) is a useful catalyst in the oxidation of SO$_2$ to SO$_3$ to produce sulfuric acid. Other important oxidations include the conversion of naphthalene or orthoxylene to phthalic anhydride. The addition of small amounts (0.02 percent) of V$_2$O$_5$ to glass reduces infiltration of ultraviolet rays, and larger amounts of the oxide produce a yellow-green glass. The oxide and metavanadates have been used to assist with the formation of black pigments from oils. The textile industry has also used small amounts of the oxide to assist the oxidation of aniline, which provides black dyes with increased color fastness.

Tunicates have been found to concentrate vanadium in green blood cells. Vanadium has also been reported to inhibit development of dental caries and cholesterol synthesis in humans, although the body levels are very low. Like many elements that are essential micronutrients, vanadium and vanadium compounds are toxic in larger amounts.

BIBLIOGRAPHY

CARLSON, O. N.; STEVENS, E. R. "Vanadium and Vanadium Alloys." In *Encyclopedia of Chemical Technology,* 2nd ed.; R. E. Kirk, D. F. Othmer, A. Standen, eds.; New York: Wiley-Interscience, 1970; Vol. 21, pp. 157–167.

GREENWOOD, N. N.; EARNSHAW, A. *Chemistry of the Elements;* New York: Pergamon, 1984; pp. 1138–1143.

LIDE, D. R., ed. *Handbook of Chemistry & Physics,* 71st ed.; Cleveland: Chemical Rubber Co., 1990–1991.

ALTON J. BANKS

Van der Waals, Johannes Diderik (1837–1923)

Johannes Diderik van der Waals was of humble origins and something of a latecomer to the world of scientific research. He was the son of a carpenter, born in Leiden, the Netherlands, on November 23, 1837. After spending his early career as a teacher in primary schools, he entered Leiden University in 1862 to study physics and obtained a doctoral degree at age 36. During his university studies van der Waals supported himself as a secondary-school physics teacher; in 1866 he became headmaster of a school at The Hague. In 1877, he became a professor of physics at the University of Amsterdam, where he remained until his retirement in 1907, when he was succeeded by his son. During his first twenty years at the University of Amsterdam, he was the only teacher of physics there. The Royal Academy of Sciences in Amsterdam benefited greatly from work done by van der Waals. In 1896, he became secretary of the academy, a post he held until 1912. After long service to the fields of physics and physical chemistry, van der Waals died in Amsterdam on March 9, 1923.

The body of van der Waals's scientific work grew out of the topic of his doctoral thesis, "On the Continuity of the Gaseous and Liquid States." Using simple mathematics, van der Waals gave a molecular explanation for the phenomena that Thomas Andrews and others had observed in vapors and liquids. Of special interest was the existence of a critical temperature, above which a gas or vapor cannot be liquefied by pressure alone, no matter how great. This explanation was one of the first descriptions of a collective molecular effect.

Some years later, van der Waals expanded on the work of his thesis to develop the law of corresponding states. This allows a calculation of the state of any gas or liquid at any temperature and pressure—if the state at the critical temperature is known. This equation of state attempts to explain the behavior of real gases, as opposed to the "ideal" gas laws of Robert BOYLE, Jacques Charles, and Joseph GAY-LUS-SAC. It still utilizes pressure (P), volume (V), absolute temperature (T), and the universal gas constant R, but it adds two other constants, a and b, to give the equation

$$(P + a/V)(V - b) = RT$$

The term a/V accounts for intermolecular attraction, and the constant b accounts for the nonoverlapping of molecules and their finite size, with a and b being different for different gases. In addition, the cohesive attraction between molecules in a liquid and in molecular crystals such as graphite and naphthalene became known as van der Waals forces.

For this body of work concerning the equation of state of gases and liquids, Johannes van der Waals was awarded the 1910 Nobel Prize in physics. This relatively unassuming man from humble beginnings,

who was known for the simple life he led, had a profound impact on twentieth-century science. Though he was physically of small stature, he cast an enormous intellectual shadow.

See also VAN DER WAALS EQUATION; VAN DER WAALS RADIUS.

BIBLIOGRAPHY

ABBOTT, D., ed. "Johannes van der Waals." In *Chemists: The Biographical Dictionary of Scientists;* New York: Peter Bedrick Books, 1983; pp. 140–141.

FARBER, E., ed. "Johannes Diderik van der Waals." In *Great Chemists;* New York: Interscience Publishers, 1961; pp. 751–755.

PRINS, J. A. "Johannes Diderik van der Waals." In *Dictionary of Scientific Biography;* C. C. Gillispie, ed.; New York: Scribner, 1975; Vol. 14, pp. 109–111.

WEBER, R. "Johannes Diderik van der Waals." In *Pioneers of Science: Nobel Prize Winners in Physics;* Bristol and London: The Institute of Physics, 1980; pp. 40–41.

KAY REAT

Van der Waals Equation

The Van der Waals equation is a relatively simple equation that expresses the real behavior of gases better than the IDEAL GAS LAW. In addition, Johannes Diderick VAN DER WAALS arrived at his corrections to the ideal gas equation ($PV = nRT$) through a particularly straightforward chain of reasoning. The KINETIC MOLECULAR THEORY of gases, from which one can derive the ideal gas law, assumes that gas molecules are point masses (they have no volume) and that there are no forces of attraction between them. The assumption of volumeless molecules also is implicit in CHARLES'S LAW ($V = kT$, with pressure and amount held constant), and the assumption of no forces of attraction is implied by DALTON'S LAW OF PARTIAL PRESSURES. But if gas molecules have no volume, then a liquid—which has molecules in contact—should be virtually volumeless as well, and it clearly is not. And if there were no forces of attraction between molecules, then liquids would not exist, since the merest agitation would be sufficient to send all the molecules flying off into the gas phase. Therefore, the volume of molecules of a gas and intermolecular forces of attraction must be considered.

In 1873, van der Waals proposed that both the calculated pressure and volume and the ideal gas law are in error. The error in the volume is that the ideal gas law assumes that the volume of the container is the free volume available for the molecules to move around in. But the molecules themselves exclude some small part of the volume of the container; this volume is unavailable for molecules to move through, since it already is occupied by other molecules. If the volume occupied by 1 mol of gas molecules is symbolized by b—the molar excluded volume—then the free volume for the ideal gas equation becomes $V_{free} = V_{container} - nb = V - nb$, since we assume V to be the container's volume. With this modification, the ideal gas law becomes $P(V - nb) = nRT$.

Van der Waals proposed that the error in the pressure arises from the molecules attracting each other. Hence, a molecule in the body of the gas is attracted to molecules in all directions. These attractions balance out, and the molecule feels no net force of attraction. On the other hand, a molecule that is about to hit the container wall has no attracting molecules between itself and the wall. It experiences no force of attraction toward the wall. Yet it is still attracted by molecules in the body of the gas. (All other attractions—above and below, to the right and the left—are balanced.) Thus, a molecule that is about to hit the container wall experiences a net force of attraction back into the body of the gas. This molecule will hit the wall with a diminished force, producing a smaller pressure than if there were no forces of attraction. To compensate, a term should be added to the ideal-gas pressure. This term should depend on the nature of the gas and also on the concentration (n/V) of the gas molecules, since concentration determines how many gas molecules are close enough to attract the molecule back into the gas. In fact, concentration should appear twice, the second time to account for the number of molecules that are about to exert pressure by hitting the wall.

The van der Waals equation therefore is $[P + (an^2/V^2)](V - nb) = nRT$. E. S. Swinbourne (1955) has pointed out that molecules of the gas are attracted to the walls of the container as well as back into the body of the gas. The pressure correction actually is required because collisions are not instantaneous, owing to the mutual attraction, or "stickiness," of the molecules. But the results of van der Waals's reasoning still are valid: The b represents the volume excluded by the molecules, whether this exclusion is conceived as being that of small, impenetrable spherical molecules or produced by strong but short-range repulsive forces between molecules. The a term is the result of forces of attraction between molecules. Consequently, in other contexts, the van der Waals radius of a molecule is taken to be its nonbonded radius. The common nonbonding intermolecular forces are referred to as van der Waals forces: dipole-dipole

attractions, exchange forces (also called London forces, or instantaneous-dipole-induced-dipole forces), and sometimes hydrogen bonds.

The values of a and b for a particular gas usually are obtained by fitting the van der Waals equation to pressure-volume isotherms (that is, lines of constant temperature), such as those in Figure 1 for carbon dioxide. The data for this figure are values obtained by Michels et al. (1937) for carbon dioxide. The observed behavior of carbon dioxide is indicated by the data points, which are joined by the heavier lines. The light wavy line represents the behavior predicted by the van der Waals equation for 29.93°C and 30.41°C. For carbon dioxide under these conditions, the only place where the van der Waals prediction differs significantly from actual behavior is in the region (outlined by a dome) in which both liquid and vapor are present. To the left of the dome, only liquid is present. To the right and above, only gas and vapor exist. Interestingly, even the incorrect predictions of the van der Waals equation (the predictions within the dome) are related to observed phenomena. They represent metastable states, states that revert to a stable situation when slightly perturbed. The van der Waals prediction is above the isotherm on the right side of the domed region. This corresponds to a vapor that has been compressed beyond the condensation pressure, a supersaturated vapor. A practical example is supersaturated water vapor sometimes

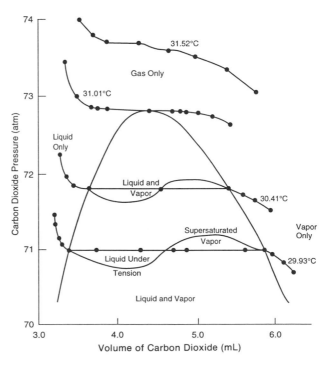

Fig. 1. Pressure-volume diagram for carbon dioxide.

found in clouds, a sitution that is perturbed by "seeding" the clouds to initiate rain. On the left side of the domed region, the van der Waals prediction is below the isotherm. This corresponds to a liquid under reduced pressure, under tension. This occurs as a propeller spins in the water. Bubbles of vapor form almost explosively as small perturbations in the propeller's surface relieve the liquid tension. These minature explosions can actually tear chunks from the metal propeller, a phenomenon known as *cavitation*. Because cavitation can cause premature failure, propeller designs are modified to minimize its occurrence.

Van der Waals proposed his equation of state for gases in his doctoral dissertation of 1873, "On the Conformity of the Gaseous and Liquid States." Van der Waals spent the rest of his scientific career investigating the behavior of fluids and explaining the behavior of liquids. He received the Nobel Prize in 1910 "for his work concerning the equation of state of gases and liquids."

See also GASES, BEHAVIOR OF.

BIBLIOGRAPHY

MICHELS, A., et al. "The Isotherms of CO_2 in the Neighbourhood of the Critical Point and Round the Coexistence Line." *Proc. Roy. Soc.* (*London*) 1937, *A160*, 358.

SWINBOURNE, E. S. "The Van der Waals Gas Equation: A Simple Kinetic Treatment." *J. Chem. Educ.* 1955, *32*, 366–369.

ROBERT K. WISMER

Van der Waals Radius

Johannes Diderik van der Waals (1837–1923) contributed to the development of the KINETIC MOLECULAR THEORY of gases, and he is memorialized by having some of his contributions named for him: the VAN DER WAALS EQUATION of state for gases and the van der Waals radius of atoms, for example. J. D. van der Waals developed an equation of state for gases that fits the actual behavior of real gases better than the IDEAL GAS LAW by including two factors to account for the fact that real gas molecules attract each other and the real gas molecules take up space. The space that gas molecules occupy can be deduced from the volume of liquid that is produced when a gas is sufficiently cooled to condense.

The van der Waals radius of an atom is the volume the atom would occupy if it were in contact with other atoms but not bound to any other atom. The van der Waals radius of the heavier noble gas elements can be determined as one-half the internuclear distance of atoms in the solid form of the element. The internuclear distance is typically determined using X-RAY CRYSTALLOGRAPHY. For other elements, in which atoms in species in solid form are chemically bound to other elements, determining the van der Waals radius is not as straightforward. When a solid consists of molecules in which atoms of the same element in different molecules are in contact (e.g., the chlorine atoms of neighboring molecules of carbon tetrachloride in solid carbon tetrachloride), one half the internuclear distance between the like atoms gives the van der Waals radius. The internuclear distance is not directly obtained from X-ray crystallography but is estimated from the electron distribution pattern that results from the analysis of the diffraction data.

For many elements, there are no compounds in which atoms of the same element are in nonbonded contact in the solid state, but there are compounds in which these elements are in nonbonded contact in the solid state with an element whose van der Waals radius is known. In these cases, the unknown van der Waals radius has been determined by subtracting the known van der Waals radius from the internuclear distance between the two atoms. Using a table of van der Waals radii, a crystallographer can readily infer whether two atoms in a crystal are bonded to each other—which gives a shorter internuclear distance than the sum of van der Waals radii of the two atoms—or whether the two atoms are constituents of separate molecules, in which case the internuclear distance is very close to the sum of van der Waals radii.

BIBLIOGRAPHY

HUHEEY, J. E.; KEITER, E. A.; KEITER, R. L. *Inorganic Chemistry;* New York: Harper Collins, 1993.

DAVID LAVALLEE

Van't Hoff, Jacobus Henricus (1852–1911)

Jacobus Henricus van't Hoff, born on August 30, 1852, in Rotterdam, the Netherlands, was the third of seven children of a physician of the same name and Alida Jacoba van't Hoff (*née* Kolff). After receiving his diploma in technology from the Delft Polytechnic School in 1871, he studied mathematics for a year at the University of Leiden. From the fall of 1872 to the spring of 1873, he worked with August KEKULÉ at the University of Bonn. In 1873 he passed his doctoral examination in chemistry at the University of Utrecht, from which institution he received his doctorate on December 22, 1874.

In early 1874 van't Hoff studied chemistry with Charles-Adolphe Wurtz in Paris, where he met Joseph Achille LE BEL, considered, with van't Hoff, to be a cofounder of stereochemistry. In September 1874 at the age of twenty-two, he published his eleven-page classic paper in Dutch, entitled *A Suggestion Looking to the Extension into Space of the Structural Formulas at Present Used in Chemistry and a Note upon the Relation between the Optical Activity and the Chemical Constitution of Organic Compounds*. This short but influential booklet, which is considered to mark the founding of STEREOCHEMISTRY and which preceded Le Bel's similar but independently conceived proposal of the asymmetric carbon atom by only two months, explained the existence of organic optical isomers by postulating that the carbon atom possesses a tetrahedral configuration.

In 1876 van't Hoff became lecturer in physics at the State Veterinary School in Utrecht, and in 1877 he became lecturer in theoretical and physical chemistry at the University of Amsterdam, where from 1878 to 1896 he was successively professor of chemistry, mineralogy, and geology and chemistry department head. His inaugural lecture of October 11, 1878, based on studies of biographies of famous scientists, was titled "The Power of Imagination in Science." In 1878 he married Johanna Francisca Mees, by whom he had two sons and two daughters.

In 1877 van't Hoff began to work on chemical thermodynamics and affinity, and in 1884 he proposed his principle of mobile equilibrium. From 1885 to 1890 his studies of OSMOTIC PRESSURE and the analogy between dilute solutions and gases appeared. These were among the earliest papers to be published in the *Zeitschrift für physikalische Chemie*, which he had cofounded in 1887 with Wilhelm OSTWALD. In 1896 he became honorary professor at the University of Berlin, where he lectured only once a week, allowing him to devote himself entirely to research. Here he began a completely new field of research, a study of the equilibria involved in the marine salt deposits at Stassfurt according to Gibbs's phase rule (see PHASE RULE), which not only was of great value to the German potash industry but also made him the founder of the science of petrology (the study of the origin, structure, chemical composition, and classification of rocks). The author of numerous books and articles and the recipient of many honors, in 1901 van't Hoff received the first Nobel Prize for chemistry "for the discovery of the laws of chemical dynamics and osmotic pressure in solutions." Like the other two "ionists," Wilhelm Ostwald and Svante August ARRHENIUS, van't Hoff visited the United States, lecturing at the University of Chicago in 1901 and describing his travels in great detail. He died in Steglitz (now Berlin) on March 1, 1911, of pulmonary tuberculosis.

BIBLIOGRAPHY

COHEN, E. "Jacobus Henricus van't Hoff, 1852–1911." In *Great Chemists;* Ed. and trans. by E. Farber; New York: Interscience, 1961; pp. 947–958.

HOLLEMAN, F. "My Reminiscences of van't Hoff." *J. Chem. Educ.* 1952, 29, 379–382.

KAUFFMAN, G. B. "The Centenary of Physical Chemistry." *Educ. Chem.* 1987, 24, 168–170.

SNELDERS, H. A. M. "Jacobus Henricus van't Hoff." In *Dictionary of Scientific Biography;* C. C. Gillispie, ed.; New York: Scribner, 1976; Vol. 13, pp. 575–581.

VAN KLOOSTER, H. S. "Van't Hoff in Retrospect." *J. Chem. Educ.* 1952, 29, 376–379.

VAN'T HOFF, J. H., *Imagination in Science;* trans. and introduced by G. F. Springer; New York: Springer-Verlag, 1967.

GEORGE B. KAUFFMAN

Vapor Pressure

When a pure gas is in equilibrium with a condensed phase of the same substance, the pressure of the gas is called the vapor pressure of the substance. If the condensed phase is a solid, the term *sublimation pressure* is also used. A pure element or compound has a fixed vapor pressure at a given temperature. The equilibrium is dynamic, with equal rates of molecular condensation and evaporation across a planar phase boundary.

Vapor pressures are a result of the balance between the energy due to attractive intermolecular forces (favoring the condensed phase) and thermal translational energy of the molecules (favoring the gas phase). Solids usually have extremely low vapor pressures at normal temperatures. Liquids have higher vapor pressures, and if continuous removal of the vapor results in complete evaporation within a short time, the liquid is termed volatile. Vapor pressures for some common substances at room temperature are shown in Table 1.

Vapor pressure increases sharply with temperature up to the critical point (Figure 1). A liquid boils when its vapor pressure reaches the external pressure (usu-

Table 1. Vapor Pressures for Some Substances at 20°C

Substance	p(torr)
mercury (liquid)	0.002
white phosphorus (solid)	0.03
naphthalene (solid)	0.05
iodine (solid)	0.25
water (liquid)	18
ethanol (liquid)	44
diethyl ether (liquid)	442
carbon dioxide (liquid)	4.5×10^4

ally 1 atm). The vapor pressure line, as shown in Figure 1, is part of the phase diagram describing equilibria between all three phases of the substance. Quantitatively, vapor pressure (p) and temperature (T) are related by the Clapeyron equation:

$$dp / dT = \Delta S / \Delta V \qquad (1)$$

where ΔS and ΔV are the changes in molar entropy and molar volume, respectively, in going from liquid to gas. Although equation 1 is general, a more useful approximate form, the Clausius-Clapeyron equation, is easily derived from it by integration:

$$\ln p = -\Delta H_{vap} / RT + \text{constant} \qquad (2)$$

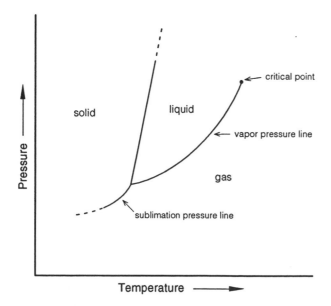

Fig. 1. Typical phase diagram for a pure substance.

Here ΔH_{vap} is the enthalpy of vaporization of the liquid. Equation 2 is valid for limited temperature ranges, and a linear plot of ln p versus $1/T$ can be used to find ΔH_{vap}. Literature listings of vapor pressures frequently utilize equation 2, and tabulate constants a and b for an empirical expression of the form:

$$\log_{10} p \text{ (atm)} = a / T + b$$

Techniques used for determining vapor pressure depend on the volatility of the substance. For a volatile liquid, the pressure can be measured directly by introducing an excess of the liquid into an evacuated volume attached to a pressure-measuring device. The simplest such volume is the space above a mercury barometer column. Other methods include measuring the weight loss in a liquid when a known volume of inert gas is bubbled through, and using an isoteniscope, in which the vapor pressure is balanced against a variable external pressure. Low vapor pressures, such as for mercury or for some organic crystals at room temperature, can be measured using a Knudsen cell, which is a small evacuated vessel with a pinhole to the surroundings through which vapor is continuously evacuated. The rate of loss of weight of a solid or liquid by effusion from such a vessel is related to the vapor pressure through kinetic theory.

In the presence of a foreign gas, the vapor pressure of a substance increases, although the effect is small except at very high pressure. The quantitative relationship is:

$$\ln p'/p = V_1(P - p)/RT \qquad (3)$$

where p is the normal vapor pressure, and p' the vapor pressure at a total pressure P. V_1 is the molar volume of the liquid. For example, the vapor pressure of water increases about 8 percent in the presence of 100 atm of argon at 25°C.

Vapor pressure over a curved interface is greater than over a planar boundary if the curvature is convex (such as in a droplet) and is less if the curvature is concave (such as in a bubble). The Kelvin equation gives the change:

$$\ln p_r/p = 2\sigma V_1/rRT \qquad (4)$$

In equation 4, σ is the surface tension and p_r is the vapor pressure over a convex surface with radius r (the right-hand side of the equation is negative for a concave surface). Once again the effect is small except for highly curved surfaces: for a water droplet with radius 100 A at 25°C, the ratio p_r/p is about 1.1. The Kelvin equation is important in understanding the mechanism of condensed phase formation. If a vapor is cooled to its saturation pressure, remote from any surface, then no spontaneous condensation can occur because the small nuclei of molecular clusters formed have high initial curvature and would therefore immediately re-evaporate. Spontaneous homogeneous nucleation requires a much higher vapor pressure. This is supersaturation or supercooling. The ratio of the supersaturation pressure to the normal vapor pressure is about 5:1 for many fluids. In practice, condensation usually occurs at much lower supersaturations due to nucleation at walls or on impurities such as dust particles (heterogeneous nucleation). For the same reason, bubble formation in a heated liquid requires a degree of superheating, and sometimes violent "bumping" occurs: Added boiling chips provide the heterogeneous nucleation centers required for smooth boiling at the normal boiling point. Nucleation theory uses statistical thermodynamics to calculate critical nuclei sizes and hence supersaturation ratios. It is moderately successful for pure materials.

The study of vapor pressures of solutions has many practical applications. Examples include distillation, molar mass determination, activity coefficient measurement, and gas solubility. An IDEAL SOLUTION can be defined as one in which all components have about the same molecular dimensions and intermolecular interactions. That is, for two components, the A-A, A-B, and B-B forces are comparable. Under these conditions, the escaping tendency of a molecule in the solution, as determined by the force field around it, is independent of composition; so the vapor pressure (p_i) is determined solely by the mole fraction (x_i) in solution:

$$p_i = x_i p_i^o \qquad (5)$$

Here p_i^o is the vapor pressure of pure component i. The total vapor pressure is the sum of the individual partial pressures, in accordance with DALTON'S LAW:

$$p_{total} = \Sigma\, p_i \qquad (6)$$

Equation 5 is Raoult's law. Originally empirical, it can be regarded as another way of defining solution ideality. (It can also be shown that for a solution in which Raoult's law is obeyed, the thermodynamic quantities $\Delta H_{mix} = 0$, $\Delta V_{mix} = 0$, and $\Delta S_{mix} = -R\Sigma x_i \ln x_i$, which is a third way of defining an ideal solution.) Ideal solutions are formed from substances that are chemically similar, such as benzene and toluene, but in most cases significant deviations from Raoult's law occur. Vapor pressures for an ideal two-component system are shown in Figure 2.

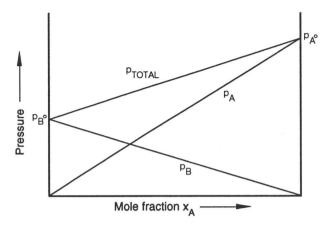

Fig. 2. Total vapor pressure and partial vapor pressure in an ideal two-component system. In this figure and successive figures, component A is more volatile.

A more common situation is that A-B interactions are significantly different from A-A and B-B interactions, and/or the components are of different sizes so that statistical mixing does not occur. The vapor pressure for each component and the total vapor pressure are sketched in Figure 3, for a case in which A-B interactions are smaller than A-A, B-B interactions (positive deviation from Raoult's law). The total vapor pressure shows a maximum. Exactly similar considerations apply when A-B interactions are larger than A-A and B-B, and the total vapor pressure curve shows a minimum (negative deviation). Examples are carbon tetrachloride-acetone and carbon tetrachloride-methyl chloride, respectively. When the solution in Figure 3 contains mostly component A (that is, it is a dilute solution of solute B in solvent A),

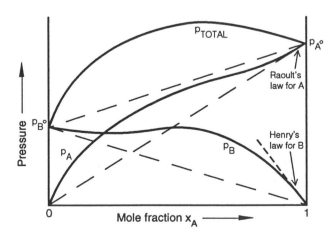

Fig. 3. Vapor pressure-composition diagram for a two-component system showing positive deviation from Raoult's law.

molecules of A are escaping from a nearly uniform A-A environment, and A obeys Raoult's law. Component B is escaping from a nearly uniform A-B environment, so the vapor pressure is still proportional to its concentration; however, the proportionality constant is not $p_B°$:

$$p_B = x_B K_{AB} \qquad (7)$$

Equation 7 expresses Henry's law, and the constant K_{AB} (in pressure units) is the Henry's law constant for the solute B in solvent A. Solutions for which the solute obeys Henry's law and the solvent obeys Raoult's law are ideal dilute solutions. Henry's law is important in studying the solubility of gases in liquids and has many applications of environmental and physiological significance. Modifications for gas solubility at curved surfaces such as in bubbles or aerosols have been developed. Henry's law is sometimes used in reciprocal form:

$$c_i = \kappa_{ij} \, p_i$$

where the constant κ_{ij} is typically expressed in units of mol L^{-1} atm^{-1}.

Separation of volatile liquid mixtures by distillation depends on the fact that at equilibrium the vapor is richer than the liquid in the more volatile components. From equations 5 and 6, for a two-component system:

$$p_{total} = p_A° \, p_B° \, / \, [p_A° + (p_B° - p_A°) \, y_A] \qquad (8)$$

where y_A is the mole fraction of A in the *vapor*. The total vapor pressure is *not* a linear function of vapor composition. Figure 4 shows p_{total} in relation to both liquid (x_A) and vapor (y_A) composition. As the exter-

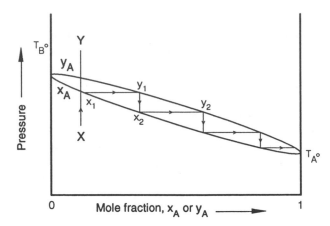

Fig. 5.   Boiling point–composition diagram for an ideal two-component system, illustrating fractional distillation.

nal pressure is lowered along the vertical line XY, vaporization starts at point x_1, and the composition of the vapor first formed is y_1. The liquid composition follows the path $x_1 x_2$, while the vapor follows $y_1 y_2$. At y_2 the liquid disappears. At any pressure between x_1 and y_2 a vapor enriched in A can be separated from the remaining liquid. In practice such separations are more usually carried out by distillation—that is, boiling the liquid at a constant external pressure (usually 1 atm). Distillation is most readily understood by transforming the vapor pressure/composition diagram to a boiling point/composition diagram. In Figure 5, heating the liquid is represented by the vertical line XY. At point x_1 boiling occurs and the vapor first appearing has composition y_1. In a fractional distillation this vapor is condensed on a vertical column above the liquid, and re-evaporated to form a vapor at a lower temperature (y_2 in Figure 5). The process continues in a series of steps as shown until pure A emerges at the top of the fractional column.

For solutions displaying positive or negative deviations from Raoult's law, the boiling point diagrams show a minimum or maximum, respectively. Complete separation by fractionation is not possible: The mixture emerging from the column (an azeotrope) attains a constant composition. Examples are ethanol/water, minimum boiling point 78°C at 96 percent ethanol; and hydrochloric acid, maximum boiling point 109°C at 20 percent acid.

The total vapor pressure above two immiscible liquids is the sum of the individual vapor pressures. The mixture therefore boils at a temperature lower than the boiling point of either component. This is the principle of steam distillation for the purification of

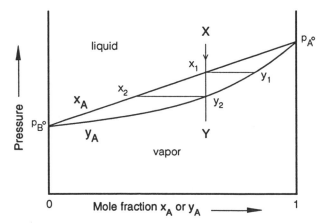

Fig. 4.   Pressure versus vapor (y), and liquid (x) mole fractions for an ideal two-component system.

organic compounds that might be unstable if distilled at their normal boiling point.

In a dilute solution, lowering of the solvent vapor pressure according to Raoult's law is a measure of the mole fraction of solvent according to equation 5, hence also of the mole fraction of solute. This is the basis for molar mass determination by vapor pressure measurement: the solute molar mass is obtained by relating its measured mole fraction to its concentration in g/L. Relative vapor pressure lowering of the solvent depends only on the *number* or particles of solute, not on their chemical nature, and is therefore called a COLLIGATIVE PROPERTY. The other three colligative properties are boiling point elevation, freezing point depression, and osmotic pressure. Historically all four were important in the early development of ionic dissociation theories, and they are sometimes still exploited, especially in nonaqueous systems, to investigate chemical equilibria (such as dimerization) in solution.

Quantitative thermodynamic treatment of non-ideal solutions is accomplished by *defining* the activity (a_i) of a component in solution, in terms of its vapor pressure above that solution. The vapor phase is usually assumed to behave ideally:

$$a_i = p_i/p_i^o \qquad (9)$$

In an ideal solution, or for the solvent in an ideal dilute solution, Raoult's law is obeyed, so that $a_i = x_i$, otherwise x_i and a_i are related by an activity coefficient γ_i:

$$a_i = \gamma_i x_i \qquad (10)$$

Thus activity coefficients in solution can be determined from vapor pressure measurements:

$$\gamma_i = p_i/(p_i^o x_i) \qquad (11)$$

With this definition of activity, $a_i \rightarrow 1$ and $\gamma_i \rightarrow 1$ for the pure component. In the solution, γ_i can be greater or less than unity, depending on whether deviation from Raoult's law is positive or negative. On an alternative scale, which relates activity to molality rather than mole fraction ($a_i = \gamma_i m_i$), vapor pressure measurements of volatile components can be similarly used to find activity coefficients of both volatile and nonvolatile components. In practice, because precise vapor pressure measurements are difficult, the isopiestic method is used for determining activity coefficients. The sample and a standard solution are placed in separate dishes and allowed to equilibrate together in an evacuated vessel. The unknown activity can be calculated from the weight changes due to solvent distillation.

BIBLIOGRAPHY

ATKINS, P. W. *Physical Chemistry*; New York: Freeman, 1990; Chapters 6 and 7.

LEWIS, G. N.; RANDALL, M.; PITZER, K. S.; BREWER, L. *Thermodynamics*; New York: McGraw-Hill, 1961.

WEISSBURGER, A., ed. *Techniques of Chemistry*; New York: Wiley, 1971; Vol. 1, Part 5.

PETER G. BOWERS

Vibrational Spectroscopy and Analysis

The temperature of any material is a manifestation of the internal motion of atoms in the molecules. The amplitudes of this internal motion increase with temperature, and as temperature reaches high values, the motion is sufficient to break solids into liquids, liquids into gases, and gaseous molecules into smaller molecules and, eventually, into atoms. If one could look at a molecule, one would find that these motions are very fast—a typical movement of an atom might bring it back into its original position within 10^{-14} seconds. Furthermore, at first glance, the motions of atoms in a molecule—or of the different molecules in a solid or liquid—might appear totally random. However, the thermal motions of the atoms in a molecule follow some distinct laws of physics, and knowledge of these laws permits the determination of the shape and structure of molecules. This can be accomplished by a branch of science known as vibrational spectroscopy. For the remainder of this discussion, only the vibrations of atoms in a molecule will be considered, although the motions of atoms in a solid can be treated by the same formal scheme.

SPECTROSCOPY is the study of the interaction of light with matter. This interaction is a result of the quantized energy levels of the molecules and the quantum character of the PHOTONS. The quantized energy levels of a molecule can be visualized as an energy ladder on which molecules can only exist at certain rungs, not at any levels in between. To advance from one rung to the next, energy has to be absorbed by the molecule. This energy is usually provided by a light particle, or photon, which must possess the exact amount of energy the molecule requires. If the photon matches this energy difference, it is absorbed by the molecule and annihilated in the process.

The energy levels of a molecule can be due to different distances between electrons and the nuclei of atoms. This gives rise to the electronic spectra of atoms, which occur mostly in the visible and ultraviolet part of the spectrum (see ULTRAVIOLET-VISIBLE WAVELENGTH). The energy levels of a molecule can also be due to different vibrational states. Just as a string can vibrate at a given frequency, or at twice the frequency (the harmonic or the octave), a molecule can oscillate in a ground and several excited states, and transitions between these states can occur when energy is absorbed in the form of a photon.

A molecule consists of atoms that are connected by chemical bonds. A molecule of N atoms, in general, has 3N modes of motion. Six of these are translations and rotations; the others (3N − 6) are vibrations. These can be visualized as distinct and mathematically well-defined patterns along which the atoms move. In these motions, all atoms move with identical frequencies and phases, but different amplitudes. A simple example may be useful to explain these concepts. Water (H_2O) is a TRIATOMIC, bent molecule with both H atoms attached to the oxygen atom. It has three vibrational modes as shown in Figure 1. The arrows attached to the atoms depict the relative motion of the atoms with respect to each other.

In vibrational spectroscopy, one observes the increase in vibrational frequency and amplitude in one of these vibrational modes when a photon of the proper energy is absorbed. The wavelengths of the photons with the particular energies are listed below the modes in Figure 1 for water. Often, the wavelength range is expressed as the inverse of the wavelength, also known as the wave number. A wave number expresses how many light waves there are in one centimeter; thus, its units are 1/cm, or cm^{-1}. The advantage of wave number units is that they are directly proportional to energy, whereas wave length is indirectly proportional to energy. The wave numbers

of the photons responsible for the three transitions in water are also listed in Figure 1.

Finally, the bottom of Figure 1 shows schematically the vibrational spectrum of water. In such a vibrational spectrum, the intensity of light transmitted by a thin layer of sample is plotted versus the wave number of wavelength of the light. The energy of the transitions responsible for molecular vibrations is generally in the infrared range of the electromagnetic spectrum, or between 4 and 50 μm. Thus, the spectrum shown in Figure 1 is also referred to as the infrared spectrum of water.

Since a molecule has 3N − 6 vibrational modes, a larger molecule such as glucose ($C_6H_{12}O_6$) exhibits a vibrational spectrum with 66 absorption peaks. One of the goals of vibrational spectroscopy is to assign a set of atomic displacements for each of the 66 absorption peaks in the spectrum. This is by no means a trivial task, but certain simplifications can be made to aid in this assignment. The foremost simplification is the principle of group frequencies. It has been shown experimentally that certain chemical groups of the molecule produce characteristic absorptions in the vibrational spectrum. A C—C single-bond stretching vibration, for example, produces a weak absorption around 800 cm^{-1}, a C=C stretching vibration absorbs at about 1,650 cm^{-1}, and a C≡O double-bond stretching vibration at 1,700 cm^{-1}. Similarly, all vibrations involving the stretching motion of C—H bonds occur about 3,000 cm^{-1}. These group frequencies can often be used to qualitatively distinguish molecules, and can be used to obtain a coarse picture of the motions involved in a molecular vibration. The two isomeric species, ethanol (CH_3CH_2OH) and dimethyl ether (CH_3OCH_3), can be distinguished by

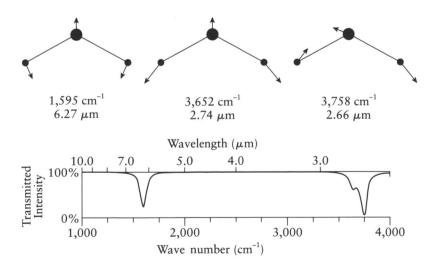

1,595 cm^{-1}
6.27 μm

3,652 cm^{-1}
2.74 μm

3,758 cm^{-1}
2.66 μm

Fig. 1.   Vibrational modes of water.

the absence of an O—H stretching frequency in dimethyl ether.

Vibrational spectra can be observed via two complementary experimental methods, infrared absorption spectroscopy (mentioned above) and Raman spectroscopy, which is a form of laser light scattering. Both techniques are widely applied in qualitative and quantitative monitoring of chemical reactions, pollutants, and structural determination of molecules.

BIBLIOGRAPHY

Bruno, T. H. *Spectroscopic Methods;* Englewood Cliffs, NJ: Prentice-Hall, 1993.

Goldman, S. *Vibration Spectrum Analysis: Basic Problem Solving;* Brooklyn, NY: Industrial Press, 1990.

Harris, D. C. *Symmetry and Spectroscopy;* New York: Dover, 1989.

Sverdlov, L. M., et al. *Vibrational Spectra of Polyatomic Molecules;* Philadelphia: Coronet Books, 1974.

MAX DIEM

Vinyl Functional Group

Vinyl is a common term in organic chemistry that describes the monosubstituted alkene functional group. The original, and formally correct, definition of the vinyl functional group is the monosubstituted ethene, Figure 1, structure 1. *Vinyl* is not, however, a term used in systematic organic nomenclature, and partly for this reason its definition has expanded over time to include that of the monosubstituted alkene, structure 2. In common usage, *vinyl* refers to the alkenyl segment that is paired with the substituent X. Thus, the informal name of the compound corresponding to structure 3 (structure 1 where X equals —Cl) is vinyl chloride (the formal name of 3 is chloroethene). Because of the common and distinctive nature of the substituted alkene, *vinyl* is often used to direct attention to the alkenyl group in an otherwise complex organic structure, as in the easily identified two vinyl groups of the protoporphyrin IX cofactor of the oxygen-carrying protein hemoglobin (structure 4). The most common use of the term *vinyl*, outside of organic chemistry, is as an adjective referring to an object made from one of the important polymers obtained by the polymerization of vinyl monomers.

The use of *vinyl* in organic chemistry is complicated yet further by the fact that the term is used most often with those substituents X where it has a particular historical legacy or where other yet more common alternative terms do not supersede (see Figure 2). Thus, vinyl is used with substituents such as vinyl

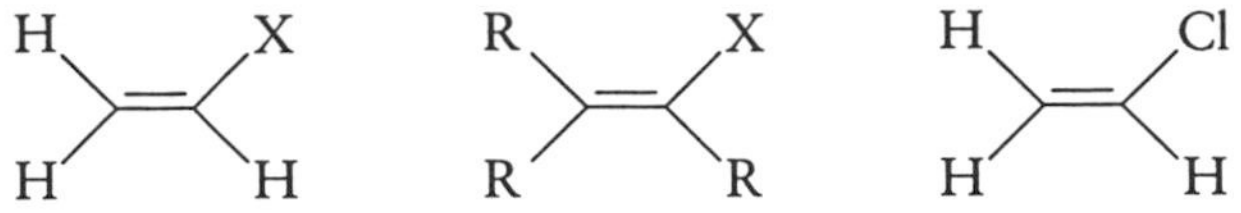

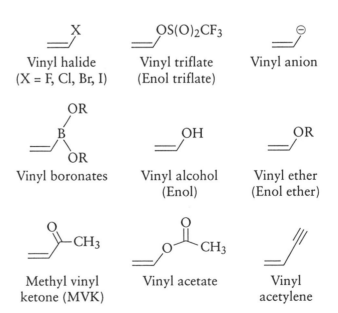

Fig. 1. (Top) Three forms of vinyl and (bottom) the protoporphyrin IX cofactor of hemoglobin, which itself contains vinyl groups.

halides (2, X = F, Cl, Br, I); vinyl alcohols (or more commonly enols, where X = OH); vinyl triflates (X = $OS(O)_2CF_3$); vinyl ethers (or enol ethers, X = OR′); vinyl silylethers (X = $OSiR′_3$); vinyl esters (X = OC(O)R′); and with organometallics such as

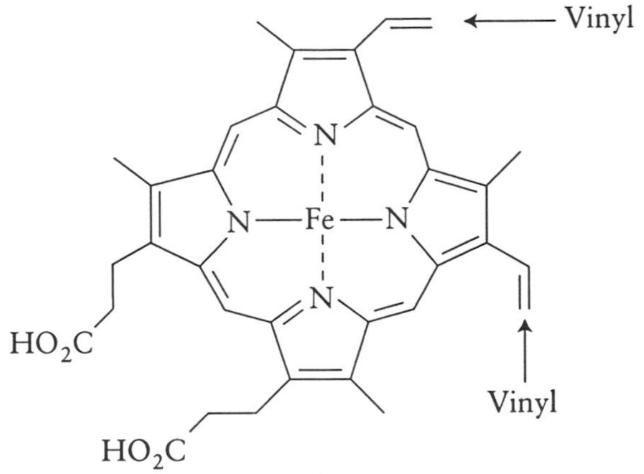

Vinyl halide
(X = F, Cl, Br, I)

Vinyl triflate
(Enol triflate)

Vinyl anion

Vinyl boronates

Vinyl alcohol
(Enol)

Vinyl ether
(Enol ether)

Methyl vinyl
ketone (MVK)

Vinyl acetate

Vinyl
acetylene

Fig. 2. Some common vinyl functional groups.

1489

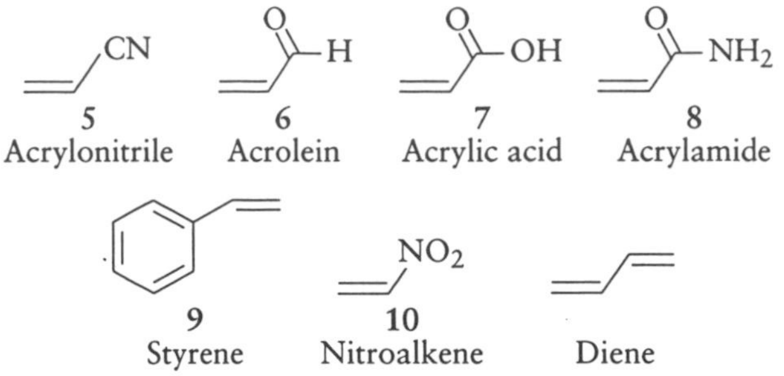

Fig. 3. Some common vinyl functional groups that are not routinely referred to as vinyl.

vinyl stannanes (X = SnR′$_3$) or vinyl Grignards (X = MgBr). Vinyl also is used to describe the oxidation states of an alkenyl segment (vinyl cation, vinyl radical, vinyl carbanion). It is not used for the ethenyl parent 1 where X is a carbonyl, carboxyl, or aryl substituent, as other terms take precedence. See Figure 3. Thus, the informal name for the nitrile 5 (X = CN) is acrylonitrile; for the aldehyde 6 (X = C(O)H), acrolein; for the carboxylic acid 7 (X = CO$_2$H), acrylic acid; for the amide 8 (X = C(O)NH$_2$), acrylamide; and for the vinylarene 9 (X = Ph, a phenyl substituent), styrene. The vinyl-substituted ketone is referred to occasionally as a vinyl ketone (although the simplest vinyl ketone is known universally as MVK, for methyl vinyl ketone) but rather more often as the more commonly used "enone" or "α,β-unsaturated ketone." The structure 3, where X = NO$_2$, is inexplicably never referred to as a "vinyl nitro" but rather as a nitroalkene or nitro olefin 10.

There are two terms, related to vinyl, that also are common terms in informal organic nomenclature. *Vinylidene* (see Figure 4) refers to a 1,1-disubstituted alkene and is exemplified by vinylidene chloride 11 (the systematic name is 1,1-dichloroethene). The adjective *vinylogous* (the noun is *vinylog* or *vinylogue*) describes two functional groups linked directly, through RESONANCE, by an alkene functional group.

Thus the methoxy-substituted cyclohexenone (Figure 5, structure 12) does not behave as having independent methoxy and ketone functional groups but rather behaves as having a single *vinylogous ester* functional group. This distinction has mechanistic importance: substitution of the methoxy of a vinylogous ester by a nucleophile proceeds by a conjugate addition/elimination mechanism that is similar to substitution at the carbonyl of an ester. A wonderful example of vinylogous behavior is provided by the mitomycins (a class of antitumor agents). Inspection of mitomycin A 13 reveals a vinylogous ester, a vinylogous amide, and a carbinolamine (the addition intermediate of an amine to a carbonyl). Carbinolamines are usually unstable intermediates, but the carbinolamine of the mitomycins is stable due to the participation of the nitrogen of the vinylogous amide in vinylogous resonance. The electron-withdrawing character of the carbonyl removes sufficient electron density from the nitrogen, such that loss of the methoxide anion (CH$_3$O−) from the carbinolamine is energetically unfavorable. The interplay of these functional groups is illustrated by the following reaction sequence of the mitomycines (Figure 6). Dissolution of mitomycin A in aqueous ammonia results in the transformation of the vinylogous ester to a second vinylogous amide, by an addition/elimination sequence, to give mitomycin C 14. Reduction of the quinone of mitomycin C gives a hydroquinone 15, with loss of the vinylogous amide. The carbinolamine of 15 is unstable and undergoes immediate loss of methoxide to give the imine 16.

Because of the exceptional importance of the substituted alkene, *vinyl* is a frequently encountered term in organic chemistry. Its customary use is illustrated by the four reactions of Figures 7 through 10. Figure 7 illustrates, in conformational perspective, the divinylcyclopropane 17 that is released by the female gametes of saltwater brown algae as an attractive

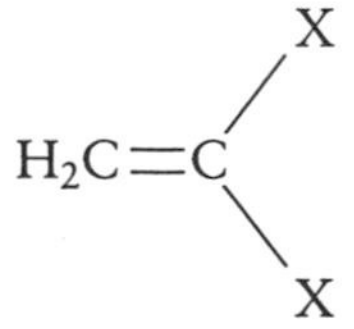

Fig. 4. Vinylidine.

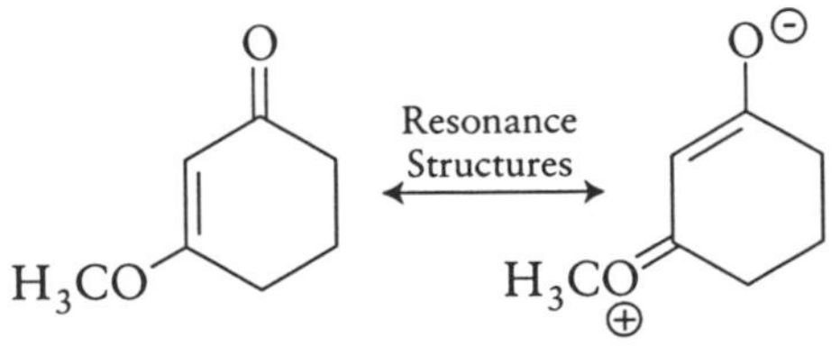

Fig. 5. Methoxy-substituted cyclohexenone, which has a single vinylogous ester functional group.

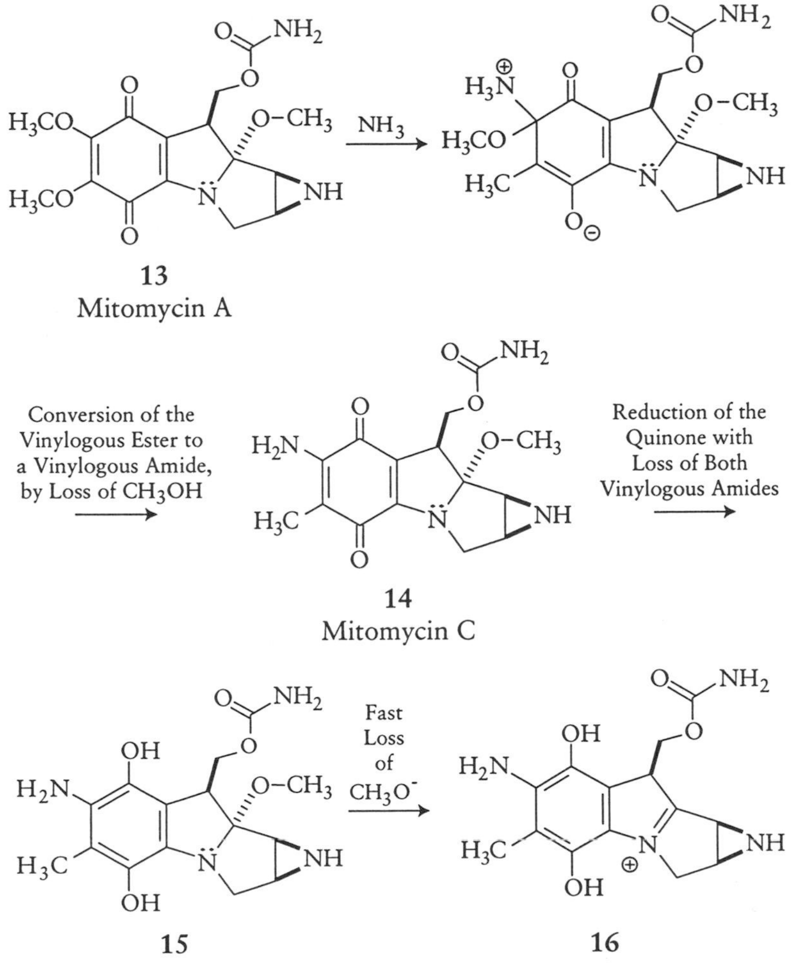

13
Mitomycin A

Conversion of the
Vinylogous Ester to
a Vinylogous Amide,
by Loss of CH₃OH
⟶

Reduction of the
Quinone with
Loss of Both
Vinylogous Amides
⟶

14
Mitomycin C

Fast
Loss
of
CH₃O⁻
⟶

15

16

Fig. 6. Mitomycin reaction sequent.

secretion to the motile male sex cells of the algae. Clearly, it is important that these "release factors" have a transitory existence such that they do not persist, as they are carried by the water currents and mislead the male. Appropriately, these divinylcyclopropanes undergo spontaneous inactivation by a

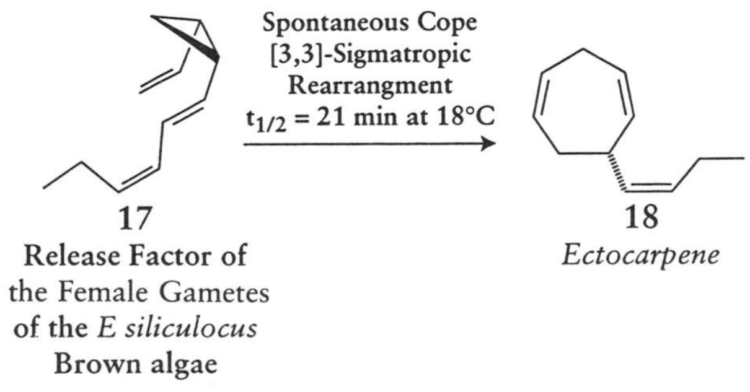

Spontaneous Cope
[3,3]-Sigmatropic
Rearrangment
$t_{1/2}$ = 21 min at 18°C
⟶

17
**Release Factor of
the Female Gametes
of the *E siliculocus*
Brown algae**

18
Ectocarpene

Fig. 7. Divinylcyclopropane viewed from a conformational perspective.

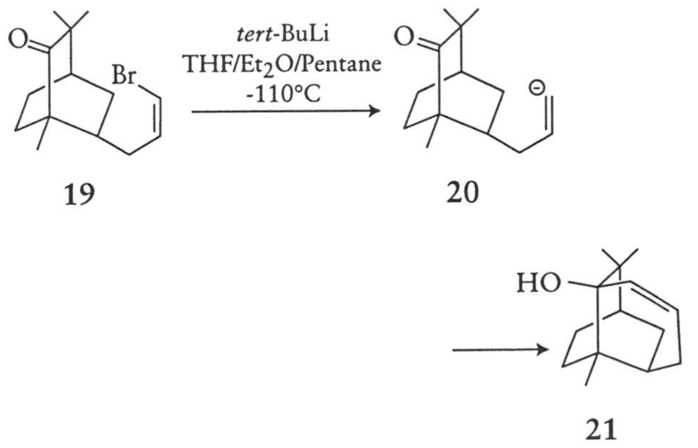

tert-BuLi
THF/Et₂O/Pentane
-110°C
⟶

19

20

21

Fig. 8. Conversion of vinyl bromide to a vinyl anion, which reacts further to produce a substance related to patchouli alcohols.

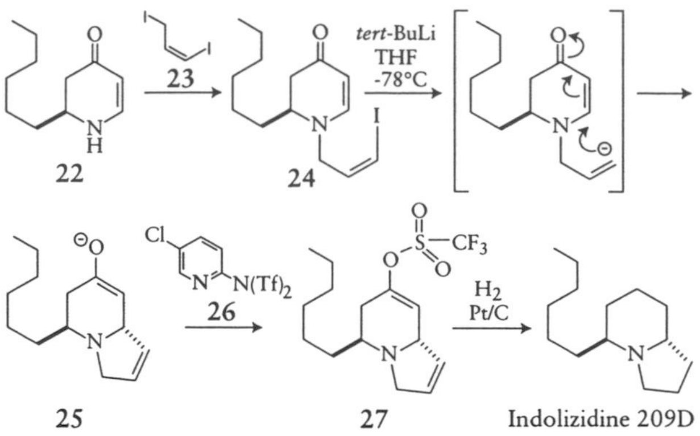

Fig. 9.   Reactive path from vinylogous amide to vinyl triflate.

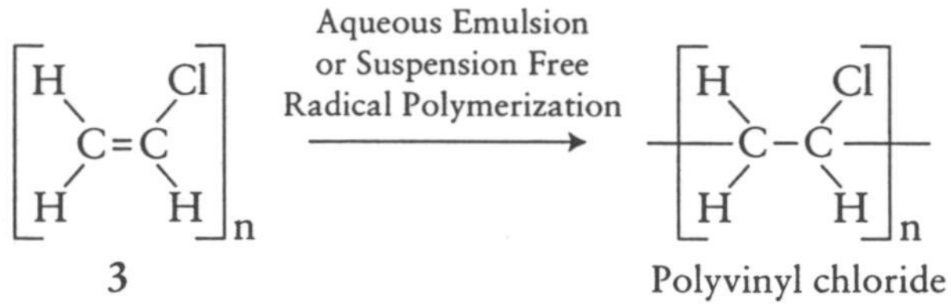

Fig. 11.   Polymerization of vinyl chloride.

Cope pericyclic reaction to give the alkenylcycloheptadiene 18. In Figure 8 the vinyl bromide 19 is metalated (converted by an organolithium to a vinyl anion). The vinyl anion reacts immediately by intramolecular addition to the ketone to give a product (21) related to the highly fragrant patchouli alcohols. In Figure 9 the vinylogous amide 22 undergoes N-alkylation by the diiodoalkene 23. This diiodoalkene has both vinyl iodide and allyl iodide functional groups: the vinyl iodide is completely unreactive to nucleophiles, while the allyl iodide is highly reactive, reacting at the nitrogen to give 24. The very poor reactivity of vinyl halides toward nucleophiles is a consequence of the greater electronegativity of the sp^2 hybridization, which destabilizes the accumulation of negative charge on the halide that must occur in the transition state for nucleophilic displacement. Following metalation, vinyl iodide 23 undergoes intramolecular conjugate addition to give enolate 25, which is trapped by the triflate donor 26 to give the vinyl triflate 27. Catalytic hydrogenation reduces the alkene and vinyl triflate functional groups to complete a synthesis of the indolizidine 209D, one of several structurally related neurotoxins secreted by the neotropical poison dart frogs. In Figure 10, the enolate of the methoxy enone (a vinylogous ester) 28 is trapped by trimethylsilylchloride to give the diene

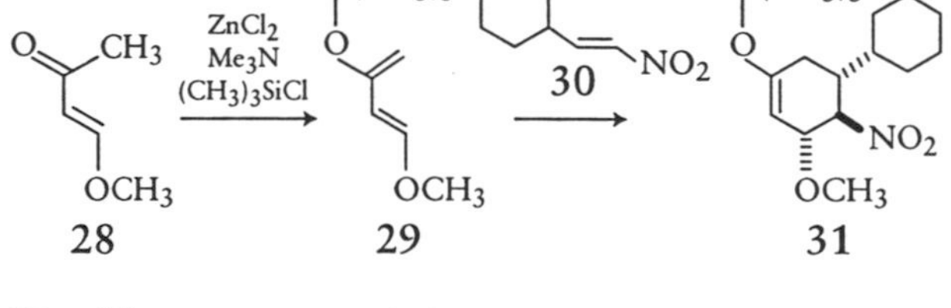

29. This diene undergoes DIELS-ALDER cycloaddition with the nitroalkene 30 to provide the vinyl silyl ether 31.

The important vinyl polymers include polyvinyl acetate, polyvinyl alcohol, polyvinyl ether, polyvinylidene chloride, and the polyvinyl halides. The best known vinyl polymer is polyvinyl chloride (PVC), a thermoplastic polymer. Vinyl chloride is polymerized (Figure 11), typically with a radical initiator, to provide a powder that upon heating in a mold fuses (cross-links) to a tough, durable plastic. The addition of plasticizers, or copolymerization with (for example) vinyl acetate or acrylonitrile, gives equally durable but malleable PVC polymers. Polyvinyl acetate is used as a binder for paints and as adhesives; polyvinyl alcohol is also an important component of adhesives and, as it has properties similar to starch, is used as an emulsifier and thickener.

BIBLIOGRAPHY

BEYER, H.; WALTER, W. Handbook of Organic Chemistry; New York: Prentice-Hall Europe, 1996; pp. 73–89, 95–98, 329.

JED F. FISHER

Fig. 10.   Reactive path from the enolate of the methoxy enone to vinyl silyl ether.

Vision, Chemistry of

Vision is a specialized form of light detection. The eye is the organ of vision. It consists of optical elements, such as the lens, and a network of sensory cells and neurons known as the retina. There are two types of photoreceptor cells in the retina: rods and cones, named because of their characteristic shapes. Rod cells are responsible for dim-light black-and-white vision, and cone cells are responsible for bright-light color vision. Light striking a photoreceptor cell is converted into an electrical signal by a photochemical reaction. The signal is processed by other cells in the retina and sent to the brain. Key elements of vision in humans include high acuity, color sensitivity, and the capability for a large dynamic range of signal intensity.

Photoreceptor Molecule

VITAMIN A AND RHODOPSIN

A *chromophore* is a chemical group that produces color by absorbing light. The chromophore in nearly all vertebrate visual pigments is derived from vitamin A_1, a fat-soluble terpene alcohol of the formula $C_{20}H_{30}O$. Vitamin A_1 is an example of a compound with a highly conjugated system of alternating single and double bonds, a polyene. It has five carbon-carbon double bonds. The overlapping *p*-orbitals of adjacent carbon atoms form a π-bond. Each double bond between adjacent carbon atoms consists of one σ- and one π-bond.

Vitamin A_1 is enzymatically oxidized in the retina to an aldehyde, retinal. Retinal is covalently attached to a specific lysine residue in the AMINO ACID sequence of the opsin protein to form the visual pigment rhodopsin. Rhodopsin is the photoreceptor molecule of the rod cell. It consists of a protein (opsin) embedded into the specialized disk membrane of the rod cell and a chromophore (retinal). The protein-chromophore linkage in rhodopsin is an unusually stable protonated Schiff base bond.

The bonding of the retinal chromophore polyene cannot be adequately described by a single electron structure. Therefore, the bonding is delocalized and displays resonance, a weighted average of canonical structural forms.

The retinal in the inactive, or dark-adapted, state of rhodopsin is a specific geometric isomer, 11-*cis*-retinal. Geometric isomers are possible because there is hindered rotation about carbon-carbon double bonds. The 11-*cis* notation refers to the fact that the carbon atom chains attached to carbon atoms 11 and 12 lie on the same side of the C_{11}–C_{12} double bond. This results in a bend in the carbon backbone in addition to a slight twist because of steric effects of

the methyl group on carbon 13. The single carbon-carbon double bond in the ionone ring of retinal (C_5–C_6) is also in the *cis* orientation in rhodopsin. The three other carbon-carbon double bonds in the retinal polyene chain are in the *trans* orientation.

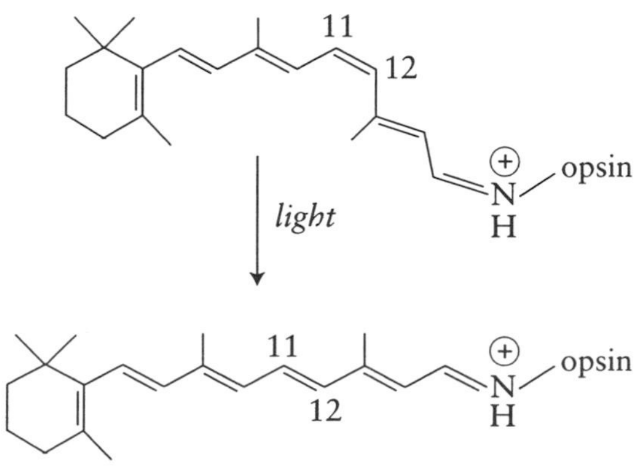

RHODOPSIN PHOTOACTIVATION

Visual excitation is triggered by photoisomerization of 11-*cis*-retinal to the 11-*trans* isomer within the retinal binding pocket of rhodopsin. The absorption of a photon, a quantum of energy, by rhodopsin causes the photoisomerization. Although rhodopsin can absorb a fairly wide range of light energies, green light is most effective in activating rhodopsin. The wavelength of light that is maximally absorbed (λ_{max} value) is 500 nm.

Rhodopsin has an unusually high quantum yield for a biological system. The quantum yield is defined as the fraction of absorbed light that results in a particular outcome. The primary quantum yield of rhodopsin is about 0.67, meaning that about two of three rhodopsin molecules that absorb a photon undergo retinal photoisomerization.

The retinal photoisomerization is the only light-dependent event in vision. Retinal isomerization very rapidly converts light energy into molecular motion, or thermal energy. The newly formed 11-*trans*-retinal isomer can then interact with specific amino acids of rhodopsin to cause a change in receptor conformation. The active receptor conformation binds to a specific signal-transducing protein in the cytoplasm of the rod cell called transducin. In this way, a photon of a specific wavelength is absorbed by rhodopsin and converted into a biochemical signal.

Biochemistry of Vision

G PROTEIN CASCADE

The rod cell protein transducin transmits a biochemical signal from rhodopsin to a cellular effector molecule. Transducin is a guanine nucleotide-binding regulatory protein, or G protein. Photoactivated rhodopsin catalyzes the exchange of guanosine 5′-diphosphate (GDP) for guanosine 5′-triphosphate (GTP) by multiple transducin molecules. The activated transducin, with GTP in its nucleotide-binding pocket, then transmits a chemical signal to the effector molecule in the signaling cascade. The effector molecule in the vertebrate visual system is the enzyme guanosine 3′ : 5′-cyclic monophosphate (cyclic GMP) phosphodiesterase. A single transducin molecule activates a single phosphodiesterase molecule.

Phosphodiesterase catalyzes the hydrolysis of cyclic GMP to guanosine 5′-monophosphate (GMP)

Fig. 1.   The chemical process of vision.

and intracellular levels of cyclic GMP drop. The lower levels of cyclic GMP cause a rod cell plasma membrane cation channel to close. Since the plasma membrane is selectively permeable to ions, which are electrically charged, an electrical potential difference exists between the inside and the outside of the rod cell. The potential increases as cation channels are closed and the influx of sodium and calcium ions, which carry positive charge, is slowed. The result is that the rod cell becomes hyperpolarized in response to light. The increase in potential varies proportionally with the strength of the light signal. The change in membrane potential is sent as an electrical signal from the plasma membrane to the synaptic terminal of the rod cell, where it is transmitted to other specialized cells of the retina.

Signal Amplification

A properly dark-adapted rod cell can detect a single photon. This extreme sensitivity is possible in part because of two important features of the visual system: (1) the stability of rhodopsin in darkness to thermal activation and (2) the high degree of signal amplification by the biochemical cascade of vision. It has been estimated that the spontaneous thermal isomerization of 11-cis-retinal in rhodopsin in darkness takes place roughly once in 300–1,000 years. In contrast, a single photon can theoretically cause as many as 1,000 cation channels in a rod cell membrane to close.

The light signal is turned off by a number of biochemical mechanisms. Light-activated rhodopsin becomes phosphorylated by a specific rhodopsin kinase enzyme. The phosphorylated form of rhodopsin can no longer interact with transducin. The active GTP-bound form of transducin has an intrinsic enzymatic GTP hydrolysis activity. Over time, the bound GTP is converted to GDP, and transducin returns to its inactive state. In addition, intracellular calcium levels drop after the cation channel of the rod cell closes. A fall in intracellular calcium concentration mediates photoreceptor cell recovery and adaptation.

The process of adaptation allows the sensitivity of the retina to adjust based on the level of light in a visual scene. Adaptation is extremely important for a useful visual system since the magnitude of light varies widely in the environment. For example, bright sunlight and dim starlight differ in luminance by at least several orders of magnitude.

Color Vision

Humans possess trichromatic color vision, or trichromacy. Most people can match any given reference color by combining the three primary colors. The three primary colors for additive color mixtures are red, green, and blue. In 1802, Thomas Young hypothesized that trichromacy resulted from humans having three separate color-sensing mechanisms. It is now known that the retina contains three classes of cone photoreceptor cells. Each class of cone cell is sensitive to a specific wavelength of light.

At the molecular level, human trichromatic color vision requires the presence of three cone pigments with broad overlapping spectral absorption. Each specific cone class contains only one type of photoreceptor molecule. The three types of cone photoreceptor molecules (red, green, and blue) are homologues of rhodopsin. The amino acid sequences of these opsins are about 40 percent identical to that of human rhodopsin. The green and red opsins are about 96 percent identical to each other and about 43 percent identical to the blue opsin.

The spectral properties of human cone pigments have been studied by a variety of techniques ranging from psychophysical color matching to microspectrophotometry. Using techniques of molecular biology, the human cone pigment genes were expressed in tissue culture cells, reconstituted with 11-cis-retinal chromophore, and studied by ultraviolet-visible absorption spectroscopy. The λ_{max} values of the pigments were estimated to be 425 nm (blue cone), 530 nm (green cone), and 560 nm (red cone). These studies confirmed the previous assignments of the cloned pigment genes, which were based on a genetic analysis.

Rhodopsin and the three cone pigments all use the same chromophore, 11-cis-retinal. This single chromophore allows all visible wavelengths of light, which range from about 400 nm (violet) to 600 nm (deep red), to be detected. A spectral tuning mechanism exists so that a particular opsin protein can modulate the absorption spectrum of its retinylidene chromophore. Spectral tuning is possible because specific amino acid side chains of each opsin can interact with the chromophore and shift its λ_{max} value.

Among the fifteen differences between the 364 amino acid human green and red pigments, seven amino acid changes are responsible for the observed 30-nm spectral shift in going from the green to the red pigment. Most of this shift is caused by amino acid side chains that contain a hydroxyl group: tyrosine, serine, and threonine.

Genetic variations in color vision may result from a mutation in one of the genes encoding a cone opsin. If the amino acid change affects spectral tuning in one of the cone pigments, then one of the cone types will contain a pigment with an anomalous absorption spectrum. Individuals with this genetic variation are called anomalous trichromats. The most common type of anomalous trichromacy is called red-green color blindness. Red-green color vision variations affect males predominantly since the genes for the red and green opsin genes are found only on the X-chromosome.

Severe color vision defects may result from complete deficiencies of one or more of the three cone opsin genes. Individuals who lack a functional red or green photoreceptor are called dichromats.

Related Signaling Systems

Rhodopsin is a member of a family of related seven-helical membrane receptor proteins. Other notable members of this family include receptors for the hormones adrenaline and glucagon. Transducin is also a member of a family of related signal-transducing G proteins. The same general system of related membrane receptors and G proteins is used by all eukaryotic organisms—from humans to yeast.

Membrane receptor molecules undergo a conformational change in response to the specific binding of a hormone ligand in the case of hormone receptors, or the absorption of light in the case of rhodopsin. This conformational change allows the portion of the receptor on the cytoplasmic surface of the plasma membrane to interact with specific proteins in the cytoplasm of the cell.

G proteins are one class of cytoplasmic proteins that interact with the activated membrane receptor. The active receptor catalyzes the exchange of GDP for GTP by multiple G protein molecules. The activated G protein then transmits a chemical signal to a cellular effector molecule.

Effector molecules may be enzymes or ion channels. An effector enzyme produces a small bioactive compound known as a second messenger molecule. For example, adrenaline is secreted into the bloodstream and binds to the seven-helical adrenergic receptor. The hormone-receptor complex activates a specific G protein, which in turn activates the effector enzyme adenylyl cyclase to produce adenosine $3':5'$-cyclic monophosphate (cAMP), which affects cellular physiology. In the visual system, the effector enzyme is cyclic GMP phosphodiesterase, and the second messenger molecule is cyclic GMP.

BIBLIOGRAPHY

HUNT, D. M.; DULAI, K. S.; BOWMAKER, J. K.; MOLLON, J. D. "The Chemistry of John Dalton's Color Blindness." *Science* 1995, *267*, 984–988.

NATHANS, J. "The Genes for Color Vision." *Sci. Amer.* 1989, *260*, 42–49.

O'BRIEN, D. F. "The Chemistry of Vision." *Science* 1982, *218*, 961–966.

SAKMAR, T. P. "Opsins." In *Handbook of Receptors and Channels*; Boca Raton, FL: CRC Press, 1994; pp. 257–276.

SCHNAPF, J. L.; BAYLOR, D. A. "How Photoreceptor Cells Respond to Light." *Sci. Amer.* 1987, *256*, 40–47.

STRYER, L. "The Molecules of Visual Excitation." *Sci. Amer.* 1987, *257*, 42–50.

STRYER, L. "Signal Transduction Cascades." In *Biochemistry*, 4th ed.; New York: Freeman, 1995; pp. 332–339.

WALD, G. "The Molecular Basis of Visual Excitation." *Nature* 1968, *219*, 800–807.

THOMAS P. SAKMAR

Vitamins

Vitamins are substances that are essential for life in addition to the proteins, minerals, fats, and carbohydrates in foods. To be considered a vitamin, a substance must meet the following criteria:

- It is an organic compound distinct from fats, carbohydrates, and proteins; it must be a natural component of foods, where it is usually present in minute amounts.
- It is essential, also usually in minute amounts, for normal physiological function (i.e., maintenance, growth, development, and/or production).
- Its absence or underutilization must cause a specific deficiency syndrome.
- It is not synthesized by the host in amounts adequate to meet normal physiological needs.

This definition is not perfect in that many species can synthesize at least some of these substances. For example, most species can synthesize ascorbic acid (vitamin C), cholecalciferol (vitamin D), and niacin. Only the few (e.g., guinea pigs, humans) who lack the enzyme L-gulonolactone oxidase require a dietary source of vitamin C, and only individuals not exposed to sunlight cannot perform the photolysis of its precursor metabolite, thus requiring vitamin D in their diets. Only those species (e.g., cats, ducks, and fishes) who are very inefficient in converting the AMINO ACID tryptophan to niacin require the latter as a vitamin. Thus the definition of a vitamin should be understood as having specific reference to animal species, stage of development, diet or nutritional status, and physical environmental conditions. Some compounds are vitamins for one species and not another; some are vitamins only under specific dietary or environmental conditions.

General Properties of Vitamins

Vitamins comprise a group of chemically heterogeneous, organic, low-molecular-weight substances that have key roles in metabolism. (See Table 1.) Most are families of chemically related substances—vitamers, sharing qualitative biological activities. Vitamers and provitamins within a vitamin family may vary in their relative biologic activities—that is, biopotency. Such differences are considered by quantifying those vitamins in terms of biopotency-based units (international units, IU) instead of mass.

Vitamins A, D, E, and K are soluble in nonpolar organic solvents and are thus referred to as fat-soluble vitamins. Vitamin C, thiamin, riboflavin, niacin, vitamin B_6, biotin, pantothenic acid, folate, and vitamin B_{12} are soluble in polar solvents and are thus referred to as water-soluble vitamins. The fat-soluble vitamins are comprised of multiple five-carbon isopentyl units. The water-soluble vitamins have few structural similarities.

In general, the fat-soluble vitamins are unstable to oxidation and susceptible to oxidative attack in the presence of heat, oxygen, metal ions, and UV light. Therefore their amounts in natural foods are highly variable, being greatly affected by the conditions of food production and processing. The water-soluble vitamins tend to be more stable under most conditions, exceptions being riboflavin; vitamins B_6 and B_{12}, which are degraded by light; and thiamin, which is sensitive to mildly alkaline conditions.

Vitamin Absorption and Transport

The enteric absorption of the fat-soluble vitamins (see Table 2) depends on micellar dispersion in the intestinal lumen. These vitamins are not soluble in the aqueous environment of the alimentary tract. In the upper tract, they dissolve in the bulk lipid phases of emulsions formed there. Pancreatic lipase converts emulsified triglycerides to free fatty acids and β-monoglycerides, which combine to form small (10–50 Å) particles called mixed micelles. The hydrophobic cores of micelles solubilize fat-soluble vitamins and other nonpolar lipids, facilitating their diffusion across the intestinal mucosa. Thus dependent on micellar dispersion, the absorption of fat-soluble vitamins requires dietary fat. Vitamins that are soluble in the intestinal lumen can be taken up by the absorptive surface of the gut directly. Some are absorbed by simple diffusion, others via specific carriers. Some (e.g., vitamin C, vitamin B_{12}, thiamin, and folate) are absorbed via carrier-dependent mechanisms at low doses and by simple diffusion at high doses.

Solubility in the aqueous environments of the blood plasma and lymph is a major determinant of ways in which the vitamins are transported to the liver and peripheral organs (see Table 3). Thus the fat-soluble vitamins depend on carriers. After absorption from the gut they associate with lipid-rich chylomicra elaborated in intestinal mucosal cells. Vitamins E and K are transferred in these particles to lipoproteins in the liver. Vitamins A and D are transported from the liver to peripheral tissues by specific carriers of hepatic origin. Some water-soluble vitamins are transported by protein carriers in the plasma. Some (riboflavin, vitamin B_6) are carried via weak, nonspecific binding to albumin; others are tightly associated with certain immunoglobulins (riboflavin) or proteins (riboflavin, vitamins A, D, E, and B_{12}). Several vitamins (e.g., vitamin C, thiamin, niacin, riboflavin, pantothenic acid, biotin, and folate) are transported in the free solution in the plasma.

In general, the fat-soluble vitamins are well retained; they tend to be stored in lipid-rich tissues (e.g., adipose, liver) in appreciable amounts such that variable patterns of food consumption may be accommodated by depositing and mobilizing tissue stores according to dietary intake. The water-soluble vitamins tend to be excreted rapidly and not retained well, the notable exception being vitamin B_{12}, which can be accumulated in the liver.

Vitamin Metabolism

Only a few vitamins function in biological systems without some metabolic activation or linkage to a cofunctional species (e.g., an enzyme). The conversion to metabolically active forms may involve substantial chemical modification and/or combination with another metabolically active species, including enzymes (see Table 4). The fat-soluble vitamins are generally excreted with the feces via the bile, while the water-soluble vitamins are excreted in the urine either intact (riboflavin, pantothenic acid) or as water-soluble metabolites (vitamin C, thiamin, niacin, riboflavin, pyridoxine, biotin, folate, vitamin B_{12}).

Table 1. Physical Properties of the Vitamins

Vitamin	Vitamer	MW	Solubility org.[a]	H$_2$O[b]	Absorption max, nm	Melting pt., °C	Color-form
Vitamin A	Retinol	286.4	+	−	325	62–64	Yellow crystal
	Retinal	284.4	+	−	373	61–64	Orange crystal
	Retinoic acid	300.4	+	sl	351	180–182	Yellow crystal
Vitamin D	Vitamin D$_2$	396.6	+	−	265	115–118	White crystal
	Vitamin D$_3$	384.6	+	−	265	84–85	White crystal
Vitamin E	α-Tocopherol	430.7	+	−	294	2.5	Yellow oil
	γ-Tocopherol	416.7	+	−	298	−2.4	Yellow oil
Vitamin K	Vitamin K$_1$	450.7	+	−	242, 248, 260, 269, 325		Yellow oil
	Vitamin K$_{2(35)}$	649.2	+	−	243, 248, 261, 270, 325–328	54	Yellow crystal
	Vitamin K$_3$	172.2	+	−		105–107	Yellow crystal
Vitamin C	Free acid	176.1	−	323	245	190–192	White crystal
	Na salt	198.1	−	620	245	218[c]	White crystal
Thiamin	Disulfide form	562.7	−	sl		177	Yellow crystal
	Hydrochloride	337.3	−	1000			White crystal
	Mononitrate	327.4	−	27		196–200[c]	White crystal
Riboflavin[d]		376.4	−	0.33	220–225, 266, 371, 444, 475	278[c]	Orange-yellow crystal
Niacin	Nicotinic acid	123.1	−	16	263	237	White crystal
	Nicotinamide	122.1	−	1000	263	128–131	White crystal
Vitamin B$_6$	Pyridoxal	167.2	−	500	293	165[c]	White crystal
	Pyridoxol(HCl)	205.6	−	220	255, 326	160	White crystal
Biotin	d-Biotin	244.3	−	0.4		167	White crystal
Pantothenic acid	Free acid	219.2	−	freely			Clear oil
	Ca salt	476.5	−	356		195[c]	White crystal
Folate	Monoglutamate	441.1	−	0.0016	256, 283, 368	250[c]	Orange-yellow crystal
Vitamin B$_{12}$	Cyanocobalamin	1,355.4	−	12.5	278, 361, 550	>300	Red crystal

[a] Mg/ml at 25°C.

[b] In organic solvents, fats, and oils.

[c] Decomposes at this temperature.

[d] Fluoresces.

Metabolic Functions of the Vitamins

The thirteen families of vitamins comprise two to three times that number of practically important vitamers and function in metabolism in four general (and not mutually exclusive) ways: as a membrane stabilizer; as H$^+$/e$^-$ donors/acceptors; as hormones; and as coenzymes (see Table 5). The type of metabolic function of a particular vitamer is dependent on its tissue/cellular distribution and its chemical activity, both of which are direct or indirect functions of its chemical structure.

Table 2. Enteric Absorption of the Vitamins

Vitamer	Digestion	Site[a]	Enterocytic Metabolism	Efficiency %	Conditions of Potential Malabsorption
			Micelle-Dependent Diffusion		
Retinol	—	D, J	Esterification	80–90	Pancreatic insufficiency (pancre-atitis,
Retinyl esters	Deesterified	D, J	Reesterification		Se-deficiency, cystic fibrosis, cancer) β-carotene
		D, J	Esterification	50–60	biliary atresia, obstructive jaun-dice, celiac disease, very low fat diet
Vitamins D	—	D, J	—	~50	Pancreatic or biliary insuffi-ciency
Tocopherols	—	D, J	—	20–80	Pancreatic or biliary insuffi-ciency
Tocoph. esters	Deesterified[b]	D, J	—	20–80	Pancreatic or biliary insuffi-ciency
MKs	—	D, J	—	10–70	Pancreatic or biliary insuffi-ciency
Menadione	—	D, J	—	10–70	Pancreatic or biliary insuffi-ciency
			Active Transport		
Phylloquinone	—	D, J	—	~80	Pancreatic or biliary insuffi-ciency
Ascorbic acid	—	I	—	70–80	D-isoascorbic acid
Thiamin	—	D	Phosphorylation		Pyrithiamin, excess ethanol
Thiamin di-P	Dephosphor[b]	D	Phosphorylation		Pyrithiamin, excess ethanol
Riboflavin	—	J	Phosphorylation		
FMN, FAD	Hydrolysis[b]	J	Phosphorylation		
Flavoproteins	Hydrolysis[b]	J	Phosphorylation		
Folylmono-glu	—	J	Glutamation		Celiac sprue
Folylpoly-glu	Hydrolysis[b]	J	Glutamation		Celiac sprue
Vitamin B_{12}	Hydrolysis[b]	I	Adenosinylation, methylation	>90	Intrinsic factor deficiency (pernicious anemia)
			Facilitated Diffusion[c]		
Nicotinic acid	—	J		>90[d]	
Nicotinamide	—	J		~100[d]	
Niacytin	Hydrolysis[b]	J			
NAD(P)	Hydrolysis[b]	J			
Biotin	—	J			Biotinidase deficiency, avidin
Biocytin	Hydrolysis[b]	J			Biotinidase deficiency, avidin
Pantothenate	—				
Coenzyme-A	Hydrolysis[b]				

Table 2. (Continued)

Vitamer	Digestion	Site[a]	Enterocytic Metabolism	Efficiency %	Conditions of Potential Malabsorption
			Simple Diffusion		
Ascorbic acid[e]	—	D, J, I	—	<50	
Thiamin[e,f]	—	J	Phosphorylation		
Nicotinic acid	—	J	—		
Nicotinamide	—	J	—		
Pyridoxol	—	J	Phosphorylation		
Pyridoxal	—	J	Phosphorylation		
Pyridoxamine	—	J	Phosphorylation		
Biotin	—	D, J	—	>95	Raw egg white (avidin)
Pantothenate	—		—		
Folylmono-glu[e]	—	J	Glutamation		
Vitamin B_{12}^{e}	—	D, J	Adenosination, methylation	~1	

[a] Duodenum, jejunum, or ileum.
[b] Yields vitamin in absorbable form.
[c] Na⁺-dependent saturable processes.
[d] Estimate may include contribution of simple diffusion.
[e] Simple diffusion important only at high doses.
[f] Symport with Na⁺.

Table 3. Postabsorptive Transport of the Vitamins in the Body

Vehicle	Vitamin	Form Transported	Distribution
		Lipoprotein-Bound	
Chylomicra[a]	Vitamin A	Retinyl esters	Lymph[a]
	Vitamin A	β-carotene	Lymph[a]
	Vitamin D	Vitamin D[b]	Lymph[a]
	Vitamin E	Tocopherols	Lymph[a]
	Vitamin K	Phylloquinones, mena-quinones, menadione	Lymph[a]
VLDL[c]/HDL[d]	Vitamin E	Tocopherols	Plasma
	Vitamin K	Mainly menaquinone-4	Plasma
		Associated Nonspecifically with Proteins	
Albumin	Riboflavin	Free riboflavin	Plasma
		Flavin mononucleotide	Plasma
	Vitamin B_6	Pyridoxal	Plasma
		Pyridoxal phosphate	Plasma
Immunoglobulins[a]	Riboflavin	Free riboflavin	Plasma

Table 3. (Continued)

Vehicle	Vitamin	Form Transported	Distribution
Bound to Specific Binding Proteins			
Retinol BP (RBP)	Vitamin A	Retinol	Plasma
Cellular RBP (CRBP)	Vitamin A	Retinol	Intracellular
Cellular RBP, type II (CRBPII)	Vitamin A	Retinol	Enterocytic
Interstitial RBP (IRBP)	Vitamin A	Retinol	Interstitial spaces
Cellular retinal BP (CRALBP)	Vitamin A	Retinal	Intracellular
Cellular retinoic acid (CRABP)	Vitamin A	Retinoic acid	Intracellular
Transcalciferin	Vitamin D	D_2; D_3; 25-OH-D; 1,25-$(OH)_2$-D; 24,25-$(OH)_2$-D	Plasma
Vitamin D receptor	Vitamin D	1,25-$(OH)_2$-D	Enterocyte
Vitamin E BP	Vitamin E	Tocopherols	Intracellular
Riboflavin BP	Riboflavin	Riboflavin	Plasma
Flavoproteins	Riboflavin	Flavin mononucleotide	Intracellular
	Riboflavin	Flavin adenine dinucleotide	Intracellular
Transcobalamin I	Vitamin B_{12}	Vitamin B_{12}	Intracellular
Transcobalamin II	Vitamin B_{12}	Methylcobalamin	Plasma
Transcobalamin III	Vitamin B_{12}		Plasma
Erythrocyte-Carried			
Erythrocyte membranes	Vitamin E	Tocopherols	Blood
Erythrocytes	Vitamin B_6	Pyridoxal phosphate	Blood
Erythrocytes	Pantothenic acid	Coenzyme A	Blood
Free in Plasma			
—	Vitamin C	Ascorbic acid	Plasma
—	Thiamin	Free thiamin	Plasma
—	Thiamin	Thiamin pyrophosphate	Plasma
—	Riboflavin	Flavin mononucleotide	Plasma
—	Pantothenic acid	Pantothenic acid	Plasma
—	Biotin	Free biotin	Plasma
—	Niacin	Nicotinic acid	Plasma
—	Niacin	Nicotinamide	Plasma
—	Folate	Pteroylmonoglutamates[f]	Plasma

[a] In mammals, lipids are absorbed into the lymphatic circulation, where they are transported to the liver and other tissues as large lipoprotein particles called *chylomicra;* in birds, reptiles, and fishes, lipids are absorbed into the hepatic portal circulation.

[b] Vitamin D without a subscript refers to both major forms of the vitamin: D_2 and D_3.

[c] Very-low-density lipoproteins.

[d] High-density lipoproteins.

[e] E.g., IgG, IgM, and IgA.

[f] Especially 5-CH_3-tetrahydrofolate.

Table 4. Vitamins That Must Be Activated Metabolically

Vitamin	Active Form(s)	Activation Step	Conditions Increasing the Need for Vitamins
Vitamin A	Retinol	Retinal reductase Retinol hydrolase	Protein insufficiency
	11-*cis*-retinol	Retinyl isomerase	
	11-*cis*-retinal	Alcohol dehydrogenase	Zinc insufficiency
Vitamin D	1,25-$(OH)_2$-D	Vitamin D 25-hydroxylase	Hepatic failure
		25-OH-D 1-hydroxylase	Renal failure, exposure to lead, estrogen deficiency, anticonvulsant drugs
Vitamin K	All forms	Dealkylation of phylloquinones, menaquinones	Hepatic failure
		Alkylation of phylloquinones, menaquinones, menadione	
Thiamin	Thiamin-diP[a]	Phosphorylation	High carbohydrate intake
Riboflavin[b]	FMN, FAD	Phosphorylation Adenosylation	
Vitamin B_6	Pyridoxal-P	Phosphorylation Oxidation	High protein intake
Niacin	NAD(H) NADP(H)	Amidation (nicotinic acid)	Low tryptophan intake
Pantothenic acid	Coenzyme A	Phosphorylations, decarboxylation, ATP-condensation, peptide bond formation	
	ATP	Phosphorylation, peptide bond formation	
Folate	C_1FH_4[c]	Reduction; addition of C_1	
Vitamin B_{12}	Methyl-B_{12}	Cobalamin methylation	Folate, CH_3 insufficiency
	5'-deoxy-adenosyl-B_{12}	Adenosylation	

[a] I.e., thiamin pyrophosphate.
[b] Formerly known as vitamin B_2.
[c] I.e., tetrahydrofolic acid.

Table 5. Metabolic Functions of the Vitamins

Vitamin	Activities
Membrane Stabilizer	
Vitamin E	Antioxidant protection of polyunsaturated phospholipids
Hormone	
Vitamin D	Several metabolites important in Ca homeostasis
H^+/e^- Donors/Acceptors	
Vitamin E	Quenching of free radicals
Vitamin K	Conversion to the epoxide form in carboxylation of peptide glutamyl residues
Vitamin C	Oxidation to dehydroascorbic acid in hydroxylations
Niacin	Interconversion of NAD^+/NADH and $NADP^+$/NADPH couples in dehydrogenases
Riboflavin	Interconversion of $FMN/FMNH/FMNH_2$ and $FAD/FADH/FADH_2$ in oxidases
Pantothenic acid	Oxidation of CoA in the synthesis/oxidation of fatty acids
Coenzymes	
Vitamin A	Rhodopsin conformational change in light-induced bleaching; gene expression
Vitamin K	Vitamin K-dependent peptide-glutamyl carboxylase
Vitamin C	Cytochrome P_{450}-dependent oxidations (drug and steroid metabolism)
Thiamin	Cofactor of α-keto acid decarboxylases and transketolase
Niacin	NAD(H)/NADP(H) used by dehydrogenases in the metabolism of carbohydrates, lipids, and protein; Krebs cycle, rhodopsin synthesis
Riboflavin	FMN; L-amino acid oxidase, lactate dehydrogenase, pyridoxine 5′-phosphate oxidase FAD: D-amino acid and glucose oxidases, succinic and acetyl CoA dehydrogenases; glutathione, vitamin K, and cytochrome reductases
Vitamin B_6	Metabolism of amino acids, porphyrins, glycogen, and epinephrine
Biotin	Carboxylations and transcarboxylations
Pantothenic acid	Fatty acid synthesis/oxidation
Folate	Single-carbon metabolism, purine synthesis, methyl-group synthesis
Vitamin B_{12}	Methylmalonyl CoA mutase, N^5-CH_3-FH_4:homocysteine methyl transferase

The Individual Vitamins

VITAMIN A

Vitamin A-active compounds occur as derivatives of retinol and as plant pigments called carotenoids, which yield retinol (Figure 1) upon metabolism. The most important pro-vitamin A carotenoid is β-carotene, which is a dimer of retinol. Retinol is oxidized in the body to the aldehyde form retinal, which functions as a chromophore in the visual pigments, and then to retinoic acid, which regulates gene expression to affect growth and cell differentiation. Accordingly, the most prominent signs of vitamin A deficiency include night blindness and squamous metaplasia. Core foods for vitamin A include green and yellow vegetables, liver, oily fishes, and vitamin A-fortified foods (e.g., margarine).

VITAMIN D

Vitamin D-active sterols (e.g., vitamin D_3, cholecalciferol; see Figure 2) are derived in vivo ultimately by photolysis of 7-dehydrocholesterol in skin exposed to sunlight. The metabolically active form is a ring- and side-chain-hydroxylated derivative, 1,25-dihydroxycholecalciferol, which functions as a steroid hormone in the regulation of calcium homeostasis. That process involves the control of the uptake of calcium from the gut, the mobilization of calcium from the bone, and the reabsorption of calcium across the renal tubule. The most prominent sign of vitamin D deficiency is undermineralization of bone, which manifests as rickets in children and osteomalacia in adults. Core foods for vitamin D include fish liver oils and fortified foods (e.g., milk).

VITAMIN E

Vitamin E-active compounds have the biological activity of α-tocopherol (1,3,5-trimethyltocol), with side chains consisting of three isopentyl units that are fully saturated (tocopherols; see Figure 3), or each contain a double bond (tocotrienols). The vitamers E function as membrane-resident, FREE RADICAL scavengers to protect polyunsaturated membrane lipids from oxidative damage; they participate in this function with the essential trace element selenium and the antioxidant vitamin C. The most prominent

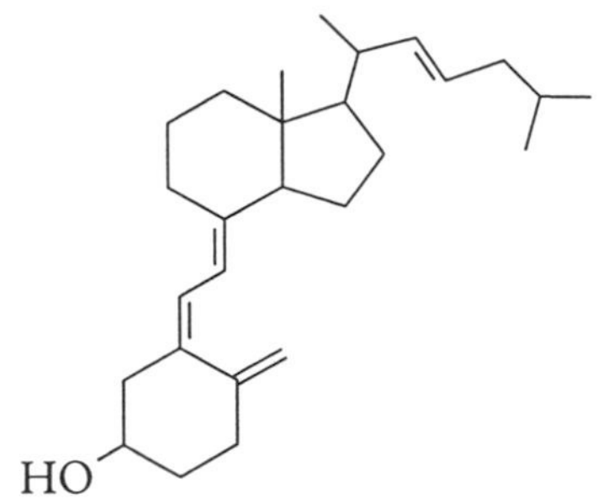

Fig. 2. Vitamin D_3 (cholecalciferol).

sign of vitamin E deficiency is increased erythrocyte hemolysis. Core foods for vitamin E are plant oils.

VITAMIN K

Vitamin K describes 2-methyl-1,4-naphthoquinone and its derivatives exhibiting the antihemorrhagic activity. The phylloquinones (Figure 4) in plants have phytyl or further-alkylated side chains, with a double bond on the proximal isoprene unit. The menaquinones, from bacteria, have side chains of varying numbers of isopentyl units, each with a double bond. The synthetic compound with no side chain, menadione, can be alkylated by humans and animals to produce menaquinones. Vitamin K functions in the posttranslational carboxylation of several calcium-binding proteins important in blood clotting and bone calcification. The most prominent sign of vitamin K deficiency is impaired clotting. Core foods for vitamin K include green, leafy vegetables, meats, and dairy products.

VITAMIN C

Vitamin C, also called ascorbic acid (Figure 5), is a strong reducing agent and is oxidized metabolically to dehydroascorbic acid via the radical intermediate semidehydroascorbic acid. The three species thus comprise a reversible redox system, making the vitamin an effective quencher of free radicals. Ascorbic acid maintains iron in the reduced (ferrous) state by undergoing oxidization to monodehydroascorbic

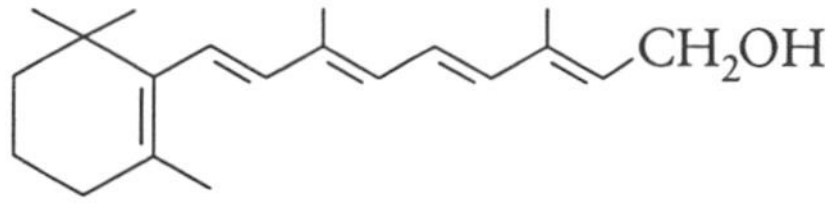

Fig. 1. All-*trans*-retinol.

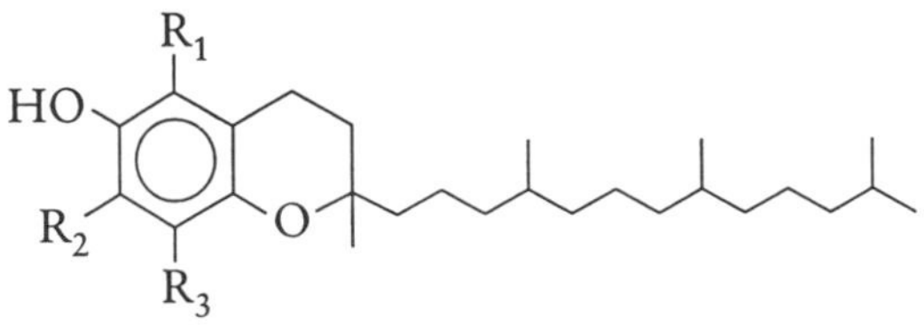

Fig. 3. Tocopherols.

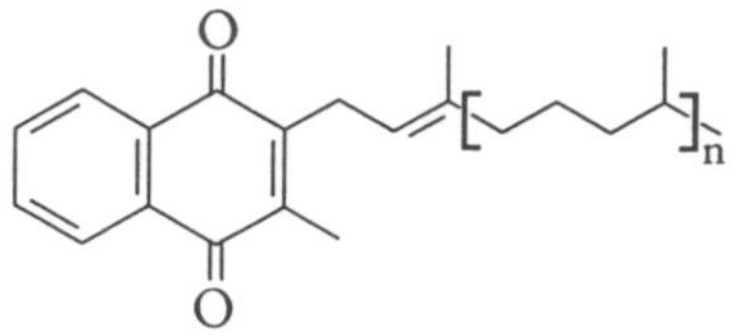

Fig. 4. Phylloquinones.

Fig. 6. Thiamin.

acid. In this way the vitamin serves as a cofactor for several ferrous-requiring enzymes in which it acts by maintaining iron in the reduced state. The most prominent signs of vitamin C deficiency are capillary fragility and hemorrhage, as seen in scurvy. Core foods for vitamin C include fruits, vegetables, and organ meats.

THIAMIN

Thiamin (Figure 6) is the designation of the compound known formerly as vitamin B_1. The thiamin thiazole hydroxyethyl group is phosphorylated in vivo to form thiamin mono-, di-, and triphosphates. Thiamin pyrophosphate "cocarboxylase" is metabolically active as a coenzyme for decarboxylations of keto-acids and transketolations. The most prominent sign of thiamin deficiency is peripheral neuropathy, as seen in human beriberi. Core foods for thiamin include whole grains, liver, and yeasts.

RIBOFLAVIN

Riboflavin was known formerly as vitamin B_2. Its metabolically active forms are flavin mononucleotide (FMN), shown in Figure 7, and flavin adenine dinucleotide (FAD). These coenzymes are versatile redox cofactors participating in either one- or two-electron redox reactions, thus serving as switching sites between two-electron donors (e.g., NADH, succinate) and one-electron acceptors (e.g., iron-sulfur proteins, heme proteins). FMN and FAD are thus involved in many enzymatic steps in the metabolism of carbohydrates, amino acids, and lipids. The most prominent signs of riboflavin deficiency are dermatologic lesions. Core foods for riboflavin include meats; dairy products; and green, leafy vegetables.

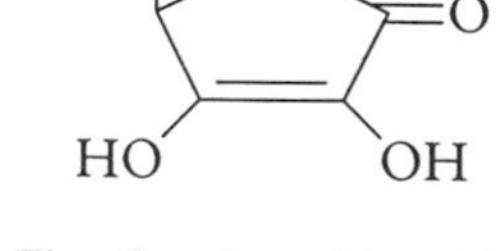

NIACIN

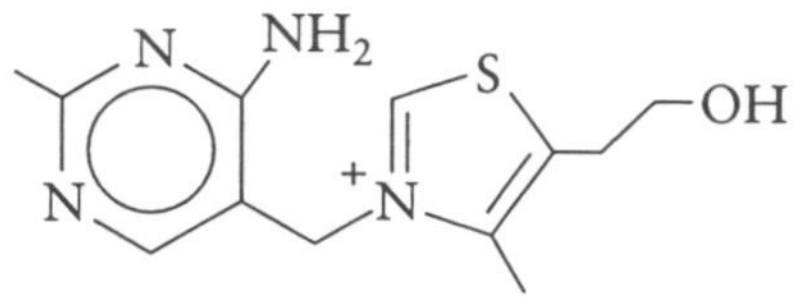

Niacin describes pyridine 3-carboxylic acid derivatives with the biological activity of nicotinamide (Figure 8). The coenzyme forms are the pyridine nucleotides NAD and NADP, which transfer hydrogen by a two-electron step, thus allowing them to serve as coenzymes for numerous reactions in the metabolism of carbohydrates, fatty acids, and amino acids. The most prominent signs of niacin deficiency are dermatologic lesions, as in human pellagra. Core foods for niacin include whole grains, meats, and yeasts.

VITAMIN B_6

Vitamin B_6 describes 2-methyl-3,5-dihydroxymethylpyridine derivatives with the biological activity of pyridoxine (Figure 9). The metabolically active form is the phosphorylated form of the aldehyde, pyridoxal phosphate, the aldehyde group which can react with primary amino groups in numerous reactions of amino acid metabolism. The most prominent signs of deficiency are dermatologic lesions. Core foods for vitamin B_6 include meats, whole grains, vegetables, and nuts.

BIOTIN

Biotin (see Figure 10) consists of conjoined ureido and thiophane rings. It bonds its enzymes covalently

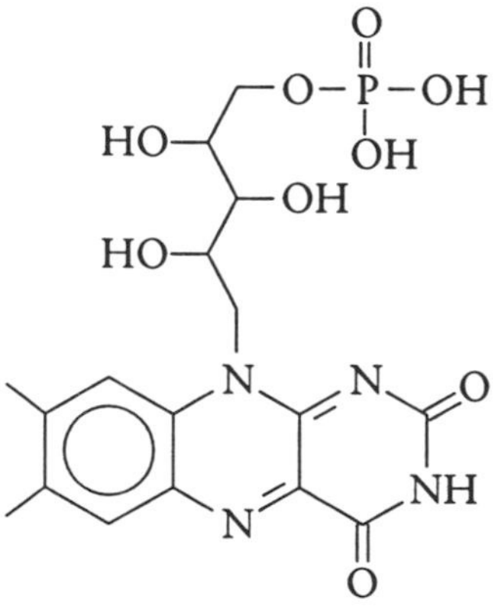

Fig. 7. Flavin mononucleotide (FMN).

Fig. 5. Ascorbic acid.

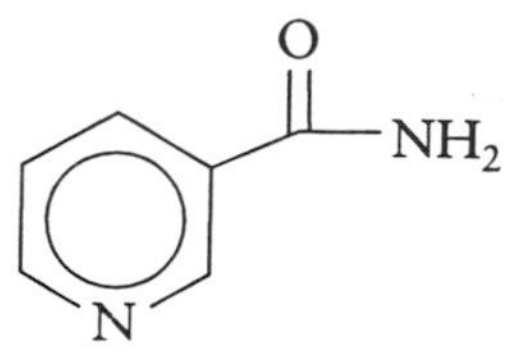

Fig. 8. Nicotinamide.

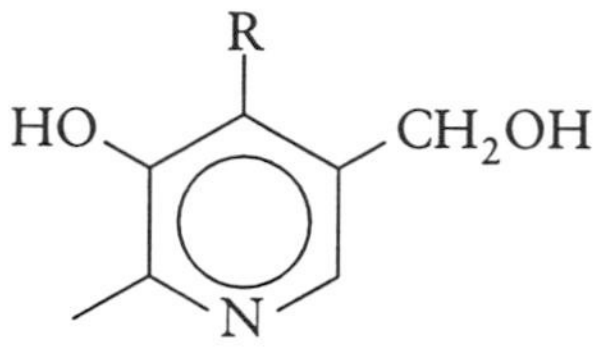

Fig. 9. Pyridoxine.

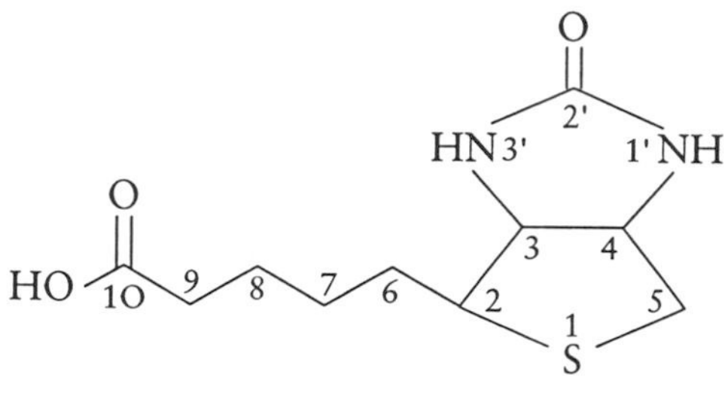

Fig. 10. Biotin.

via an amide linkage between its thiophane C-2 and a lysyl ε-amino group of the protein. Its covalent bonding of the carboxy group of carbonyl phosphate (formed from bicarbonate and ATP) allows biotin to serve as a transporter of CO_2 in 4-carboxylase involved in the metabolism of lipids, glucose, some amino acids, and energy. The most prominent signs of biotin deficiency are dermatologic lesions. Core

foods for biotin include milk, liver, egg yolk, and a few vegetables.

PANTOTHENIC ACID

Pantothenic acid designates the compound dihydroxy-β,β-dimethylbutyryl-β-alanine. It has two metabolically active forms, coenzyme A (Figure 11) and acyl carrier protein, in which the vitamin is linked via phosphodiester groups with adenosine-3′,5′-diphosphate or a serine residue, respectively. In these forms, pantothenic acid serves as a carrier of acyl groups in the metabolism of fatty acids, amino acids, and carbohydrates. The most prominent signs of pantothenic acid deficiency are dermatologic lesions. Core foods for pantothenic acid include meats, mushrooms, some vegetables, and yeasts.

FOLATE

Folate describes pteroylmonoglutamic acid and related compounds. The term *folates* is also used for this group with one or more L-glutamic acid residues. Tissues contain "conjugase" enzymes that hydrolytically remove glutamyl residues to release the monoglutamyl form, folic acid. The reduced compound is called tetrahydrofolic acid (Figure 12): its single-carbon derivatives are named according to the specific carbon moiety bound. Therefore the folates include a large number of chemically related species, each differing with respect to both the number of glutamyl residues and the nature of the single-carbon unit bound at three possible sites on the pteroylglutamic acid structure. Most naturally occurring folates also have a single-carbon unit; these forms participate in the "single-carbon pool" in the metabolism of amino acids and nucleotides. The most prominent sign of folic acid deficiency is anemia. Core foods for folates include liver; mushrooms; and green, leafy vegetables.

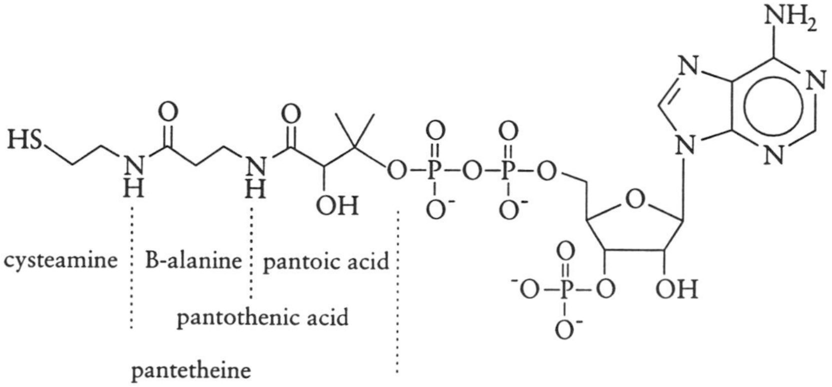

Fig. 11. Coenzyme A (showing its constituent parts).

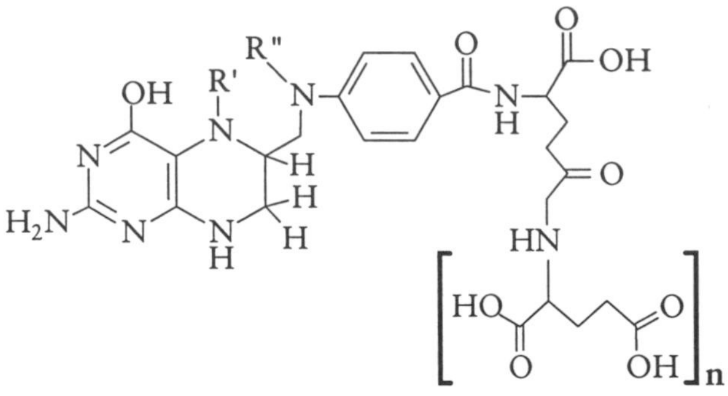

Fig. 12. Tetrahydrofolic acid and
its derivatives.

VITAMIN B12

Vitamin B_{12} is the descriptor for corrinoids (compounds containing the porphyrinlike, cobalt-centered, corrin nucleus) exhibiting the biological activity of cyanocobalamin (Figure 13), which has a cyano-ligand (CN^-) bound to the cobalt atom. The corrin cobalt atom can bind a nucleotide below the plane of the ring in addition to a small ligand above the plane. Vitamin B_{12} enzymes use the vitamin with either methyl or adenosyl ligands; these play key roles in the metabolism of propionate, amino acids, and single carbons. The most prominent sign of vitamin B_{12} deficiency is anemia. Core foods for vitamin B_{12} include meats, poultry, and bacterially fermented foods.

Table 6. Recommended Dietary Allowances

Age (yrs) or Condition	Vit. A	Vit. D	Vit. E	Vit. K	Vit. C	Thiamin	Ribo-flavin	Niacin	Vit. B_6	Folate	Vit. B_{12}
	μg^a	μg	μg^b	μg	mg	mg	mg	mg^c	mg	μg	μg
Infants											
0–0.5	375	7.5	3	5	30	0.3	0.4	5	0.3	25	0.3
0.5–1	375	10.0	4	10	35	0.4	0.5	6	0.6	35	0.5
Children											
1–3	400	10.0	6	15	40	0.7	0.8	9	1.0	50	0.7
4–6	500	10.0	7	20	45	0.9	1.1	12	1.1	75	1.0
7–10	700	10.0	7	30	45	1.0	1.2	13	1.4	100	1.4
Males											
11–14	1,000	10.0	10	45	50	1.3	1.5	17	1.7	150	2.0
15–18	1,000	10.0	10	65	60	1.5	1.8	20	2.0	200	2.0
19–24	1,000	10.0	10	70	60	1.5	1.7	19	2.0	200	2.0
25–50	1,000	5.0	10	80	60	1.5	1.7	19	2.0	200	2.0
51+	1,000	5.0	10	80	60	1.2	1.4	15	2.0	200	2.0
Females											
11–14	800	10.0	8	45	50	1.1	1.3	15	1.4	150	2.0
15–18	800	10.0	8	55	60	1.1	1.3	15	1.5	180	2.0
19–24	800	10.0	8	60	60	1.1	1.3	15	1.6	180	2.0
25–50	800	5.0	8	65	60	1.1	1.3	15	1.6	180	2.0
51+	800	5.0	8	65	60	1.0	1.2	13	1.6	180	2.0
Pregnant	800	10.0	10	65	70	1.5	1.6	17	2.2	400	2.2
Lactating											
0–6 mos.	1,300	10.0	12	65	95	1.6	1.8	20	2.1	280	2.6
6–12 mos.	1,200	10.0	11	65	90	1.6	1.7	20	2.1	280	2.6

a Retinol equivalents.
b α-tocopherol equivalents.
c Niacin equivalents.
Source: National Academy of Sciences, *Recommended Dietary Allowances*, 10th ed.; Washington, DC: National Academy Press, 1989.

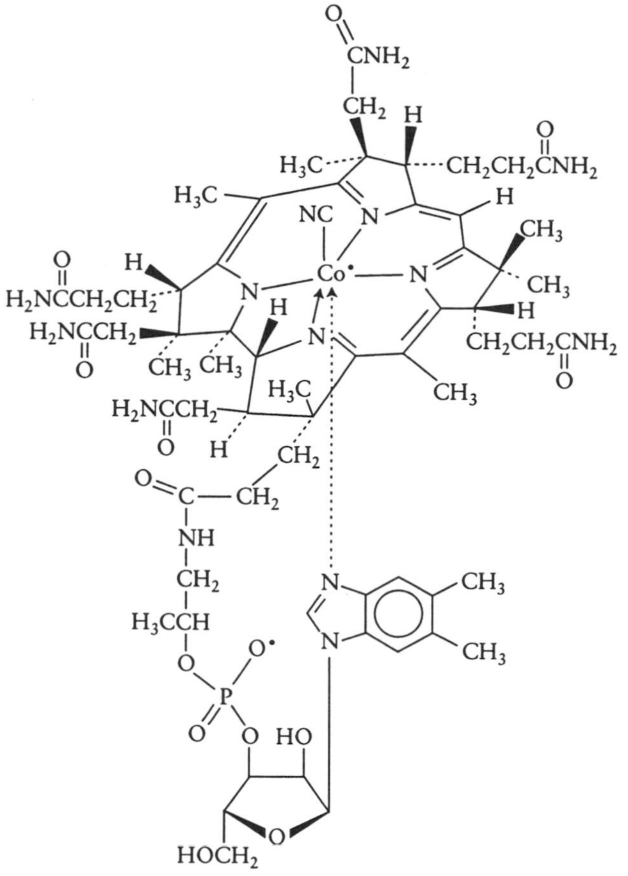

Fig. 13. Cyanocobalamin.

Vitamin Requirements

The quantitative estimates of vitamin requirements are established by prominent scientific advisory groups and reviewed periodically. In the United States these values are the Recommended Dietary Allowances (RDAs). They are amounts of vitamins required to prevent known clinical deficiency signs in healthy people eating mixed diets.

BIBLIOGRAPHY

COMBS, G. F., JR. *The Vitamins: Fundamental Aspects in Nutrition and Health;* New York: Academic Press, 1992.

MACHLIN, L. J., ed. *Handbook of Vitamins,* 2nd ed.; New York: Marcel Dekker, 1991.

National Academy of Sciences. *Recommended Dietary Allowances,* 10th ed.; Washington, DC: National Academy Press, 1989.

GERALD F. COMBS, JR.

Von Laue, Max Theodor Felix

See LAUE, MAX THEODOR FELIX VON.

VSEPR Theory

See VALENCE SHELL ELECTRON PAIR REPULSION.

Waste Treatment

Waste consists of the unwanted by-products and residuals of all activities, including chemical manufacturing and chemical consumption. Waste can occur in three common thermodynamic states—i.e., solid, liquid, and gas. Construction debris, bath water, and landfill gas from aging disposal sites are examples. Waste is also classified as hazardous or nonhazardous. Empty pesticide containers are hazardous solid waste common to household trash. Benzene is a hazardous substance released as a gas via normal operations at many gasoline filling stations and as a liquid from leaking petroleum-product storage tanks. Sanitary wastewater is a universal, dual-phase (liquid/solid) nonhazardous waste. CARBON DIOXIDE is a generally nonhazardous gaseous waste from human and microbial respiration.

Nonhazardous Solid Wastes

Construction debris, bulky household trash including newsprint and plastics, unwanted appliances, and cardboard, glass, or metal containers are types of nonhazardous solid wastes. This material is disposed of by recycling, landfilling, and/or incineration. Recycling may constitute pretreatment prior to the other processes. Incineration may be coupled to energy production by recovery of heat from combustion gases with a steam boiler system. The primary residual from incineration is hearth ash, which arises from noncombustible material, such as glass and metal, contained in the municipal solid waste (MSW) stream. Fly ash consists of fine particles that arise from noncombustibles and condensation processes in cooled combustion gases; fly ash is removed by an air pollution control (APC) system located between the combustion unit(s) and the boiler (if any) or exhaust stack.

Chlorine, nitrogen, and sulfur in the incinerator feed stream usually appear as inorganic acids in combustion gas. If hydrochloric, nitric, and sulfuric acids are neutralized and removed in a dry scrubber, a solid saltlike waste results. If the acids are treated in a wet scrubbing, a predominantly liquid slurry waste results. Ashes and dry-scrubber effluent are landfilled; wet-scrubber effluent must be dewatered before landfilling (see also LANDFILLS). The overall incineration process is described by the following simplistic mass balances:

Solid Waste → Noncombustibles + Combustible + Moisture
Noncombustibles → Ashes
Combustibles → Products of Incomplete Combustion (PICs) + Carbon Dioxide + Inorganic Acids (Scrubber Residues) + Steam (Energy)

High-molecular-weight PICs, carbon monoxide, acid gases, and residual particulate matter are present in small quantities in stack gases. Concentrations of species in stack gases are related in very complex fashion to vapor pressure, condensed phase formation and particle size, adsorption onto uncontrolled inert particulate matter, and APC system design and efficiency. Stack-gas composition is determined by sampling and analyses; both are difficult to carry out and interpret in the context of National Ambient Air Quality Standards (NAAQS). Moisture will condense during boiler heat exchange, scrubbing, and particle separation; saturation can be expected at the point of discharge.

Landfilling has been one of the primary methods of disposal for nonhazardous wastes. Combustor ash, as-collected MSW, baled and compacted MSW, construction debris, and unwanted appliances are interred in landfills. At the end of daily operations, daily cover is placed over the accumulated mass to limit odor, rodents, birds, and infiltration of precipitation. Landfilling is less and less attractive as a means of disposal because of residential and/or commercial land development, negative public attitudes (the "not in my backyard" [NIMBY] philosophy), and strict regulation of gas and liquid emissions (see EMISSION ABATEMENT). Oxygen diffusion through fill and cover is slow. With the exception of arid areas, precipitation and moisture contributed by compressed fill yield an abundant source of water, generally present as a leachate mound. Internal conditions are suitable for anaerobic fermentation. Microbial metabolism of organic carbon leads to acetic acid and methane in sequential acetogenesis and methanogenesis reactions. With glucose as the model source, these reactions can be expressed simply as

Acetogenesis
$$C_6O_6H_6 + 2 H_2O \rightarrow 2 C_2O_2H_4 + 2 CO_2 + H_2$$

Methanogenesis
$$C_2O_2H_4 \rightarrow CH_4 + CO_2$$

The combination of gases generated by the fermentation process is termed landfill gas; it forms flammable and/or explosive mixtures with air, in the fill, and at points of discharge through fissures and pores. Controlled venting is mandatory in many instances. Dilute acetic acid participates in leaching metals and sparingly soluble organic species from filled matter. This process was the basis of the Environmental Protection (EP) Toxicity Test required by the Resource Conservation and Recovery Act (RCRA). A leachate plume can lead to pollution of neighboring surface water and groundwater. Thus, leachate control by impermeable barriers, collection, and treatment is generally required.

Wastewater

Sanitary wastewater is the most common nonhazardous liquid waste. It includes discharges from residential and commercial (industrial) sanitary facilities, household cooking and washing activities, industrial wash water, and some industrial process water. In urban areas, wastewater is collected and transported in sewers to a central treatment plant. In areas of low population density, wastewater is treated in on-site septic systems or "turkey" mounds. Wastewater generation rates vary greatly with time of day, season, and the intensity of residential and nonresidential development. Coupled storm-water management and/or combined sewers for storm-water and wastewater management add greatly to the variability of flow rate. Equalization basins are used to collect and store wastewater from irregular sources, permitting steady (metered) flow to the treatment system.

The first step in treatment is physical; mechanical screening is utilized to separate floating solids and suspended debris. Liquid, with some finely divided particulate matter, passes to primarily microbial (secondary) treatment. Dispersed biomass or a biofilm (trickling filter) metabolizes dissolved and suspended carbonaceous matter, under aerobic conditions. Nitrogen and phosphorus are necessary macronutrients. As a rule of thumb, the carbon : nitrogen : phosphorus ratio should be approximately 100 : 10 : 1. In the dispersed biomass reactor, air is introduced via spargers or large sweeps to enhance mass transfer rates at the air/water interface. Air flow in trickling filters may be passive or forced.

Two gross parameters are used to describe oxygen requirements in aerobic microbial treatment processes—i.e., biochemical oxygen demand (BOD) and chemical oxygen demand (COD). BOD is the oxygen uptake due entirely to microbial respiration and growth; with glucose as a simple substrate, BOD is expressed by

Respiration
$$C_6O_6H_6 + (9/2) O_2 \rightarrow 6 CO_2 + 3 H_2O$$

Growth
$$C_6O_6H_6 + O + N + P \rightarrow \text{New Cell Mass}$$

On the other hand, COD is the oxygen uptake for strongly oxidizing nonmicrobial processes. The following examples illustrate this quantity for typical inorganic and refractory organic species:

$$2 FeO + (1/2) O_2 \rightarrow Fe_2O_3$$
$$CCl_4 + O_2 \rightarrow CO_2 + 2 Cl_2$$

For readily biodegradable species, COD is nearly equal to BOD, and the COD : BOD ratio is approximately 1. A COD : BOD ratio greater than 3 is taken to imply a toxic or refractory nature for the substrate(s). Conventional microbial treatment will not work well for wastewater with this characteristic.

Overflow from secondary treatment with dispersed biomass is settled; treated settler overflow is discharged, after disinfection, denitrification, and/or trace-contaminant removal, to a receiving water. Most thickened underflow is recycled, to return active biomass to the aeration tank. Because some growth occurs during aeration, a portion of the recovered microbial must be "wasted." Excess sludge is dewatered and disposed of by landfilling or incineration. Excess trickling filter biofilm is removed by backwashing with a portion of product water, dewatered, and disposed of in an identical manner.

In aeration processes, volatile organic compounds (VOCs) are stripped into the airstream and released to the atmosphere. Sorption onto biomass slows this process to some extent; however, many VOCs are refractory and build up in the aeration tank with recycled biomass. Pure oxygen and enriched airstreams can be treated for VOC removal and recycled to the aeration tank. Volatile species can be removed from one-pass air also; however, the cost of cleaning a large volume of inert gas (nitrogen) is high and difficult to justify.

Small household waste treatment systems—i.e., septic tanks and turkey mounds—are intended for the digestion of bulk and suspended solids. Microbial digestion proceeds by aerobic or anaerobic pathways; both processes can occur in some systems, as functions of depth, decomposition rate, and oxygen diffusion in surrounding soil or cover material. Well-balanced treatment systems discharge clarified, nutrient-rich effluent to soil and groundwater. Disposal sites are often identified by luxuriant surface vegetation. New biomass and undigested solids may accumulate in a septic tank, requiring periodic physical removal of settled sludge. Similarly, biomass and undigested solid matter may build up in void spaces of mounds and, over time, may limit infiltration and wastewater flow.

Nonhazardous Gas Treatment

Massive discharges of gases, not commonly viewed as hazardous, can still have impacts on human health and the environment. Sulfur dioxide and oxides of nitrogen emitted by electric-generating stations are closely regulated under the Clean Air Act (CAA), as amended in 1990. Acid gases produced during the incineration of MSW have the potential for undesirable effects and are controlled. High concentrations of unburned hydrocarbons, carbon monoxide, and finely divided particulate matter occur in urban areas, not in association with massive individual sources, but owing to the large numbers of cars, buses, and trucks in use.

Treatment technologies used to control gaseous emissions include thermal destruction, chemical and catalytic reactions at lower temperatures, sorption, and particle separation by filtration or electrostatic precipitation. Temperature control in two or more combustion stages contributes to essentially complete destruction of carbonaceous matter in incinerators. Flares, some utilizing supplementary fuel and air, are used in petroleum refining, landfills, and industrial processes to complete destruction of both flammable and nonflammable gases. Sulfur dioxide is *adsorbed* and reacted with lime (dry) or *absorbed* and reacted with caustic or lime slurry (wet) to yield poorly soluble, nonvolatile calcium sulfate (gypsum) or a sodium sulfate liquor:

$$\text{Typical}$$
$$SO_2 + CaO \text{ or } Ca(OH)_2 \rightarrow Ca(SO_4)$$

Acid gases are scrubbed with water or dilute caustic; acid gases may also be sorbed and neutralized.

Unburned hydrocarbons and carbon monoxide are oxidized catalytically in all new gasoline-powered vehicles. Oxides of nitrogen can be adsorbed initially onto a solid substrate and/or reacted directly with ammonia or some other cost-effective chemical reducing agent. Particulate matter is removed from a gas stream by filtration on fabric or by electrostatic charge addition and precipitation on counter-charged surfaces.

Hazardous Wastes

Hazardous substances have been identified in legislation, by judicial interpretation, and by agency actions taken under authority derived from legislation. Sources of legislative intent include (but are not limited to) the Resources Conservation and Recovery Act (RCRA), the Comprehensive Environmental Response, Compensation, and Liability Act (CERCLA), the Superfund Amendments and Reauthorization Act (SARA), the Clean Air Act (CAA), as amended, and the Occupational Safety and Health Act (OSHA). Agencies such as the Environmental Protection Agency (EPA), the Occupational Safety and Health

Administration (OSHA), and the Department of Transportation (DOT) have created standards and criteria pursuant to these congressional directives.

Probably the most comprehensive, generic identification of hazardous substances is contained in the RCRA. Materials are characterized by specific chemical properties: toxicity, ignitability, corrosivity, and reactivity. A substance is toxic if it is fatal to humans or small animals at low doses or if it contains any of 358 tabulated constituents. A material is corrosive if it has a pH less than 2 (highly acidic) or greater than 12 (highly alkaline). In addition, substances are listed as hazardous by origin; residues, containers, and off-specification material from a large number of industrial or commercial activities are also listed as hazardous.

The preferred methods for treatment of hazardous wastes are destruction, detoxification, or reduction in volume. The goal is always to minimize threats to human health and/or the environment. A clear distinction must be made between organic and inorganic substances with respect to destruction. High temperatures and strongly oxidizing conditions can destroy organic substances, by conversion to carbon dioxide, water, and mineral acids. It is not certain, however, that oxides of nitrogen and polychlorinated dibenzo(p)dioxins/furans (PCDDs and PCDFs), respectively, are not the outcome of some destructive processes. Inorganic species include metals and salts that cannot be destroyed. Thus, detoxification and reduced volume (often interpreted as reduced mobility or availability in the environment) are important considerations.

In some instances, prohibitions and strict restrictions on manufacture, sale, and use have been invoked. Asbestos, asbestos-containing materials (ACMs), DDT and other pesticides, and polychlorinated biphenyls (PCBs) have been banned. Chlorofluorocarbons (CFCs) and other chlorine-containing species are strictly regulated. The use of lead in plumbing fixtures, solder, and coatings is limited and closely monitored.

Hazardous solids include contaminated solids and sediments—e.g., the dioxin-laden soil from Times Beach (Missouri) and the PCB-laden sediment in the Hudson River (New York). The most common treatment methods for solid hazardous wastes that are concentrated organic matter are conventional incineration, thermal desorption with subsequent destruction in the gas phase, and strong oxidation. Incineration does not work well with solids and sediments because of the large mass of inert material that must be treated; thermal desorption and strong oxidation can be applied cost effectively to excavated soils and sediments. Elemental (inorganic) hazardous wastes, such as weakly radioactive matter, leachable mine tailings, and industrial plating or metalworking residues, may require concentration (reduction in volume) of contaminants prior to commercial recovery or long-term, stable storage.

Hazardous gases are a primary focus of the CAA Amendments of 1990. This law sets standards for massive emissions and hazardous air pollutants (HAPs) that require extensive monitoring and emission controls for both stationary and nonstationary sources (vehicles). Adsorption, absorption, catalytic oxidation, conventional combustion, and micron-size particle removal (filtration) are technologies that have proven utility.

Pollution Prevention

Some or all hazardous waste treatment can be avoided by in-plant and in-home pollution prevention. Substitution of benign or more environmentally friendly chemicals for solvents, degreasing agents, air conditioning and refrigerant fluids, and agricultural chemicals can contribute. Recovery and recycling to return unavoidable process streams for reuse within a process reduces demand for manufactured, virgin materials and minimizes waste treatment requirements. In some instances, process redesign or integration of hazardous waste reduction techniques at an early stage of process design and development can lead to substantial reduction in hazardous waste generation.

BIBLIOGRAPHY

HENRY, J. G.; HEINKE, G. W. *Environmental Science and Engineering,* 2nd ed.; Upper Saddle River, NJ: Prentice Hall, 1996.

Metcalf and Eddy, Inc. *Wastewater Engineering: Treatment, Disposal, Reuse,* 2nd ed.; New York: McGraw-Hill, 1979.

SINCERO, A. P.; SINCERO, G. A. *Environmental Engineering: A Design Approach;* Upper Saddle River, NJ: Prentice Hall, 1996.

ROBERT C. AHLERT

Water

See ATMOSPHERE; FREEZING AND MELTING; HEAVY WATER; LIQUID; SOLUBILITY; SOLUTIONS; WATER PURIFICATION.

Water Purification

Raw water for various consumer uses is obtained from two primary natural sources—i.e., fresh surface waters and groundwater. Desalination is employed in the western United States to convert brackish surface water to a product suitable for public or industrial applications. In the Middle East, seawater is converted to water for irrigation and industrial process water, largely via membrane separations. In the southwestern United States and South Africa, wastewater is treated and recycled for various nonpotable applications.

Large-scale or river basin recycling is of very common occurrence. In the United States, the Passaic, Delaware, Potomac, and Colorado rivers are well-documented examples of multiple water use. Multiple use is more the rule, rather than the exception, in both urbanized and developing countries. Rivers, lakes, and groundwater systems serve as "receiving" waters for untreated, partially treated, and fully treated domestic and industrial wastewaters. In addition, freshwater systems receive eroded soil, animal and vegetative debris, and agricultural chemicals, in response to rainfall runoff and snowmelt. Some removal of organic constituents occurs via microbial degradation, vaporization, photochemical reactions, and abiotic oxidation/reduction reactions. However, residual organic solutes, inorganic salts, and suspended matter (turbidity) must be removed by treatment of raw water, prior to use. After treatment, product water must be disinfected to minimize proliferation of microorganisms in storage, distribution, and end-use systems.

Purification is interpreted as the treatment and disinfection of raw water to meet quality requirements appropriate to specific applications. Quality criteria vary with intended use. The three dominant uses are (1) drinking water for human consumption, (2) industrial process and cooling water, and (3) water for laboratory dilution, reagent, and specialty chemical services.

Unit cost, effectiveness of monitoring and quality control, and process reliability are primary considerations in the selection of a purification process and the methods of process operation. Large volumes of high-quality water must be available at low cost for domestic consumption. Reliability and control must be adequate at all times to protect public health. Large volumes of water with specific quality characteristics are necessary for general industrial process and cooling applications. Low cost is important, although the cost of the water supply is a factor that is "value added" to energy or material product. Limited volumes of very high-quality water are required in the last category. Reliability, quality control, and extreme purity are necessary, at higher costs that are often small compared to the unit prices of products or services.

Drinking Water

Purification of drinking water is carried out to meet three classes of goals: public acceptance, protection of public health, and regulatory requirements. The user community expects drinking water to have acceptable taste, odor, and appearance. Many relatively harmless species can contribute to taste or odor that is offensive. Excessive suspended solids or color are sources of poor appearance; high concentrations of iron, manganese, or turbidity in raw water, and the inability of a treatment process to cope with variations in these quantities, can be responsible for poor appearance. Disinfection is the primary tool used to prevent the spread of water-borne disease. Water at the tap must have some residual disinfectant; however, in the case of chlorine, excess "free" disinfectant can lead to taste and odor problems and/or the presence of compounds deemed dangerous to public health—i.e., trihalomethanes (THMs). Thus, acceptance and health protection can be competing quality goals.

Federal regulations are intended to provide uniform, national standards for drinking water to protect human health and the environment. There is much overlap between federal regulatory standards and the operating criteria established by local water districts and purveyors. The Safe Drinking Water Act (as amended in 1986) created maximum contaminant levels (MCLs) for eighty-three specific constituents potentially present in raw or processed drinking water. Also, the amended act created maximum contaminant level goals (MCLGs); MCLGs are not enforceable but establish ultimate levels, as technology, treatment, and cost permit. National primary drinking water regulations are revised annually and are based on current MCLs. MCLs for inorganic species are cited in the Code of Federal Regulations (40 CFR Section 141, July 1993a), as follows:

Contaminant	MCL (mg/L)
Arsenic	0.05
Asbestos	7 million long (>10 μm) fibers/liter
Barium	1.0
Cadmium	0.005

Contaminant	MCL (mg/L)
Chromium	0.05
Fluoride	4.0
Lead	0.05
Mercury	0.002
Nitrate (as N)	10.0
Nitrite (as N)	1.0
Nitrate + Nitrite (as N)	10.0
Selenium	0.01

The administrator of the U.S. Environmental Protection Agency (USEPA) identifies the "best technology, treatment technique or other means available for achieving" (BATs) desired levels (40 CFR Section 141, July 1993b). In this context, ion exchange (IE) and reverse osmosis (RO) are recommended for nitrate/nitrite removal.

MCLs cited for endrin and total trihalomethanes (TTHMs) are 0.0002 and 0.1 mg/L, respectively (40 CFR Section 141, July 1993c). Eighteen volatile organic compounds (VOCs) have drinking water MCLs, driven by health considerations (USEPA, 1991). Thirteen of these species are chlorinated hydrocarbons (four aliphatic, six olefinic, and three aromatic); the remaining five include benzene and methyl-, ethyl-, and vinyl-substituted benzene. MCLs are low—i.e., 0.002 mg/L for vinyl chloride and 0.005 mg/L for benzene (40 CFR Section 141, July 1993d). Seven species have MCLGs of zero; these compounds are known or highly suspect carcinogens.

Eighteen agricultural chemicals are cited (40 CFR Section 141, July 1993d); nine have zero MCLGs by reason of highly suspect carcinogenicity. The MCL of ethylene dibromide is 0.00005 mg/L; most others are less than 0.004 mg/L. BATs recommended by USEPA are tabulated for both classes of organic compounds (40 CFR Section 141, July 1993e). Adsorption from water onto granular activated carbon (GAC) is suggested generally. Packed tower aeration (PTA) is recommended for many of the target species, especially vinyl chloride.

COAGULATION

Colloids, suspended solids, microorganisms, and viruses are characterized by relatively strong surface charges, often structured as an electric double layer. Electrostatic repulsion gives stability to dispersions of fine particles. Coagulation is intended to destabilize suspensions through the addition of coagulant chemicals such as aluminum sulfate, ferric chloride or sulfate, and synthetic polymer and/or natural polymer (conchiolin from shellfish shell). Coagulant solution or suspension is mixed rapidly with raw water that has received an initial disinfection to prevent bioslime formation in process vessels, and screening to remove coarse solids and debris. Mixing is completed in a few seconds. The combination of coagulant chemicals and the sequence of coagulant addition, for optimum reduction of turbidity and suspended iron, is usually determined by laboratory tests with typical raw water samples.

FLOCCULATION

Settleable and filterable solids are separated after flocculation. The latter incorporates gentle, slow mixing to promote particle interactions that lead to agglomeration of destabilized particulate matter into larger aggregates.

SEDIMENTATION

After particle-size growth has taken place, the process stream passes to a clarification basin or tank. The aqueous suspension is separated into supernatant clarified water and an underflow suspension of thickened, concentrated particulate matter (sludge). Sedimentation removes larger, dense, nonfloccuble particulate matter—i.e., sand and silt. Larger aggregates formed during flocculation are settled also. Clarification devices generally depend on gravity settling; geometry is variable—i.e., rectangular, circular, or conical. Clarifiers may utilize a sludge rake or scraper and may be operated in a one-pass or a sludge recirculation mode. A "sludge blanket" clarifier employs a siphon for sludge removal.

FILTRATION

Clarifier retention time and volume are generally insufficient for gravity settling to achieve desired suspended solids concentrations. Filters and centrifuges, to polish the clarified process stream, are available in many configurations. Some filters employ a single medium—e.g., graded sand—as packing; multimedia filters that use coal and/or sand layers of differing particle sizes are also common. A filter screen or porous retaining plate, together with a precoat layer or filter aid (diatomaceous earth) added to the process stream, are an effective approach; in this design, the suspended solids form the bulk of the filter medium. In all cases, periodic back washing with a portion of the product water is an inexpensive and reliable method of sludge removal.

The sequence of operations that includes primary screening, coagulant addition, mixing, flocculation, clarification, and filtration reduces raw water turbidity, as well as color and color-forming potential. This traditional treatment process contributes to acceptability and attainment of secondary standards. Microorganisms and viruses are present as dispersed solids in raw water; thus, reduction of suspended solids levels is a step toward protection of public health.

DISINFECTION

This process is required to minimize viable microorganisms, control water-borne disease, and avoid formation of bioslime in distribution and end-use systems. Disinfection can be accomplished with strong oxidizing chemicals, biocides, or radiation. Strong oxidizers include chlorine, hypochlorite or permanganate salts, chloramines, chlorine dioxide, or ozone.

Chlorine is popular in the United States; residual "free" chlorine, as hypochlorous acid, remains to protect against postcontamination of downstream systems. Chlorine has been found to contribute to THM formation, with an impact on public health. Excess chlorine and products of chlorine in reaction with trace organic solutes, especially phenol and substituted phenolic species, lead to taste and odor problems. Ozone is popular in Europe; it is an effective disinfectant, but the absence of a residual provides no protection from postcontamination. The absence of a residual is likewise the major problem with radiation sterilization. Chloramines, chlorine dioxide, and oxidizing salts are less effective disinfectants, more expensive and substantially less likely to yield THMs, as compared to chlorine. The latter benefit—i.e., avoidance of THM formation—is a significant consideration.

TRACE TOXINS

To some extent, the basic drinking water treatment train will limit inorganic contaminant levels. However, to achieve reliable control on these species, supplementary treatment may be necessary. GAC, IE, and RO are advanced technologies that have been identified as BATs.

THMs, VOCs, and agricultural chemicals, in raw or process water at concentrations in excess of primary drinking water standards, are less likely to be reduced to designated levels by conventional treatment. The favored advanced treatment for these three classes of compounds is GAC. The inclusion of a carbon adsorption unit in the process must be integrated with end-of-process disinfection, since the carbon can yield soluble chlorinated species in the presence of residual chlorine. Activated carbon will remove THM precursors and avoid THM formation during or after clorination, if the unit is located upstream of disinfection.

Industrial Process and Cooling Water

Relatively clean water is required for process purposes in very large quantities. Once-through cooling water can represent 90–95 percent of the total industrial liquid "waste" stream. Potentially toxic trace constituents are not a dominant factor. Operational issues, peculiar to the industrial mission, dictate the design of water pretreatment processes.

ONCE-THROUGH COOLING

Water cooling without reuse is practiced by many power-generating stations. If raw water frequently has elevated levels of suspended solids, some pretreatment to remove particulate matter is necessary to prevent physical obstruction of cooling equipment. Similarly, "shock" treatment with biocide is necessary to prevent formation of bioslime on heat exchange surfaces. Fouling of heat exchange surfaces can occur as a consequence of corrosive attack. One or more of the following pretreatment steps may be justified economically to minimize losses in efficiency and direct costs of cleaning: pH control, elimination of dissolved oxygen, and reduction in total or selective dissolved salt concentrations. Acid/base addition, IE, RO, and/or chemical reduction may be indicated.

CONSUMPTIVE COOLING

Consumptive cooling minimizes the volume of raw water to be pretreated. Cooling water is recirculated through a cooling system (tower), and latent heat is exchanged for sensible heat, via simultaneous heat and mass transfer. Raw water is required as makeup for evaporative losses and the blow-down stream used to prevent accumulation of dissolved solids in the circulating water. Biocide levels must be maintained in the recirculating system to prevent biological fouling; disinfection is continuous in-process, rather than shock, pretreatment. Total or selective dissolved solids removal, with IE or RO, must be utilized to avoid concentrating solutes and to minimize blow-down volumes.

PROCESS WATER

Conditioning of process water is heavily dependent on the nature of the specific process. Soluble and suspended process wastes may overwhelm raw water impurities, such that wastewater treatment takes precedence over raw-water pretreatment. Species present in raw water may catalyze or enter into unwanted chemical reactions.

High-Quality Water

LABORATORY PURPOSES

Comparison of water quality against the stringent MCLs imposed on drinking water makes it necessary to have ultrapure water available for dilution, control, and blanks in analytical laboratories. In short, the laboratory must have a reasonably absolute standard against which product water quality can be measured. Laboratory water supplies utilize single or multiple distillation, membrane separation, and deionization to prepare ultrapure product from feed of initial drinking water quality.

SPECIAL MANUFACTURING

Steam generation in high-pressure boilers, pharmaceutical manufacturing, and processing of electronic devices require water of ultrahigh purity. Boiler water must be free of species that can foul heat transfer surfaces, decreasing steam production and increasing operating temperatures at a risk to boiler integrity.

Poor quality surface coating and ion implantation for electronic devices are results of impurities in water. Ultrahigh purity is required in rinsing, cleaning, and nonconducting background. Preliminary deionization, pH control, dissolved gas removal, RO, and final IE are appropriate. Treatment costs may be high, but become a trivial part of product value. Restrictions on chlorofluorocarbon (CFC) manufacture and use have resulted in an increased interest in aqueous-based cleaning practices in plating and related electronics manufacturing activities.

Drugs and health-related materials must be free of latent or viable microbial activity, viruses, or substances capable of eliciting allergic or immune system responses from human receptors. Since chemical addition is not desirable, organism and biochemical removals utilize radiation and microfiltration. The latter is a physical process that functions at a size scale that includes colloids and some large molecular species.

Measurements and Control

Instrument technology provides on-line measurement systems for monitoring and control of process operations. Flow, pH, conductivity or resistivity, turbidity, and oxidation/reduction state can be sensed and reported or recorded remotely. Overall product quality is assured by product samples and laboratory assays, at two levels. The primary level includes traditional measurements that relate to acceptability and human health—i.e., suspended solids, microbial plate count, nitrate, etc. Analytic techniques are described in detail in the manual *Standard Methods* (Clesceri et al., 1989).

BIBLIOGRAPHY

CLESCERI, L. S. (Water Environmental Association); GREENBERG, A. E. (American Public Health Association); TRUSSELL, R. R. (American Water Works Association). *Standard Methods for the Examination of Water and Wastewater*, 19th ed.; Washington, DC: American Public Health Association, 1995.

Code of Federal Regulations. *40 CFR Subpart B—Maximum Contaminant Levels—Sections 141.11(b), 141.11(c) and 141.11(d) and 40 CFR Subpart F—National Revised Primary Drinking Water Regulations: Maximum Contaminant Levels—Section 141.62(b);* Washington, DC: Office of the Federal Register, July 1993.

———. *40 CFR Subpart B—Maximum Contaminant Levels—Section 141.12;* Washington, DC: Office of the Federal Register, July 1993.

———. *40 CFR Subpart F—National Revised Primary Drinking Water Regulations: Maximum Contaminant Levels—Section 141.62(c);* Washington, DC: Office of the Federal Register, July 1993.

———. *40 CFR Subpart G—National Revised Primary Drinking Water Regulations: Maximum Contaminant Levels—Section 141.61(a) and 141.61(c);* Washington, DC: Office of the Federal Register, July 1993.

———. *40 CFR Subpart G—National Revised Primary Drinking Water Regulations: Maximum Contaminant Levels—Section 141.61(b);* Washington, DC: Office of the Federal Register, July 1993.

KAWAMURA, S. *Integrated Design of Water Treatment Facilities;* New York: Wiley, 1991.

LORCH, W. *Handbook of Water Purification;* New York: Wiley, 1987.

U.S. Environmental Protection Agency, Office of Drinking Water. *Drinking Water Health Advisory: Volatile Organic Compounds;* Chelsea MI: Lewis Publishers, 1991.

VAN HALL, C. E., ed. *Measurement of Organic Pollutants in Water and Wastewater;* ASTM Special Technical Publication 686; Philadelphia: ASTM, 1979.

ROBERT C. AHLERT

Wave Function

See PSI FUNCTION.

Weights and Measures

See INTERNATIONAL SYSTEM OF UNITS.

Werner, Alfred (1866–1919)

Alfred Werner, the founder of COORDINATION chemistry, was born on December 12, 1866, in Mulhouse, France, the fourth and last child of Jean-Adam Werner, an iron worker, and his second wife, Salomé Jeanette Werner (*née* Tesché). Although Alsace was annexed to the second German empire in 1871, the Werner family continued to speak French at home, and their sympathies remained with France. Young Werner's spirit of rebellion and resistance to the authority of the German occupation forces may have contributed to the revolutionary and iconoclastic theory with which his name is associated. Despite his reverence for German science (most of his articles appeared in German journals), Werner's political and cultural ties were with France.

At the age of six, Werner was enrolled in the Catholic École Libre des Frères, where his tremendous self-confidence and stubborn independence that kept him from submitting blindly to authority became evident. He attended the École Professionelle, a technical school where he studied chemistry (1878–1885). During his year of compulsory military service in the German army (1885–1886), he audited chemistry courses at the Technische Hochschule in Karlsruhe. He then attended the Eidgenössisches Polytechnikum (now the Eidgenössische Technische Hochschule [ETH]) in Zürich, from which he received a degree in technical chemistry on August 3, 1889. He received his doctorate on October 13, 1890.

Within three years (1890–1893), Werner produced his three most important theoretical papers. His doctoral dissertation (1890), one of the classics of stereochemistry, coauthored with his mentor Arthur Hantzsch, extended LE BEL and VAN'T HOFF's concept of the tetrahedral carbon atom (1874) to the nitrogen atom, explained many puzzling cases of geometrically isomeric organic trivalent nitrogen compounds, and placed the stereochemistry of nitrogen on a firm theoretical basis. In his *Habilitationsschrift* (1891), an original research paper required in order to teach at a university, Werner attacked August KEKULÉ, the supreme patriarch of structural organic chemistry, by replacing Kekulé's rigidly directed valences by his more flexible approach of viewing affinity as a variously divisible, attractive force emanating from the center of an atom and acting equally in all directions.

Werner spent the winter semester of 1891–1892 working on thermochemical problems with Marcellin Berthelot at the Collège de France, but returned from Paris to Zürich in 1892 to become *Privatdozent* (unsalaried lecturer) at the Polytechnikum. In the fall of 1893 he became associate professor at the University of Zürich, where he remained for a quarter-century. In 1894 he married Emma Wilhelmine Giesker and became a Swiss citizen. This union produced a son and a daughter. In 1895 he was promoted to professor.

Werner was appointed to the university largely because of the almost overnight fame produced by the publication of his third most important theoretical paper (1893), in which he proposed the basic postulates of his revolutionary and controversial coordination theory, which had occurred to him in a classic example of the "flash of genius." At 2 A.M., Werner, who was primarily an organic chemist whose knowledge of inorganic chemistry was very limited, awoke with the solution to the riddle of what were then called "molecular compounds." It had come to him like a flash of lightning. He arose from his bed and wrote so quickly and steadily that by 5 P.M. he had finished his most important paper.

Werner's coordination theory, with its concepts of COORDINATION NUMBER, primary and secondary valence, addition and intercalation compounds, and octahedral, square planar, and tetrahedral configurations, not only provided a logical explanation for known molecular compounds but also predicted series of unknown compounds, whose discovery by Werner and his students in a quarter-century tour de force of research activity confirmed his controversial ideas. In 1913 Werner became the first Swiss chemist to be awarded the Nobel Prize "in recognition of his work on the linkage of atoms in molecules, by which he has thrown fresh light on old problems and opened new fields of research, particularly in inorganic chemistry." Soon afterward chronic arteriosclerosis of the brain, aggravated by excessive drinking, destroyed Werner's physical and mental health, and he died in Zürich on November 15, 1919.

Werner's ideas encompassed almost the whole of systematic inorganic chemistry and even found application in organic chemistry. He was the first to show that stereochemistry is a general phenomenon not limited to carbon compounds, and his views of valence and chemical bonding stimulated research on these fundamental issues. Today, when the theoretical and practical importance of coordination chemistry is unquestioned, it is clear that the foundations of modern structural inorganic chemistry were erected largely by one man, Alfred Werner, who has justly been called the inorganic Kekulé.

BIBLIOGRAPHY

KAUFFMAN, G. B. *Alfred Werner, Founder of Coordination Chemistry;* New York: Springer Verlag, 1966.

———. "Alfred Werner's Coordination Theory: A Brief Historical Introduction"; *Educ. Chem.* 1967, *4*, 11–18.

———. *Classics in Coordination Chemistry:* The *Selected Papers of Alfred Werner,* Part 1; New York: Dover, 1968.

———, ed. *Werner Centennial;* Washington, DC: American Chemical Society, 1967.

MORGAN, G. T. "Alfred Werner." *J. Chem. Soc.* 1920, *117*, 1639–1648.

GEORGE B. KAUFFMAN

Wieland, Heinrich (1877–1957)

Heinrich Wieland was part of a progeny of German chemists spanning three generations—his father (Theodor), sons (Otto and Theodor), and son-in-law (Feodor Lynen). It is not difficult to infer that this lineage had, in addition to traditional family ties, significant scientific interaction. Heinrich's father did thesis work (but did not submit a thesis) under Adolf Strecker, who had discovered cholic acid in 1948. Wieland's proposal for the structure of cholic acid, as a complex steroid, was recognized by the Nobel Prize committee in 1928. His son-in-law Lynen, also intrigued with the subject of steroids, received the Nobel Prize in 1964.

Wieland's breadth of interests may have been influenced by the diversity of his training, that is excerpted here from his own account: "began Fall semester 1896, University of Munich; began Fall semester 1897, First Laboratory of Berlin under Dr. P. Cory; 1898–1899, Technische Hochschule at Stuttgart under Professor Hell; 1899–1901, University of Munich." He received his Ph.D. ("Attempts to Synthesize Phenylated Alleles"), under Johannes Thiele, from the University of Munich in July 1901. He was the private assistant to Professor Harries in Berlin for the winter semester of 1901–1902, and for Professor Baeyer beginning in spring 1902.

His earliest publications focused on nitrogen-containing compounds. The subjects included alkene additions by nitrogen oxides, the mechanism of aromatic nitration, fulminic acid polymerization, and the discovery of the first nitrogen free radical (prepared from hydrazines). The prominence of his studies at the University of Munich was acknowledged when he was given a promotion to a senior lectureship in 1913.

It is interesting to juxtapose the history Wieland's scientific achievements with the political circumstances in Germany during the First and Second World Wars. In 1917–1918 he did chemical warfare research at the Kaiser Wilhelm Institute. Between the wars, his career developed rapidly. In 1918 he moved to the position at Technische Hochschule in Munich for which he had been accepted but had taken a leave of absence. In 1921 he became a department chair at the University of Freiburg. In 1925 he was nominated as director of the prestigious University of Munich, where he spent the rest of his career. His replacement of the resigning director, Richard Willstätter, was under unfortunate circumstances since it is believed to have been influenced by anti-Semitic trends, which were already beginning. Willstätter had personally recommended Wieland to be his replacement.

By the time of Wieland's promotion, he had already expanded considerably the scope of his research and was being nominated by prominent German scientists for the Nobel Prize. His contributions in the area of bile acids and toad venoms provided the basis for his nomination. Cholic acid, a bile salt found in the gut, was found to solubilize hydrophobic compounds such as aromatic compounds and fatty acids. This discovery is called the cholic acid principle, for which Wieland is given credit. Wieland's award of the 1928 Nobel Prize, however, was based on his structural determination of cholic acid. The process of elucidating the structure of a new natural product at that time was very tedious, requiring the identification of fragments from chemical degradation, from which a proposal for the entire structure would be made. Some have found it amusing that the structure of cholic acid required correction, based on improved X-ray crystallographic analysis techniques available in 1932.

With the new director position, the size of Wieland's group continued to grow. He had almost 600 students and postdoctoral assistants in his laboratories during his career. He made significant contributions in many areas of chemistry. His pioneering natural product chemistry of nitrogen-containing ALKALOIDS, peptides, and STEROIDS focused on strychnine and morphine, the cobra and mushroom (death cap) toxins, methylcholanthrene (a carcinogenic toxin), and many others. While these topics appear to have been influenced by government interests in wartime applicability, these complex natural products were instrumental in providing tools for the study of biological systems and in providing significant challenges to synthetic chemistry methodology. It is indeed tragic to note the famous Dachau concentration camp is not far from Munich. Human experimentation with some of these substances may have occurred.

Wieland had over 350 publications over the span of his career. Another measure of his great teaching ability is the accomplishments of those he trained. His son-in-law Feodor Lynen, for example, received the Nobel Prize in 1964 for the discovery of how activated acetic acid is used for the biosynthesis of fatty acids and terpenes. This work followed Wieland's early interest with yeast metabolism of acetic acid and steroid biosynthesis.

This pioneering work with acetic acid metabolism led to Wieland's proposal that biological oxidation could occur by dehydrogenation. Wieland was openly criticized for his views during a dispute with a contemporary, Otto Warburg. One of Warburgs's students, Krebs, realized that both were partially correct, leading to Krebs's formulation of the citric acid cycle for which he received the 1953 Nobel Prize. Perhaps Wieland would have devoted more attention to this topic if political circumstances were different. Nonetheless, Wieland is sometimes called the "founder of modern biochemistry."

Wieland's ability to conduct laboratory work abruptly ended when Allied bombing destroyed the institute in 1944. Wieland also lost his library when he was evicted from his personal residence by the U.S. Army for their needs. The house issue was resolved in 1947 just prior to his seventieth birthday, due to pressure by American citizens and an article in the New York Times. He remained editor of LIEBIG'S *Annalen der Chemie* until shortly before his death.

BIBLIOGRAPHY

JONES, D. P. In *Dictionary of Scientific Biography;* C. C. Gillispie, P. Abailard, L. Berg, eds.; New York: Scribner, 1981; Vol. 1, pp. 334–335.

WITKOP, B. "Remembering Heinrich Wieland (1877–1957): Portrait of an Organic Chemist and Founder of Modern Biochemistry." *Medicinal Research Reviews* 1992; 12, 195–274.

DWIGHT M. PETERSON

Wilkinson, Geoffrey (1921–1996)

British inorganic chemist Sir Geoffrey Wilkinson was born in Todmorden, Yorkshire, on July 14, 1921. In 1946 Wilkinson received his Ph.D. from Imperial College in London and began his professional career in the United States working on the development of the atomic bomb. He was a research associate at the University of California from 1946 until 1950 and at the Massachusetts Institute of Technology from 1950 to 1951. Wilkinson then began teaching at Harvard University, remaining there from 1951 until 1956.

While in the United States he began research on ferrocene. Wilkinson showed that this molecule has a structure with an iron atom sandwiched between two flat five-carbon rings. For his studies of METALLOCENES, Wilkinson shared the 1973 Nobel Prize in chemistry with E. O. Fischer, who had independently carried out similar research.

In 1956 Wilkinson returned to the University of London to assume a faculty position as professor of inorganic chemistry. During the 1960s he focused his studies on the use of complexes containing a metal-hydrogen bond. The rhodium complex known as Wilkinson's catalyst was the first to be used as a homogeneous catalyst for adding hydrogen to the double bonds of ALKENES.

Wilkinson's awards and honors include an American Chemical Society Award in inorganic chemistry, the Lavoisier Medal from the French Chemical Society as well as the 1973 Nobel Prize in chemistry. Wilkinson was knighted in 1976. He died in 1996.

BIBLIOGRAPHY

DAINTITH, J.; MITCHELL, S.; TOOTILL, E. *A Biographical Encyclopedia of Scientists;* New York: Facts on File, 1981.

MILLAR, D.; MILLAR, I.; MILLAR, J.; MILLAR, M. *The Cambridge Dictionary of Scientists;* Cambridge, U.K.: Cambridge University Press, 1996.

Who's Who of British Scientists (1969/1970), London: Longman, 1970.

ANNA R. BERGSTROM

Wöhler, Friedrich (1800–1882)

Friedrich Wöhler was born in Eschersheim, Germany, and died in Göttingen, Germany. Chemistry was a boyhood pastime for Wöhler that he never outgrew. While attending high school in Frankfort beginning in 1814, he performed experiments in a home laboratory. At Marburg University, where his father had studied, he began studies to be a physician and continued his experiments in a laboratory he built in his room. His interest in chemistry was well known; one professor chastised him for wasting time with experiments when he should have been studying.

Nevertheless, his early research was productive; in 1821, he published the first of nearly 300 papers. The chemist L. Gmelin attracted him to Heidelberg, where he completed his studies and received the degree of doctor of medicine in surgery and midwifery in 1823. He never practiced medicine but, encouraged by Gmelin, went to Sweden to study chemistry with J. BERZELIUS. Berzelius found his new student enthusiastic but lacking in good laboratory techniques. Wöhler was a quick learner, and over the year spent with Berzelius, he matured into a deliberate, exacting, and reflective scientist who disliked speculation. In 1825, he returned to Germany to teach at the Berlin Technical School. In 1828, he was appointed professor by royal decree. After moving his family to Cassel due to a cholera outbreak, he joined them in 1831, accepting a position at a new industrial school. In 1836, Wöhler received a university appointment at the George Augustus University in Göttingen, where he established one of the great teaching laboratories of Germany, training many chemists until his retirement in 1880. A very gentle, warm man, he was much loved by his students.

Owing to his syntheses of urea and systematic study of the benzoyl radical with J. LIEBIG, Wöhler is best remembered as one of the founders and developers of organic chemistry. Although others (J. Davy in 1812; M. Vauquelin in 1818) had unwittingly prepared urea, Wöhler unequivocally synthesized urea from cyanogen and aqueous ammonia in 1824 and communicated a more effective method, from ammonium chloride and silver cyanate, in 1828. As the first recognized synthesis of an organic compound, it overthrew the theory of vitalism, which stated that compounds made by plants or animals could not be synthesized in the laboratory. At the 1837 meeting of the British Association for the Advancement of Science, Liebig announced that Wöhler's urea synthesis "must be considered one of the discoveries with which a new area of science has commenced."

It was through his investigations of cyanates while still working with Berzelius that Wöhler met Liebig. They were introduced through J. GAY-LUSSAC, who had noted that the reported compositions of cyanic acid by Wöhler and fulminic acid by Liebig were identical, constituting a then-rare example of what would become known as isomerism (see ISOMERS). Liebig and Wöhler worked well together and collaborated in numerous areas. Most notable was their report of the chemistry of benzaldehyde, which detailed unanticipated, systematic REACTIVITY patterns of organic compounds. Demonstrating that the benzoyl radical C_7H_5O could remain intact throughout a series of chemical reactions, the work introduced the first significant data leading to the concept of functional groups. In other, independent work, Wöhler prepared beryllium, aluminum, phosphorus, crystalline silicon, and boron, as well as calcium carbide, silicon nitride, silicon hydride, and the oxides of phosphorus. Later in his career (1863), he noted the now well-known analogy of carbon and silicon compounds. In conjunction with the Liebig laboratory at Giessen, his laboratory at Göttingen was largely responsible for establishing the German dominance in organic chemistry that prevailed well into the twentieth century.

BIBLIOGRAPHY

GORDON, N. E. "Editor's Outlook: Friedrich Wöhler." *J. Chem. Educ.* 1928, *5*, 1537–1538.

KLOOSTER, H. S. "Friedrich Wöhler and His American Pupils." *J. Chem. Educ.* 1944, *21*, 158–170.

LEICESTER, H. M. *The Historical Background of Chemistry,* New York: Dover, 1971.

LEICESTER, H. M.; KLICKSTEIN, H. S., eds. *A Source Book in Chemistry, 1400–1900;* Cambridge, MA: Harvard University Press, 1968.

SMITH, E. F. "Some Experiences of Dr. Edgar W. Smith as a Student under Wöhler." *J. Chem. Educ.* 1928, *5*, 1554–1557.

WARREN, W. H. "Contemporary Reception of Wöhler's Discovery of the Synthesis of Urea." *J. Chem. Educ.* 1928, *5*, 1539–1553.

MARTIN HULCE

Wood

Wood—a natural, cellular, composite material of botanical origin—possesses unique structural and chemical characteristics that render it desirable for a broad variety of end uses. Wood is obtained from two broad categories of plants known as softwoods (gymnosperms) and hardwoods (angiosperms). The wood tissue, including cells and intercellular substance, is a composite material constructed from three principal polymers: CELLULOSE is the skeleton, hemicelluloses the matrix, and lignin the encrusting substance. Additionally, heartwood in particular contains many low-molecular-weight organic compounds known as extractives or extraneous substances. A few extractives, such as tannins, are polymeric. Inorganic constituents (ash) generally amount to no more than 0.1–0.5 percent of the wood. The distribution of these polymers and extractives in wood cannot be defined precisely for a given tree species or even for a given tree. The wood of the conifers (softwood) differs in its chemistry from that of the arboreal angiosperms (hardwood), and among the latter there are differences between those native to the tropics and to the temperate zones. Even within a tree, chemical composition varies with roots, stems, and branches. Juvenile wood differs from mature wood, sapwood from heartwood, and earlywood from latewood. Compression wood in conifers and tension wood in hardwoods have chemical compositions vastly different from normal wood. Tracheids and ray cells in softwood have dissimilar chemical compositions, and this applies to the fibers, vessels, and ray cells of hardwoods. On the ultrastructural level, the middle lamella, the primary wall, and the secondary wall have significantly different chemical compositions. Within the secondary wall there are also differences in composition between the S_1, S_2, and S_3 layers (Figure 1). The percentages of the principal chemicals in softwood and hardwood are summarized in Table 1. The distribution of these polymers, in general, across a typical cell wall is illustrated in Figure 2.

Cellulose, the skeleton of the cell walls, is the main constituent of wood. Approximately 40–45 percent of the dry wood is cellulose, located predominantly in

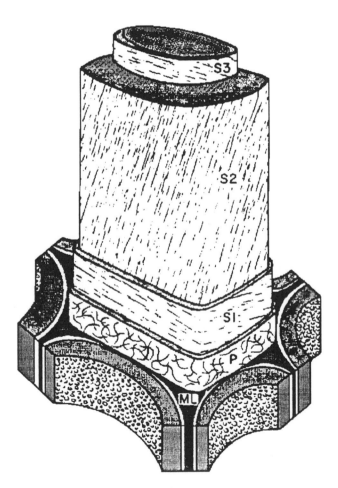

Fig. 1. Simplified structure of a wood cell. Keys: middle lamella (ML), primary wall (P), the outer (S1), middle (S2), and inner (S3) layers of the secondary wall.

the secondary cell wall. Cellulose is a polysaccharide that is made up of D-glucose units linked successively through 1,4-glycosidic bonds. It constitutes the microfibrils and is responsible for the strength of fiber because of its high degree of polymerization and linear orientation. Cellulose possesses a ribbonlike, flat molecular structure. Through inter- and intrahydrogen bondings, cellulose molecules form a two-phase structure that contains crystalline and amorphous structures. While the crystalline portion of the microstructure contributes to the strength of the molecule, the amorphous portion provides the softness and pliability. Plant cells can experience very high osmotic pressures, and so the cell walls must be strong—both to withstand pressure and to support the weight of the plant. Cellulose in wood is imbedded in an amorphous matrix of lignin, which functions as a cement to glue cellulose fibers together. Lignin also holds cellulose molecules together within the fiber cell wall. The cellulose resists tension while the matrix helps resist compression.

Lignin is a complex, highly branched, three-dimensional polymer composed of phenylpropane units with a somewhat different composition in softwoods and hardwoods. It encrusts the intercellullar space and any openings in the cell wall after the cellulose and hemicelluloses have been deposited. Hence, the concentration of lignin is high in the middle lamella (70 percent) and low in the secondary wall. However, due to the thickness, at least 70 percent of the lignin in softwoods is located in the secondary wall. Lignin is biosynthetically derived from three precursors: p-coumaryl alcohol, coniferyl alcohol, and sinapyl alcohol (Figure 3). The coupling reactions take place among the reactive sites of these units and a rather complicated structure is formed. P-coumaryl alcohol is a minor precursor of softwoods and hardwoods; coniferyl alcohol is the predominant precursor of softwood lignin; and coniferyl alcohol and sinapyl alcohol are both precursors of hardwood lignin. A typical softwood lignin contains the following functional groups per 100 building units: methoxyl (90–95), phenolic hydroxyl (20), phenolic ether (80), aliphatic hydroxyl (90), benzyl alcohol or ether (40), and carbonyl (20). Approximately one-third of the interunit linkages in lignin are carbon-carbon bonds and the remainder carbon-oxygen. The most important interunit linkage is the arylglycerol β-aryl ether bond (50). Other major bonds are in phenylcoumaran, the pinoresinol, and the diphenyl types of linkages. A partial picture of the molecular structure is shown in Figure 4. Due to its three-dimensional molecular structure, although lignin is amorphous in nature, it is a rigid material.

Table 1. Chemical Composition of Various Lignocellulosic Materials

Lignocellulosic Source	Cellulose (%)	Hemicellulose (%)	Lignin (%)	Extract (%)
Hardwood	43–47	25–35	16–24	2–8
Softwood	40–44	25–29	25–31	1–5
Abaca	63.72	5–10	21.83	1.6
Bagasse	40	30	20	10
Coir	32–43	10–20	43–49	4.5
Corn cobs	45	35	15	5
Corn stalks	35	25	35	5
Cotton	95	2	0.9	0.4
Flax (retted)	71.2	20.6	2.2	6.0
Flax (unretted)	62.8	12.3	2.8	13.1
Hemp	70.2	22.4	5.7	1.7
Henequen	77.6	4–8	13.1	3.6
Istle	73.48	4–8	17.37	1.9
Jute	71.5	13.6	13.1	1.8
Kenaf	36.0	21.5	17.8	2.2
Ramie	76.2	16.7	0.7	6.4
Sisal	73.1	14.2	11.0	1.7
Sunn	80.4	10.2	6.4	3.0
Wheat straw	30	50	15	5

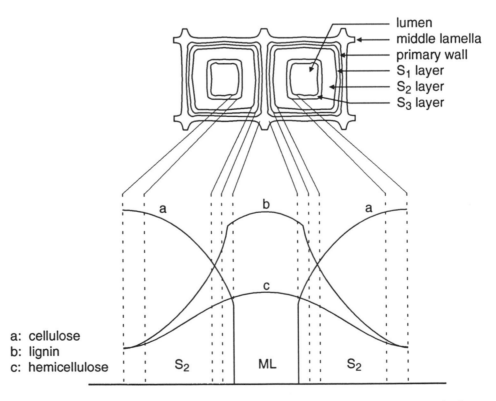

Fig. 2. Chemical components across wood cell walls. (Reproduced from *Journal of Applied Polymer Science*, with permission of John Wiley & Sons.)

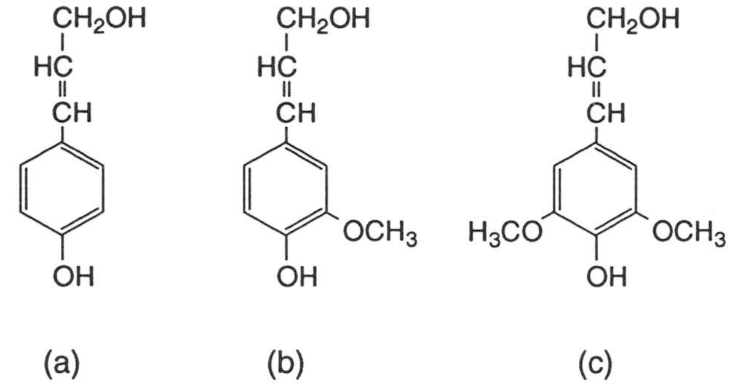

Fig. 3. Lignin precursors: (a) P-coumaryl alcohol, (b) coniferyl alcohol, and (c) sinapyl alcohol.

Cellulose and lignin make up the major portion of wood. Both of these polymers provide strength and rigidity to the wood. However, in trees they are too rigid to sustain mechanical stress; hence, the presence of hemicelluloses provides flexibility for the tree or wood. Hemicelluloses also increase the packing density of the cell wall. They are made up of a mixture of carbohydrates with a degree of polymerization of about 250 (i.e., made up of 250 units of monomeric units). Some of the hemicelluloses are also branched. These carbohydrates are glucose, mannose, galactose, xylose, arabinose, 4-O-methylglucuronic acid, and galacturonic acid residues. Some hardwoods contain trace amounts of rhamnose. The predominant hemicelluloses in softwood are a family of galactoglucomannans, which together account for 15 to

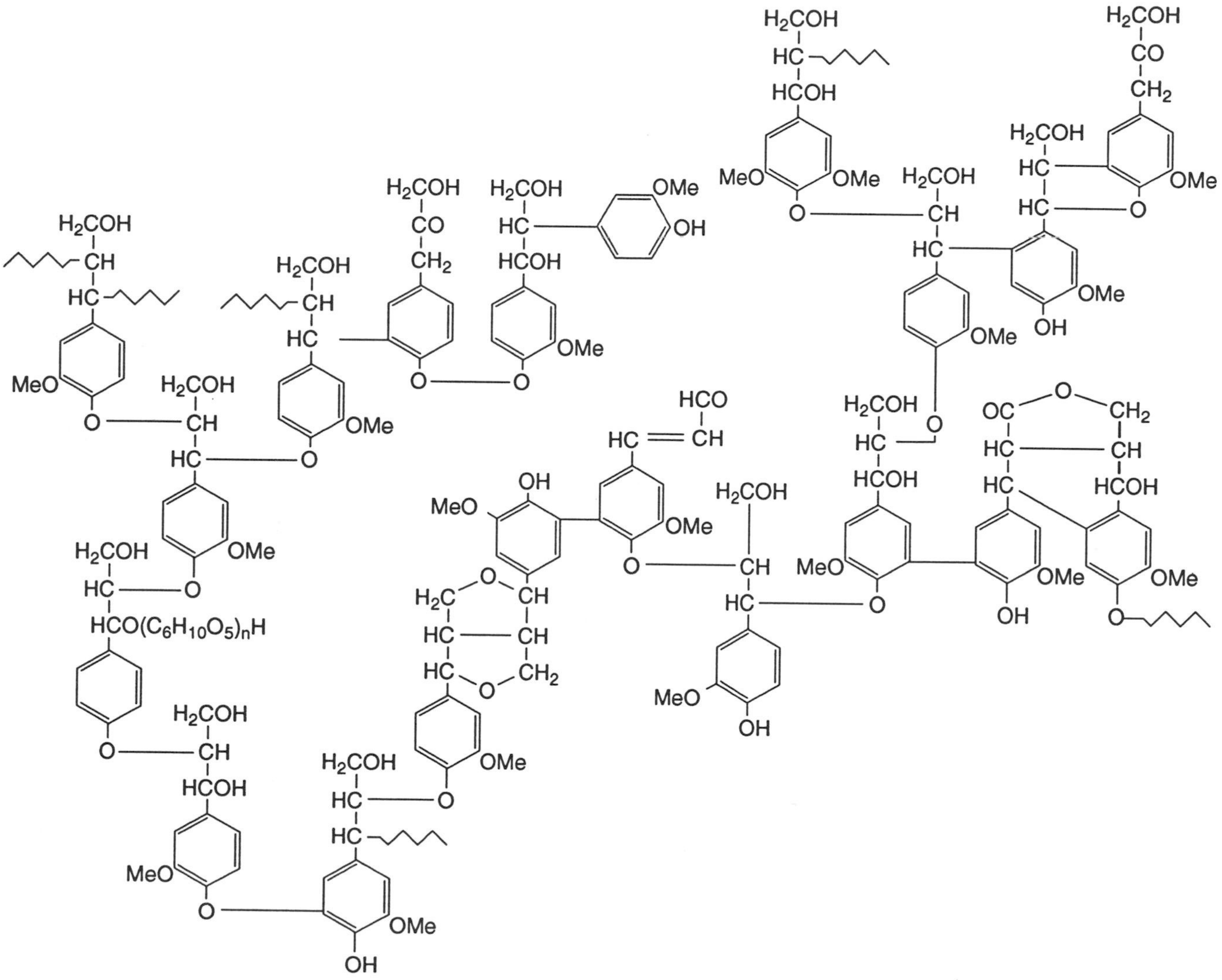

Fig. 4. A possible polymeric structure of softwood lignin.

20 percent of this type of wood. They consist of a main chain of $(1 \rightarrow 4)$-linked β-D-glucopyranose and β-D-mannopyranose residues, some of which have a single side chain of α-D-galactopyranose attached to their 6-position. The glucose to mannose ratio is about $1:3$ but the ratio of galactose to glucose can vary from $1:1$ to $1:10$. Acetyl groups are also attached to some of the residues in the glucomannan backbone. The predominant hemicellulose in all hardwoods is an acidic xylan. The xylan consists of a linear, main chain of $(1 \rightarrow 4)$-linked β-D-xylopyranose residues. Single-unit side chains, consisting of a $(1 \rightarrow 2)$-linked 4-O-methyl-α-D-glucuronic acid residue, are randomly distributed along the xylan backbone, usually with an average number of one side chain per 10 xylose units. There are also 7 acetyl groups per 10 xylose residues attached to the hydroxyl groups of the xylan chain.

Wood also contains small quantities of extraneous materials called extractives. The nature of the extractives is frequently typical of a tree species. They are chemicals not directly involved in the strength of the wood, formed primarily at the transition from sapwood to heartwood. The vast majority of the extractives in wood are of a low molecular weight and are located outside the cell wall. A few of them are reserve food materials and others are protective agents, but most of them seem to serve no specific function. Many of them are located only in the heartwood. The larger number of different extractives in wood can be classified as terpenes, resin acids, tannins, polyphenols, lignans, tropolones, fats, waxes, and carbohydrates. The inorganic materials are normally 0.1 to 0.5 percent of the oven-dry weight of wood. The alkali earths—i.e., calcium, potassium, and magnesium—usually account for 70 percent of the total inorganic materials present. Occasionally, manganese, silicon, and silica are present in appreciable amounts depending on the wood species. Although extractives play little role in strength, they are totally responsible for many of wood's other properties, such as color, odor, taste, and decay resistance.

BIBLIOGRAPHY

FENGEL, D.; WEGENER, G. *Wood: Chemistry, Ultrastructure, Reactions*; Berlin and New York: Walter de Gruyter, 1984.

HON, D. N.-S. *Polymer News* 1988, *13*, 134–140.

HON, D. N.-S.; SHIRAISHI, N., eds. *Wood and Cellulosic Chemistry*; New York: Marcel Dekker, 1991.

SJÖSTRÖM, E. *Wood Chemistry: Fundamentals and Applications*; New York: Academic Press, 1981.

DAVID N.-S. HON

Woodward, Robert Burns (1917–1979)

Robert Burns Woodward was born on April 10, 1917, in Boston, Massachusetts. He is recognized by the worldwide scientific community as the leading figure of this century in organic chemistry. Woodward began his scientific career at the age of eight and before becoming a teenager had already completed much of the experimental work that would normally be carried out in college. Indeed, when Woodward was a freshman at the Massachusetts Institute of Technology, he was given his own research laboratory at the age of sixteen. In only four years he had graduated with both a bachelor's and doctor's degree. After a short stay at the University of Illinois, Woodward returned to Cambridge, Massachusetts, and to Harvard University, where he was to spend the rest of his life and his career.

Chemists have made contributions of monumental proportions to humankind's progress in the past two centuries. In 1838 WÖHLER and LIEBIG, two leading chemists of their time, stated, "the philosophy of chemistry will draw the conclusion that the synthesis of all organic compounds, as long as they are not part of an organism, must be seen as not merely probable but certain." This prediction was admirably fulfilled by Woodward.

In 1944 at the age of twenty-seven, Woodward, then an assistant professor in the chemistry department at Harvard, succeeded in synthesizing quinine. Quinine, the most important antimalarial at the time, had not been available since the beginning of World War II. The research was supported by Edwin Land and the Polaroid Corporation of Cambridge, who sought to replace quinine, which they used as a polarizer. Woodward's synthesis of quinine did not of itself provide a commercial route to quinine. Rather, it completed the goal that generations of chemists had strived for since the 1850s and initiated the modern area of organic synthesis. His synthesis of quinine and the work that he performed on the then-miracle antibiotic penicillin established Woodward as the leader of what history may well look back and describe as the "golden age" of organic chemistry.

Organic chemists plan their synthetic strategies both forward, from readily available starting materials, and backward, from their ultimate goal the finished product. This exercise requires an intimate knowledge of the three-dimensional (3D) structure of each intermediate in the synthetic scheme and an understanding and intuition of how 3D properties affect reactivity of molecules. Woodward was a master of these skills. Not only was he a key figure in most of the major events that directed organic chemistry during this century, but he was also a leader in the use of spectroscopic methods that chemists use to aid their exploration into the detailed structure of molecules. Early in his career he developed theories that related the structure of a molecule and the way in which it absorbs ultraviolet and visible light. He also helped explain how the absorption of polarized light could be used to predict the 3D structure of molecules by his codevelopment of the octant rule. Woodward was a pioneer in the use of infrared and NUCLEAR MAGNETIC RESONANCE spectroscopy, and helped establish HIGH-PERFORMANCE LIQUID CHROMATOGRAPHY (HPLC), which is still the most versatile method to separate complex mixtures into their individual components.

Throughout his career Woodward focused his attention on determining the structure of and synthesizing molecules of biological importance. These achievements are too extensive to list in detail but include molecules such as strychnine, antibiotics, cholesterol, cortisone, lysergic acid, ferrocene, reserpine, tetracyclines, and chlorophyll. This work resulted in 1965 in the award of the Nobel Prize in chemistry to Woodward for "contributions to the art of organic synthesis." Woodward's most spectacular and difficult synthesis was that of nature's most complex nonpolymeric molecule, vitamin B_{12}. Woodward and his colleague Albert Eschenmoser completed the synthesis in 1976; the experimental work involved more than one hundred chemists from twenty nations.

During the early steps of the vitamin B_{12} synthesis Woodward was surprised and no doubt dismayed that the predicted outcome of one of his reactions was wrong. Woodward never allowed the complexities of nature to daunt him. Indeed, he always interpreted the message that nature gave him and used this knowledge to achieve his goals. B_{12} provides the paradigm for this, since this one reaction that had the audacity to defy and confuse him resulted in a collaboration with Roald Hoffmann (who was awarded the Nobel Prize in 1981 for this work). Together they developed a set of theoretical rules, using MOLECULAR ORBITAL THEORY to explain what, at the time, as viewed as a wide range of (apparently) unrelated organic reactions, and to predict many others. Indeed, so powerful and all-encompassing are the Woodward-Hoffmann rules that when Woodward and Hoffmann considered violations of these rules they concluded, "there are none."

Woodward died in Cambridge on July 8, 1979, at the age of sixty-two. However, he slept for only three hours a day the whole of his adult life, so even in his relatively short life he spent more hours awake than the average person who lives to be one hundred.

BIBLIOGRAPHY

BOWDON, M. E.; BENFEY, T. *Robert Burns Woodward and the Art of Organic Synthesis;* Philadelphia: Chemical Heritage Foundation, 1992.

DAVID DOLPHIN

Work Function

The *work function A* for a closed thermodynamic system is defined as $A = U - TS$, where U is the internal energy of the system, T is the temperature, and S is the entropy. Because U, T, and S depend only on the state of the system, A is also classified as a state function. H. Von Helmholtz originally called A the "free energy", but because the Gibbs (free) energy G is often called the free energy, the work function is usually known as the *Helmholtz energy* (Helmholtz energy function or Helmholtz free energy). The Helmholtz energy is related to the Gibbs energy by $G = H - TS = U + PV - TS = A + PV$, where H is the enthalpy, P is the pressure of the system, and V is the volume of the system.

For an infinitesimal change in the thermodynamic system, $dA = dU - T\,dS - S\,dT$, which becomes $dA = đa + đw - T\,dS - S\,dT$ upon substitution of the First Law of Thermodynamics. If the process is reversible, $đq = T\,dS$ and $đw$ will be equal to the maximum amount of work involved in the process. Under these conditions, $dA = T\,dS + đw_{max} - T\,dS - S\,dT = đw_{max} - S\,dT$.

Under isothermal conditions $dT = 0$, and the expression for the infinitesimal change in the system shows that the Helmholtz energy is equal to the maximum work, $dA = đw_{max}$. If the maximum work consists of a useful work component (e.g., electrochemical, gravitational, or mechanical) in addition to the reversible press-volume (expansion) work component, $đw_{max} = đw_{useful} + đw_{expansion} = đw_{useful} - P\,dV$. Under constant volume conditions, $dA = đw_{useful}$. The direct connection of A to the work involved in a thermodynamic process is the basis for the original name of work function and the symbol A, which is derived from *Arbeit*, the German word for work.

If only reversible pressure-volume work is involved in a thermodynamic process, the expression for the infinitesimal change in the system becomes $dA = đw_{max} - S\,dT = -P\,dV - S\,dT$. From this result, the volume dependence of A is given by $(\partial A/\partial V)_T = -P$, and the temperature dependence is given by $(\partial A/\partial T)_V = -S$.

For a constant volume process carried out under isothermal conditions, the process will continue until A reaches a minimum. Processes that are thermodynamically spontaneous have values of $\Delta A < 0$ and will proceed until reaching thermodynamic equilibrium at which $\Delta A = 0$.

Theoretical values of A can be calculated for molecular systems at a given temperature by considering the translational, rotational, vibrational, electronic, and nuclear contributions to the canonical ensemble partition function Q using $A = -k_B T \ln Q$, where k_B is the Boltzmann constant.

In addition to the thermodynamic applications described above, the term *work function* also represents the energy required to remove an electron from the surface of a metal during the photoelectric effect and is represented by the symbol Φ.

See also PHOTOELECTRIC EFFECT; STATISTICAL MECHANICS; THERMOCHEMISTRY.

CLYDE METZ

X

X-ray Crystallography

X-ray crystallography is the use of X rays to determine structures of molecules as they exist in single crystals. X-ray crystallography was born in 1913 when W. L. BRAGG demonstrated that placing a crystalline material in an X-ray beam produces a diffraction pattern on photographic film. This demonstrated that X rays are electromagnetic radiation and that a crystal is a repeating pattern of atoms or molecules that we call a unit cell. For large objects we use microscopes in which the lens carries out a Fourier transformation of the scattered light to produce an enlarged image of the object. Since it is not possible to make a lens for X rays, we cannot easily reconstruct the enlarged image. The regular repeating pattern of a crystal acts as a three-dimensional diffraction grating for X rays, which produces unique diffraction spots, or reflections. The diffraction data can be interpreted to produce the image of the molecules within the crystal since X-ray wavelength is on the scale of atoms and molecules.

To carry out an X-ray structure determination we need appropriate single crystals (0.1–1.0 mm on a side), a source of X rays, a means of positioning the crystal, a means of detecting the diffracted X rays, and computational facilities for processing the diffraction data for solving and analyzing the structure.

X rays are produced by bombarding a metal target with high-energy electrons, producing a characteristic wavelength radiation. A second method uses synchrotron radiation produced as a consequence of the acceleration of electrons. In both cases the X rays are monochromatized. The preferred method of monochromatizing is the use of a single-crystal monochromater placed in the primary beam; its diffraction is used for the further diffraction experiment. The beam is then passed through a collimator, which limits it to a fine, uniform beam directed toward the crystal.

If they are not air sensitive, the crystals are mounted on a glass fiber. If they are air sensitive, they are mounted in a sealed thin-walled capillary tube with a controlled environment. The latter is typically used with biological macromolecules. The crystal sample is then mounted on a device for positioning in the X-ray beam.

The earliest means of X-ray detection was photographic. Today photon counters are now commonly used and are moved about to count the diffracted X rays at the observing position. Today an X-ray diffractometer uses a movable photon counter and a goniostat to systematically move the crystal for the collection of data. Most recently, several types of X-ray area detectors have been developed that are sensitive to both the X-ray intensity and position. Unlike film they can be read electronically. For large molecules area detectors have become essential.

The crystallographic unit cell and its orientation must be determined by observing the location of a dozen or more diffraction spots. The unit cell geometry, along with intensities of certain reflections, provides information as to the space group of the crystal.

1527

At this point a systematic collection of the intensity data is undertaken.

A complete set of data is collected, anywhere from a few hundred to tens of thousands of reflections depending on the size of the molecule. Intensities, or I's, are converted to structure factor amplitudes, or F's, correcting for background scattering, crystal decay, polarization of the X-ray beam, and geometrical factors.

The theory of structure determination states that the electron density, $\rho(\mathbf{r})$, is the Fourier transform of the structure factors, $F(\mathbf{s})$. That is, the electron density at a point $\mathbf{r}$ is given by

$$\rho(\mathbf{r}) = (1/V) \, \Sigma \, F(\mathbf{s}) \, e^{-\pi i \mathbf{r} \cdot \mathbf{s}}$$

where V = the volume of the unit cell, i = the square root of -1, and $F(\mathbf{s})$ = the structure factor at the scattering position $\mathbf{s}$. The italics indicate a complex number, bolding indicates three-space vectors, and a bullet, ·, indicates a dot product.

The structure factor, $F(\mathbf{s})$, is a complex number having an amplitude and a phase:

$$F(\mathbf{s}) = |F(\mathbf{s})| \, e^{-i\varphi(\mathbf{s})} = F(\mathbf{s}) \, e^{-i\varphi(\mathbf{s})}$$

where $|F(\mathbf{s})| = F(\mathbf{s})$, the amplitude of the structure factor, and $e^{-i\varphi(\mathbf{s})}$ is the phase. The corrected intensities, $I(\mathbf{s})$'s, are proportional to the square of the amplitude of the structure factor giving no information about the phase. The problem of solving the structure is that of finding the phase of each of the structure factors.

A corollary of the equation for $\rho(\mathbf{r})$ is the inverse equation

$$F(\mathbf{s}) = \int \rho(\mathbf{r}) \, e^{2\pi i \mathbf{r} \cdot \mathbf{s}} \, d\mathbf{r}$$

where the integral is over the volume of the unit cell and can be rewritten as

$$F(\mathbf{s}) = \Sigma \, f_j \, e^{2\pi i \mathbf{r}_j \cdot \mathbf{s}}$$

where the summation is over the atoms in a unit cell and f_j, the atomic scattering factor, is the scattering by a give atom at position $\mathbf{r}_j$ and is proportional to the number of electrons in the atom.

The earliest and simplest method of solving a structure is trial and error. If correct, the calculated and observed structure factor amplitudes will show a reasonable agreement as commonly represented by the R value,

$$R = \Sigma \, ||F_{obs}| - |F_{calc}||/\Sigma|F_{obs}|$$

In 1935 A. L. Patterson demonstrated that the Patterson function, $P(\mathbf{r})$, the Fourier transform of the intensities,

$$P(\mathbf{r}) = (1/V) \, \Sigma \, I(\mathbf{s}) \, e^{-2\pi i \mathbf{r} \cdot \mathbf{s}}$$

contains information about the weighted vectors from each atom to each other atom. Analysis of $P(\mathbf{r})$ allows for the solution of many small molecule structures as well as more complex structures where a few heavy atoms dominate.

The next advance, "direct methods," by J. Karl, H. Hauptman, and others in the late 1950s and 1960s is based on the severe restriction placed on the phases by the facts that the electron density is everywhere positive and is built up from a number of nearly spherical atoms. The method uses iterative analysis to find sets of phases that are consistent with a structure having the known number and types of atoms. These phase sets are then used to calculate electron density maps, which are examined for reasonable model structures.

M. Perutz and coworkers in the 1950s developed the method of isomorphous replacement for larger molecules. Here data are obtained from a native and from several isomorphous heavy-atom derivative structures, where all except a few atoms are in the same position as the native. The changed atoms can be located from the differences in intensities. Using several such sets allows for the determination of the native phases.

The molecular replacement method starts with a model structure, possibly a related molecule of known structure. The scattering of the isolated model is used to calculate the scattering that would be obtained by placing the model in different orientations and positions in the observed crystal's unit cell. Agreement between the observed and model F's is used to select possible starting models.

The above methods result in either a phase set or a model from which a phase set may be calculated. An iterative procedure is then carried out for the purpose of calculating electron density maps, improving the model, and calculating new phases. A least-squares refinement procedure follows in which the model is adjusted to improve the agreement between the observed and calculated F's. For structures where insufficient data are available the data are supplemented with constraining equations on bond distances, angles, and other geometrical data.

A correct structure produces an R value of below 10 percent for a small molecule and below 20 percent for macromolecules.

The information obtained from crystal structures is a function of the size of the molecule. From very small molecules information about the wave functions of the atoms and molecules is obtained. From small molecules details of the molecular geometry, bond distances, angles, torsion angles, and chiralities are obtained in addition to insights into inter- and intramolecular packing. Conformations of macromolecules are observed along with the active sites of enzymes, with substrates and/or substrate analogs, and with drugs bound to macromolecules. The interactions between the elements of macromolecular complexes such as protein-DNA complexes are also visualized.

The closely allied field of X-ray fiber diffraction analysis applies many of the same techniques. These studies are carried out on materials—normally, polymers—which have a helical repeat strcuture rather than the unit cell.

Powder pattern analysis is an analytical technique that uses the diffraction produced by a finely ground powder to identify the material and or impurities present. The powder contains crystals in all orientations; diffraction rings are observed rather than diffraction spots, where the radius and intensity of these rings are characteristic of the crystal. Impurities will show up as aberrant lines, which may be identifiable in terms of the kind and amount of impurity.

BIBLIOGRAPHY

GLUSKER, J. P.; TRUEBLOOD, K. N. *Crystal Structure Analysis: A Primer,* 2nd ed.; New York: Oxford University Press, 1985.

LADD, M. F.; PALMER, R. A. *Structure Determination by X-Ray Crystallography,* 2nd ed.; New York: Plenum, 1986.

STOUT, G. H.; JENSEN, L. H. *X-Ray Structure Determination,* 2nd ed.; New York: Wiley, 1989.

GARY J. QUIGLEY

X-ray Diffraction

The structure of matter is one of the most important properties of any chemical system. By *structure* we mean the nature of chemical bonds in a compound and how electrons are shared. Structural properties of considerable interest include bond distances, BOND ANGLES, overall symmetry, and COORDINATION NUMBERS.

X-ray diffraction is a method used to determine the structure of chemical compounds. X-ray diffraction involves the diffraction of X rays from various planes of atomic layers. X rays are known to penetrate matter and consequently are used in dental and medical applications. X-ray diffraction is governed by Bragg's law as given in equation 1:

$$n\lambda = 2d \sin \theta \qquad (1)$$

where n is an integer, λ is the wavelength of the X rays, d is the spacing between different parallel planes, and θ is the angle of incidence of the X rays with the planes of atoms from which diffraction occurs.

Bragg's law can be derived from the diagram shown in Figure 1, which shows the relationship between θ and d. There must be an integral number (n) of wavelengths for constructive interference and diffraction to occur. The extra distance that wave 2 travels with respect to wave 1 is CB + BD. Both CB and BD are equivalent to sin θ based on geometrical principles. Therefore, the extra distance CB + BD equals $2d$ sin θ.

X-ray diffraction is commonly used to study either single crystals or powders. Diffraction occurs when solids having long-range order interact with X rays. For a powder, it is very important that the sample is prepared so that all of the different planes of atoms can interact in a random way so that all are observed in the resultant powder diffraction pattern. This means that pressing a pellet or pressing a powder onto a surface should usually be avoided.

The crystalline aluminosilicate mineral zeolite Y exhibits a characteristic X-ray powder diffraction pattern that contains very sharp and intense peaks. Zeolites are catalytic agents that are widely used in a variety of commercial and industrial applications, such as the cracking of petroleum, a process whereby complex hydrocarbons are broken down in the presence of heat and the zeolite into lighter products like gasoline. Zeolites are also used as builders in the manufacture of SOAPS AND DETERGENTS, as abrasives in toothpaste and dental powders, and in many other applications. Zeolite Y has a high degree of structural symmetry and belongs to the cubic crystal system (see CRYSTAL HARDNESS and CRYSTAL SYSTEMS). The particular pattern of diffraction lines exhibited by a given material can be a unique representation of that material. This means that all materials with the same structure have an X-ray powder diffraction pattern that can be used to identify the material.

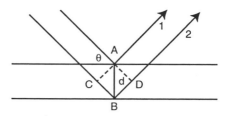

Fig. 1. Diffraction of X ray from planes in a solid.

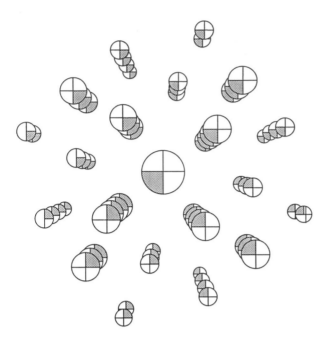

Fig. 2.   Structure of cryptomelane, $KMn_8O_{16} \cdot n\ H_2O$, showing various atoms. Atoms are represented as thermal ellipsoids. K is in the center; the small atoms are MN, which are surrounded by O atoms.

Particular applications of X-ray powder diffraction include identification of unknown substances by comparison to known databases of X-ray patterns, phase identification, determination of the structure of unknowns, and the purity of materials. There are several complications in such analyses. One particular problem is that amorphous or noncrystalline materials do not show a powder diffraction pattern. Another problem is that X-ray powder diffraction is not very sensitive with respect to other analytical methods such as atomic absorption, which can be used to determine concentrations as low as the part-per-million and -billion range.

The width of a peak in X-ray powder diffraction is related to particle size. At particle sizes less than about 200 Å, peaks begin to broaden. The average sizes of such small particles can often be measured in this way.

The intensities of the diffraction peaks are related to electron densities in a material and, therefore, to the specific types of atoms in a molecule. There are several factors that affect intensity, such as the specific structure, geometrical factors related to the instrumentation that is used, angular dependencies of X rays, thermal vibrations of atoms, absorption of X rays by the sample, and other factors.

The other major application of X-ray diffraction is in the determination of the exact location of atoms and how they are bound in a molecule. Single crystals are most often needed for this purpose. In this case, a single crystal is mounted on a device known as a goniometer, which consists of a series of arcs in different directions so that the crystal can be turned to different orientations with respect to the incident X rays. In such experiments, the crystal is rotated to different orientations, and intensities of diffracted X rays are measured for different planes of atoms.

Once the intensities for various planes are collected, the experimenter then uses a computer and knowledge of the chemical composition to try to solve the structure of that material. A three-dimensional electron density (probability) map is created by such mathematical manipulations, which is necessary because X rays cannot be refocused. Statistical and refinement methods are used to obtain a best fit to the data.

An example of a representation of a crystal structure of a mineral known as cryptomelane [$KMn_8O_{16} \cdot n\ H_2O$] is shown in Figure 2. The atoms are represented as probability ellipsoids. From the positions of of the different atoms, it is possible to obtain information about bond distances, bond angles, local coordination numbers, and regularity of the different units in the structure.

X-ray diffraction is one of the most powerful analytical methods that can be used to undestand chemical systems. There are several available books and articles on this fascinating subject.

BIBLIOGRAPHY

BRAGG, W. H.; BRAGG. W. L. *X-rays and Crystal Structure;* London: G. Bell and Sons, 1925.

GLUSKER, J. P.; TRUEBLOOD, K. N. *Crystal Structure Analysis;* London: Oxford University Press, 1972.

STOUT, G. H.; JENSEN, L. H. *X-ray Structure Determination;* New York: Macmillan, 1968.

VAINSHTEIN, B. K. *Modern Crystallography;* New York: Springer-Verlag, 1981.

WYCKOFF, R. W. G. *Crystal Structures;* New York: Wiley, 1971; Vols. 1–6.

STEVEN L. SUIB

X-ray Fluorescence Spectroscopy

X RAYS are a short-wavelength form of electromagnetic radiation discovered by Wilhelm Roentgen just before the turn of the century. X-ray photons are produced following the ejection of an inner orbital electron from an excited atom. Like all other forms of electromagnetic radiation, most of the properties of

X-radiation can be grouped into three areas: scatter, diffraction, and fluorescence. Scattering occurs where a beam of X-radiation interacts with the loosely bound outer electrons of an element leading to a change in direction of the incident beam. Such scatter can be coherent (scattered wavelength same as incident wavelength), or incoherent (scattered wavelength longer than incident wavelength). X-RAY DIFFRACTION is a combination of two phenomena: coherent scatter and interference. At any point where two or more waves cross one another, they may interfere. Interference does not imply the impedance of one wave train by another, but rather describes the effect of superposition of one wave upon another. A suitable single crystal can act as a "diffraction grating" since there are layers of atoms (scatterers) at unique distances (d) from each other in the crystal lattice. Thus, under certain geometric conditions, wavelengths (λ) being scattered from the crystal, that are exactly in phase, may add to one another, giving diffraction maxima at an angle (Θ). This relationship is described by Bragg's law:

$$n\lambda = 2d \sin \Theta$$

The diffraction property is a useful way of dispersing a polychromatic beam of radiation since, by placing the crystal in the path of the beam, each unique wavelength will diffract at a unique diffraction angle. Fluorescence occurs when the primary X-ray photons are energetic enough to create electron vacancies in the specimen, leading in turn to the generation of secondary (fluorescence) radiation. The secondary radiation is characteristic of the elements making up the specimen. The technique used to isolate and measure individual characteristic wavelengths is following excitation by primary X-radiation, which is called X-ray fluorescence spectroscopy.

X-ray fluorescence spectroscopy provides the means of the identification of an element by measurement of its characteristic X-ray emission wavelength or energy (Jenkins, 1976). The method allows the quantization of a given element by first measuring the emitted characteristic line intensity and then relating this intensity to elemental concentration. The first analytical use of the X-ray spectroscopic method dates back to the early 1920s. Thus, while the roots of the method go back to the early part of this century, it is only during the last thirty years or so that the technique has gained major significance as a routine means of elemental analysis (Birks, 1976). Modern X-ray spectrometers use either the diffracting power of a single crystal to isolate narrow wavelength bands or a proportional detector to isolate narrow energy bands from the polychromatic radiation excited in the sample. The two methods are referred to as wavelength dispersive spectroscopy and energy dispersive spectroscopy. Because the relationship between emission wavelength and atomic number is known, isolation of individual characteristic lines allows the unique identification of an element to be made, and elemental concentrations can be estimated from characteristic line intensities. Both types of spectrometer lend themselves admirably to the qualitative and quantitative analysis of solid materials and solutions. Over the past thirty years or so, the X-ray fluorescence method has become one of the more valuable methods for the qualitative and quantitative analysis of materials (Bertin, 1975; Jenkins et al., 1981). In particular, its speed, accuracy, and versatility have made it the method of choice in more than 20,000 laboratories all over the world.

Relationship between Wavelength and Atomic Number

Since all emitted X-ray photons have energies proportional to the differences in the energy states of atomic electrons, the lines from a given element will be characteristic of that element. The relationship between the wavelength λ of a characteristic X-ray photon and the atomic number Z of the excited element was first established by H. G. J. Moseley. MOSELEY'S LAW is written

$$1/\lambda = K[Z - \sigma]^2$$

in which K is a constant that takes on different values for each spectral series. σ is a shielding constant that has a value of just less than unity. Since there is a simple relationship between the characteristic X-ray photon and the atomic number of the element from which the characteristic emission line occurs, X-ray fluorescence spectroscopy is ideally suited for qualitative elemental analysis. Thus by measuring the wavelengths, or energies, of a given series of lines from an unknown material, the atomic numbers of the excited elements can be established. The inherent simplicity of characteristic X-ray spectra makes the process of allocating atomic numbers to the emission lines relatively easy, and the chance of making a gross error is rather small. There are only one hundred or so elements, and within the range of the conventional spectrometer, each element gives, on average, only a half dozen lines. A further benefit of the X-ray emission spectrum for qualitative analysis is that because transitions do arise from inner orbitals, the effect of chemical combination or valence state is almost negligible.

Instrumentation for X-ray Fluorescence Analysis

The basic function of the spectrometer is to separate the polychromatic beam of radiation coming from the specimen in order that the intensities of each individual characteristic line can be measured. A spectrometer should provide sufficient resolution of lines to allow such data to be taken, at the same time providing a sufficiently large response above the background to make the measurements statistically significant, especially at low analyte concentration levels. It is also necessary that the spectrometer allow measurements over the wavelength range to be covered. There is a wide variety of instrumentation available today for the application of X-ray fluorescence techniques, but each of the systems comprises three main parts: the source, the detector, and the actual spectrometer itself. Several different types of source have been employed for the excitation of characteristic X-radiation, including those based on electrons, X rays, gamma rays, protons, and synchrotron radiation. By far the most common source today is the X-ray photon source. An X-ray detector is typically a transducer for converting X-ray photon energy into voltage pulses. Detectors work through a process of photon-ionization in which interaction between the entering X-ray photon and the active detector material produces a number of electrons. The current produced by these electrons is converted to a voltage pulse by a capacitor and resistor, such that one digital voltage pulse is produced for each entering X-ray photon, in addition to being sensitive to the appropriate photon energies—i.e., being applicable to a given range of wavelengths or energies.

A wavelength dispersive spectrometer may be a single-channel instrument in which a single crystal and a single detector are used for the measurement of a series of wavelengths sequentially or a multichannel spectrometer in which many crystal/detector sets are used to measure elements simultaneously. Of these two basic types, the sequential systems are the most common. The energy dispersive spectrometer consists of the excitation source and the spectrometer/detection system. The spectrometer/detector is typically a Si(Li) detector, which is a proportional detector of high intrinsic resolution. A multichannel analyzer is used to collect, integrate, and display the resolved pulses. Most systems can be equipped with multisample-handling facilities and can be automated by use of a computer, which is then also available for spectral stripping, peak identification, quantitative analysis, and other useful functions. Typical spectrometer systems are capable of precisions of the order of a few tenths of one percent with sensitivities down to the low ppm level. Single-channel wavelength dispersive spectrometers are typically employed for both routine and nonroutine analysis of a wide range of products, including ferrous and nonferrous alloys, oils, slags and sinters, ores and minerals, thin films, and so on. These systems are very flexible but, relative to multichannel spectrometers, are somewhat slow. The multichannel wavelength dispersive instruments are used almost exclusively for routine, high throughput, analyses where the great need is for fast accurate analysis, but where flexibility is of no importance. Energy dispersive spectrometers have the great advantage of being able to display information on all elements at the same time. They lack somewhat in resolution compared with the wavelength dispersive spectrometer, but the ability to reveal elements absent as well as elements present makes the energy dispersive spectrometer ideal for general troubleshooting problems. They have been particularly effective in the fields of scrap alloy sorting, in forensic science, and in the provision of elemental data to supplement X-ray powder diffraction data.

The modern wavelength dispersive system was introduced commercially in the early 1950s, and energy dispersive spectrometers in the early 1970s. While each of these instrument types provides the same basic type of information, the characteristics of the two methods differ mainly in their relative sensitivities and the way in which data is collected and presented. Generally speaking, the wavelength dispersive system is roughly one to two orders of magnitude more sensitive than the energy dispersive system. Against this, however, the energy dispersive spectrometer measures all elements within its range at essentially the same time, whereas the wavelength dispersive system identifies only those elements for which it is programmed.

Quantitative Analysis

The correlation between the characteristic line intensity of an analyte element and the concentration of that element is typically nonlinear over wide ranges of concentration, due mainly to interelement effects between the analyte element and other elements making up the specimen matrix. Nevertheless, because the relationship between characteristic line intensity and elemental composition is reasonably well understood, good quantitative data can be obtained. Today, conventional X-ray fluorescence spectrometers allow the rapid quantitation of all elements in the periodic table from fluorine (atomic number 9) and upward. Recent advances in wavelength dispersive spectrometers have extended this element range down to carbon (atomic number 6). Over most of the measurable range, accuracies of a few tenths of one percent are possible, with detection limits down to the low ppm level. Typical analysis times vary from about ten seconds to three minutes per element. The minimum sample size required is of the order of a few milligrams, although typical sample sizes are probably about several grams. Good accuracy is obtainable, and in favorable cases standard deviations of the order of a few tenths of one percent are possible. The sensitivity of the X-ray fluorescence technique is fair, and determinations down to the low ppm level are possible for most elements.

The versatility of the X-ray fluorescence has been more recently enhanced by the development of special purpose instruments. As an example, the total reflection X-ray fluorescence (TRXRF) technique (Schwenke and Knoth, 1982) uses a special sample support and employs glancing-angle primary beam optics. The TRXRF method has found great application in the analysis of natural waters (West and Nurrenberg, 1988). The concentration levels of, for example, transition metals in rain, river, and sea waters are normally too low to allow estimation by standard X-ray fluorescence techniques, unless preconcentration is employed. Using TRXRF concentration levels down to less than 10 μg/l are achievable. While the TRXRF method is most applicable to homogeneous liquid samples in which the sample is evaporated onto the optical flat, success has also been achieved in the application of the method to solids including particulates, sediments, air dusts, and minerals.

BIBLIOGRAPHY

BERTIN, E. P. *Principles and Practice of X-Ray Spectrometric Analysis*, 2nd ed.; New York: Plenum, 1975.

BIRKS, L. S. *History of X-Ray Spectrochemical Analysis*; American Chemical Society Centennial Volume; Washington, DC: American Chemical Society, 1976.

JENKINS, R. *An Introduction to X-Ray Spectrometry*; London: Wiley, 1976; Chapter 4.

————. *X-Ray Fluorescence Spectrometry*; New York: Wiley, 1988.

JENKINS, R.; GOULD, R. W.; GEDCKE, D. *Quantitative X-Ray Spectrometry*, 2nd ed.; New York: Dekker, 1993.

WEST, T. S.; NURRENBERG, H. W., eds. *The Determination of Trace Metals in Natural Waters*; Oxford: Blackwell, 1988.

RONALD JENKINS

X Rays

X rays are a form of electromagnetic radiation that are produced when an accelerated electron beam strikes a metal target enclosed in an evacuated glass tube. The radiation so produced travels at the speed of light ($c = 2.99 \times 10^8$ m sec^{-1}) and has wavelengths in the range 10^{-7}–10^{-9} m, which is more commonly given as 0.1–10 Å (angstroms). X rays were discovered in a serendipitous fashion by Wilhelm Röntgen in Germany on November 8, 1895, and, indeed, are still called Röntgen rays in Germany. The type of apparatus used by Röntgen is shown in Figure 1. A photographic plate placed accidentally near the glass tube was blackened indicating that some form of high-energy radiation had been generated.

A typical emission spectrum recorded when electrons strike a metal target is shown in Figure 2. Radiation is produced over a wide range of wavelengths. This polychromatic radiation is often described as "white" radiation. The intensity and short wavelength limit are dependent on the voltage inducing the electron beam toward the metallic target. Superimposed on the white radiation are sharp lines that correspond to X-ray emission, the wavelengths of which are determined by the element being used as the target. The wavelengths, and hence energies of these "characteristic" X rays, are unaffected by changes in the voltage generating the electrons although the intensities of the characteristic X rays are increased with increasing voltage. The characteristic line with the highest energy and shortest wavelength is called the K_β line and the more intense line with longer wavelength is known as the K_α line.

After the discovery of the existence of X rays in 1895, there were many investigations into their properties, and it was quickly recognized that this highly penetrating radiation had medical applications al-

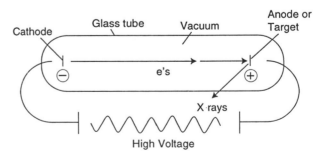

Fig. 1.   Drawing of the type of apparatus used to generate X rays.

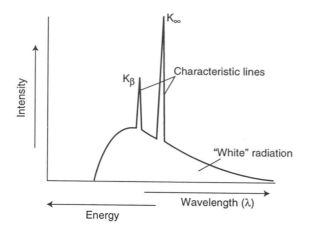

Fig. 2. X-ray emission spectrum showing the "white" radiation and the characteristic K lines.

lowing one to view bone structure and diagnose fractures. In 1913, the British physicist Henry MOSELEY carried out the investigations that revealed the origin of X rays. Moseley recognized that the energy and wavelength of a particular characteristic X ray (for example, the K_α line) could be related to the atomic number of the target material by the equation

$$\nu^{1/2} = C(Z - \sigma)$$

where

ν = frequency of the X-radiation ($\nu = c/\lambda$)

Z = atomic number of the target material

and C and σ are constants for a particular characteristic line—that is, the values would be different for the K_α and the K_β lines.

Moseley's equation paved the way for the recognition of the origin of the characteristic X rays and also provided experimental validation of the Rutherford and Bohr theories of atomic structure. The incident electrons have a range of energies, and when they encounter the electrons of the target material, they are scattered producing radiation of varying wavelength corresponding to the white radiation. However, when the accelerated electrons striking the target material have sufficient energy to remove an electron from a particular orbital or "shell," in the terminology of X-ray investigations, then the vacancy so created by the lost electron will be filled by an electronic transition from an outer shell. The energy, and hence the wavelength and frequency, of this transition will be governed by the nuclear charge or atomic number of the target element. Figure 3 shows a simplified picture of an atom with electron shells,

which are designated K, L, M, and so forth. The K-, L-, and M-shells correspond to atomic orbitals with principal quantum numbers 1, 2, and 3, respectively. If an electron is ejected from the K-shell, then an electronic transition from the L- or M-shells will return the atom to a lower energy state. A transition from the L-shell will give rise to a K_α X ray and one from the M-shell will result in a K_β X ray. The L → K transition is more probable, so the intensity of the K_α line is greater, but the M → K transition involves a greater change in energy, so the wavelength of the K_β line is shorter. As the energies of the electrons in both the L- and the M-shells of an atom are not exactly the same, the K_α emission is in fact a doublet (K_{α_1} and K_{α_2}), and the K_β emission consists of four closely spaced lines, as was confirmed by later, more detailed studies.

Most X rays used in industrial, academic, and medical applications are produced by modern versions of the equipment used by Röntgen. The X-ray tube consists of an evacuated chamber enclosed by heavy glass and having two electrodes attached to a source of high voltage. The chemical nature of the anode or target determines the wavelength of the characteristic radiation. The heat generated by the electron beam striking the target requires an efficient water cooling system. Typical applied voltages are in the 30 to 60 kV range with tube current from 15 to 35 mA. In many applications, very intense X rays are required. These more intense X-ray sources can be generated by rotating the target or anode so that no one spot on the electrode is receiving the electron beam at all times, thus dispersing the heat produced over a much greater area of the target.

Very intense X-ray sources are produced in storage ring particle accelerators known as synchrotons. The

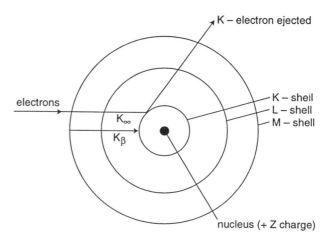

Fig. 3. Schematic drawing of Bohr atom, with K-, L-, and M-shells.

X rays so produced are polychromatic (i.e., white) and of very high intensity (10^3–10^4 times that of a conventional source). Monochromatic radiation is obtained by X-ray scattering from a particular plane in a crystal according to Bragg's law (see below). Suitable adjustments of the crystal, known as a monochromator, can allow the user to select X rays of almost any desired wavelength.

X rays have many applications in chemistry. In this section, three such uses will be discussed briefly: X-ray absorption, X-ray fluorescence, and X-ray diffraction.

When X rays strike a particular element, the absorption of the radiation generally follows the equation

$$I_t = I_o \exp [-\mu \ell]$$

where

I_t = transmitted intensity of X rays

I_o = incident intensity of X rays

ℓ = thickness of absorbing element

μ = linear absorption coefficient, which is a constant for a given element and for X rays of a given wavelength

For a particular element, increasing the energy or shortening the wavelength of the incident X-radiation normally causes less absorption, thus lowering the value of μ, the absorption coefficient. However, if the absorption of X rays for a given element is examined over a range of incident wavelengths, a more complex pattern is observed (Figure 4). The curve from A to B follows the pattern expected; increasing the energy of the incident X-rays increases transmission. However, at a particular wavelength, known as an absorption edge, the transmitted intensity drops

by a large amount (B → C), and then at higher energies the previous pattern resumes. At the absorption edge the energy of the incident X rays is exactly that required to eject an electron from a particular shell of the absorbing element. There are different absorption edges for the K-, L-, and the M-shells. The wavelengths of these absorption edges are characteristic of the absorbing material. In this way, the measurement of X-ray absorption edges in a given sample provides a means of identifying elements present therein.

X-ray fluorescence analysis consists of exposing a sample of unknown composition of an electron, X-ray, or gamma-ray beam. Any of these sources will produce X rays from the sample. The characteristic wavelengths produced and their intensities can provide both the nature and amounts of the elements in the sample.

The most important application of X rays to the advancement of chemical knowledge has come from the results of X-ray diffraction studies. X rays are scattered by the electrons contained in atoms, ions, and molecules. Crystals are made up of ordered arrangements in three dimensions of molecules, ions, or atoms, and therefore of electron density surrounding these ions or atoms. The basic repeating pattern of this ordered arrangement is said to be contained in the unit cell of the crystal. Unit cells have dimensions from 5 to 100 Å (10^{-10} m) in each direction. The regular repeating pattern of electron density in these unit cells serves as a diffraction grating for X rays. The first X-ray diffraction pattern from a crystal was obtained in the laboratory of the German physicist Max von LAUE in 1912. The principles involved were developed by von Laue and the English father-son combination of W. H. BRAGG and W. L. BRAGG. From W. L. Bragg (later Sir Lawrence Bragg) came the Bragg law of diffraction,

$$2d \sin \theta = \lambda$$

where d is the perpendicular spacing between identical planes in adjacent unit cells, θ is the angle of incidence and scattering from these crystal planes, and λ is the wavelength of the X rays.

Applications and extension of these principles led to the determination of the electron density and hence the structure of the molecules that were contained in the crystal unit cell. X-ray diffraction studies have revealed the structures of complex organic and organometallic compounds, metal alloys, proteins, nucleic acids, and viruses. The development of the principles and the techniques of X-ray diffraction of crystals has resulted in awarding of Nobel Prizes to W. H. Bragg, W. L. Bragg, Max Perutz, and John

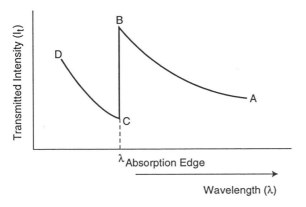

Fig. 4. Transmission of X rays as a function of wavelength.

Kendrew, Francis Crick, James Watson, Dorothy HODGKIN, Jerome Karle, and Herbert Hauptman, among other scientists studying X-ray diffraction.

BIBLIOGRAPHY

CULLITY, B. D. *Elements of X-Ray Diffraction;* Reading, MA; Addison-Wesley, 1956; Chapter 1.

JENKINS, R. *X-Ray Fluorescence Spectrometry;* New York: Wiley, 1988.

KONIGSBERGER, D. C.; PRINS, R., ed. New York: Wiley, 1988.

STOUT, G. H.; JENSEN, L. H. *X-Ray Structure Determination;* New York: Wiley, 1989.

IAIN C. PAUL

Z

Zeeman Effect

The influence in the line spectrum of an element by a magnetic field is known as the *Zeeman effect*. In 1896 Pieter Zeeman observed that the yellow doublet *D* lines found in the flame emission spectrum of sodium were broadened when the flame was placed between the poles of a magnet. High-resolution spectroscopy studies showed that the spectral lines were actually split in the presence of the applied magnetic field into series of closely spaced lines.

The line splitting results from the interaction between the applied magnetic field and the MAGNETIC MOMENT of the atom resulting from the electron motion. Consider the transition shown in Figure 1, in which the lower energy level consists of one energy state and the higher energy level consists of three energy states. In the absence of an external magnetic field, all of the states in a given level are degenerate—have the same energy. An applied field splits the level into its states, each state having a slightly different energy. Figure 1 shows the splitting that occurs for the system in the presence of a weak, homogeneous magnetic field. The separation is proportional to the strength of the applied field and is several orders of magnitude less than the energy corresponding to the unresolved transition.

In the absence of an applied field, only one energy change is involved for the transition shown in the figure and only one spectral line, known as a singlet, is observed. In the presence of an applied field, three slightly different energy changes are involved in the transition, and three closely spaced spectral lines, known as a triplet, are observed. Detailed calculations indicate that the optical polarization of the various components for a given transition will be either perpendicular or parallel to the applied field, and studying these polarizations is a useful tool for spectral analysis.

The total electron spin of an atom is determined by the sum of the spin quantum numbers of all of the electrons. For atoms in which the total spin is zero, the normal Zeeman effect is observed in which all single spectral lines form triplets with identical splittings. For atoms in which the total spin is nonzero, the splitting of each energy level by the applied field is different, and the energies of the transitions between these states will all be different. The line spectra for these atoms will be much more complex than those described above, and this phenomenon is known as the anomalous Zeeman effect.

As the strength of the applied field increases, the separation between the Zeeman energy states increases. At very high field strengths, the separation between the energy states can become greater than the separation between the original energy levels. Under these conditions all atoms show the normal Zeeman triplets, each of which contains closely spaced components. This limiting set of conditions is known as the Paschen-Back effect.

A nuclear Zeeman effect is observed for atoms having nonzero nuclear spin when an external field is

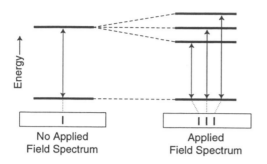

Fig. 1.   Normal Zeeman effect.

applied. Because the nuclear magnetic moment is about one-thousandth of that for an electron, the field strength needed to resolve the Zeeman lines will be strong enough to produce nuclear Paschen-Back conditions.

See also ANGULAR MOMENTUM; ATOMIC ABSORPTION, EMISSION, AND FLUORESCENCE SPECTROSCOPY.

CLYDE METZ

Ziegler-Natta Reactions

The Ziegler-Natta catalysts are formed from reactions involving transition metal compounds of Groups IV–VIII (mainly titanium, vanadium, zirconium) with alkyls, aryls, or hydrides of Groups I–IV, under inert conditions (Tait, 1989; Corradini et al., 1989; Porri et al., 1989). These catalysts have great industrial relevance, since they are used for the production of several polymeric materials: thermoplastics like high-density polyethylene (HDPE), linear low-density polyethylene (LLDPE), isotactic polypropylene (and, in much smaller volumes, isotactic poly-1-butene and poly-4-methyl-pentene), and rubbers like polybutadiene, polyisoprene and ethylene-propylene copolymers (see ELASTOMERS).

The catalytic systems were discovered by Ziegler in 1953 (Ziegler et al., 1955). The discovery of how to synthesize stereoregular POLYMERS was made by Natta and his coworkers in 1954 by using the same catalysts (Natta, 1965).

We define as a *regular polymer* a linear polymer whose molecules can be substantially described by a unique species of constitutional units in a unique sequential arrangement. We define a polymer as *stereoregular* if the succession of configurations is regular too (we mean by configuration the spatial arrangement of the various bonds, without considering the multiplicity of arrangements that arise from rotation around single bonds).

Cellulose, guttapercha, and natural rubber are examples of stereoregular polymers, which nature is able to synthesize. The research conducted in the three decades after 1922, when Staudinger first proposed the term *macromolecule*, led to the synthesis of semicrystalline polymers—the nylons by polycondensation and low-density polyethylene by polyaddition—all polymers of great practical importance. These are regular polymers, whose constitutional units have only one configuration; consequently, they do not present the problem of stereoregularity. On the other hand, the vinyl polymers already known at the time—poly(vinyl chloride) or polystyrene, obtained by radical polymerization processes—were found amorphous on roentgenographic examination, even when they had a quite regular constitution. The researchers of those years, in general, did not recognize that the lack of crystallinity was related to the lack of regularity in the succession of configurations.

In the early 1950s, there was the quite contemporaneous discovery—in three different laboratories—of processes for the polymerization of ethylene at low pressures using solid catalysts: the catalyst used by Standard Oil of Indiana was molybdenum(VI) oxide supported on aluminum oxide; the one used by Phillips Petroleum was chromium(VI) oxide, still supported on silica/alumina; the one studied by Ziegler and his coworkers at the Max Planck Institute at Mühlheim resulted from the reaction between triethylaluminum and titanium tetrachloride. Nowadays it is hypothesized that the mechanism of polymerization (insertion of a monomer molecule in a metal-carbon bond) is quite similar for all three catalysts.

The polyethylene obtained (the already cited HDPE) is more highly crystalline, is more rigid and dense, and has a much more regular structure than the one previously known that was obtained at very high temperature and pressure and that had been industrially produced in the previous fifteen years (now called low-density polyethylene, LDPE). The latter shows macromolecules with both long and short branches and is consequently less crystalline than the almost completely linear polyethylene obtained with the catalytic processes mentioned above. Linear polyethylene does not have a stereogenic carbon in its constitutional unit, and hence the problem of stereoisomerism is not present.

In the laboratories of Professor Natta in Milan it was found that the Ziegler catalysts could polymerize (besides ethylene) propylene, styrene, and several α-olefins to high linear polymers; these polymers appeared crystalline when examined by X-ray diffraction techniques and were able to give oriented fibers.

In less than one year since the preparation of the first polymer of propylene, Natta was able to communicate, in the meeting of the Accademia dei Lincei of December 1954, that a new chapter had been opened in the field of macromolecular chemistry owing to the discovery of processes that yielded polymers with an extraordinary regularity in their structure from the perspective of their chemical constitution and of the configuration of the successive monomeric units along the chain of each macromolecule.

This regularity is particularly impressive considering that a propene monomer unit can be inserted in the polymer chain in four (constitutionally and/or configurationally) different ways and that a chain with 1,000 monomer units, for instance, could present as many as $4^{1000} \approx 10^{600}$ different stereoisomers.

X-ray examination allowed the determination of the lattice constants of several new crystalline polymers like polypropylene, polybutene-1, and polystyrene. The identity period that resulted along the chain axis—as determined by fiber diffraction patterns—was of the order of 6.5 Å and could be attributed to a chain segment containing three monomeric units. Therefore, the idea that the crystallinity could have originated from a regular alternation of monomeric units characterized by enantiomeric steric configurations was ruled out. On the contrary, it was evident that the polymeric chains had to be constituted by regular successions of monomeric units with the same steric configuration (Natta, 1965). This kind of structure was designated as *isotactic* (from the Greek words *isos*, "the same," and *tasso*, "to put in order"). The examination of the crystal structure of isotactic polymers showed that the chain conformation of these polymers is always helical.

Figure 1 shows the different types of stereoisomerism present in vinyl polymers, viewing the main chains as lying down on a plane: (a) the isotactic sequence of configurations; (b) the syndiotactic sequence of configurations; (c) no order present, the atactic sequence of configurations.

It can be useful to discuss the possible configurations for a vinyl polymer in terms of relative local configurations. As seen in Figure 2, two monomeric units in sequence, constituting a "diad," can have two nonequivalent relative configurations; analogous to the classic case of tartaric acids, these relative configurations are indicated as *meso* (abbreviation *m*) and *racemic* (abbreviation *r*), respectively.

A polymer will be fully isotactic if all the diads are *m*; it will be fully syndiotactic if all the diads are *r*. In general, and according to the catalytic system, intermediate situations can occur.

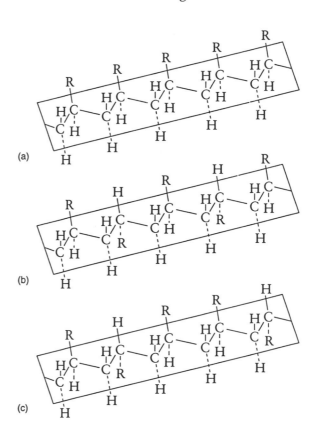

Fig. 1. Schematic representation of the configuration of (a) isotactic, (b) syndiotactic, and (c) atactic vinyl polymers.

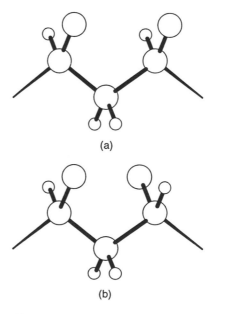

Fig. 2. (a) Meso (*m*) and (b) racemic (*r*) relative configurations in a head-to-tail vinyl polymer.

Figure 3 shows a model of the chain of polypropylene, in two orthogonal projections, along the axis and perpendicularly to the axis of the chain. It is seen that a helix with three monomeric units per repeating unit is formed by the alternation of carbon-carbon bonds in *gauche* and *trans* conformations. Helicoidal chain structures in a polymer allow the repetition of identical configurational units in such a way that they take equivalent conformations in respect to an axis. It may be interesting to recall that at the beginning of the 1950s the application of similar principles led PAULING to suggest the α-helix model for polyaminoacid and led Watson and Crick to the double helix model for DNA.

Early on, it was also found that the molecular masses of the polymers obtained by Ziegler-Natta catalysis can be easily regulated by adding hydrogen as a transfer agent. This is the procedure that is now followed to regulate the molecular masses in the industrial production of polyolefins.

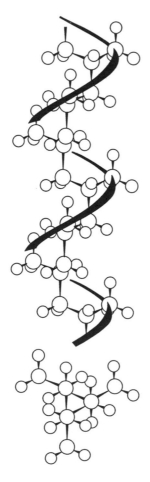

Fig. 3. Chain conformation of isotactic polypropylene.

Some years ago, the opinion was that no further improvement in the control of the polymerization stereospecificity could be achieved from research on the catalysts. The art (if not the science) of preparing $TiCl_3$ catalysts with the highest possible surface activity appeared to be mature.

At the end of the 1960s, however, new catalytic systems for the polymerization of ethylene were implemented in which the titanium chloride was supported on a matrix, as, for instance, magnesium oxide or chloride. These new catalysts showed a very high activity in the polymerization of ethylene with yields of the order of 10^6 instead of 10^4 grams of polymer per gram of titanium. In the second-generation plants for the production of high-density polyethylene, the use of these catalysts avoided the expensive process of separating the catalyst from the polymer.

The new supported catalysts, however, were unsatisfactory for the polymerization of propylene, where control of the succession of the *m* versus the *r* configurations along the polymer chain is necessary. The isotacticity index (the insoluble fraction in boiling *n*-heptane) of polypropylenes, obtained with high yields using catalysts for ethylene polymerization of the above kind, was in the range of 30 to 60 percent. This means that approximately one half of the macromolecules have a fraction of relative *m* configurations lower than 95 percent, so that the most interesting technological characteristics of the polymer are heavily damaged. More recently, the research to design a high-yield-supported catalyst, also for the isotactic polymerization of propylene, has led, with the help of the previous experience in the field, to the implementation in the industrial research laboratories of a new catalytic system capable of yields so high (thousands of kilograms of polymer per gram of titanium) that the depuration (purification) process could be eliminated in the production plants with such high isotacticity that the heavy cost of the extraction of the amorphous fraction could be considerably reduced (Barbé et al., 1986).

In the 1980s, the discovery of the homogeneous stereospecific catalysts for the polymerization of 1-alkenes opened up new prospects for research on stereospecific polymerization and on stereoregular polyolefins. The discovery has been achieved by Ewen (Ewen, 1984; Ewen et al., 1988) on the basis of earlier researches on metallocenes in combination with alkyl-Al-oxanes by Sinn and Kaminsky (Sinn and Kaminsky, 1980).

Depending on the specific metallocene π-ligands used, these systems present completely different stereospecific behaviors. For instance, catalytic systems

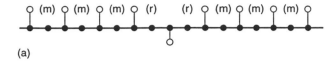

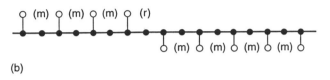

(a)

(b)

Fig. 4. Possible stereodefects along the chain of a prevailingly isotactic polymer.

containing the metallocene stereorigid ligand ethylene-*bis*–(1-indenyl) or ethylene-*bis*–(4,5,6,7-tetrahydro-1-indenyl) polymerize α-olefins to isotactic polymers (Ewen, 1984), while catalytic systems constituting the metallocene stereorigid ligand isopropyl(cyclopentadienyl-1-fluorenyl) instead polymerize α-olefins to syndiotactic polymer (Ewen et al., 1988). Hence, it has been possible to tune the structure of these catalysts to produce a series of new stereoregular polymers, in particular a series of new crystalline syndiotactic polymers.

The active site for these homogeneous catalysts is cationic, as shown by the synthesis of a wide series of group 4 metallocene cations able to polymerize ethylene and propylene without any aluminum cocatalyst. The finding by Ewen that ethylene-*bis*–(1-indenyl)ZrCl$_2$/methyl/alumoxane and ethylene-*bis*–(1-indenyl)ZrCH$_3^+$B(C$_6$F$_5$)$_4^-$ produce isotatic polypropylene with the same microstructural defects is the best available proof of the cationic nature of the active species (Ewen et al., 1991). Hence, the polymerization is thought to occur on a cation (at a transition metal, as Ti or Zr) combined with a compatible noncoordinating anion.

A main feature of the new homogeneous catalytic systems is that they can be "single site," that is, they can include all identical catalytic sites. This can be a great advantage with respect to the heterogeneous catalytic systems, for which several sites with different reactivities and regio- and stereospecificities are present. In particular, single-site catalytic systems can allow also a better control of the molecular mass distribution as well as, for copolymers, a better control of the comonomer composition and distribution (Kaminsky and Arndt, 1997). Relevant industrial applications may be forthcoming (Horton, 1994).

It may be of interest now to discuss briefly the possible polymerization mechanism and the features

that lead to the extremely high stereoregularity observed for polypropylenes.

First of all, it is worth noting that the Ziegler-Natta systems have to be considered catalysts rather than initiators. In fact the *initiator* is defined as a substance that starts a chain reaction and is consumed in the reaction, in contrast with a catalyst, which is not consumed. Many industrially important Ziegler-Natta systems can produce many polymer molecules per transition metal atoms, with all but the first resulting from chain transfer.

It is widely accepted that, in the presence of Ziegler-Natta catalysts, poly-1-alkene chain growth consists of successive insertions of 1-alkene molecules into a transition metal-carbon bond. A *cis* opening of the 1-alkene double bond has been proved conclusively.

An insertion leading to the formation of a M-CH$_2$-CHR-P sequence (where M = metal; P = growing polymer chain) is said to be *primary* while one leading to the formation of a M-CHR-CH$_2$-P sequence is said to be *secondary*. Studies of the microstructure of the polymers, through nuclear magnetic resonance techniques (Zambelli and Tosi, 1974), have shown, for instance, that for the isotactic polymerization of the 1-alkenes the insertions tend to be nearly always primary.

For some catalytic systems based on layered TiCl$_3$ (as well as for typical homogeneous isospecific metallocene-based systems) it has been found—again by means of nuclear magnetic resonance studies—that the resulting polypropylene macromolecules may show a fraction of *m* dyads higher than 99 percent and that in such polymers the *r* dyads appear always to be in pairs. With reference to Figure 4, the few inversions of configuration happen according to model *a* rather than according to the example of model *b*.

This indicates that for these catalytic systems the steric control is dictated by the chirality of the organometallic complex of which the active site is part, rather than by the configuration of the asymmetric tertiary C atom of the last inserted monomer unit (because this hypothesis would be in agreement with model *b* and not with model *a*).

It is generally accepted that the differences in the rates of insertion of units having different configurations are mainly owing to nonbonded interactions at the catalytic site (Cossee, 1967; Corradini et al., 1982; Guerra et al., 1994).

Let us consider first the homogeneous metallocene-based catalysts, for which the stereospecificity mechanism is relatively well understood. The polymerization takes place principally through migratory insertion at a cation according to Figure 5 (Ewen, 1984; Cossee, 1967):

A tendentially isotactic polymer results if, at every other step, isomorphous faces of the olefin are inserted preferably; a tendentially syndiotactic polymer results if, at every other step, enantiomorphous faces of the olefin are inserted preferably. Hence, for instance, an isotactic or a syndiotactic polymer is obtained when a binary symmetry axis C_2 (as in scheme 1) or a symmetry plane C_s characterizes the π-ligand (Ewen, 1984; Ewen et al., 1988).

As an example, the calculated minimum energy coordination intermediates, corresponding to two successive polymerization steps, are sketched in Figure 5 for the metallocene catalysts with the chiral (R,R)-ethylene-*bis*–(1-indenyl) ligand (with C_2 symmetry and homotopic coordination positions for olefin and growing chain) and with the isopropyl(1-fluorenyl-cyclopentadienyl) ligand (with C_s symmetry and enantiotopic coordination positions for olefin and growing chain). All these minimum energy intermediates present the growing chain far from the bulky substituents of the cyclopentadienyl rings and the methyl group of the propene monomer on the opposite side with respect to the second (and following) atom of the growing chain. The results of Figure 6 are able to rationalize the isospecificity of the catalytic systems based on the *rac*-ethylene-*bis*–(1-indenyl) ligand and the syndiospecificity of the catalytic systems based on the isopropyl(1-fluorenyl-cyclopentadienyl) ligand. A similar mechanism could work for the heterogeneous Ziegler-Natta catalysts (Cossee, 1967; Corradini et al., 1982).

Hence, molecular mechanics studies have suggested that the chiral environment of the metal atom

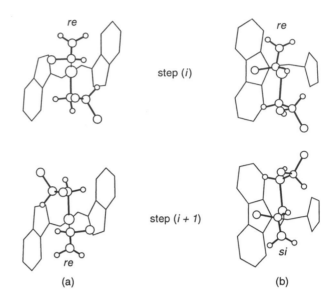

Fig. 6. The calculated minimum energy coordination intermediates, corresponding to two successive polymerization steps, for the metallocene catalysts with: (a) the chiral (R,R)-ethylene-*bis*–(1-indenyl) ligand (isospecific catalyst) (b) the isopropyl(1-fluorenyl-cyclopentadienyl) ligand (syndiospecific catalyst). Close to the propene monomer, its chirality of coordination (*re* or *si*) is indicated.

forces the growing chain to a chiral orientation, which in turn favors the insertion of the coordinated alkene with only one of the two prochiral faces (Corradini et al., 1982; Guerra et al., 1994).

In this respect, it may be of interest to recall that a relevant experimental proof of the importance of the role played by the growing chain in determining the steric course of the insertion reaction was given by the observation of the stereospecificity in the first step of polymerization. In fact, analyses by ^{13}C NMR techniques of the polymer endgroups have shown that when, in the first step of polymerization, the alkyl group bonded to the metal is a methyl group, the insertion of the monomer is essentially nonenantioselective, while when the alkyl group is an isobutyl group, the first insertion is enantioselective as are the successive insertions (Zambelli et al., 1982; Longo et al., 1987).

Besides the stereoregular polymers of olefins, the thorough investigation of the possibilities offered by the Ziegler-Natta catalysts in particular and of the stereospecific polymerization catalysis in general has led to the acquisition of new classes of polymers—for instance, those of dienes—too numerous to be discussed here (Porri et al., 1989; 1991). Let us only

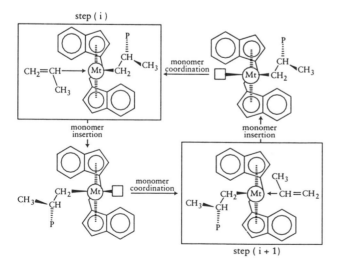

Fig. 5. Scheme 1.

recall that Ziegler-Natta catalysts are used in industrial plants for the production of cis-1,4-polybutadiene and of cis-1,4-polyisoprene rubbers.

BIBLIOGRAPHY

BARBÉ, P. C.; CECCHIN, G.; NORISTI, L. "The Catalytic System Ti-Complex/MgCl$_2$." *Adv. Polym. Sci.* 1986, *81*, 1–83.

BRINTZINGER, H. H.; FISHER, D.; MULHAUPT, R.; RIEGER, B.; WAYMOUTH, R. M. "Stereo-specific Polymerization with Chiral Metallocene Catalysts." *Angew. Chem. Int. Ed. Engl.* 1995, *34*, 1143–1170.

CORRADINI, P.; BARONE, V.; FUSCO, R.; GUERRA, G. "Steric Control in Ziegler-Natta Catalysts: An Analysis of Nonbonded Interactions at Model Catalytic Sites" *J. Catal.* 1982, *77*, 32–42.

CORRADINI, P.; BUSICO, V.; GUERRA, G. "Monoalkene Polymerization: Stereospecificity." In *Comprehensive Polymer Science;* G. C. Eastmond, A. Ledwith, S. Russo, P. Sigwalt, eds.; Oxford: Pergamon, 1989; Vol. 4, Chapter 3.

COSSEE, P. "The Mechanism of Ziegler-Natta Polymerization II: Quantum Chemical and Crystal-Chemical Aspects." In *The Stereochemistry of Macromolecules;* A. D. Ketley, ed.; New York: Marcel Dekker, 1967; Vol. 1, Chapter 3.

EWEN, J. A. "Mechanism of Stereochemical Control in Propylene Polymerizations with Soluble Group 4B Metallocene/Methylalumoxane Catalysts." *J. Am. Chem. Soc.* 1984, *104*, 6355–6364.

EWEN, J. A.; ELDER, M. J.; JONES, R. L.; HASPESLAGH, L.; ATWOOD, J. L.; BOTT, S. G.; ROBINSON, K. "Metallocene/Polypropylene Structural Relationships: Implications on Polymerization and Stereochemical Control Mechanisms." *Makromol. Chem. Symp.* 1991, *48/49*, 253–295.

EWEN, J. A.; JONES, R. L.; RAZAVI, A.; FERRARA, J. D. "Syndiospecific Propylene Polymerizations with Group 4 Metallocenes." *J. Am. Chem. Soc.* 1988, *110*, 6255–6256.

GUERRA, G.; CAVALLO, L.; MOSCARDI, G.; VACATELLO, M.; CORRADINI, P. "Enantioselectivity in the Regioirregular Placements and Regiospecificity in the Isospecific Polymerization of Propene with Homogeneous Ziegler-Natta Catalysts." *J. Am. Chem. Soc.* 1994, *116*, 2988–2995.

HORTON, A. D., "Metallocene Catalysis: Polymers by Design?" *Trends in Polym. Sci.* 1994, *2*, 158–166.

KAMINSKY, W.; ARNDT, M. "Metallocenes for Polymer Catalysis." *Adv. Polym. Sci.* 1997, *127*, 143–187.

LONGO, P.; GRASSI, A.; PELLECCHIA, C.; ZAMBELLI, A., "^{13}C-Enriched End Group of Isotactic Polypropylene and Poly(1-Butene) Prepared in the Presence of Ethylenediindenyldimethyltitanium and Methylalumoxane." *Macromolecules* 1987, *20*, 1015–1018.

NATTA, G., "From the Stereospecific Polymerization to the Asymmetric Autocatalytic Synthesis of Macromolecules." *Science* 1965, *147*, 261–272.

PORRI, L.; GIARRUSSO, A. "Conjugated Diene Polymerization." In *Comprehensive Polymer Science;* G. C. Eastmond, A. Ledwith, S. Russo, P. Sigwalt, eds.; Oxford: Pergamon, 1989; Vol. 4, Chapter 5.

PORRI, L.; GIARRUSSO, A.; RICCI, G. "Recent Views on the Mechanism of Diolefin Polymerization with Transition Metal Initiator Systems." *Progr. Polym. Sci.* 1991, *16*, 405–441.

SINN, H.; KAMINSKY, W. "Ziegler-Natta Catalysis." *Adv. Organomet. Chem.* 1980, *18*, 99–149.

TAIT, P. J. T. "Monoalkene Polymerization: Ziegler-Natta and Transition Metal Catalysts." In *Comprehensive Polymer Science;* G. C. Eastmond, A. Ledwith, S. Russo, P. Sigwalt, eds.; Oxford: Pergamon, 1989; Vol. 4, Chapter 1.

ZAMBELLI, A.; SACCHI, M. C.; LOCATELLI, P.; ZANNONI, G. "Isotactic Polymerization of α-Olefins: Stereoregulation for Different Reactive Chain Ends." *Macromolecules* 1982, *15*, 211–212.

ZAMBELLI, A.; TOSI, C. "Stereochemistry of Propylene Polymerization." *Adv. Polym. Sci.* 1974, *15*, 31–60.

ZIEGLER, K.; HOLZKAMP, E.; BREIL, H.; MARTIN, H. "The Mühlheim Low-Pressure Polyethylene Process." *Angew. Chem.* 1955, *67*, 541–547.

PAOLO CORRADINI
GAETANO GUERRA

Zinc

30 Zn 65.38

Melting Point: 693 K; **Boiling Point:** 1180 K; **Density:** 7.14 g/cm^3; **Most Common Ion:** + 2

Zinc is the twenty-fourth most abundant element in the Earth's crust with an average abundance of 76 ppm. The name *zinc* is thought to be derived from the German word *Zinke* (spike). Zinc is not found free in nature, and typically occurs as sulfide deposits, frequently associated with sulfides of lead, cadmium, iron, or copper. Zinc has been an important component of brass from our earliest history. Brass from Palestine dates from between 1400 B.C. and 1000 B.C. Zinc was not isolated until the thirteenth century A.D. in India when it was obtained by reducing calamine (zinc carbonate). The concurrent deposits of several metals with zinc prevented isolation of the metal for years. Reduction of the metal oxide is hampered by the relatively low boiling point of the metal. The metallurgy of zinc moved from India to China, where sinc coins were used in the period 1370–1640. Europeans learned of the techniques, and zinc production began in Europe about 1750.

Zinc is a blue-white, widely distributed metal whose major ores are ZnS, known as zinc blende or sphalerite, and $ZnCO_3$, known as calamine or smithsonite. Large deposits of the metal are located in Australia, Canada, and the United States. Once separated from coincident metals, zinc is recovered from these minerals by roasting to form the oxide followed by reduction by charcoal or carbon. The metal is subsequently distilled. World production of zinc is approximately six million tons per year.

The relative importance of the uses of zinc has changed over time. Presently about 40 percent of the zinc recovered is used for prevention of corrosion. Galvanized products are formed in several processes. Brasses continue to be an important alloy of zinc. A number of other alloys, with aluminum, copper, or magnesium, are used for diecasting. Alloys of zinc are used for certain solders. Dry cell BATTERIES—for example, the Leclanché cell—continue to use zinc.

Zinc compounds are important to industry. Zinc oxide, ZnS, and $ZnCO_3$ are used in the manufacture of paints. The oxide is used as an antiseptic and as a mordant in printing and dyeing of textiles. The phos-phor zinc sulfide is used in making luminous dials, X rays, and television screens. Zinc sulfate is used to promote crenellation of rayon during manufacture. The sulfate is also used in trace quantities as a fertilizer and for the control of certain plant diseases.

Zinc is an element essential to humans, and the average human contains about 2 grams of zinc. There are several important zinc enzymes: carboxypeptidase A and carbonic anhydrase, to name but two. Zinc ions are also required for thermolysin, and both DNA and RNA polymerases.

BIBLIOGRAPHY

GREENWOOD, N. N.; EARNSHAW, A. *Chemistry of the Elements;* New York: Pergamon, 1984; pp. 1395–1402.

LIDE, D. R., ed. *Handbook of Chemistry & Physics,* 71st ed.; Cleveland: Chemical Rubber Co., 1990–1991.

SCHLECHTEN, A. W.; THOMPSON, A. P. "Zinc and Zinc Alloys." In *Encyclopedia of Chemical Technology,* 2nd ed.; R. E. Kirk, D. F. Othmer, A. Standen, eds.; New York: Wiley-Interscience, 1970; Vol. 22, pp. 555–603.

THOMPSON, A. P. "Zinc Compounds." In *Encyclopedia of Chemical Technology,* 2nd ed.; R. E. Kirk, D. F. Othmer, A. Standen, eds.; New York: Wiley-Interscience, 1970; Vol. 22, pp. 604–613.

ALTON J. BANKS

NOBEL PRIZE WINNERS IN CHEMISTRY

1901 Jacobus Henricus van't Hoff

1902 Hermann Emil Fischer

1903 Svante August Arrhenius

1904 William Ramsay

1905 Johann Friedrich Wilhelm Adolf von Baeyer

1906 Henri Moissan

1907 Eduard Buchner

1908 Ernest Rutherford

1909 Wilhelm Ostwald

1910 Otto Wallach

1911 Marie Curie

1912 Victor Grignard
Paul Sabatier

1913 Alfred Werner

1914 Theodore William Richards

1915 Richard Martin Willstätter

1916 *No award*

1917 *No award*

1918 Fritz Haber

1919 *No award*

1920 Walther Hermann Nernst

1921 Frederick Soddy

1922 Francis William Aston

1923 Fritz Pregl

1924 *No award*

1925 Richard Adolf Zsigmondy

1926 The (Theodor) Svedberg

1927 Heinrich Otto Wieland

1928 Adolf Otto Reinhold Windaus

1929 Hans Karl August Simon von Euler-Chelpin
Arthur Harden

1930 Hans Fischer

1931 Carl Bosch
Friedrich Bergius

1932 Irving Langmuir

1933 *No award*

1934 Harold Clayton Urey

1935 Frédéric Joliot
Irène Joliot-Curie

1936 Petrus (Peter) Josephus Wilhelmus Debye

1937 Walter Norman Haworth
Paul Karrer

1938 Richard Kuhn [declined by his government but later reinstated]

1939 Adolf Friedrich Johann Butenandt
Leopold Ruzicka

1940 *No award*

1941 *No award*

1942 *No award*

1943 George De Hevesy [Georg von Hevesy]

1944 Otto Hahn

1945 Arturri Ilmari Virtanen

1946 John Howard Northrop
Wendell Meredith Stanley
James Batcheller Sumner

1947 Robert Robinson

1948 Arne Wilhelm Kaurin Tiselius

1949 William Francis Giauque

1950 Kurt Alder
Otto Paul Hermann Diels

1951 Edwin Mattison McMillan
Glenn Theodore Seaborg

1952 Archer John Porter Martin
Richard Laurence Millington Synge

1953 Hermann Staudinger

1954 Linus Carl Pauling

1955 Vincent du Vigneaud

1956 Cyril Norman Hinshelwood
Nikolai Nikolaevitch Semenov

1957 Alexander R. Todd

1958 Frederick Sanger

1959 Jaroslav Heyrovsky

1960 Willard Frank Libby

1961 Melvin Calvin

1962 John Cowdery Kendrew
Max Ferdinand Pertutz

1963 Giulio Natta
Karl Ziegler

1964 Dorothy Crowfoot Hodgkin

1965 Robert Burns Woodward

1966 Robert S. Mulliken

1967 Manfred Eigen
Ronald George Wreyford Norrish
George Porter

1968 Lars Onsager

1969 Derek H. R. Barton
Odd Hassel

1970 Luis F. Leloir

1971 Gerhard Herzberg

1972 Christian B. Anfinsen
Stanford Moore
William Howard Stein

1973 Ernst Otto Fischer
Geoffrey Wilkinson

1974 Paul J. Flory

1975 John Warcup Cornforth
Vladimir Prelog

1976 William N. Lipscomb

1977 Ilya Prigogine

1978 Peter D. Mitchell

1979 Herbert C. Brown
Georg Wittig

1980 Paul Berg
Walter Gilbert
Frederick Sanger

1981 Kenichi Fukui
Roald Hoffmann

1982 Aaron Klug

1983 Henry Taube

1984 Robert Bruce Merrifield

1985 Herbert A. Hauptman
Jerome Karle

1986 Dudley R. Herschbach
Yuan T. Lee
John C. Polanyi

1987 Donald J. Cram
Jean-Marie Lehn
Charles J. Pedersen

1988 Johann Deisenhofer
Robert Huber
Hartmut Michel

1989 Sidney Altman
Thomas R. Cech

1990 Elias James Corey

1991 Richard R. Ernst

1992 Rudolph A. Marcus

1993 Kary B. Mullis
Michael Smith

1994 George A. Olah

1995 Paul Crutzen
Mario Molina
F. Sherwood Rowland

1996 Robert F. Curl
Harold W. Kroto
Richard E. Smalley

INDEX

Page numbers in **boldface** refer to the main entry on a subject.
Page numbers in *italics* indicate figures. Page numbers followed by *t* indicate tables.

calcium oxide formation, 674
carbon dioxide role in producing, 304–305
corrosion, 422
crystal polymorphs, 433
freshwater buffer, 274
as inorganic compound, 298, 777
as representative polymorph, 1238t
Calcium channel blockers, 486
Calcium chloride. *See* Calcium
Calcium fluoride (fluorspar), 639–640, 1413
Calcium-40, 1289
Calcium hydride, 950
Calcium nitrate fertilizers, 620
Calcium oxide, 674
Calcium phosphate, 259
Calcium sulfate, 391, 674
California
    boron deposit, 264
    Gold Rush of 1849, 690
California Institute of Technology
    Millikan presidency, 948, 949
    X-ray crystallography, 268
Californium
    atomic number and symbol, 18, 1455
    first synthesis, 1030–1031
    number of radioactive isotopes, 19
Calixarenes, 429–430, 1180
Calomel electrode, 589, 1002
Calorie. *See* Dietetic calorie; Nutrition
Calorimetry, 713
Calphostin 26, biosynthesis, 1286, *1287*
Calutron (instrument), 813
Calvin-Benson cycle, 1211
Calvin cycle, 306
Cambridge University
    Aston association, 175
    Bragg (William Lawrence) association, 268
    Chadwick association, 333
    Dirac association, 477
    Hodgkin association, 743–744
    Lowry association, 873
    Rayleigh association, 1300–1301

Rutherford association, 1324
Sanger association, 1330
Thomson association, 1446–1447
Camera lens. *See* Photography, chemistry of
cAMP. *See* Cyclic adenosine monophosphate
Camphor, 376, *555*
CAM plants. *See* Crassulacean acid metabolism
Camptothecin, 152, *153*
Canada
    asbestos deposits, 173, 174
    coal reserves and consumption, 369t
    cobalt deposits, 372
    copper deposits, 421
    gold deposits, 690
    heavy water nuclear reaction technology, 718
    Herzberg, Gerhard, 730
    lead deposits and production, 845
    platinum deposits, 1215
    potassium mining, 1245
    selenium production, 1334
    silver production, 1346
    zinc peosits, 1544
Cancer
    asbestos-related, 174
    carcinogenic process, 339–340
    carcinogens, 160, 174, 339–342, 992–996, 1216, 1224, 1291
    cell damage, 150
    diet link, 644
    estrogen and, 1395–1396
    improper DNA alkylation link, 74–75
    nicotine link, 1008
    see also Anticancer agents; *specific cancers*
Candela (unit of measure), 795
Candle wax, 385
Cannabis sativa. See Marijuana
Cannizzaro, Stanislao, **282–283**, 940
    defense of Avogadro's hypothesis, 184, 198–199, 282, 360, 478, 940, 973, 977
    list of definitive atomic masses, 184, 282, 887, 973, 977, 1157

Cannizzaro reaction, 651
Cans, aluminum, 879
Capillarity, defined, 861
Caprolactam, 1065
Captopril, 486, 895, *896*
Carat (gem unit), 887
Carbaboranes. *See* Carboranes
Carbamate protecting groups, 1256t
Carbamic acids, 97
Carbamoyl intermediates, 1004
Carbamoyl phosphate synthetase I, 119
Carbanions, 81–82, **283–286**, 315–316
    alkylation reactions, 74
    carbon nucleophiles, 942
    crown ethers and cryptands uses, 428
Carbenes, 203, **286–292**, 539, 1107, 1108, 1109, 1110
Carbenicillin, 139
Carbenoids, 290, 291–292
Carbinolamines, 546, *548*, 1490
Carbocations, **292–298**, 412, 543–544, *550*
    of alcohols, 33
    of alkenes, 70, 72, 297, 411
    in alkylation reaction, 74, 75
    of alkynes, *84*
    tert-butyl, 412
    carbanions vs., 283
    direct observation of, 295, 296–297
    of enols, 1427
    nonclassical, 296–297
    as reactive intermediates, 293–294, 295
    structure and stability, 293
    superacid use in observing, 295, 296–297, 1415
Carbohydrates, 1404–1408
    aldehydic behavior, 41–45
    biochemistry of, 1406–1407
    caloric content, 474
    carbon cycle, 301
    carbonyl groups in, 312
    commercial uses, 1408
    (D)-series, 1385
    ethanol fermented from, 590
    Fischer (Emil) studies, 626
    Gay-Lussac role in discovering, 682
    as macronutrient, 1059–1060

Mass spectrometry (*cont.*)
  peptide characterization and study, 1148
  relative atomic mass determinations, 194
  resolving powers, 503
  specific conditions, 1369
Mass-to-charge ratio (*m/z* values), 889, 890
Mass transfer, 1463
Mastication. *See* Chewing
Matchbooks, 1194
Maté leaves, 279
Material Safety Data Sheets, 869
Materials science and engineering, **890–892**, 927
  aqua regia use, 155
  coke production, 370–371
  computer applications, 392
  gas phase reactions and, 674
  metallocenes uses, 923–926
  microscopy etchants, 943
  qualitative analysis, 1267–1269
  separation processes, 1343, 1344
  smelting, 1347–1349
  stereochemistry, 1386
  sulfuric acid uses, 1413
  unit operations, 1464
  zinc processes, 1543
  see also Industrial chemistry
Mating stance, 1188
Matrix mechanics, 478, 718, 1332
Matrix receptors, 1306
Matrix vesicles, 259–260
Matsen, Albert, 1276
Matter
  atom, classical, 181–185
  Boyle theory of, 182, 264–265
  Dalton theory of, 952–953
  energy manifestations, 1434
  energy transformations, 559–565
  intermolecular forces, 581, 790–793
  Lavoisier theory of, 183, 842–843
  mass as constant property, 413–414, 886
  particulate nature of, 1140–1142

  phase, 1170–1171
  properties, 1248–1249
  see also Gases; Liquid; Solids; States of matter; *specific components*
Matte smelting, 1347–1348
Matthias rule, 1417
Mauna Loa Observatory (Hawaii), 306, *307*, 572
Mauve dye (mauveine), 500, 745
Maxam, A. M., 479
Maximum contaminant levels (MCLs), drinking water, 1513–1514
Max Planck Institute
  Butenandt and, 277
  Heisenberg and, 719
  Planck and, 1214
  Von Laue and, 842
  Ziegler and, 1538
  see also Kaiser Wilhelm Institute
Maxwell, James Clerk
  on energy distribution, 831, *832*, 1300
  and Faraday's work, 614
  kinetic molecular theory, 827–828, 976–977, 1142
  light wave theory, 185, 1207–1208
Maxwell distribution. *See* Statistical mechanics
Mayo Clinic, 132, 146
Mayrow, John, 359
Maytansine, 50
Mazurs, E. G., 1163
MBE. *See* Molecular beam epitaxy
MBS. *See* Minimum basis sets
MBS SCF method, 1279, 1280, 1281
M-C bonds. *See* Organometallics
McClellan, A. L., 476
MC excited state, 606
McGill University, 1324, 1353
McGraw-Hill Encyclopedia of Science & Technology, 869
McLean, A. D., 1273
MCLT excited state, 106
McMillan, E., 1030
MCPA (herbicide), 724, 728*t*

MDMA (methylenedioxymethamphetamine; ecstasy), 495
Meadow saffron, 52, 57
Mean free path
  kinetic theory of gases, 675, 676
  Maxwell estimation of molecular, 976–977
Measure, units of. *See* International system of units
Measurement
  of atmospheric pressure, 807
  atomic force microscopy, 193
  classical stoichiometric laws, 359–361
  colligative properties, 373–376
  of crystal hardness, 434
  of density, 467
  of dipole moments, 476
  of electrical conductivity, 505
  of energy absorption of emission. *See* Spectroscopy
  of infrared spectra, 785
  of kinetic isotope effects, 825
  mole concept, 951–955
  of osmotic pressure, 1119–1120
  of partial pressure, 454
  of pressure, 885, 1247
  qualitative analysis, 1267–1269
  of radiation, 684–685, 1290
  of rapid reaction processes, 830–831
  of temperature, 1431
  titration, 1448–1451
  of water purification, 1516
  see also Barometer; International System of Units; Manometer; Thermometer
Meat tenderizers, 579
Mebenzadole, 488
Mechanical energy
  defined, 559
  energy transformations, 561
Mechanical (hearth) roasting, 1322
Medicinal chemistry, **893–897**
  alkaloids and, 49, 53–58
  alkynes and, 79–80
  analgesics, 125–127
  anesthetics, 130–135

Tetrahydropyran ring system, 597
Tetraiodothyronine, 745
Tetralin, 997
Tetramethyltetraselenafulvalene (TMTSF), 1417
Tetramethyl tin, 778
1,3,5,7-Tetranitro-1,3,5,7-tetrazocine (HMX), 608–609
Tetrapeptides, 1146
Tetrapropylammonium perruthenate (TPAP), 31
Tetrapyrroles, 248, *998*, *999*, 1240
Tetratomic, **1431–1432**
Tetrodotoxin, 57, *58*, 1217
Teutsch, Georges, 417
Textile Fiber Products Identification Act (TFPIA), 622, 623, 624
Textiles. *See* Fibers
Thailand, 1447
Thalén, Tobias Robert, 170
Thalidomide, 1387
Thallium, 671
Thebaine, 50, 52, 125
Thénard, Louis-Jacques, 263, 455, 682, 1345, 1408
Thénard's blue, 381
Theobromine, 50
Theophilline, 50
Theoretical chemistry, **1432–1433**
Born-Oppenheimer approximation, 963, 1272, 1433
see also Bioenergetics; Kinetics; Quantum mechanical computations; Quantum theory; Statistical mechanics; Thermochemistry
Theoretische Grundlagen der organischen Chemie (Hückel), 750
Therapeutic drugs. *See* Drugs, beneficial
Thermal conductivity
for deuterium analysis, 468
diffusion, 1341
of gases, 681–684, 714
heat transfer, 714–715
of metals, 714, 927
Thermal conductivity detector, 356

gas-liquid chromatography, 679
Thermal cracking, 62
Thermal energy. *See* Thermochemistry
Thermal radiation, 714–715
Thermal reactions
of batteries, 216
defined, 830
dissociation, 478
elimination (pyrolysis), 546
entropy, 1437–1438
of gases. *See* Gas laws
kinetics and, 830
Woodward-Hoffmann rules, 1152t, 1155
Thermochemistry, **1433–1440**
bar unit, 211
Boltzmann studies, 252–253
Born-Haber cycle, 261–263
clusters, 364–365
colligative properties, 335, 373–376, 394
confinement effect, 1435
defined, 1433
elimination reactions, 546, 547
energy transformations, 559–565
entropy, 1377–1378, 1437–1439, 1525
equation of state, 580
equilibria, 580–587
explosives and propellants, 607–610, 873
of freezing and melting. *See subhead* melting and freezing process *below*
heat capacity, 713, 1434
heat transfer, 713–717
ideal gas law, 770, 1371
ionic strength effects, 802–803
ionization constant, 804
Kelvin contribution, 818
kinetics, 831, *832*, 834
Laws of Thermodynamics, 561, 564, 1002, 1435–1439
Le Châtelier studies, 847
magnetic behavior, 882
melting and freezing process, 660–661, 899–900
Nernst heat theorem, 1002
phase rule, 1171

phases and phase diagrams, 1171–1176
phosphoric acid preparation, 1193
Planck studies, 1214
quantum chemical computations, 1271–1276
quantum theory, 1282
reactivity, 1303
real solutions, 781
reversible and irreversible operation, 582–583
statistical mechanics, 1142, 1372–1379, 1434–1435
theoretical chemistry, 1432
three common states. *See* Gases; Liquid; Solids
work function, 1525–1526
see also Crystal lattice energy; Fusion; Statistical thermodynamics; Thermal reactions
Thermodynamics. *See* Thermochemistry
Thermodynamics, First Law of, 1435, 1436
Thermodynamics, Second Law of, 561, 564, 1438–1439; *see also* Entropy
Thermodynamics, statistical. *See* Statistical thermodynamics
Thermodynamics, Third Law of, 1002, 1437
Thermogenesis, 748
Thermoluminescence, **1440–1441**
Thermolysin, 1544
Thermometer
gas, 680
mercury, 907
scales, 1431
Thermonuclear reaction, deuterium, 468
Thermonuclear weapons. *See* Hydrogen bomb
Thermoplastics, 504, 624–625, 1230, 1538
Thermosets (heat-set plastics), 1235
Thermotropic liquid crystals, 862t, 864
THF. *See* Tetrahydrofuran
Thiamin, 1062, *1504*
biosynthetic pathway, 999, 1000
deficiency, 1063

Uranium-236, as fission product, 1041

Uranium-238 radioactive decay series, 459, 1027–1028, 1289
  duration of decay steps, 1291
  half-life, 1289
  as nuclear reactor fuel, 1033, 1043
  radon gas, 1290

Uranium-239 radioactive decay series, 1030

Urate, 149

Urbain, Georges, 985

Urea. *See* Urine chemistry

Urea cycle. *See* Krebs cycle

Urease, 1413–1414

Urethanes, 97

Urey, H. C., 468, 1327

Uric acid, 1467
  as antioxidant, 659
  blood testing, 362
  compound symnthesis, 626

Uridine diphosphoglucose (UDPglu), 916

Urinalysis, 1464, 1465–1466

Urine chemistry, **1464–1471**
  abnormal constituents, 1468–1471
  catecholamine elimination, 326
  chemical composition, 1466–1468
  clinical, 361–362
  evaluation, 361, 362, 1465–1466
  formation, 1464–1465
  insulin effects, 748
  osmosis and, 1121, 1122
  phosphorus and, 1194, 1195
  urea fertilizers, 619
  urea as fixed nitrogen, 1017
  urea synthesis, 119–122, 298, 809, 1014, 1080, *1081*, 1520

Urobilinogen, 1468

Urokinase, 487, 579

Uronic acids, 1406

Uroporphyrinogen-III (XI), 1240

Usanovich acid-base concepts, 172, 873

USSR (former). *See* Commonwealth of Independent States

UV radiation. *See* Ultraviolet spectroscopy; Ultraviolet (UV) radiation

UV-visible wavelength. *See* Ultraviolet-visible wavelength

UV-Vis spectroscopy (UV/visible spectrometry), 1269, 1462

V

Vaccines, 234, 237, 1143–1144
  antibodies, 144–145

Vaccum, Boylean, 264

Vacuum gauge. *See* Manometer

Vacuum pump, 266

Vacuum tube, 849

Vacuum ultraviolet lamp (VUV), 1203

Vacuum UV region, 1461

Valence, **1473–1474**
  classical molecular theory and, 975, 976
  concept uses, 1473
  Erlenmeyer introduction of concept, 589
  lone pair, 872
  Lowry studies, 873
  nonstoichiometric compounds, 1024–1025
  periodicity, 940, 1159, 1161, 1167
  quantization, 1161, 1473
  Werner approach, 419, 1517
  see also Atomic mass unit

Valence bond theory, 817, 979, **1474–1475**
  Couper structures, 423–424, 975, 1474
  covalent bonds, 980, 1312
  electron-deficient bond, 522–523
  Kekulé structures, 156, 423
  Lewis structures, 424, 522, 849–850, 977–978, 1312, 1319, 1475–1476
  localized bonding, 871
  molecular orbital theory comparison, 337–339
  orbital overlap, 1079–1080
  Pauling and, 337, 852, 979, 980, 1319, 1474
  resonance theory, 1319
  standard references, 1475
  see also Molecular orbital theory

Valence-Bond Theory and Chemical Structure (eds. Klein and Trinajstic), 1475

Valence electron, **1475**
  defined, 1079, 1475
  electrical conduction, 507
  electron-deficient bond and, 522
  ionization energy, 806
  isoelectronic, 808

Valence orbitals. *See* Valence electron

Valence shell electron pair repulsion, 978–979, **1475–1479**
  lone pairs, 872, 978, 1477
  molecular geometry, 958
  multiple bonds, 1477–1478
  polyatomic ionic structures and, 1223

Validation of method, 130

Valine, 117, 123, 616*t*, 1060, 1146
  nonpolar side chain, 112, *113*

Valinomycin, 1147

Valium (diazepam), *132, 133,* 219, 220–221, 485, 492, 893

Vallate papilla, 1423, *1424*

Vanadium, **1479**
  biological function, 928, 1454
  electron configuration, *521*
  Ziegler-Natta reactions, 1538

Vanadium oxide, 1479

Van Bommel, A. J., *554, 556,* 1384

Vancomycin
  cyclic ether linkages, 597, *598*
  fermentation organism, *616*
  hydrogen bonding, 318
  mode of action, 142
  as organochlorine drug, 1097
  selective toxicity, 1217

Van de Graaff accelerators, 1027

Van der Waals, Johannes Diderik, **1479–1480**

Van der Waals equation, **1480–1482**
  atomic force, 192
  critical constants, 425
  and ideal gas law, 770
  and kinetic molecular theory, 829, 1481
  and nonideal gases, 580, 1023–1024

ISBN 0-02-897224-4

90000